Chemical Signals in Vertebrates 13

Bruce A. Schulte • Thomas E. Goodwin
Michael H. Ferkin
Editors

Chemical Signals in Vertebrates 13

Springer

Editors
Bruce A. Schulte
Department of Biology
Western Kentucky University
Bowling Green, KY, USA

Thomas E. Goodwin
Department of Chemistry
Hendrix College
Conway, AR, USA

Michael H. Ferkin
Department of Biological Sciences
The University of Memphis
Memphis, TN, USA

ISBN 978-3-319-22025-3 ISBN 978-3-319-22026-0 (eBook)
DOI 10.1007/978-3-319-22026-0

Library of Congress Control Number: 2015955742

Springer Cham Heidelberg New York Dordrecht London

Printed on acid-free paper

Springer International Publishing AG Switzerland is part of Springer Science+Business Media (www.springer.com)

Foreword and Dedication

The 13th volume of *Chemical Signals in Vertebrates* (CSiV) emanates from the first joint meeting of CSiV and the International Society of Chemical Ecology (ISCE), which was held at the University of Illinois Urbana-Champaign in June 2014. Dr. May Berenbaum was the local host and CSiV greatly appreciates the marvelous job done by May, the organizing committee, and everyone involved. The editors of CSiV 13 invited Dr. Berenbaum to write an overview paper on CSiV-ISCE shared areas of interest and future prospects for collaboration (see the contribution by Dr. May Berenbaum in this volume).

In this Foreword we wish to pay homage to two pioneering giants in the CSiV community, Dr. Dietland Müller-Schwarze and Dr. Robert Johnston. In December 2014, CSiV lost Dr. Johnson, a dear friend, wonderful colleague, and a scientific leader in the realm of chemical communication. Drs. Michael Ferkin and Aras Petrulis have composed a fitting remembrance to Bob and this volume is produced in his honor (see their remembrance in this volume).

Dr. Dietland Müller-Schwarze was an editor of the first 10 of the past 12 volumes in the CSiV series. Essentially from the formal conception of the study of chemical signals in vertebrates, Dr. Müller-Schwarze has been an exemplary leader and he continues to be a guiding force in this field. Thus, we asked him to write a historical and future perspective for CSiV 13. The CSiV meeting in 2014 marked the first one that Dietland could not attend, so we look forward to seeing you at the next one, Dietland!

From what we can determine, this also was the first meeting in which none of the original attendees at the 1976 symposium at the Gideon Putnam Hotel in Saratoga Springs, NY, was in attendance. Thus, CSiV has truly been handed off to the next generation of researchers studying chemical signals in vertebrates. Consequently, in this volume you will find contributions from individuals who are just beginning on this quest and those who have been studying vertebrate chemical signals for many years.

With gratitude we dedicate this volume to Dietland Müller-Schwarze and to the memory of Robert Johnston for their contributions that have advanced the study of Chemical Signals in Vertebrates. Thank you Dietland and Bob!

Bowling Green, KY — Bruce A. Schulte
Memphis, TN — Michael H. Ferkin
Conway, AR — Thomas E. Goodwin

In Memory: A Tribute to Dr. Robert (Bob) E. Johnston

Dr. Robert Elliott (Bob) Johnston, a pioneer and leader in the study of odor communication for nearly five decades, passed away at the age 72 on December 20, 2014. Bob was an active contributor to the field of chemical communication in vertebrates, a CSiV volume editor, and the host of a CSiV meeting held at Cornell University. Bob was born in Philadelphia on April 16, 1942, the oldest child of Robert Henry and Margaret Elliott (Riley) Johnston. Bob earned his B.A. from Dartmouth College in 1964. He began his doctoral work at Brown University and then moved to Rockefeller University. Bob studied under the supervision of Donald Pfaffman and earned his Ph.D. from Rockefeller University in 1970. After graduating Bob joined the psychology faculty at Cornell University. For both of us and for many others, this is where Bob's legacy began. This book is published in memory of Bob Johnston with great appreciation of his many contributions to the growth of our field.

Bob studied chemical and other forms of communication in a variety of animals, including humans. Indeed, his collaboration with Bob Kraut on the evolutionary function of human smiling produced a very influential paper on human evolutionary psychology (Kraut and Johnston 1979). However, his main study animal was the golden hamster. Bob was one of the first to study odor communication in a truly systematic way. His early work examined scent marking behavior and the responses of golden hamsters to the scent marks of conspecifics. Among many findings, Bob found that flank marking was aggressive or competitively motivated in both males and females. He also discovered that vaginal marking was used by females as a sexual solicitation behavior directed towards male conspecifics (Johnston 1974, 1977, 1979). From these and many other studies, Bob built a program spanning 44 years that studied odor communication from the level of the cell to the level of the whole organism. Bob and coworkers developed and tested hypotheses that addressed important mechanistic and functional questions about individual (e.g., Johnston 1993; Johnston et al. 1993a, b), kin (e.g., Heth et al. 1998; Mateo and Johnston 2000), and species recognition (e.g., delBarco-Trillo and Johnston 2013) as well as those surrounding scent over-marking (e.g., Johnston et al. 1994, 1997; Wilcox and

Johnston 1995; Ferkin et al. 1999) and sexual communication (e.g., Ferkin and Johnston 1995). Bob and his students also conducted a series of experiments examining how the nervous system controls communication behavior, social memory formation, and sexual recognition (e.g., Pfeiffer and Johnston 1994; Petrulis and Johnston 1999; Petrulis et al. 1999; Lai et al. 2005).

Bob was a talented researcher, writer, and empiricist. One of Bob's strengths as a scientist was his ability to use a synthetic approach to developing testable hypotheses for understanding odor communication. Bob's approach to experiments on rodent communication and individual recognition through odors has influenced researchers worldwide. He developed connections with German and Turkish researchers that led to studying wild hamsters in Turkey. Another of Bob's strengths was his incredible skill as a mentor. Anyone that has had the pleasure of working with Bob, and there are over 150 undergraduates, graduate students, postdocs, and visiting faculty with whom he collaborated, valued highly his support, positive attitude, and sage advice.

What we found most gratifying was Bob's caring, thoughtfulness, and willingness to share his ideas and expertise. Bob had a calmness that allowed him to be a great listener and a remarkable sounding board for research ideas. He had a great mind and could design an experiment on the fly. Bob would challenge you to think outside the box. He encouraged you to go to the next step and design a "cool" experiment and he helped you do it. In the competitive field of academia, Bob was unselfish in sharing his knowledge and expertise with you. After being with Bob, you realized that you learned something special. It was important to him that you understood how to conduct science and interact with others. In doing so, Bob left a strong, positive imprint on many lives, including our own.

There were other sides of Bob that were known to close friends and family. Bob was a former collegiate athlete. He used his 6 ft 4 in. frame to play basketball for Havertown High School and Dartmouth College. Bob continued to play pick-up games of basketball at Cornell University. He also enjoyed squash, cross-country skiing, ice skating, hiking, and dancing. In addition, Bob was a talented artist, accomplished photographer, and sculptor. Much of this work was influenced by Bob's love of nature. Bob was too modest about his skills as an artist. Finally, Bob had a great sense of humor and was fun to be around. Ask any of his students and postdocs, as well as the unsuspecting that sat on the swing in his backyard that took you soaring over a deep gorge.

Bob was a great and kind man. He was an excellent teacher, leading researcher, and outstanding mentor. For many of us he became a colleague and more importantly a dear friend. Bob will be sorely missed by those whose lives he touched. He was a loving father to two sons, Aiden and Alexander, and devoted and loving husband to Joan for 44 years. Joan and Bob traveled extensively to attend professional meetings in different countries and to observe animals in their natural habitat. Their travels took them to the South America, Australia, China, the Republic of Dagestan, Turkey, and "accidently" Syria. Bob was a great storyteller. It was exciting to hear stories about their exploits and adventures and the neat animals they saw. He mentioned that one of his most interesting and satisfying experiences was to finally be able to watch the hamsters he studied in the lab for over 40 years in their natural habitat. Bob's study site was in Turkey near its border with Syria. Bob would watch the Syrian and Turkish soldiers look at each other and at him. Despite the international intrigue, Bob discovered that free-living golden hamsters are not active at night (Gatterman et al. 2008), a surprise to pet hamster owners everywhere.

Memphis, TN Michael H. Ferkin
Atlanta, GA Aras Petrulis

References

delbarco-Trillo J, Johnston RE (2013) Interactions with heterospecific males do not affect how female *Mesocricetus* hamsters respond to conspecific males. Anim Behav 86:151–156

Ferkin MH, Johnston RE (1995) Meadow voles, *Microtus pennsylvanicus*, use multiple sources of scent for sex recognition. Anim Behav 49:37–44

Ferkin MH, Dunsavage J, Johnston RE (1999) Meadow voles can discriminate between the top and bottom scent of an over-mark and prefer the top scent. J Comp Psych 113:43–51

Gattermann R, Johnston RE, Yigit N, Fritzsche P, Larimer S, Ozkurt S, Neumann K, Song Z, Colak E, Johnston J, McPhee ME (2008) Golden hamsters are nocturnal in captivity but diurnal in nature. Biol Lett 4:253–255

Heth G, Todrank J, Johnston RE (1998) Kin recognition in golden hamsters: evidence for phenotype matching. Anim Behav 56:409–417

Johnston RE (1974) Sexual attraction function of golden hamster vaginal secretion. Behav Biol 12:111–117.

Johnston RE (1977) The causation of two scent-marking behaviour patterns in female hamsters (*Mesocricetus auratus*). Anim Behav 25:317–327

Johnston RE (1979) Olfactory preferences, scent marking, and "proceptivity" in female hamsters. Horm Behav 13:21–39

Johnston RE (1993) Memory for individual scent in hamsters (*Mesocricetus auratus*) as assessed by habituation methods. J Comp Psych 107:201–207

Johnston RE, Chiang G, Tung C (1994) The information in scent over-marks of golden hamsters. Anim Behav 48:323–330

Johnston RE, Sorokin ES, Ferkin MH (1997) Female voles discriminate males' over-marks and prefer top-scent males. Anim Behav 54:679–690

Johnston RE, Derzie A, Chiang G, Jernigan P, Lee H-C (1993) Individual scent signatures in golden hamsters: evidence for specialization of function. Anim Behav 45:1061–1070

Kraut RE, Johnston RE (1979) Social and emotional messages of smiling: An ethological approach. J Pers Soc Psych 37:1539–1553

Lai W-S, Ramiro L-LR, Yu HA, Johnston RE (2005) Recognition of familiar individuals in golden hamsters: a new method and functional neuroanatomy. J Neurosci 25:11239–11247

Mateo JM, Johnston RE (2000) Kin recognition and the 'armpit effect': evidence of self–referent phenotype matching. Proc Royal Soc Biol Sci 267:695–700

Petrulis A, Johnston RE (1999) Lesions centered on the medial amygdala impair scent-marking and sex-odor recognition but spare discrimination of individual odors in female golden hamsters. Behav Neurosci 113:345–357

Petrulis A, Peng M, Johnston RE (1999) Effects of vomeronasal organ removal on individual odor discrimination, sex-odor preference, and scent marking by female hamsters. Physiol Behav 66:73–83

Pfeiffer CA, Johnston RE (1994) Hormonal and behavioral responses of male hamsters to females and female odors: roles of olfaction, the vomeronasal system, and sexual experience. Physiol Behav 55:129–138

Wilcox RM, Johnston RE (1995) Scent counter-marks: specialized mechanisms of perception and response to individual odors in golden hamsters (*Mesocricetus auratus*). J Comp Psych 109:349–356

Contents

Part I Overviews of Chemical Ecology (in) Vertebrates

1 **Are Mammals Just Furry Bugs with Fewer Legs? Convergences in Mammalian and Insect Chemical Ecology** 3
May R. Berenbaum

2 **Chemical Signals in Vertebrates 13: Where We Stand and What Might Be Next** 11
Dietland Müller-Schwarze

3 **Assessing the Role of Olfactory Cues in the Early Life History of Coral Reef Fish: Current Methods and Future Directions** 17
Rohan M. Brooker and Danielle L. Dixson

4 **A Small Expose on Bovine Pheromones: with Special Reference to Modifications of the Reproductive Cycle** 33
Kristina Nordéus, Renée Båge, Hans Gustafsson, Robert Glinwood, and Lennart Söderquist

Part II Microbiomes and Chemical Signals

5 **Coding of Group Odor in the Subcaudal Gland Secretion of the European Badger *Meles meles*: Chemical Composition and Pouch Microbiota** 45
Christina D. Buesching, H. Veronica Tinnesand, YungWa Sin, Frank Rosell, Terry Burke, and David W. Macdonald

6 **The Role of Bacteria in Chemical Signals of Elephant Musth: Proximate Causes and Biochemical Pathways** 63
Thomas E. Goodwin, Innocent H. Harelimana, Laura J. MacDonald, Daniel B. Mark, Aline Umuhire Juru, Qin Yin, James A. Engman, Randall A. Kopper, Cheryl F. Lichti, Samuel G. Mackintosh, James D. Shoemaker, Mark V. Sutherland, Alan J. Tackett, and Bruce A. Schulte

7 Age-Related Variation in the Scent Pouch Bacterial Communities of Striped Hyenas (*Hyaena hyaena*) 87
Kevin R. Theis, Arvind Venkataraman, Aaron P. Wagner, Kay E. Holekamp, and Thomas M. Schmidt

8 Bacterial Communities Associated with Junco Preen Glands: Preliminary Ramifications for Chemical Signaling 105
Danielle J. Whittaker and Kevin R. Theis

Part III Intraspecific Signaling

9 Age-Related Effects on Individual Discrimination Among Meadow Voles, *Microtus pennsylvanicus* 121
Michael H. Ferkin, Christian T. Vlautin, and Lyndsey M. Pierson

10 Putative Pheromone of the Indian Crestless Porcupine, *Hystrix brachyura* 137
Biswatosh Ghosh, Mousumi Poddar-Sarkar, Sibdas Ray, and Ratan Lal Brahmachary

11 Chemical Cues, Hibernation and Reproduction in Female Short-Beaked Echidnas (*Tachyglossus aculeatus setosus*): Implications for Sexual Conflict 145
Rachel L. Harris, Elissa Z. Cameron, Noel W. Davies, and Stewart C. Nicol

12 Chemical Analyses Reveal Family-Specific Nest Odor Profiles in Zebra Finches (*Taeniopygia guttata*): A Pilot Study 167
Sarah Kohlwey, E. Tobias Krause, Markus C. Baier, Caroline Müller, and Barbara A. Caspers

13 Exploring the Use of Olfactory Cues in a Nonsocial Context in Zebra Finches (*Taeniopygia guttata*) 177
E. Tobias Krause, Johanna Kabbert, and Barbara A. Caspers

14 Variation in Urinary Amino Acids in the Mozambique Tilapia: A Potential Signal of Dominance or Individuality? 189
Olesya Kutsyna, Zélia Velez, Adelino V.M. Canário, Tina Keller-Costa, and Peter C. Hubbard

15 A Low Phytoestrogen Diet Reduces the Proceptivity But Not the Attractivity of Meadow Voles (*Microtus pennsylvanicus*) 205
Lyndsey Pierson, Matthew Hetherington, and Michael H. Ferkin

16 Dominance Hierarchy in Indian Blackbuck (*Antelope cervicapra* L.): Sources, Behavior and Role of Pheromone Signals 217
Thangavel Rajagopal and Govindaraju Archunan

17 **Asian Elephant Reflections: Chirality Counts** ... 229
L.E.L. 'Bets' Rasmussen, David R. Greenwood, Thomas E. Goodwin, and Bruce A. Schulte

Part IV Interspecific Signaling

18 **Detection of Fish and Newt Kairomones by Ovipositing Mosquitoes** ... 247
Lauren L. Eveland, Jason R. Bohenek, Alon Silberbush, and William J. Resetarits Jr.

19 **Evolutionary Aspects of the Use of Predator Odors in Antipredator Behaviors of Lumholtz's Tree-Kangaroos (*Dendrolagus lumholtzi*)** ... 261
Sigrid R. Heise-Pavlov

20 **Small Emissions with Major Consequences: Specialized Malodorous Defenses in Birds** ... 281
Gregory Röder, Vittorio Baglione, Diana Bolopo, José M. Marcos, Alfréd Trnka, and Daniela Canestrari

Part V Frog Chemosignals

21 **A Review of Chemical Defense in Poison Frogs (Dendrobatidae): Ecology, Pharmacokinetics, and Autoresistance** ... 305
Juan C. Santos, Rebecca D. Tarvin, and Lauren A. O'Connell

22 **Semiochemicals in Anurans: Testing Different Categories with One Poison Frog Species** ... 339
Lisa M. Schulte

23 **Chemical Communication in Archaic New Zealand Frogs** ... 351
Bruce Waldman

Part VI Studies of Domestic and Zoo-Housed Animals

24 **Chemical Signals in Giant Panda Urine (*Ailuropoda melanoleuca*)** ... 363
Martin Dehnhard, Thomas B. Hildebrandt, Cathleen Meerheim, Iain Valentine, and Frank Göritz

25 **Semiochemical Communication in Dogs in the Context of Reproductive Behavior** ... 381
Michał Dzięcioł, Antoni Szumny, Wojciech Niżański, Marcin Wrzosek, Ewa Stańczyk, and Joanna Polikowska

26 **Hormonal and Behavioral Responses to Odor Cues in Zoo-Housed African Painted Dogs (*Lycaon pictus*)** ... 391
Michelle L. Rafacz, Matthew R. Heintz, and Rachel M. Santymire

27 Responses to Domestic Cat Chemical Signals in the House Mouse Are Modulated by Early Olfactory Experience 401
Vera V. Voznessenskaya, Ilya G. Kvasha, Artyom B. Klinov, and Tatiana K. Laktionova

Part VII Field and Analytical Approaches

28 Does Deconvolution Help to Disentangle the Complexities of Mammal Odors? 415
Peter Apps

29 Guidelines for Collecting and Extracting Avian Odors in a Remote Field: Case Study of a Subantarctic Seabird 435
Marianne Gabirot, Jérôme Mardon, Sylvie Campagna, Nigel West, Francesco Bonadonna, and Sandra M. Saunders

30 Pair-Specific Scents in African Wild Dogs, *Lycaon pictus*, and an Example of a Potential Method to Identify Signals Within Complex Mixtures 461
Neil R. Jordan, Peter J. Apps, Krystyna A. Golabek, and J. Weldon McNutt

31 Automated Headspace Solid-Phase Microextraction of Urinary VOCs from Eleven Maned Wolves (*Chrysocyon brachyurus*): A Recursive Workflow for GC–MS Analysis 477
Marieke E. Kester, Elizabeth W. Freeman, Nucharin Songsasen, and Thomas B. Huff

32 Learning and Applications of Chemical Signals in Vertebrates for Human–Wildlife Conflict Mitigation 499
Bruce A. Schulte

33 An Experimental Test of the Effect of Diet on Preen Wax Composition in New Zealand Silvereyes (*Zosterops lateralis*) 511
Laura Azzani, Justin L. Rasmussen, Steven Gieseg, and James V. Briskie

Index 527

Contributors

Peter Apps Paul G. Allen Family Foundation Laboratory for Wildlife Chemistry, Botswana Predator Conservation Trust, Maun, Botswana

Govindaraju Archunan Center for Pheromone Technology, Department of Animal Science, Bharathidasan University, Tiruchirappalli, Tamil Nadu, India

Laura Azzani School of Biological Sciences, University of Canterbury, Christchurch, New Zealand

Renée Båge Department of Clinical Sciences, Swedish University of Agricultural Sciences (SLU), Uppsala, Sweden

Vittorio Baglione Department of Agro-forestry, University of Valladolid, Valladolid, Spain

Sustainable Forest Management Research Institute, Palencia, Spain

Markus C. Baier Department of Chemical Ecology, Bielefeld University, Bielefeld, Germany

May R. Berenbaum Department of Entomology, University of Illinois at Urbana-Champaign, Urbana, IL, USA

Jason R. Bohenek Department of Biology, The University of Mississippi, University, MS, USA

Diana Bolopo Department of Agro-forestry, University of Valladolid, Valladolid, Spain

Francesco Bonadonna CEFE UMR 5175, CNRS – Université de Montpellier – Université Paul-Valéry Montpellier – EPHE, Montpellier, France

Ratan Lal Brahmachary Biology Division, Indian Statistical Institute, Kolkata, India

James V. Briskie School of Biological Sciences, University of Canterbury, Christchurch, New Zealand

Rohan M. Brooker School of Marine Science and Policy, University of Delaware, Newark, DE, USA

School of Biology, Georgia Institute of Technology, Atlanta, GA, USA

Christina D. Buesching Recanati-Kaplan Centre, Wildlife Conservation Research Unit, Department of Zoology, University of Oxford, Tubney, Oxford, UK

Terry Burke NERC Biomolecular Analysis Facility – Sheffield, Department of Animal & Plant Sciences, University of Sheffield, Sheffield, UK

Elissa Z. Cameron School of Biological Sciences, University of Tasmania, Hobart, TAS, Australia

Sylvie Campagna Université de Nimes, Nimes, France

Adelino V.M. Canário Centro de Ciências do Mar, Universidade do Algarve, Faro, Portugal

Daniela Canestrari Department of Biology of Organisms and Systems (BOS), University of Oviedo, Oviedo, Spain

Research Unit of Biodiversity (UMIB, CSIC), University of Oviedo, Oviedo, Spain

Barbara A. Caspers Department of Animal Behaviour, Bielefeld University, Bielefeld, Germany

Noel W. Davies Central Science Laboratory, University of Tasmania, Hobart, TAS, Australia

Martin Dehnhard Leibniz Institute for Zoo and Wildlife Research, Berlin, Germany

Danielle L. Dixson School of Marine Science and Policy, University of Delaware, Newark, DE, USA

School of Biology, Georgia Institute of Technology, Atlanta, GA, USA

Michał Dzięcioł Department of Reproduction and Clinic of Farm Animals, Wrocław University of Environmental and Life Sciences, Wrocław, Poland

James A. Engman Department of Biology, Henderson State University, Arkadelphia, AR, USA

Lauren L. Eveland Department of Biology, The University of Mississippi, University, MS, USA

Michael H. Ferkin Department of Biological Sciences, The University of Memphis, Memphis, TN, USA

Elizabeth W. Freeman New Century College, George Mason University, Fairfax, VA, USA

Center for Species Survival, Smithsonian Conservation Biology Institute, National Zoological Park, VA, USA

Marianne Gabirot Organisms and Environment Research Division, Cardiff School of Biosciences, Cardiff University, Cardiff, UK

CEFE UMR 5175, CNRS – Université de Montpellier – Université Paul-Valéry Montpellier – EPHE, Montpellier, France

Biswatosh Ghosh Department of Zoology, Bidhannagar College, Salt Lake, Kolkata, India

Steven P. Gieseg School of Biological Sciences, University of Canterbury, Christchurch, New Zealand

Robert Glinwood Department of Plant Production Ecology, SLU, Uppsala, Sweden

Krystyna A. Golabek Botswana Predator Conservation Trust, Maun, Botswana

Department of Zoology, Wildlife Conservation Research Unit, University of Oxford, Oxford, UK

Thomas E. Goodwin Department of Chemistry, Hendrix College, Conway, AR, USA

Frank Göritz Leibniz Institute for Zoo and Wildlife Research, Berlin, Germany

David R. Greenwood The New Zealand Institute for Plant & Food Research Limited, Auckland, New Zealand

School of Biological Sciences, University of Auckland, Auckland, New Zealand

Hans Gustafsson Växa Sverige, Uppsala, Sweden

Department of Clinical Sciences, Swedish University of Agricultural Sciences (SLU), Uppsala, Sweden

Innocent H. Harelimana Department of Chemistry, Hendrix College, Conway, AR, USA

Rachel L. Harris School of Biological Sciences, University of Tasmania, Hobart, TAS, Australia

Matthew R. Heintz Lincoln Park Zoo, Conservation and Science Department, Chicago, IL, USA

Sigrid R. Heise-Pavlov Centre for Rainforest Studies at the School for Field Studies, Yungaburra, QLD, Australia

Matthew Hetherington Department of Biological Sciences, The University of Memphis, Memphis, TN, USA

Thomas B. Hildebrandt Leibniz Institute for Zoo and Wildlife Research, Berlin, Germany

Kay E. Holekamp Department of Zoology, Michigan State University, East Lansing, MI, USA

BEACON Center for the Study of Evolution in Action, Michigan State University, East Lansing, MI, USA

Peter C. Hubbard Centro de Ciências do Mar, Universidade do Algarve, Faro, Portugal

Thomas B. Huff Potomac Environmental and Research Education Center, George Mason University, Fairfax, VA, USA

Neil R. Jordan Botswana Predator Conservation Trust, Maun, Botswana

Centre for Ecosystem Science, School of Biological, Earth and Environmental Sciences, University of New South Wales, Sydney, NSW, Australia

Taronga Conservation Society Australia, Taronga Western Plains Zoo, Wildlife Reproduction Centre, Dubbo, NSW, Australia

Aline Umuhire Juru Department of Chemistry, Hendrix College, Conway, AR, USA

Johanna Kabbert Department of Animal Behaviour, Bielefeld University, Bielefeld, Germany

Tina Keller-Costa Centro de Ciências do Mar, Universidade do Algarve, Faro, Portugal

Marieke E. Kester Department of Environmental Science and Policy, George Mason University, Fairfax, VA, USA

Center for Species Survival, Smithsonian Conservation Biology Institute, National Zoological Park, VA, USA

Artyom B. Klinov A. N. Severtsov Institute of Ecology and Evolution, Russian Academy of Sciences, Moscow, Russia

Sarah Kohlwey Department of Animal Behaviour, Bielefeld University, Bielefeld, Germany

Randall A. Kopper Department of Chemistry, Hendrix College, Conway, AR, USA

E. Tobias Krause Department of Animal Behaviour, Bielefeld University, Bielefeld, Germany

Institute of Animal Welfare and Animal Husbandry, Friedrich-Loeffler-Institut, Celle, Germany

Olesya Kutsyna Centro de Ciências do Mar, Universidade do Algarve, Faro, Portugal

Ilya G. Kvasha A. N. Severtsov Institute of Ecology and Evolution, Russian Academy of Sciences, Moscow, Russia

Tatiana K. Laktionova A. N. Severtsov Institute of Ecology and Evolution, Russian Academy of Sciences, Moscow, Russia

Cheryl F. Lichti Department of Pharmacology and Toxicology, University of Texas Medical Branch at Galveston, Galveston, TX, USA

David W. Macdonald Recanati-Kaplan Centre, Wildlife Conservation Research Unit, Department of Zoology, University of Oxford, Tubney, Oxford, UK

Laura J. MacDonald Department of Chemistry, Hendrix College, Conway, AR, USA

Samuel G. Mackintosh Department of Biochemistry and Molecular Biology, University of Arkansas for Medical Sciences (UAMS), Little Rock, AR, USA

José M. Marcos Department of Agro-forestry, University of Valladolid, Valladolid, Spain

Jérôme Mardon CEFE UMR 5175, CNRS – Université de Montpellier – Université Paul-Valéry Montpellier – EPHE, Montpellier, France

Atmospheric and Environmental Chemistry Research (AECR) Group, School of Chemistry and Biochemistry, The University of Western Australia, Crawley, WA, Australia

Daniel B. Mark Department of Chemistry, Hendrix College, Conway, AR, USA

J. Weldon McNutt Botswana Predator Conservation Trust, Maun, Botswana

Cathleen Meerheim Leibniz Institute for Zoo and Wildlife Research, Berlin, Germany

Caroline Müller Department of Chemical Ecology, Bielefeld University, Bielefeld, Germany

Dietland Müller-Schwarze College of Environmental Science and Forestry, State University of New York, Syracuse, NY, USA

Stewart C. Nicol School of Biological Sciences, University of Tasmania, Hobart, TAS, Australia

Wojciech Niżański Department of Reproduction and Clinic of Farm Animals, Wrocław University of Environmental and Life Sciences, Wrocław, Poland

Kristina Nordéus Department of Clinical Sciences, Swedish University of Agricultural Sciences (SLU), Uppsala, Sweden

Lauren A. O'Connell Center for Systems Biology, Northwest Labs, Harvard University, Cambridge, MA, USA

Lyndsey Pierson Department of Biological Sciences, The University of Memphis, Memphis, TN, USA

Mousumi Poddar-Sarkar Department of Botany, University of Calcutta, Kolkata, India

Joanna Polikowska Department of Chemistry, Wroclaw University of Environmental and Life Sciences, Wrocław, Poland

Michelle L. Rafacz Science and Mathematics Department, Columbia College Chicago, Chicago, IL, USA

Thangavel Rajagopal Center for Pheromone Technology, Department of Animal Science, Bharathidasan University, Tiruchirappalli, Tamil Nadu, India

Department of Zoology, Thiagarajar College, Madurai, Tamil Nadu, India

L.E.L. 'Bets' Rasmussen Department of Environmental & Biomolecular Systems, OGI School of Science & Engineering, Oregon Health & Science University, Beaverton, OR, USA

Justin L. Rasmussen School of Biological Sciences, University of Canterbury, Christchurch, New Zealand

Sibdas Ray Department of Chemistry, University of Calcutta, Kolkata, India

William J. Resetarits Jr. Department of Biology, The University of Mississippi, University, MS, USA

Gregory Röder Laboratory of Fundamental and Applied Research in Chemical Ecology (FARCE), Institute of Biology, University of Neuchâtel, Neuchâtel, Switzerland

Frank Rosell Faculty of Arts and Sciences, Department of Environmental and Health Studies, Telemark University College, Telemark, Norway

Juan C. Santos Department of Zoology, Biodiversity Research Centre, University of British Columbia, Vancouver, BC, Canada

Rachel M. Santymire Lincoln Park Zoo, Conservation and Science Department, Chicago, IL, USA

Sandra M. Saunders Atmospheric and Environmental Chemistry Research (AECR) Group, School of Chemistry and Biochemistry, The University of Western Australia, Crawley, WA, Australia

Thomas M. Schmidt Department of Internal Medicine, University of Michigan, Ann Arbor, MI, USA

Bruce A. Schulte Department of Biology, Western Kentucky University, Bowling Green, KY, USA

Lisa M. Schulte Department of Biology, East Carolina University, Greenville, NC, USA

Department of Biogeography, Trier University, Trier, Germany

James D. Shoemaker E. A. Doisy Department of Biochemistry and Molecular Biology, Saint Louis University School of Medicine, St. Louis, MO, USA

Alon Silberbush Department of Life Sciences, Center for Biological Control (CBC), Ben-Gurion University of the Negev, Beer-Sheva, Israel

YungWa Sin Recanati-Kaplan Centre, Wildlife Conservation Research Unit, Department of Zoology, University of Oxford, Tubney, Oxford, UK

NERC Biomolecular Analysis Facility – Sheffield, Department of Animal & Plant Sciences, University of Sheffield, Sheffield, UK

Lennart Söderquist Department of Clinical Sciences, Swedish University of Agricultural Sciences (SLU), Uppsala, Sweden

Nucharin Songsasen Center for Species Survival, Smithsonian Conservation Biology Institute, National Zoological Park, Front Royal, VA, USA

Ewa Stańczyk Department of Reproduction and Clinic of Farm Animals, Wrocław University of Environmental and Life Sciences, Wrocław, Poland

Mark V. Sutherland Department of Biology, Hendrix College, Conway, AR, USA

Antoni Szumny Department of Chemistry, Wroclaw University of Environmental and Life Sciences, Wrocław, Poland

Alan J. Tackett Department of Biochemistry and Molecular Biology, University of Arkansas for Medical Sciences (UAMS), Little Rock, AR, USA

Rebecca D. Tarvin University of Texas at Austin, Austin, TX, USA

Kevin R. Theis Department of Zoology, Michigan State University, East Lansing, MI, USA

BEACON Center for the Study of Evolution in Action, Michigan State University, East Lansing, MI, USA

H. Veronica Tinnesand Faculty of Arts and Sciences, Department of Environmental and Health Studies, Telemark University College, Telemark, Norway

Alfréd Trnka Department of Biology, University of Trnava, Trnava, Slovakia

Iain Valentine Edinburgh Zoo, Royal Zoological Society of Scotland, Edinburgh, Scotland, UK

Zélia Velez Centro de Ciências do Mar, Universidade do Algarve, Faro, Portugal

Arvind Venkataraman Department of Internal Medicine, University of Michigan, Ann Arbor, MI, USA

Christian T. Vlautin Department of Biological Sciences, The University of Memphis, Memphis, TN, USA

Vera V. Voznessenskaya A. N. Severtsov Institute of Ecology and Evolution, Russian Academy of Sciences, Moscow, Russia

Aaron P. Wagner BEACON Center for the Study of Evolution in Action, Michigan State University, East Lansing, MI, USA

Bruce Waldman Laboratory of Behavioral and Population Ecology, School of Biological Sciences, Seoul National University, Seoul, South Korea

Nigel West ChemCentre, Resources and Chemistry Precinct, Curtin University, Bentley, WA, Australia

Danielle J. Whittaker BEACON Center for the Study of Evolution in Action, Michigan State University, East Lansing, MI, USA

Marcin Wrzosek Centre for Experimental Diagnostics and Biomedical Innovations, University of Environmental and Life Sciences, Wrocław, Poland

Qin Yin Department of Chemistry, Hendrix College, Conway, AR, USA

Part I
Overviews of Chemical Ecology (in) Vertebrates

Chapter 1
Are Mammals Just Furry Bugs with Fewer Legs? Convergences in Mammalian and Insect Chemical Ecology

May R. Berenbaum

1.1 CSiV vs. CSiI

The 2014 meeting of the International Society of Chemical Ecology (ISCE) was held on the campus of the University of Illinois at Urbana-Champaign. The 30th annual meeting of the society, it was the first to be held in the United States since the 2008 meeting at Pennsylvania State University and the first ever held jointly with the Chemical Signals in Vertebrates (CSiV) group, which held its 13th triennial meeting. That these two groups have never before held co-located meetings is a bit surprising given that they share a lengthy common history. The idea for founding an international society of chemical ecology was proposed at the first Gordon Conference on Plant-Herbivore Interactions, held in 1980 in Santa Barbara, California, co-organized by Paul Feeny of Cornell University and Gerald Rosenthal of the University of Kentucky. After the second Gordon Conference on Plant-Herbivore Interactions, in 1983, the Executive Committee convened to plan the first meeting of the new Society in Austin, TX, for 1984. At that meeting, vertebrates featured prominently; among the 130 participants at the first meeting was Dietland Müller-Schwarze, a founding member of CSiV.

Müller-Schwarze (this volume) has provided a description of the trajectory of studies in Chemical Signals in Vertebrates, from the first CSiV meeting in 1976 to the 13th meeting in 2014 in Urbana, IL, along with a projection for future studies. If past is prologue, many of the same prognostications should apply to invertebrate (particularly insect) chemical ecology as well as vertebrate chemical ecology. This volume provides an excellent opportunity to review the commonalities in chemical signals between vertebrates, as exemplified by mammals, and invertebrates, as exemplified by insects. The last common ancestor of mammals and insects probably

M.R. Berenbaum (✉)
Department of Entomology, University of Illinois at Urbana-Champaign,
Urbana, IL 61801-3795, USA
e-mail: maybe@illinois.edu

B.A. Schulte et al. (eds.), *Chemical Signals in Vertebrates 13*,
DOI 10.1007/978-3-319-22026-0_1

dates back at least 550 million years, but, despite the evolutionary distance, there are remarkable similarities between two seemingly wildly disparate taxa, likely arising from selective constraints imposed by a terrestrial lifestyle.

1.2 Common Ecology

Relatively few animal phyla have achieved dominance in terrestrial habitats. Among invertebrates, the arthropods, particularly insects, are arguably the animals that have achieved the greatest success; the vast majority of the 900,000+ described species are terrestrial. Among vertebrates, the mammals, which comprise approximately 5500 species, also underwent an adaptive radiation in terrestrial environments, which, although not as large as that of reptiles (with 9500 species) or birds (with upwards of 9800 species), share certain ecological similarities with insects. The Mammaliaformes, comprising modern mammals and their extinct close relatives, appeared in the Triassic, contemporaneously with archosaurs; their success was likely linked to adoption of nocturnal lifestyles, in part to reduce predation risk from the dominant archosaurs (Hall et al. 2012). Nocturnal habits to escape predation characterize many insect taxa; possibly due to release of constraints imposed by visually orienting diurnal predators, night-active insects tend to exhibit larger body sizes than day-active insects in the same communities (Guevara and Aviles 2013). Nocturnal activity, in turn, would have reduced reliance on visual signals and increased reliance on the chemical senses, olfaction, and gustation.

The acquisition of hair, as well, may have initially been an adaptation to facilitate nocturnal living by mammals, both by providing insulation and by serving as sensory structures for detecting non-visual environmental signals (Hall et al. 2012). In insects, hairs (setae or sensilla) can be similarly innervated and serve as sensory structures (as well as insulating material) (Steinbrecht 1996). In both insects and mammals, hairs also serve an important function in dispersing chemical signals (e.g., tail hair tufts in Asian elephants, Raha et al. 2013, and hair pencils in male noctuid moths, Birch et al. 1989).

Other similarities between mammals and insects relate to certain aspects of their behavior. Terrestrial life places a premium on intromittent insemination (to protect sperm viability), so there is a need to bring males and females into close (closest) proximity for mating (Cloudsley-Thompson 1988). Chemical signals in both groups are important in marking territories or protecting resources from competitors, consistent with nocturnal activity (during which time visual signals would be ineffective). Both groups also include species that live in family groups; eusociality (i.e., social systems with reproductive division of labor, cooperative brood care, and overlapping generations) is essentially restricted to arthropods and mammals (Sherman et al. 1995). Social living in both groups is mediated by chemicals and in both groups these signals are invaluable for mediating recognition of individuals or group members (Wilson 1970).

1.3 Common Signal Detection Patterns with Evolutionarily Independent Proteins

Terrestrial living has undoubtedly contributed to similarities between mammals and insects. In both groups, membranes of chemosensory cells are bathed in liquid and odorants must travel through a liquid-filled space to reach receptors (Hildebrand and Shepherd 1997). As well, membranes in both groups are modified to increase surface area and cells are positioned in proximity to a number of accessory cells. There is neuronal conservation, with small bipolar neurons with an arborescent terminus and with a direct axonal connection to the central nervous system conveying information about chemical stimuli (Hildebrand and Shepherd 1997).

With the vast evolutionary gulf between mammals and insects, however, it is not altogether surprising that chemosensory proteins, and the molecular underpinnings of olfaction and gustation, have evolved independently. Mammalian olfactory receptors are 7-transmembrane G-protein-coupled receptors; insect olfactory receptors represent an entirely different family of transmembrane proteins, oriented differently in the cell membrane and representing a phylogenetically independent line of transmembrane proteins (Benton et al. 2006), which form ligand-gated ion channels. Moreover, insect odorant receptors comprise a heteromeric complex with both an obligate noncanonical Orco unit and a canonical Or required for functionality (Ray et al. 2014). Mammals and insects also utilize groups of chemosensory proteins with no orthologues in the other group; in vertebrates there are v1r and v2r G-protein-coupled receptor proteins associated with vomerolfaction, and in insects there are ionotropic receptors that have apparently evolved from ionotropic glutamate receptors, conserved ligand-gated ion channels (Rytz et al. 2013).

1.4 Common Signal Processing Pathways with Evolutionarily Independent Proteins

In both mammals and insects (indeed, in vertebrates and invertebrates generally), only one olfactory receptor protein (from a large inventory of such proteins) is typically expressed in each neuron and each receptor protein is expressed in only a few neurons (Fuss and Ray 2009, Sim et al. 2012). Although this general pattern is strikingly similar across taxa, the olfactory receptor gene families are independently evolved and the regulatory process by which neurons regulate expression of a particular protein also differs between mammals and insects (Ray et al. 2008). Structural elements are similar beyond peripheral perception of chemical signals as well (Hildebrand and Shepherd 1997). In both mammals and insects, neurons expressing the same protein synapse in the same glomerulus. In both groups as well, olfactory cells project directly to the brain. The olfactory tract leads to the primary olfactory cortex; from there the primary projection goes to the thalamus and then to the neocortex. In mammals, each olfactory receptor protein is expressed in a subset of sensory neurons; axons of these, as they do in insects, extend directly to the brain to contact glomerular neuropils in the olfactory bulb. From these glomeruli in mammals, from 1 to 100 projection neurons transmit information to the higher brain centers, such as the olfactory cortex. In insects, olfactory receptor cells project to the deutocerebrum, where they form synapses in glomerular neuropils. Interneurons then communicate with the mushroom bodies (corpora pedunculata) of the protocerebrum; from there, descending pathways carry out behavioral responses (Hildebrand 1995; Hildebrand and Shepherd 1997). Thus, the "olfactory relay" is organizationally similar in insects and mammals. Based on the shared structure and function of olfactory sensory systems, Hildebrand (1995) has speculated that olfactory systems based on glomerular organization arose coincident with colonization of dry land (and hence contact with airborne signals) 500 million years ago.

Similarly, perireceptor phenomena share similarities in mammals and insects. Cytochrome P450s, for example, are involved in both taxa in processing xenobiotics, including those bound to receptors. These P450s evolved in the context of a terrestrial environment, where lipophilic toxins must be oxidized to facilitate export out of the body (Gonzalez and Nebert 1990). Although P450s involved in biosynthesis of endogenous compounds (such as sterols and steroids) tend to be conserved to some degree across vertebrates and invertebrates, the P450s associated with detoxification have evolved independently in vertebrates and invertebrates (Kawashima and Satta 2014). Intriguingly, though, there is one P450 Clan, comprising CYP4 genes, that is represented in both insect and mammalian genomes; in a few insect species, CYP4 genes are associated with pheromone degradation in sensory structures and in humans some CYP4 genes encode enzymes that carry out detoxification reactions (Hsu et al. 2007)

In terms of taste, there are many commonalities between mammals and insects in how gustatory systems are organized (Yarmolinsky et al. 2009). Both systems are dedicated to recognizing nutritionally important dietary constituents with a fairly limited inventory of labeled inputs—sweet, bitter, water, and carbon dioxide in insects and sweet, sour, salty, bitter, and umami in mammals. Segregated groups of peripheral receptor cells are responsible for the primary tastes. This overall organizational similarity belies molecular differences; as with olfactory receptors, gustatory receptor proteins in mammals and insects are fundamentally different.

1.5 Nature of Chemical Signals

Chemicals involved in communication in both mammals and insects are by and large secondary metabolites—that is, they are generally not molecules involved in the primary physiological processes of life, such as respiration, digestion, excretion, and in the case of green plants photosynthesis (Berenbaum and Seigler 1992). All secondary metabolic pathways derive ultimately from primary pathways. Just as variation in primary metabolic processes affects the probability of survival and is subject to natural selection, variation in secondary metabolic processes also affects the probability of survival and is subject to natural selection. Differences in selective forces, arising from diverse ecological contexts, account at least in part for the tremendous diversity of structure that is characteristic of secondary metabolites (Berenbaum and Seigler 1992).

There are limits, however, to the variation in secondary metabolism available for natural selection to act upon. These limits are imposed by, among other things, availability of raw materials for constructing secondary metabolites, suitability of those materials for effecting biological activity, phylogenetic constraints on primary metabolic pathways from which secondary pathways arise, and energetic and toxicological implications of particular pathways (Berenbaum and Seigler 1992). The starting materials for signal chemicals are almost always primary metabolites, which include amino acids (for making proteins for building structures and catalyzing reactions), nucleotide bases (for assembling genetic materials), sugars and fatty acids (for energy storage), and glycerol (for making cell membranes). These primary metabolites are universally distributed among all organisms. Thus, despite the initially dumbfounding diversity of structures of secondary metabolites produced by organisms, there are clear patterns of similarity that emerge upon close inspection (Berenbaum and Seigler 1992).

Most obvious of those similarities is the relatively conservative elemental composition of secondary metabolites. As is the case for most primary metabolites, most secondary metabolites are composed of carbon, oxygen, hydrogen, and nitrogen. These elements are relatively abundant in the lithosphere (essentially, Earth's crust) and are hence available for incorporation into molecules. That availability of elements affects secondary metabolism can be seen by comparing secondary metabolites in terrestrial and marine organisms. Compounds containing halogens such as chlorine or bromine atoms are extremely rare (although not nonexistent) in insects and mammals, which are found largely in terrestrial or freshwater environments. By comparison, seawater is comparatively rich in bromine and chlorine; accordingly, marine invertebrates frequently produce secondary metabolites containing these elements. Dibromo-indigo, for example, is a pigment produced by the sea snail *Bolinus brandaris* (Vougogiannopoulou and Skaltsounis 2012)

In addition to elemental composition, the structure of chemical signals is constrained by the medium through which they travel. Chemical signals tend to differ significantly in structure in terrestrial and aquatic environments (Wyatt 2003). In air, the rate at which molecules diffuse is to an extent a function of molecular weight; high-molecular-weight molecules tend to be relatively nonvolatile. Most insect sex pheromones, thus, are on the order of MW 200–300 (Wyatt 2003). In contrast, in aqueous environments, diffusion rates depend not on volatility but on water solubility, a physical property that is not directly associated with molecular weight (Wyatt 2003). In aquatic systems, many biologically important signal substances are large polar molecules such as proteins, which, with effectively zero vapor pressure, would be of little use in an airborne communication system (Wilson 1970).

That there are constraints on structure and composition of signal chemicals means that the same signal chemical may be produced by wildly disparate organisms. Chemical signals, then, may not be unique to particular taxa or habitats. That they serve a similar ecological function increases the likelihood that chemical signals may be shared between mammals and insects.

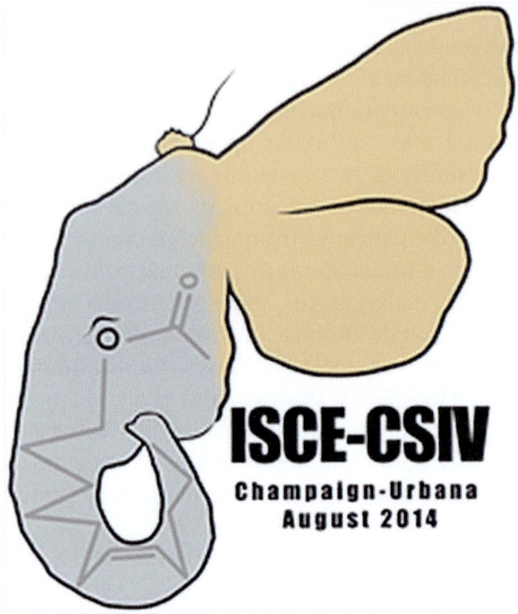

Fig. 1.1 Elemoth (designed by Joseph Wong)

The basic requirements for life are in force irrespective of the environment—obtaining nutrients, escaping predation, and reproducing. As well, most of the basic primary processes are for the most part universal in occurrence, so biosynthetic starting materials are conserved throughout the living world. Finally, all organisms are constrained and directed by the same basic ecological and evolutionary processes. Thus, even when chemical identities and transmission dynamics differ, many basic principles pertaining to chemical signals emerge. For these reasons, the logo for the 2014 joint meeting of the International Society of Chemical Ecology and Chemical Signals in Vertebrates (ISCE-CSiV 2014) was a fictional chimeric creature that could be called an "elemoth"—part Asian elephant (*Elephas maximus*) and part cabbage looper (*Trichoplusia ni*) (Fig. 1.1). The elemoth embodies the constraints on structure and composition of signal chemicals in very different ecological contexts. (*Z*)-7-dodecenyl acetate, a component of the sex pheromone of over 140 species of moths, including the cabbage looper *Trichoplusia ni*, also is produced as a urinary preovulatory female-to-male pheromone in the Asian elephant (Rasmussen et al. 1997). Subsequent analyses of urine from African elephants (*Loxodonta africana*) have detected the bark beetle aggregation pheromones frontalin, *exo*-brevicomin, and *endo*-brevicomin and the aphid alarm pheromones (*E,E*)-α-farnesene and (*E*)-β-farnesene (Goodwin et al. 2006). The elemoth seemed to us to be an apt symbol for a gathering of scientists dedicated to identifying the substances, characterizing the transmission dynamics, and elucidating the mechanisms underlying chemical mediation of interactions among organisms. Given those similarities, it would seem that future joint meetings would be in the interest of all chemical ecologists.

Acknowledgments I thank my colleague Hugh Robertson for helpful comments on the manuscript and I thank Bruce Schulte both for inviting me to contribute this chapter and for patiently waiting for me to complete it!

References

Benton R, Sachse S, Michnick SW, Vosshall LB (2006) Atypical membrane topology ad heteromeric function of *Drosophila* odorant receptors in vivo. PLoS Biol 4, e20

Berenbaum MR, Seigler D (1992) Biochemicals: engineering problems for natural selection. In: Roitberg BD, Isman MD (eds) Insect chemical ecology. An evolutionary approach. Routledge, New York, pp 84–121

Birch MC, Lucas D, White PR (1989) The courtship behavior of the cabbage moth *Mamestra brassicae* (Lepidoptera: Noctuidae), and the role of male hair-pencils. J Insect Behav 2:227–239

Cloudsley-Thompson JL (1988) Evolutionary trends in reproduction. Evolution and adaptation of terrestrial arthropods. Springer, Berlin, pp 57–79

Fuss SH, Ray A (2009) Mechanisms of odorant receptor gene choice in *Drosophila* and vertebrates. Mol Cell Neurosci 41:101–112

Gonzalez FJ, Nebert DW (1990) Evolution of the P450 gene superfamily: animal-plant 'warfare', molecular drive and human genetic differences in drug oxidation. Trends Genet 6:182–186

Goodwin TE, Eggert MS, House SJ, Weddell ME, Schulte BA, Rasmussen LEL (2006) Insect pheromones and precursors in female African elephant urine. J Chem Ecol 32:1849–1853

Guevara J, Aviles L (2013) Community-wide body size differences between nocturnal and diurnal insects. Ecology 94:537–543

Hall MI, Kamilar JM, Kirk EC (2012) Eye shape and the nocturnal bottleneck of mammals. Proc Roy Soc London B: Biol Sci 279:4962–4968

Hildebrand JG (1995) Analysis of chemical signals by nervous systems. Proc Natl Acad Sci U S A 92:67–74

Hildebrand J, Shepherd GM (1997) Molecular mechanisms of olfactory discrimination: Converging evidence for common principles across phyla. Annu Rev Neurosci 20:263–278

Hsu M-H, Savas U, Griffin KJ, Johnson EF (2007) Human cytochrome P450 family 4 enzymes: Function, genetic variation and regulation. Drug Metab Rev 39:515–538

Kawashima A, Satta Y (2014) Substrate-dependent evolution of cytochrome P450: Rapid turnover of the detoxification-type and conservation of the biosynthesis-type. PLoS One 9, e100059

Raha P, Poddar-Sarkar M, Nag UK, Bhar R, Brahmachary RL (2013) Ultrastructure and chemical composition of elephant hair in the context of chemical signals in the Asian elephant *Elephas maximus*. In: East ML, Dehnhard M (eds) Chemical signals in vertebrates 12. Springer, New York, pp 227–234

Rasmussen LEL, Lee TD, Zhang A, Roelofs WL, Daves G Jr (1997) Purification, identification, concentration and bioactivity of (Z)-7-dodecen-1-yl acetate: Sex pheromone of the female Asian elephant, *Elephas maximus*. Chem Senses 22:417–437

Ray A, van der Goes van Naters W, Carlson JR (2008) A regulatory code for neuron-specific odor receptor expression. PLoS Biol 6, e125

Ray A, van der Goes van Naters W, Carlson JR (2014) Molecular determinants of odorant receptor function in insects. J Biosci 39:555–563

Rytz R, Croset V, Benton R (2013) Ionotropic receptors (IRs): Chemosensory ionotropic glutamate receptors in Drosophila and beyond. Insect Biochem Mol Biol 43:888–897

Sherman PW, Lacey EA, Reeve HK, Keller L (1995) The eusociality continuum. Behav Ecol 6:102–108

Sim CK, Perry S, Tharada SK, Lipsick JS, Ray A (2012) Epigenetic regulation of olfactory receptor gene expression by the Myb-MuvB/dREAM complex. Genes Dev 26:2483–2498

Steinbrecht RA (1996) Structure and function of insect olfactory sensilla. Ciba Found Symp 200:174–177

Vougogiannopoulou K, Skaltsounis A-L (2012) From Tyrian purple to kinase modulators: naturally halogenated indirubins and synthetic analogues. Planta Med 78:1515–1528

Wilson EO (1970) Chemical communication within animal species. In Sondheimer E, Simeone JB (eds) Chemical Ecology. New York: Academic Press, pp. 133–156

Wyatt TD (2003) Pheromones and animal behaviour: Communication by smell and taste. Cambridge University Press, Cambridge, pp 1–22

Yarmolinsky DA, Zuker CS, Ryba NJP (2009) Common sense about taste: from mammals to insects. Cell 139:234–239

Chapter 2
Chemical Signals in Vertebrates 13: Where We Stand and What Might Be Next

Dietland Müller-Schwarze

2.1 The Beginnings

The first Chemical Signals in Vertebrates took place in 1976 at Saratoga Springs, New York. Entomologists had studied insect pheromones and regularly covered their field in published books and in meetings. I had felt then that we had reached the critical mass to fill a room with vertebrate odor communication researchers from around the world. We were able to put together the first book specifically focused on vertebrate chemical communication (Müller-Schwarze and Mozell 1977). The papers overwhelmingly dealt with intraspecific communication.

It was the time when Michael Stoddart could write in 1976: "It has not been shown hitherto that rodents perceive the odour of their predators and secondly, if it is perceived, whether it is interpreted as being associated with the predator and resulting in an observable behavioral change."

2.2 Where We Stand

While the number of participants in the CSiV meetings has oscillated around the same level over the years, CSiV13 marked a complete rejuvenation: For the first time, the meeting was carried by an entire new generation, advancing the field in new ways.

The scope of research areas keeps diversifying: The first CSiV meeting emphasized pheromones, i.e., intraspecific communication, following the examples of insect pheromones. Now, research has broadened to include a variety of interactions

D. Müller-Schwarze (✉)
College of Environmental Science and Forestry, State University of New York, Syracuse, NY, 13210, USA
e-mail: dmullers@esf.edu

B.A. Schulte et al. (eds.), *Chemical Signals in Vertebrates 13*,
DOI 10.1007/978-3-319-22026-0_2

between species. In 2014, 9 out of 25 presentations (36 %) dealt with interspecific behavior, while in 1976 under 10 % did so.

Researchers study more taxonomic groups now: About 19 % of the papers in this volume deal with birds, while none were included in the 1976 volume. After Bernice Wenzel had pioneered the field of avian olfaction, it was not until the nineties that at any one meeting more than two papers dealt with birds. In this current volume, mammals were studied in 56 % of the papers, while in the earlier volumes mammals usually accounted for 75 % or more of the papers. This reflects the greater diversity of taxonomic groups now being investigated for their chemical communication.

During the intervening decades we have learned to appreciate the complexities of vertebrate semiochemicals, both in terms of chemical composition, and the sophistication of learning and interpretation of signals by animals. As this volume shows, investigators have developed sophisticated bioassays, methods of chemical separations and identification, and statistical approaches to deal with complex mixtures of compounds that often occur in small amounts, and can be volatile, or non-volatile such as proteins. The identification of darcin, a protein in male mouse urine that attracts and informs females (Roberts et al. 2010), constituted a major milestone.

Synergistic effects of compounds in complex pheromone mixtures show that presence and ratios of components matter enormously. For instance, in amphibians, the plethodontid modulation factor (PMF) in the salamander *Plethodon shermani* from the male's mental gland on his chin influences the female's courtship behavior. While the whole secretion increases the female's receptivity, a major fraction actually decreases it (Wilburn et al. 2015). Such findings complicate the traditional search for the "active principle" and call for new interpretations. Different components in a mixture may have different functions.

2.3 What Might Be Next?

2.3.1 New Tools

The study of chemical signaling started with interactions between two individuals of the same species and has proceeded to consider ever more complex situations. We are inching ever closer to understand how chemical information flows through pieces of an ecosystem. For instance, prey species may determine danger levels by monitoring the diet of their predators.

We humans may smell a pine forest, a swamp, or a penguin rookery, but to animals, there is immensely more detail. Their odorsphere consists of many odorscapes: On the ocean we have odor patches, odor mountains, odor valleys, odor holes, and odor waves. We have every reason to assume that seabirds and other animals can read these odorscape patterns. On land there are odor lanes, odor rivers, and odor fields, depending on the lay of the land and air currents. And then there are small points in space we know as scent marks, but most of the time we humans cannot smell. We "are chemosensory idiots," as E.O. Wilson puts it in his latest book (2014). Our large body size and bipedalism forced us to rely on audiovisual communication and made us "sensory cripples" (Wilson 2014) when it comes to olfaction.

We need tools to extend our meager chemical senses and monitor the odorscapes that matter so much to animals. Enter the "electronic noses," tools that are being developed right now. For instance, the Caltech electronic nose relies on conducting polymers that swell upon exposure to vapors. Odor mixtures yield specific electronic fingerprints that can be processed with principal component analysis, for instance. The Caltech electronic nose is trainable for specific odors. The sensitivity for biogenic amines is as good as 10 parts per trillion. An electronic nose can not only distinguish rose oil from garlic odor, but even Pepsi from Coke, R- from S-enantiomers, and a 51 %:49 % mixture of two analytes from a 50:50 mixture (Lewis Research Group 2014). Currently, electronic noses are being used to detect disease, off-flavors in the food industry, and leaking nitrogen in the International Space Station (Kremer 2014).

Behavioral researchers and chemical ecologists might want to have a serious look at electronic noses as tools in their arsenal. For use in the field, the ongoing miniaturizing of the devices is promising. Eventually devices will be small enough so they can be attached to mobile phones (Covington 2014) and will be programmable to detect specific compounds.

2.3.2 *Current and Future Environmental Challenges*

Basic research has advanced sufficiently that vertebrate chemical ecologists now can help understand and even solve urgent environmental problems of our time in a variety of ways. Advances in neuroscience, genetics, physiology, behavior studies, and chemistry promise exciting new possibilities. The following areas are a few examples of opportunities for chemical ecologists.

2.3.2.1 Dealing with Disruption of Chemical Communication

Acidification of both fresh- and seawater can seriously disrupt chemical communication and predator–prey interactions in aquatic organisms. Ocean acidification and warming interferes with chemical signaling. A lower pH affects the chemosignals themselves and their interactions with the chemosensory receptor proteins (Wyatt et al. 2014). Lower pH may also affect responses to alarm pheromones: fathead minnows (*Pimephales promelas*) and finescale dace (*Chrosomus neogaeus*) no longer respond to their species' alarm pheromone if the pH is lowered from 8.0 to 6.0 because the N-oxide group on the pheromone molecule is lost (Brown et al. 2000).

In the area of predator–prey interactions, Royce-Malmgren and Watson (1987) showed that changes in pH altered preying behavior in Atlantic salmon (*Salmo salar*). Prey fish leak amino acids that attract salmon. While glycine attracts salmon, they avoid L-alanine. If the pH of their freshwater environment is lowered from 7.6 to 5.1, the fish become indifferent to glycine, but are now attracted to L-alanine. Since the latter could also be emanating from predatory fish, the response proves maladaptive.

2.3.2.2 Mitigating Wildlife Problems

In many contexts, we have to coexist with wildlife, but the challenges consist in minimizing negative impacts for humans. If we want to keep animals from certain places, plants, or situations, we can condition them to artificial signal stimuli (Schulte, 2016). In some cases, however, the conditioning may be very specific to the one place where the animal experienced the negative/painful stimulus. For example, at Yellowstone National Park, rangers shot rubber bullets at grizzly bears begging for food from tourists along roads. The bears avoided only the particular bend of the road were they had been hit, but continued to beg at other stretches of the road. However, such local effects may be all that is needed in many cases.

We might want to steer colonization by wildlife: In many situations the "push-pull" concept (Schulte 2016) is desirable. An example is habitat scenting to keep beavers out of sensitive areas. To prevent beavers from settling in attractive sites one can apply scent from male and female beavers (the "push" part), mimicking occupancy by an adult pair. Newly arriving beavers avoided these sites and settled elsewhere, provided there were other, unscented attractive sites available elsewhere (the "pull" part). We experimentally scented habitat in one saturated and two less dense populations. The artificial scent kept beavers from colonizing a particular site only if enough other sites were available (Welsh and Müller-Schwarze 1989).

2.3.2.3 Habitat Imprinting in Fish

Olfactory habitat imprinting has been tried to boost, reintroduce, or guide local populations of economically important fish such as salmon to avoid unwanted mixing of gene pools. Morpholine-imprinted Coho salmon have been successfully attracted to the Mad River in California after morpholine had been added to the river (Hassler and Kucas 1988). 2-Phenylethanol (phenylethyl alcohol) imprinting in salmon resulted in a higher percentage of neurons sensitive to this compound in the nose. This might be helpful in ascertaining that imprinting has taken place before the fish are released into the wild (Barinaga 1999). Recent experiments in coral reef fish showed that chemical imprinting is possible, but innate preferences also play a role (Dixson et al. 2014).

2.3.2.4 Conservation and Breeding of Endangered Species

Semiochemicalsmost likely will serve as very important tools to enhance captive breeding of endangered species. Priming pheromones are particularly important. Olfactory stimuli that enhance breeding by accelerating puberty or stimulating estrus cycles are sought, while those that hinder successful reproduction such as puberty-delaying or pregnancy-blocking pheromones should be avoided. Because of the various odor effects, close attention must be paid to group composition. Presence of strange vs. familiar males, dominance relationships, or grouping of females all affect reproductive behavior, often via odors. Such interfering odors are to be avoided by various means, ranging from social isolation to proper ventilation. Odors also have successfully been used in behavioral enrichment to stimulate captive animals (e.g., Rafacz et al. 2016). However, unintended consequences may occur for captive breeding: the odors provided may contain compounds that interfere with sexual maturation, estrus, or pregnancy.

After successful reproduction we have to prepare the animals for release into their natural habitat. This includes, among other issues, teaching them the correct prey odors and avoiding imprinting them on humans. Although challenging, preserving endangered species with the aid of pheromones will be one of the most important practical applications. Again Wilson (2014): "We cannot talk in the language of pheromones, but it will be well to learn more about how other organisms do it, in order better to save them and with them the majority part of the environment on which they depend."

References

Barinaga M (1999) Salmon follow watery odors home. Science 286:705–706

Brown GE, Adrian JC, Smyth E, Let H, Brennan S (2000) Ostariophysan alarm pheromones: laboratory and field tests of the functional significance of nitrogen oxides. J Chem Ecol 26:139–154

Covington J (2014) Interview. The Guardian, 2 April 2014

Dixson DL, Jones GP, Munday PL, Planes S, Pratchett MS, Thorrold SR (2014) Experimental evaluation of imprinting and the role innate preference plays in habitat selection in a coral reef fish. Oecologia 174:99–107

Hassler TJ, Kucas ST (1988) Returns of morpholine-imprinted coho salmon to the Mad River, California. North Am J Fish Manage 8:356–358

Kremer W (2014) Sniffing out cancer with electronic noses. BBC News 8 March 2014. www.bbc.com/news/magazine-26472225

Lewis Research Group (2014) nsl.caltech.edu/research: nose

Müller-Schwarze D, Mozell MM (1977) Chemical signals in vertebrates. Plenum, New York

Rafacz ML, Heintz MR, Santymire R (2016) Hormonal and behavioral responses to odor cues in zoo-housed African painted dogs (*Lycaon pictus*). New York: Springer

Roberts SA, Simpson DM, Armstrong SD, Davidson AJ, Robertson DH, McLean L, Beynon RJ, Hurst JL (2010) Darcin: a male pheromone that stimulates female memory and sexual attraction to an individual male's odour. BMC Biol 8:75

Royce-Malmgren CH, Watson WH III (1987) Modification of olfactory-related behavior in juvenile Atlantic salmon by changes in pH. J Chem Ecol 13:533–546

Schulte, BA (2016) Learning and applications of chemical signals in vertebrates for human-wildlife conflict mitigation. New York: Springer

Stoddart DM (1976) Effects of weasels (*Mustela nivalis*) on trapped samples of their prey. Oecologia 22:439–441

Welsh RG, Müller-Schwarze D (1989) Experimental habitat scenting inhibits colonization by beaver, *Castor canadensis*. J Chem Ecol 15:887–893

Wilburn DB, Eddy SL, Chouinard AJ, Arnold SJ, Feldhoff RC, Houck LD (2015) Pheromone isoform composition differentially affects female behaviour in the red-legged salamander, *Plethodon shermani*. Anim Behav 100:1–7

Wilson EO (2014) The meaning of human existence. Liveright (W.W. Norton), New York

Wyatt TD, Hardege JD, Tershak J (2014) Ocean acidification foils chemical signals. Science 346:176

Chapter 3
Assessing the Role of Olfactory Cues in the Early Life History of Coral Reef Fish: Current Methods and Future Directions

Rohan M. Brooker and Danielle L. Dixson

3.1 Introduction

Coral reefs are one of the most dynamic ecosystems on earth, home to an unparalleled diversity of marine organisms (Reaka-Kudla 1997). Due to the multitude of biotic and abiotic processes continuously occurring within these systems, the surrounding waters contain a complex mixture of chemical signals. These potential cues hold important information that could impact the ecology of reef-associated organisms. For instance, while coral reefs appear to our eyes as a kaleidoscope of visual information, many organisms primarily interact with this environment using chemosensory systems, relying on odors within the water column to gain detailed, often crucial information about their surroundings (Arvedlund and Kavanagh 2009). Many broad-scale ecological processes that drive coral reef dynamics, such as patterns of distribution and abundance, diversity, and resilience, are influenced by behavioral interactions that occur between organisms and their environments (Leis 2002; Baird et al. 2003). Therefore, empirical studies that further our understanding of how olfactory cues influence these behavioral interactions are vital.

Fishes are one of the most conspicuous and morphologically diverse groups of mobile coral reef fauna, occupying a wide variety of ecological niches and often fulfilling important functional roles (Bellwood et al. 2003; Goatley and Bellwood 2010). While many species have good visual capabilities (Siebeck and Marshall 2007), they also rely on an array of alternative sensory systems, such as acoustic, olfactory, and electrical perception to interact with their habitats (Myrberg and Fuiman 2002). In particular, many have evolved sensitive chemosensory organs, making waterborne chemicals an important source of information for these species.

R.M. Brooker • D.L. Dixson (✉)
School of Marine Science and Policy, University of Delaware, Newark, DE 19716, USA

School of Biology, Georgia Institute of Technology, Atlanta, GA 30332, USA
e-mail: dixson@udel.edu

B.A. Schulte et al. (eds.), *Chemical Signals in Vertebrates 13*,
DOI 10.1007/978-3-319-22026-0_3

Almost all demersal reef fishes have a bipartite life cycle, consisting of a planktonic larval stage, lasting from hours to months in duration (Luiz et al. 2013), that facilitates spatial dispersal of individuals, and a juvenile-adult stage following metamorphosis and settlement onto the benthos. As behavioral decisions made during this time can directly influence survival, the ability to receive, interpret, and integrate sources of relevant olfactory information into these decisions is critical.

Understanding the nature of odor molecules in the marine environment is essential when attempting to determine the ability of organisms, such as a larval fish, to detect and interpret them. Due to the relatively slow rate at which chemicals disperse via diffusion in seawater, hydrodynamic processes (Atema 2012) often primarily determine the dispersal fields of waterborne chemicals over ecologically relevant spatial scales. Therefore, most studies attempting to isolate the olfactory preferences of young reef fishes, and the influence of these preferences on behavior, have done so under controlled laboratory conditions (e.g., Atema et al. 2002; Holmes and McCormick 2010). Many developments in this area are relatively recent, such as the study of larval and juvenile cognition and the role of olfactory cues in learning post-settlement. Consequently, many important questions have been answered without any need to isolate the specific chemicals involved. However, as reefs shift away from coral dominated systems (Bell et al. 2013), the relative availability of odors and the ability of fishes to receive odors are altered (Munday et al. 2009); in this situation, a greater understanding of the chemical properties of olfactory cues will likely prove invaluable.

The aim of this chapter is not to provide a comprehensive review of olfactory cue use by larval and juvenile coral reef fishes, but rather to (1) highlight the value of incorporating chemical ecology into behavioral studies by discussing two notable areas where this has been particularly informative, larval olfactory preferences and learning in juveniles; (2) give an overview of the current state of knowledge regarding the nature of olfactory cues relevant to juvenile coral reef fishes, and the unique challenges examining these cues present; (3) evaluate the key methods used to assess the role of olfactory cues in juvenile reef fish behavior; and finally (4) provide some suggestions of future directions which will improve our understanding of these important processes.

3.2 Olfactory Cues and the Early Life History of Coral Reef Fishes

Olfactory cues can be important mediators of many behavioral processes throughout the life cycles of coral reef fishes. For instance, adults use odors to locate and select food (DeBose et al. 2008; Lönnstedt et al. 2012a), to navigate through reef habitats (Devine et al. 2012), to regulate mutualistic interactions (Dixson and Hay 2012), and to avoid detection (Brooker et al. 2015). However, sources of olfactory information appear to be particularly important throughout early life history stages, especially during the critical period when individuals transition from planktonic larvae to benthic-dwelling juveniles. This time is particularly hazardous, as individuals must first locate a suitable habitat on which to settle, and then rapidly adapt their behavior to this new environment. Globally, the community composition and biodiversity of coral reefs is in decline due to climate change and other anthropogenic impacts (Bellwood et al. 2004; Carpenter et al. 2008; De'ath et al. 2012), with resilience largely dependent on a continued larval supply. For this reason, determining the behavioral mechanisms used by reef-associated fishes before, during, and post-settlement, and what effects the rapid degradation of reef habitats will have on these behaviors, has swiftly become a priority (Lecchini et al. 2013).

3.2.1 The Role of Olfactory Cues in Habitat Recognition and Settlement Site Selection

Although pelagic larvae are subjected to strong hydrodynamic processes that can affect dispersal (Cowen et al. 2006), species often display non-random distributions (Bell et al. 1982; Shulman 1998) and high levels of self-recruitment to natal reefs (Jones et al. 1999; Taylor and Hellberg 2003; Almany et al. 2007). These patterns, along with the remarkable swimming and sensory competency of many late-stage larvae (Fisher et al. 2000), suggest that reef fishes exert substantial control over where they settle. For a larval fish, recruitment to optimal habitat is crucial for immediate survival as well as long-term fitness (Booth and Wellington 1998; Munday 2001), while at the population level, patterns of larval retention and dispersal affect connectivity and therefore genetic population structure (Gerlach et al. 2007). Not surprisingly, reef fish larvae have evolved finely tuned, multisensory mechanisms to locate and orientate towards appropriate habitats using a variety of environmental cues (Lecchini et al. 2005b). While the relative importance of these cues appears to vary temporally depending on ontogeny and relative spatial scale (Dixson et al. 2011; Huijbers et al. 2012), evidence suggests that a larvae's chemosensory system can be engaged throughout this process (Arvedlund et al. 1999; Vail and McCormick 2011; Paris et al. 2013).

Coral reef fish larvae can exhibit preferences for a variety of coral reef-based cues, including live coral (Coppock et al. 2013; Dixson et al. 2014b), seagrass (Huijbers et al. 2012; Igulu et al. 2013), conspecifics (Sweatman 1983; Lecchini et al. 2005a), and specialized habitats such as a specific anemone species (Elliott et al. 1995; Arvedlund et al. 1999). While many settlement-stage larvae and post-settlement juveniles have now been shown to exhibit olfactory preferences for reef based cues, and will orient and swim towards preferred odors under laboratory conditions or over within-reef spatial scales, the degree to which these odors influence the behavior of larvae far from the reef, relative to other sources of sensory information, has remained uncertain. However, odors may also be important over these larger spatial scales if, for instance, larvae can select between currents based on the relative strengths of a preferred odor (Gerlach et al. 2007). Recent work has revealed that larvae are indeed able to discriminate reef odors from substantial distances offshore, changing their direction and swimming speeds in response (Paris et al. 2013; Lecchini et al. 2014). In addition, some species exhibit preferences for the odor of native vegetation, which often indicates adjacent coral habitats, and may occur in strong enough concentrations to lure distant larvae (Dixson et al. 2008). Exactly how concentrated these odors must be to initiate a behavioral response is unknown, although the olfactory receptors of marine fishes often have low detection thresholds for behaviorally relevant chemicals (i.e., Hubbard et al. 2003, 2011; Yacoob and Browman 2007).

It appears as if many of these chemically mediated behaviors are regulated by the presence of specific olfactory cues. Therefore, changes to the chemical landscape of reefs following habitat degradation may negatively affect the ability of larval reef fish to locate suitable habitats if these cues are removed or modified. For instance, while larvae are attracted to the odor of native rainforest vegetation these preferences do not extend to general vegetation odor (Dixson et al. 2008), suggesting that as coastal rainforests are cleared in favor of non-endemic agricultural crops, the ability of fishes to locate adjacent reef habitats using terrestrial odors could be compromised. Other consequences of coastal development, such as sediment enrichment and re-suspension, also appear to reduce the ability of fishes to distinguish olfactory cues (Wenger et al. 2011). Globally, many coral reefs are shifting from coral-dominated to algae-dominated systems, due to the targeted removal of herbivorous fish, increased coastal eutrophication, and other confounding anthropogenic impacts (Bellwood et al. 2004; Carpenter et al. 2008). While larvae are attracted to the olfactory cues produced by healthy, coral-dominated habitats they avoid the odor of seaweed-dominated, highly fished reefs (Lecchini et al. 2013; Dixson et al. 2014b) potentially limiting how effectively larvae can replenish degraded coral habitats. Overall, it appears that the removal or dilution of key odors could have serious consequences for recruitment.

3.2.2 Predator Identification by Juvenile Reef Fishes and the Role of Odor

The few larval fish that successfully manage to navigate through the pelagic environment to a suitable habitat find themselves in an equally dangerous situation once they arrive. Largely naïve to their new surroundings, individuals must rapidly identify what organisms within their new community pose an immediate threat via predation, either by learning or through innate recognition. Odors play an important role in learning processes, often in conjunction with other sensory stimuli, and can provide detailed and precise information to the fish about its surroundings. The majority of work on predator recognition by juvenile reef fishes has examined their responses to conspecific and con-generic alarm cues, now known to exist in a variety of species (Ferrari et al. 2010). These cues originate within the epidermis where they likely serve a primary role unrelated to signaling, but are released when a fish's skin is damaged, for instance during a predation event (Chivers et al. 2007). When paired with an additional stimulus, such as the odor or visual appearance of a predator, surrounding naïve fish that receive both stimuli can quickly learn to associate the predator with risk (Ferrari et al. 2010).

While the consequences of misidentifying a potential predator can mean death, it is also energetically expensive to be vigilant against non-threatening species at the expense of other requirements such as acquiring food or defending territory from competitors. The learning mechanisms for predator recognition should therefore be accurate but also adaptable, and this is increasingly being shown to be the case. For instance, it is now known that juveniles are able to extract key components from multiple sources of information to identify relative risk (Manassa et al. 2013), socially transmit risk identification (Manassa et al. 2014), continuously update this information via latent inhibition mechanisms (Mitchell et al. 2011), and adjust this reassessment relative to background levels of risk (Chivers et al. 2014a).

While complex learning mechanisms appear to be essential for juvenile survival, the larvae and juveniles of some coral reef fishes also have the ability to innately recognize predator odors as a threat. For instance, naïve, captive breed larvae of the orange clownfish, *Amphiprion percula*, will avoid the odor of predators and they can distinguish between the odors of predatory and non-predatory fish (Dixson et al. 2010). This ability is likely due to diet-based olfactory cues released by a predator as waste, such as conspecific and heterospecific alarm cues (Brown et al. 1995), or more general metabolites that indicate a picivorous diet (Dixson et al. 2012). Innate recognition has also been demonstrated in the field, with level of recruitment by reef-naïve fishes lower on patch reefs when paired with predator odors (Vail and McCormick 2011). Interestingly, this same study identified higher relative recruitment by one species onto patch reefs paired with the odor of adult non-predators, suggesting an innate attraction to odors that may signify low predator abundance. The ability to recognize and locate threats from a distance, without prior experience, would be highly valuable for larvae as they return to the reef and need to distinguish between habitat patches. It is possible that, post-settlement, innate predator recognition is then modified by the learning mechanisms outlined above, leading to a finely tuned, habitat-specific comprehension of risk.

These findings suggest odors can play an important role in prey assessment by juvenile fishes, and so may be essential for survival. However, the predicted effects of climate change on oceanic chemistry will likely limit the ability of these fishes to receive and interpret these odors. Exposure to the increased levels of CO_2 predicted for coral reefs has been shown to disrupt the olfactory system of juvenile reef fish (Munday et al. 2009). Fishes treated to ocean acidification levels expected to occur by the end of this century have displayed decreased sensory abilities and resulting impacts on behavior due to impaired neurotransmitter function (Nilsson et al. 2012). This pathway has recently been shown to impact risk-assessment in juvenile reef fish, with treated individuals unable to distinguish between predators and non-predators directly leading to lower survival (Chivers et al. 2014b).

3.3 The Nature of Odors on Coral Reefs and Their Perception by Juvenile Reef Fishes

While the waters surrounding coral reefs are a rich source of information, the ability of a fish to access this information is dependent on a number of factors. To determine how odor can mediate the behavior of juvenile fishes, it is necessary to understand what these odors are, how fishes receive and perceive these odors, the concentration required to elicit a behavioral response, and how relevant odor molecules move through the water column.

3.3.1 Odors in a Coral Reef Context

In contrast to some areas of aquatic chemical ecology, such as the study of defensive metabolites in aquatic organisms, our knowledge of information-carrying molecules is sparse, especially concerning the chemicals relevant to coral reef fishes. This is largely a result of research objectives that emphasize understanding the role molecules play in the ecology and evolution of associated species, as opposed to their chemical structure. Nonetheless, it is likely these cues share several key properties; they are soluble in water, occur in very low concentrations, and degrade quickly (Brönmark and Hansson 2012; Chivers et al. 2013). Therefore, it is also likely that receivers on coral reefs have evolved recognition mechanisms that detect chemicals at low concentrations, similar to those observed in other marine teleosts. For example, the gilthead seabream (*Sparus auratus*) has a detection threshold for the bodily fluids of conspecifics of up to $1{:}10^{7.4}$ (Hubbard et al. 2003). Marine fishes from non-coral reef systems display sensitivity to a range of odorants including amino acids, polyamines, bile salts, prostaglandins and steroids (Hara 1994). As there are genetic similarities between the olfactory receptors of teleosts (Hashiguchi et al. 2008), it is likely coral reef fishes can also detect these chemicals. While very little is known about the odors used by juvenile coral reef fishes to locate settlement sites, fish exhibit preferences for a wide variety of habitat-specific odors including corals, vegetation, and conspecifics. For instance, Lecchini et al. (2005b) found *Chromis viridis* larvae were attracted to isolated biochemical constituents associated with juvenile conspecifics, suggesting that these could serve as settlement cues (see also Sect. 2.1). It is likely that different odors may be relevant over different spatial scales or at different stages of settlement. For instance, the odor used to orientate towards a reef from miles offshore is likely to differ from the odor used to select a microhabitat once a larvae approaches the reef matrix. However, whether the presence of a single key chemical or a complex chemical mixture drives these behaviors is unknown.

With regard to chemical alarm cues, again very little is currently known for any species of coral reef fish. However, recent work has suggested at the nature of these molecules. For instance, the mucus-based alarm cue of the freshwater zebrafish contains a glycosaminoglycan, chondriten (Mathuru et al. 2012). While exposure to isolated chondriten fragments invokes an anti-predator response, this is not as strong as when individuals are exposed to the crude alarm cue, indicating that the alarm cues of fishes may consist of complex chemical mixtures as opposed to one isolated chemical. In addition, the alarm cues of the lemon damselfish have been shown to quickly lose their potency (Chivers et al. 2013), suggesting that alarm cue molecules rapidly degrade following exposure to the external environment.

3.3.2 Odor Perception in Larval and Juvenile Fishes

In order to perceive and react to an odor, a fish must first receive it from its external environment via its chemosensory organs. In the majority of teleost fishes, this organ consists of two chambers located above, but isolated from, the mouth. Each chamber consists of an anterior and posterior nostril, allowing for water inflow and outflow respectably, bridged by an interior channel lined with olfactory epithelium-baring lamellae (Cox 2008). Both biomechanical and behavioral mechanisms are then employed to increase the rate of water flow through each chamber, and therefore enhance the potential for odor molecules to come into contact with this epithelium. For instance, the anterior nostrils of many species are raised, reducing the hydrodynamic effect of boundary layers that can limit ability of water to enter a chamber. Chamber ventilation is also facilitated by cilia located on the olfactory epithelium, as well as compression and expansion of accessory sacks (Cox 2008). These olfactory organs are often well developed in settlement stages fishes, which constantly "sniff" as a side effect of gill ventilation (Atema et al. 2002). Chemical information is received by the numerous sensory neurons located on the olfactory epithelia. These neurons project to segregated areas of the olfactory bulb, where they synapse on mitral cell dendrites. Information is transported from the olfactory bulb to the brain via olfactory tracts, where it is finally decoded, invoking a behavioral response (Laberge and Hara 2001). In addition to this "nose," many species also possess chemosensory cells on the surface of their body, the pharynx, and body cavity (Daghfous et al. 2012). To initiate a behavioral reaction, such as chemotaxis or anti-predator responses, chemical concentrations must cross a response threshold, with neurons responding best to sudden increases in concentrations (Atema 2012).

3.3.3 The Movement of Odors and the Importance of Hydrodynamic Processes

On their own, odor molecules are of limited value to an organism, as they provide no information to the receiver about direction of origin. For this reason, an individual attempting to use odors to locate a specific habitat, or identify a specific predator must often (a) rely on odor gradients, and (b) incorporate additional sources of sensory information. While many of the biotic and abiotic substances present within coral reefs constantly release odors, transmission via molecular diffusion is 10,000 times slower in water than in air, massively reducing the active space around an odor source within which a receiver can detect a signal (Rosenthal and Lobel 2005). Diffusion gradients are therefore only relevant in the immediate vicinity of the source. Instead, hydrodynamic processes such as currents, eddies, and wakes play the key role in odor dispersal over the majority of behaviorally relevant spatial scales. The generally variable nature of these hydrodynamic processes means that larvae attempting to orientate towards or away from a specific olfactory gradient are likely to be exposed to a highly variable signal with odor plumes constantly changing shape, orientation, and concentration (Atema 2012; Lecchini et al. 2014).

3.4 Assessing the Role of Odors in Behavior

Our understanding of the role odors play in the behavior of juvenile coral reef fishes has advanced rapidly over the past two decades, along with the field of larval reef fish ecology as a whole. This is largely due to the rapid development of research in this area, as pre- and post-settlement fishes became an increasingly viable experimental model. Experiments, using novel methods as well as techniques adapted from other systems, have revealed much about the ecological roles of odor. However, we still know very little about the chemical nature of these olfactory cues.

3.4.1 Collecting Larvae

Progress in this field has largely been made possible due to technological advancements, in areas such as larval rearing and larval collection, which have allowed researchers access to young, chemically naïve coral reef fishes in ample and largely predictable numbers. Innovative methods for collecting larvae, such as crest nets (Dufour and Galzin 1993) and light traps (Doherty 1987), have proven especially valuable. These devices capture a multi-species assemblage of late-stage, undamaged larvae with no prior experience of recruitment habitats. These larvae can then be used to test both pre-settlement and post-settlement behavior under laboratory and field conditions (Leis and McCormick 2002). In particular, modern versions of the light trap are often designed to be inexpensive and easily deployed under various field conditions (e.g., Stobutzki and Bellwood 1997), motivating their widespread appeal and use. However, light traps are selective, attracting a fairly standard proportion of the species assemblage with catches largely consisting of pomacentrids.

Likewise, the majority of research on captive bred fishes has also used species from this family, or other easily bred, benthic egg laying fishes. As a result, there is a bias in the literature with studies of olfactory preferences and behavior conducted using a relatively narrow range of model species. As captive rearing techniques continue to advance, it is likely a greater variety of larvae, including commercially important species, will also become viable models. Another potential drawback of using collection techniques such as crest nets and light traps is that they generally only target late-stage larvae as they approach the reef. As larval sensory competency appears to vary with ontogeny (Dixson et al. 2011), the findings of light trap studies only accurately represent the behaviors of settlement stage larvae, ignoring the rest of larval development.

3.4.2 Laboratory-Based Behavioral Experiments

The majority of studies into the role of odor in behavior have employed laboratory-based experimental methods. Given that numerous odors could influence larval behavior and habitat selection, an initial step is to establish whether juvenile reef fishes exhibit olfactory discrimination, what odors are preferred or avoided, and whether these preferences affect settlement-relevant behaviors. Choice experiments are ideally suited to this research with equipment such as flume chambers (Atema et al. 2002) and aquarium choice arenas (Lecchini et al. 2005b) proving invaluable tools. For example, choice flumes, which in the simplest sense consist of two masses of moving water separated by a sharp boundary over which a larval fish can move at will (Atema et al. 2002), allow olfactory preferences between two odor sources to be established. These types of experiments have many benefits. Preferences can be defined quickly, as can strength of responses, and detection thresholds. They also have the key benefit of allowing the researcher to isolate the cues being tested, as well as examine the effects of other aspects of water chemistry, such as pH, on preferences (i.e., Munday et al. 2009). However, as these experiments only indicate the presence or absence of a preference between paired odors, how these preferences influence a fish's behavior under natural conditions often remains unclear. The majority of predator recognition studies have measured the variation in anti-predator behavior following exposure to different odors (Holmes and McCormick 2010; Mitchell et al. 2013). These studies have modified methods established for studies in freshwater systems. Laboratory experiments have proven especially useful in the study of climate change effects on behavior (Lönnstedt et al. 2013; Dixson et al. 2014b). While choice experiments have allowed larval responses to various odors from degraded habitats to be tested, arguably their greatest value has been in determining the effects of ocean acidification on olfaction in larval fishes, and exactly how neurotransmitter impairment may affect behavior (Dixson et al. 2010; Nilsson et al. 2012; Welch et al. 2014).

3.4.3 In Situ Experiments and Field Validation

While laboratory-based experiments are important for establishing how odors can influence larval behaviors, experiments under natural field conditions that attempt to validate laboratory experiments can reveal whether those behavioral modifications are ecologically relevant. Unfortunately, this is generally difficult to achieve, as isolating the effects of odor on behavior, as opposed to other sources of environmental stimuli, is a complicated task. Some recent studies have attempted to bring olfactory cues into the field to observe their effect on wild populations (Lönnstedt et al. 2013). Other studies have taken fishes with known behavioral responses to specific odors back into a natural setting to observe survival rates (Lönnstedt et al. 2012b), effective detection distances, and the relative importance of different sensory mechanisms (Lecchini et al. 2013). In practice, field experiments have often fulfilled a supporting role, presented alongside laboratory based behavioral experiments. By combining these approaches, the effects of odor on behavior within natural systems can be inferred without isolating specific chemical cues (Paris et al. 2013; Dixson et al. 2014b). However, caution is needed as behavioral patterns observed in the laboratory can conflict with those in the field (Dixson et al. 2014a).

3.4.4 Analysis of Olfactory Cue Biochemistry

Despite the wealth of knowledge that now exists regarding the olfactory preferences of larvae, the effect of these preferences on other behaviors, the environmental source of these odors, and the influence of climate change on responses to these odors, very little is known regarding the odors themselves. Almost no studies to date have determined the chemical structure of the odor molecules used by coral reef fishes. This is largely an artifact of the field's relatively recent expansion as well as the general objective of most research to date, which has been to establish the ecological and evolutionary role of odor-influenced behavior. In this regard, the majority of studies have used bulk odor cues in the form of crude water samples either that have been collected from a specific water mass assumed to represent an odor source, or that have held a specific organism or other odor-producing item for a known length of time. While this approach allows broad inferences to be made, we do not know what key chemical or chemicals from that source are binding to a fish's chemoreceptors and causing behavioral responses. In addition, this approach may expose fish to odors at much higher concentrations than would occur naturally, bringing into question the active concentration in a field setting. Some recent work has attempted to define aspects of these chemicals, for instance their potency through time (Chivers et al. 2013) or movement through hydrodynamic processes (Paris et al. 2013). As discussed at length above, promising attempts to define the nature of chemical alarm cues used by freshwater fishes (Mathuru et al. 2012) indicate that it may indeed be possible to isolate the molecular structure of olfactory cues used by coral reef fishes.

3.5 Future Directions

Larval and juvenile coral reef fishes have highly developed chemosensory systems, and they use odors extensively to gather vital environmental information throughout their early life histories. However, many interesting and important questions remain unanswered regarding the chemical nature of these cues, the abilities of coral reef fishes to receive and interpret cues over various spatial scales and ontogenetic stages, and ultimately how odors influence the behavior of larval and juvenile coral reef fishes under natural conditions. While the chemical nature of odors that influence teleost behavior has been documented (Hara 1994), similar studies have yet to be conducted in coral reef systems, with almost nothing known regarding the various odors that elicit behavioral responses in larval and juvenile coral reef fishes. While the odors of various environmental components have been shown to attract or repel larvae, determining the molecular basis of this chemotaxis could reveal whether behavioral responses are dependent on the presence of a specific chemical component, or alternatively whether a range of habitat-specific chemicals will elicit similar responses. As it is possible that these fishes are responding to complex chemical mixes, studies should be conducted to determine whether subtly different combinations of habitat specific-chemicals produce variable reaction strengths. If specific chemicals or chemical mixes are identified as essential for drawing larvae to a particular habitat, this could have important implications for both reef management and reef regeneration, opening the potential of chemically seeding reefs to improve levels of recruitment. With regards to predator–prey interactions, determining the molecular basis of the chemical alarm cues present in coral reef fishes is an important next step in this field of research. Identifying the chemicals involved will allow us to better understand their primary function, the evolution of these cues, and how they operate within coral reef communities. To this end, techniques that have proven valuable in the study of olfaction in freshwater and marine fishes from other systems, such as mass spectrometry (Mathuru et al. 2012) and electrophysiology (Scott and Scott-Johnson 2002, Hubbard et al. 2011) will likely prove similarly useful when applied to coral reef species.

While many studies have investigated the behavioral responses of larval and juvenile coral reef fishes to odor, very little is known about the role that odor concentration plays in mediating these responses. Given the dispersive nature of odors in the ocean, it is likely that the chemosensory systems of these fishes have evolved sensitivity to low concentrations of relevant odor molecules, as is the case in other marine teleosts (Hubbard et al. 2003, 2011; Yacoob and Browman 2007). Therefore, subtle differences in odor concentration could play an important role in determining the strength of resulting behavioral responses, such as reorientation and swimming speed when searching for habitat, or reduction in feeding rates when avoiding predators. An ongoing question is what odors influence the behavior of larvae between hatching and settlement, driving larvae away from, and back towards the reef. Understanding the concentration thresholds below which specific odors fail to cause a behavioral response, as well as formulating estimates of their concentrations and dispersal patterns in nature, may help us to understand what odors are ecologically relevant at different stages of larval dispersal and settlement. In addition to this, further work is needed to determine how much response variation exists between larvae at different stages of development, especially following hatching, as well as for a greater range of ecologically disparate species. While we now have an extensive understanding of how odors can affect the behavior of coral reef fishes, future studies that examine how these odors are produced and received in greater detail will likely reveal substantially more about their ecological relevance.

References

Almany GR, Berumen ML, Thorrold SR, Planes S, Jones GP (2007) Local replenishment of coral reef fish populations in a marine reserve. Science 316:742–744

Arvedlund M, Kavanagh K (2009) The senses and environmental cues used by marine larvae of fish and decapod crustaceans to find tropical coastal ecosystems. In: Nagelkerken I (ed) Ecological Connectivity among Tropical Coastal Ecosystems. Springer Press, Netherlands, pp 135–184

Arvedlund M, McCormick MI, Fautin DG, Bildsøe M (1999) Host recognition and possible imprinting in the anemonefish *Amphiprion melanopus* (Pisces: Pomacentridae). Mar Ecol Prog Ser 188:207–218

Atema J (2012) Aquatic odor dispersal fields: opportunities and limits of detection, communication and navigation. In: Brönmark C, Hansson L-A (eds) Chemical Ecology in Aquatic Systems. Oxford University Press, Oxford, pp 1–18

Atema J, Kingsford MJ, Gerlach G (2002) Larval reef fish could use odour for detection, retention and orientation to reefs. Mar Ecol Prog Ser 241:151–160

Baird AH, Babcock RC, Mundy CP (2003) Habitat selection by larvae influences the depth distribution of six common coral species. Mar Ecol Prog Ser 252:289–293

Bell LJ, Moyer JT, Numachi K (1982) Morphological and genetic variation in Japanese populations of the anemonefish *Amphiprion clarkii*. Mar Biol 72:99–108

Bell JJ, Davy SK, Jones T, Taylor MW, Webster NS (2013) Could some coral reefs become sponge reefs as our climate changes? Glob Chang Biol 19:2613–2624

Bellwood DR, Hoey AS, Choat JH (2003) Limited functional redundancy in high diversity systems: resilience and ecosystem function on coral reefs. Ecol Lett 6:281–285

Bellwood DR, Hughes TP, Folke C, Nystrom M (2004) Confronting the coral reef crisis. Nature 429:827–833

Booth DJ, Wellington G (1998) Settlement preferences in coral-reef fishes: effects on patterns of adult and juvenile distributions, individual fitness and population structure. Aust J Ecol 23:274–279

Brönmark C, Hansson L-A (2012) Aquatic chemical ecology: new directions and challenges for the future. In: Brönmark C, Hansson L-A (eds) Chemical Ecology in Aquatic Systems. Oxford University Press, Oxford, pp 272–278

Brooker RM, Munday PL, Chivers DP, Jones GP (2015) You are what you eat: diet-induced chemical camouflage in a coral-feeding reef fish. Proc R Soc Lond B Biol Sci 282:20141887

Brown GE, Chivers DP, Smith RJF (1995) Fathead minnows avoid conspedfic and heterospedfic alarm pheromones in the faeces of northern pike. J Fish Biol 47:387–393

Carpenter KE, Abrar M, Aeby G, Aronson RB, Banks S, Bruckner A, Chiriboga A, Cortés J, Delbeek JC, DeVantier L, Edgar GJ, Edwards AJ, Fenner D, Guzmán HM, Hoeksema BW, Hodgson G, Johan O, Licuanan WY, Livingstone SR, Lovell ER, Moore JA, Obura DO, Ochavillo D, Polidoro BA, Precht WF, Quibilan MC, Reboton C, Richards ZT, Rogers AD,

Sanciangco J, Sheppard A, Sheppard C, Smith J, Stuart S, Turak E, Veron JEN, Wallace C, Weil E, Wood E (2008) One-Third of reef-building corals face elevated extinction risk from climate change and local impacts. Science 321:560–563

Chivers DP, Wisenden BD, Hindman CJ, Michalak TA, Kusch RC, Kaminskyj SGW, Jack KL, Ferrari MCO, Pollock RJ, Halbgewachs CF, Pollock MS, Alemadi S, James CT, Savaloja RK, Goater CP, Corwin A, Mirza RS, Kiesecker JM, Brown GE, Adrian JC, Krone PH, Blaustein AR, Mathis A (2007) Epidermal 'alarm substance' cells of fishes maintained by non-alarm functions: possible defence against pathogens, parasites and UVB radiation. Proc R Soc Lond B Biol Sci 274:2611–2619

Chivers DP, Dixson DL, White JR, McCormick MI, Ferrari MCO (2013) Degradation of chemical alarm cues and assessment of risk throughout the day. Ecol Evol 3:3925–3934

Chivers DP, McCormick MI, Mitchell MD, Ramasamy RA, Ferrari MCO (2014a) Background level of risk determines how prey categorize predators and non-predators. Proc R Soc Lond B Biol Sci 281:20140355

Chivers DP, McCormick MI, Nilsson GE, Munday PL, Watson S-A, Meekan MG, Mitchell MD, Corkill KC, Ferrari MCO (2014b) Impaired learning of predators and lower prey survival under elevated CO2: a consequence of neurotransmitter interference. Glob Chang Biol 20:515–522

Coppock AG, Gardiner NM, Jones GP (2013) Olfactory discrimination in juvenile coral reef fishes: Response to conspecifics and corals. J Exp Mar Biol Ecol 443:21–26

Cowen RK, Paris CB, Srinivasan A (2006) Scaling of connectivity in marine populations. Science 311:522–527

Cox JPL (2008) Hydrodynamic aspects of fish olfaction. J R Soc Interface 5:575–593

Daghfous G, Green WW, Zielinski BS, Dubuc R (2012) Chemosensory-induced motor behaviors in fish. Curr Opin Neurobiol 22:223–230

De'ath G, Fabricius KL, Sweatman H, Puotinen M (2012) The 27-year decline of coral cover on the Great Barrier Reef and its causes. Proc Natl Acad Sci U S A 109:17995–17999

DeBose JL, Lema SC, Nevitt GA (2008) Dimethylsulfoniopropionate as a foraging cue for reef fishes. Science 319:1356

Devine B, Munday P, Jones G (2012) Homing ability of adult cardinalfish is affected by elevated carbon dioxide. Oecologia 168:269–276

Dixson DL, Hay ME (2012) Corals chemically cue mutualistic fishes to remove competing seaweeds. Science 338:804–807

Dixson DL, Jones GP, Munday PL, Planes S, Pratchett MS, Srinivasan M, Syms C, Thorrold SR (2008) Coral reef fish smell leaves to find island homes. Proc R Soc Lond B Biol Sci 275:2831–2839

Dixson DL, Munday PL, Jones GP (2010) Ocean acidification disrupts the innate ability of fish to detect predator olfactory cues. Ecol Lett 13:68–75

Dixson DL, Munday PL, Pratchett M, Jones GP (2011) Ontogenetic changes in responses to settlement cues by Anemonefish. Coral Reefs 30:903–910

Dixson DL, Pratchett MS, Munday PL (2012) Reef fishes innately distinguish predators based on olfactory cues associated with recent prey items rather than individual species. Anim Behav 84:45–51

Dixson D, Jones G, Munday P, Planes S, Pratchett M, Thorrold S (2014a) Experimental evaluation of imprinting and the role innate preference plays in habitat selection in a coral reef fish. Oecologia 174:99–107

Dixson DL, Abrego D, Hay ME (2014b) Chemically mediated behavior of recruiting corals and fishes: A tipping point that may limit reef recovery. Science 345:892–897

Doherty PJ (1987) Light-traps: Selective but useful devices for quantifying the distributions and abundances of larval fishes. Bull Mar Sci 41:423–431

Dufour V, Galzin R (1993) Colonization patterns of reef fish larvae to the lagoon at Moorea Island, French Polynesia. Mar Ecol Prog Ser 102:143–152

Elliott JK, Elliott JM, Mariscal RN (1995) Host selection, location, and association behaviors of anemonefishes in field settlement experiments. Mar Biol 122:377–389

Ferrari MCO, Wisenden BD, Chivers DP (2010) Chemical ecology of predator–prey interactions in aquatic ecosystems: a review and prospectus. Can J Zool 88:698–724

Fisher R, Bellwood DR, Job SD (2000) Development of swimming abilities in reef fish larvae. Mar Ecol Prog Ser 202:163–173

Gerlach G, Atema J, Kingsford MJ, Black KP, Miller-Sims V (2007) Smelling home can prevent dispersal of reef fish larvae. Proc Natl Acad Sci U S A 104:858–863

Goatley CHR, Bellwood DR (2010) Biologically mediated sediment fluxes on coral reefs: sediment removal and off-reef transportation by the surgeonfish *Ctenochaetus striatus*. Mar Ecol Prog Ser 415:237–245

Hara TJ (1994) The diversity of chemical stimulation in fish olfaction and gustation. Rev Fish Biol Fisher 4:1–35

Hashiguchi Y, Furuta Y, Nishida M (2008) Evolutionary patterns and selective pressures of odorant/pheromone receptor gene families in teleost fishes. PLoS One 3, e4083

Holmes TH, McCormick MI (2010) Smell, learn and live: the role of chemical alarm cues in predator learning during early life history in a marine fish. Behav Processes 83:299–305

Hubbard PC, Barata EN, Canário AVM (2003) Olfactory sensitivity of the gilthead seabream (*Sparus auratus* L) to conspecific body fluids. J Chem Ecol 29:2481–2498

Hubbard P, Barata E, Ozório RA, Valente LP, Canário AM (2011) Olfactory sensitivity to amino acids in the blackspot sea bream (*Pagellus bogaraveo*): a comparison between olfactory receptor recording techniques in seawater. J Comp Phys A 197:839–849

Huijbers CM, Nagelkerken I, Lössbroek PAC, Schulten IE, Siegenthaler A, Holderied MW, Simpson SD (2012) A test of the senses: fish select novel habitats by responding to multiple cues. Ecology 93:46–55

Igulu MM, Nagelkerken I, van der Beek M, Schippers M, van Eck R, Mgaya YD (2013) Orientation from open water to settlement habitats by coral reef fish: behavioral flexibility in the use of multiple reliable cues. Mar Ecol Prog Ser 493:243–257

Jones GP, Milicich MJ, Emslie MJ, Lunow C (1999) Self-recruitment in a coral reef fish population. Nature 402:802–804

Laberge F, Hara TJ (2001) Neurobiology of fish olfaction: a review. Brain Res Rev 36:46–59

Lecchini D, Planes S, Galzin R (2005a) Experimental assessment of sensory modalities of coral-reef fish larvae in the recognition of their settlement habitat. Behav Ecol Sociobiol 58:18–26

Lecchini D, Shima J, Banaigs B, Galzin R (2005b) Larval sensory abilities and mechanisms of habitat selection of a coral reef fish during settlement. Oecologia 143:326–334

Lecchini D, Waqalevu VP, Parmentier E, Radford CA, Banaigs B (2013) Fish larvae prefer coral over algal water cues: implications of coral reef degradation. Mar Ecol Prog Ser 475:303–307

Lecchini D, Miura T, Lecellier G, Banaigs B, Nakamura Y (2014) Transmission distance of chemical cues from coral habitats: implications for marine larval settlement in context of reef degradation. Mar Biol 161:1677–1686

Leis JM (2002) Pacific coral-reef fishes: the implications of behaviour and ecology of larvae for biodiversity and conservation, and a reassessment of the open population paradigm. Environ Biol Fish 65:199–208

Leis JM, McCormick MI (2002) The biology, behavior, and ecology of the pelagic, larval stage of coral reef fishes. In: Sale PF (ed) Coral Reef Fishes. Academic, San Diego, pp 171–199

Lönnstedt OM, McCormick MI, Chivers DP (2012a) Well-informed foraging: damage-released chemical cues of injured prey signal quality and size to predators. Oecologia 168:651–658

Lönnstedt OM, McCormick MI, Meekan MG, Ferrari MCO, Chivers DP (2012b) Learn and live: predator experience and feeding history determines prey behaviour and survival. Proc R Soc Lond B Biol Sci 279:2091–2098

Lönnstedt OM, McCormick MI, Chivers DP (2013) Degraded environments alter prey risk assessment. Ecol Evol 3:38–47

Luiz OJ, Allen AP, Robertson DR, Floeter SR, Kulbicki M, Vigliola L, Becheler R, Madin JS (2013) Adult and larval traits as determinants of geographic range size among tropical reef fishes. Proc Natl Acad Sci U S A 110:16498–16502

Manassa RP, Dixson DL, McCormick MI, Chivers DP (2013) Coral reef fish incorporate multiple sources of visual and chemical information to mediate predation risk. Anim Behav 86: 717–722

Manassa RP, McCormick MI, Dixson DL, Ferrari MCO, Chivers DP (2014) Social learning of predators by coral reef fish: does observer number influence acquisition of information? Behav Ecol Sociobiol 68:1237–1244

Mathuru AS, Kibat C, Cheong WF, Shui G, Wenk MR, Friedrich RW, Jesuthasan S (2012) Chondroitin fragments are odorants that trigger fear behavior in fish. Curr Biol 22:538–544

Mitchell MD, McCormick MI, Ferrari MO, Chivers DP (2011) Friend or foe? The role of latent inhibition in predator and non-predator labelling by coral reef fishes. Anim Cogn 14:707–714

Mitchell MD, McCormick MI, Chivers DP, Ferrari MO (2013) Generalization of learned predator recognition in coral reef ecosystems: how cautious are damselfish? Funct Ecol 27:299–304

Munday PL (2001) Fitness consequences of habitat use and competition among coral-dwelling fishes. Oecologia 128:585–593

Munday PL, Dixson DL, Donelson JM, Jones GP, Pratchett MS, Devitsina GV, Døving KB (2009) Ocean acidification impairs olfactory discrimination and homing ability of a marine fish. Proc Natl Acad Sci U S A 106:1848–1852

Myrberg AA Jr, Fuiman LA (2002) The sensory world of coral reef fishes. In: Sale PF (ed) Coral Reef Fishes. Academic, San Diego, pp 123–148

Nilsson GE, Dixson DL, Domenici P, McCormick MI, Sorensen C, Watson S-A, Munday PL (2012) Near-future carbon dioxide levels alter fish behaviour by interfering with neurotransmitter function. Nat Clim Change 2:201–204

Paris CB, Atema J, Irisson J-O, Kingsford M, Gerlach G, Guigand CM (2013) Reef odor: a wake up call for navigation in reef fish larvae. PLoS One 8, e72808

Reaka-Kudla ML (1997) The global biodiversity of coral reefs: a comparison with rain forests. In: Reaka-Kudla ML, Wilson DE, Wilson EO (eds) Biodiversity II: Understanding and Protecting Our Biological Resources. Joseph Henry Press, Washington D. C, pp 83–108

Rosenthal GG, Lobel PS (2005) Communication. In: Katherine A, Sloman RRW, Sigal B (eds) Fish Physiology. Academic, New York, pp 39–78

Scott JW, Scott-Johnson PE (2002) The electro-olfactogram: a review of its history and uses. Microsc Res Techniq 58:152–160

Shulman MJ (1998) What can population genetics tell us about dispersal and biogeographic history of coral-reef fishes? Aust J Ecol 23:216–225

Siebeck UE, Marshall NJ (2007) Potential ultraviolet vision in pre-settlement larvae and settled reef fish—A comparison across 23 families. Vision Res 47:2337–2352

Stobutzki IC, Bellwood DR (1997) Sustained swimming abilities of the late pelagic stages of coral reef fishes. Mar Ecol Prog Ser 149:35–41

Sweatman HPA (1983) Influence of conspecifics on choice of settlement sites by larvae of two pomacentrid fishes (*Dascyllus aruanus* and *D. reticulatus*) on coral reefs. Mar Biol 75: 225–229

Taylor MS, Hellberg ME (2003) Genetic evidence for local retention of pelagic larvae in a Caribbean reef fish. Science 299:107–109

Vail AL, McCormick MI (2011) Metamorphosing reef fishes avoid predator scent when choosing a home. Biol Lett 7:921–924

Welch MJ, Watson S-A, Welsh JQ, McCormick MI, Munday PL (2014) Effects of elevated CO2 on fish behaviour undiminished by transgenerational acclimation. Nat Clim Change 4:1086–1089

Wenger A, Johansen JL, Jones GP (2011) Suspended sediment impairs habitat choice and chemosensory discrimination in two coral reef fishes. Coral Reefs 30:879–887

Yacoob SY, Browman HI (2007) Olfactory and gustatory sensitivity to some feed-related chemicals in the Atlantic halibut (*Hippoglossus hippoglossus*). Aquaculture 263:303–309

Chapter 4
A Small Expose on Bovine Pheromones: with Special Reference to Modifications of the Reproductive Cycle

Kristina Nordéus, Renée Båge, Hans Gustafsson, Robert Glinwood, and Lennart Söderquist

4.1 Background

During the twentieth century, dairy farming underwent great changes, developing into a regular industry. Since the 1950s milk yields have increased vastly, mainly through improved management and extensive breeding programs. At the same time decreasing fertility is a growing concern to dairy farmers. The final two decades of the last century saw calving rates to first service falling annually by 0.5 % and 1 % in the USA (Butler and Smith 1989; Beam and Butler 1999) and the UK (Royal et al. 2000), respectively, and the same trend was also observed in other developed countries (Lucy 2001).

One of the underlying mechanisms of the decreased fertility is depressed dairy cow estrous behavior, which makes it difficult for farmers to determine the optimal time for artificial insemination or AI (Dobson et al. 2008). Estrous synchronization technologies used with timed artificial insemination (TAI) protocols optimize breeding and circumvent the need for time-consuming estrus detections, but these methods include the administration of several exogenous hormones and are consequently not used in some European countries, such as Sweden. The use of synthetic pheromones may be novel reproductive management tools to manipulate estrous

K. Nordéus (✉) • R. Båge • L. Söderquist
Department of Clinical Sciences, Swedish University of Agricultural Sciences (SLU), 7054, 750 07 Uppsala, Sweden
e-mail: kristina.nordeus@gmail.com

H. Gustafsson
Växa Sverige, 7054, 750 07 Uppsala, Sweden

Department of Clinical Sciences, Swedish University of Agricultural Sciences (SLU), 7054, 750 07 Uppsala, Sweden

R. Glinwood
Department of Crop Production Ecology, SLU, 7043, 750 07 Uppsala, Sweden

B.A. Schulte et al. (eds.), *Chemical Signals in Vertebrates 13*,
DOI 10.1007/978-3-319-22026-0_4

cycles, enhance estrous behavior, accurately determine when a cow is in estrus, or facilitate collection of semen. If so identifying chemicals that cause these effects could lead to environmentally sustainable, labor-efficient, cheap technologies that would be of great interest to farmers and consumers.

4.2 Modifications of the Reproductive Cycle

Exposure to conspecifics or their excreta has been shown to cause modifications of reproductive cycles in female mice (Vanderlee and Boot 1955; Whitten 1956), rats (McClintock 1978), sheep (Radford and Watson 1957) and humans (McClintock 1971). When female rats are housed together, inter-estrus intervals decrease, an effect known as enhancement (McClintock 1983). Conversely, when female mice are housed together, longer inter-estrus intervals, or suppression, may occur (Vanderlee and Boot 1955; McClintock 1983). Exposing female mice to males or their excreta, on the other hand, may result in shorter cycles (Whitten 1956).

4.2.1 Estrous Synchrony

Estrous synchrony, i.e., synchronization of the estrous cycles of females housed together, has been widely debated over the years (Schank 2001, 2004), but there are studies to suggest that inter-female chemical communication may cause synchrony in rats (McClintock 1978) and in humans (McClintock 1971; Stern and McClintock 1998).

In 1979 Aron suggested three different mechanisms by which manipulation of female cyclicity can occur. One explanation could be that the follicular growth rate is altered. Another possible explanation is that a preovulatory surge of luteinizing hormone (LH) is induced. Finally, the neuroendocrine structures controlling the corpus luteum may be affected.

Follicular growth (via follicle-stimulating hormone, FSH) and final maturation of the dominant follicle (via LH) as well as the preovulatory LH surge are affected by the secretion of gonadotropin-releasing hormone or GnRH (Forde et al. 2011), so effects on GnRH secretion can result in both a more rapid maturation of the follicle and an induction of the preovulatory LH surge.

Effects on LH pulse frequency (Shinohara et al. 2001) as well as on the timing of the preovulatory LH surge (Stern and McClintock 1998) have been demonstrated in women exposed to female axillary secretions. Moreover, introduction of a ram into a group of anestrous ewes induces pulsatile secretion of LH within minutes after introduction and a preovulatory LH surge can be detected within 36 h (Signoret 1991).

4.2.2 Influence of Other Females on Female Cattle

Studies on both dairy cows (Hurnik et al. 1975) and beef cows (Floyd et al. 2009) indicated that estrous behavior is intensified and duration of estrus prolonged with increasing numbers of animals simultaneously in estrus. Crowding or confinement of movement of estrus-cycling cows also hastens postpartum resumption of luteal activity in beef cows (Berardinelli and Joshi 2005b). In a limited study Hradecky (1989) found that estrous cows, through the secretion of "primary attractive odors" induce the production of "secondary attractive odors" in pen-mates, which attract the interest of the bull, but to a lesser extent than the former. When exposing cyclic dairy heifers to vaginal mucus and urine from estrous cows following treatment with prostaglandins, Izard and Vandenbergh (1982b) found that the degree of estrous synchrony was higher than in control animals exposed to water. Exposure to estrous urine and vaginal mucus has also been found to influence the LH pulsatility pattern preceding the LH peak (Nordéus et al. 2012b) and there are indications that the exposure may also influence the actual LH peak (Nordéus et al. 2012a). Nishimura et al. (1991) reported that diestrous heifers smeared with their own vaginal discharge from previous cycles were mounted by herd mates more than control animals treated with water. This effect, however, did not manifest when animals were smeared with estrous vaginal discharge from another animal, indicating that the vaginal mucus may contain individual-specific volatile cues.

Farmers occasionally provide anecdotal evidence of estrous synchrony in cattle. However, since cows in general are either in postpartum or lactational anestrus or are pregnant, and the cyclicity of heifers is often much less closely monitored, opportunities to observe such an effect scientifically in the field are scarce. Although there are studies to suggest that bovine inter-female pheromones actually exist and can cause synchronization of estrous cycles or enhancement of estrous behavior, proof is still lacking.

4.2.3 Influence of Males on Female Cattle

Studies on bovine pheromones have predominantly been aimed at investigating the stimulatory effect of males on female reproductive parameters, an effect also known as biostimulation (Chenoweth 1983). Studies are mainly focussed at resumption of ovarian cyclicity after calving. For example, Izard and Vandenbergh (1982a) found that prepubertal heifers receiving oronasal treatments with bull urine reached puberty earlier than heifers receiving control treatment with water. The urine-treated heifers calved earlier and had a shorter calving season than the control heifers, but pregnancy rates were the same for the two groups.

This biostimulatory effect on ovarian activity has also been demonstrated in zebu (*Bos indicus*) cows (Rekwot et al. 2000) and in suckled beef (*Bos taurus*) cows

(Zalesky et al. 1984; Custer et al. 1990; Landaeta-Hernández et al. 2004), where bull exposure shortened the postpartum anestrus. It was further supported by Berardinelli and Joshi (2005b), who found that exposure to bulls, or their excretory products, hastened resumption of luteal activity after calving. However, there have also been reports suggesting no positive effect of bull-exposure on postpartum anestrus (Larson et al. 1994).

The biostimulatory effect seems to be greater when bulls are exchanged during the course of the experiment, compared to continuous exposure to the same bulls (Miller and Ungerfeld 2008). There seems to be an inverse linear relationship between the daily duration of bull exposure and the intervals from either calving or the start of bull exposure to resumption of ovarian activity (Tauck et al. 2010a), with longer daily exposure resulting in a shorter interval. There may also be a more rapid return to cyclicity when cows are exposed later in the postpartum anestrous period (Berardinelli and Joshi 2005a). This difference, however, was not seen by Fernandez et al. (1993).

The response to male stimuli can be modified by nutritional conditions. Monje et al. (1992) found that cows fed above their energy requirements showed a greater response than those fed below their requirements. Stumpf et al. (1992), however, demonstrated that the male effect was greater in cows subjected to a low dietary regimen preceding calving than in those subjected to a high dietary regimen. The explanation provided for the discrepancy between this and previous studies was that the exposure to males for cows with lesser body condition may help to overcome the inhibition of LH secretion caused by nutritional deficiencies, making the male effect more substantial than in well-fed cows, whereas for cows in deeper negative energy balance the male effect was not enough to overcome this inhibition.

Contrary to Custer et al. (1990), who did not see any effect of bull exposure on LH characteristics, Tauck et al. (2010b) found that the LH pulse frequency tended to be greater in anestrous suckled beef cows exposed acutely to bulls. In this study, cows were kept haltered in stalls, but bulls could come into contact with the frontal aspects of each cow. Since physical contact was possible we can view this as some sort of fence-line contact. The exposure also resulted in a lower cortisol pulse frequency. Tauck and Berardinelli (2007) demonstrated that AI pregnancy rates for cows inseminated 12 h after estrus were greater for suckled beef cows exposed to bulls or their excretory products, than for fence-line contact with bulls, when using a progestin-based protocol for synchronization of estrus. TAI pregnancy rates, however, did not differ between bull-exposed cows and unexposed cows. The opposite was seen when using a protocol including GnRH and prostaglandins (Berardinelli et al. 2007).

4.2.4 *Influence of Females on Male Cattle*

Bull bioassays have been frequently used in attempts to isolate and identify estrus-specific odorscows. Many different behavioral parameters assignable to pericopulatory behavior, such as flehmen, sniffing, licking, erection, and mounting, have been evaluated but with varying success. Several studies indicate that bulls are able to discriminate between estrous and non-estrous odor (Hart et al. 1946; Paleologou 1977). Sankar and Archunan (2004) found that the duration of flehmen behavior displayed by bulls was greater during exposure to estrous samples of vaginal mucus, saliva, feces and milk smeared on to the genital region of non-estrous cows, than during exposure to samples collected during other estrous cycle stages. Among the substances tested, the response to estrous vaginal mucus was significantly higher than for the other substances.

When exposing dairy bulls to urine from cows in various stages of cyclicity in a bowl, Houpt et al. (1989) found that the number and duration of flehmen, but not sniffing and licking, were parameters suitable to discriminate between estrous and non-estrous urine. The flehmen response to vaginal mucus, however, did not differ from that to water. Interestingly enough, only seven of the 15 bulls tested responded to estrous urine with two flehmen or more, which was the criterion for inclusion in the study. One explanation given for this finding was that sexual experience with females may be important in order for bulls to respond to estrous cues properly. Indeed, Presicce et al. (1993) found that bulls from a stud farm, with no prior contact with female cattle, displayed stronger pericopulatory behavior to urine collected from teaser bulls, than to compounds from estrous blood.

Klemm et al. (1987) and Rivard and Klemm (1989) also presented their samples to bulls in a dish. The substances studied were vaginal mucus (Klemm et al. 1987; Rivard and Klemm 1989), and serum and vulval skin gland secretions (Rivard and Klemm 1989). They found that all three of these body fluids from cows in estrus evoked a series of chained behaviors, which could be divided into three categories (Rivard and Klemm 1989). The attraction phase comprised initial responses, such as head orientation toward the sample, sniffing the air, moving toward the sample, sniffing the sample at close range, salivation, and urination. It was followed by the detection phase, which included behaviors such as licking, tongue manipulation, hypersalivation, labored breathing, flehmen and vocalization. During the phase of sexual preparation the bulls displayed penis protrusion, penile secretion, head butting, and mounting behavior. Only 8 of 23 different estrous samples induced the complete series of behaviors.

Geary et al. (1991) recorded the preference of bulls, either as total time spent adjacent to heifers or as number of flehmen towards heifers, for either estrous or diestrous heifers in a pairwise choice test, and found, contrary to the studies mentioned above, indications that bulls do not show olfactory preference for heifers in estrus.

Monitoring of heart rate could also potentially be used as a bioassay to evaluate bioactivity in different substances. Nordéus et al. (2012c) found average heart rate in bulls to be significantly higher during the first minute of exposure to estrous vaginal discharge compared to the other substances tested. A similar effect was seen for heifers when exposed to bull urine.

4.3 Potential Bovine Pheromones

According to a study by Kiddy and Mitchell (1981), in which dogs were trained to detect estrous samples of bovine vaginal fluids, estrous odor emerges from day 3 before estrus, peaks on the day of estrus, and disappears the day after. Similarly, trained rats show behavioral response towards samples collected 2 days before to 2 days after estrus (Dehnhard et al. 1991).

Ramesh Kumar et al. (2000) analyzed urine from cows in different stages of cyclicity and found that 1-iodo undecane and di-n-propyl phthalate were unique to the estrous samples. Further evidence that 1-iodo undecane is estrus-specific was provided by Sankar and Archunan (2008). When they compared volatile profiles from feces of cows in different stages of the reproductive cycle, they found three compounds, acetic acid, proprionic acid and 1-iodo undecane, that were specific for the estrous phase. As a bioassay, they smeared compounds on to the genital region of non-estrous cows. Bulls were allowed to sniff the cows for 30 min, with pre-copulatory behaviors and copulation being recorded. They found that the mixture of the three compounds resulted in significantly longer duration of flehmen behavior and increased number of mounts than the individual compounds and control, indicating that these substances may be involved in the induction of mating behaviors. Nordéus et al. (2014) found 1-hexadecanol to be present in greater amounts in estrous urine samples than in urine collected during the luteal phase. The bioactivity of this compound has not yet been investigated in cattle, but it has been suggested to have pheromonal effects in other species (Zhang et al. 2007; Hagemeyer 2010). According to Nagnan-Le Meillour et al. (2013) administration of one or more of six volatile compounds, identified from bovine estrous urine, can improve the reproductive function of a bull with effects primarily on the libido but also on semen production. The compounds are coumarin, squalene, 6-amino undecane, 2-butanone, 9-octadecenoic acid, and 1,2-dichloroethylene.

The dispersion of estrus-specific compounds in the bovine body has been demonstrated previously in swabs from the vulva and fluids from the vagina, urine, milk, and blood (Kiddy et al. 1984; Rivard and Klemm 1989). Weidong et al. (1997) investigated the presence of estrus-specific volatiles in milk, but did not find any compounds that were qualitatively different between stages. They did, however, find that there were significant quantitative differences and concluded that cycle stage can be determined by analyzing 15 compounds in milk. Klemm et al. (1994) found that blood acetaldehyde levels decreased rapidly just before, or at onset of, estrus, and suggested that estrus and ovulation could potentially be predicted by monitoring levels of acetaldehyde in milk, saliva, sweat, or breath. Acetaldehyde was also found to be estrus specific in bovine vaginal secretions (Ma et al. 1995). Increasing amounts were found followed by a drop in quantities 0–3 days before estrus. Acetaldehyde has, however, been tested in a bull bioassay previously (Presicce et al. 1993) and was not behaviorally active when tested as a singular component. Five estrus-specific compounds, trimethylamine, acetic acid, phenol 4-propyl, pentanoic acid, and proprionic acid, were identified in saliva Sankar et al. (2007). Results from the bioassay indicated that trimethylamine may be involved in attracting the bull to the estrous cow.

Several studies have focused on vaginal secretions, which are profound during estrus. Preti (1984) patented a method for detecting bovine estrus based on the quantification of methyl heptanol in vaginal secretion. Hradecky (1986) found that the concentration of free fatty acids in estrous vaginal discharge increased gradually before estrus and decreased rapidly thereafter. The concentration of free fatty acids in urine, but not in vaginal discharge, was affected by the ruminal concentration. Klemm et al. (1987) found nine estrus-specific compounds in samples that had tested positively in a bull behavioral assay. These compounds included two ketones, four amines, one alcohol, one diol, and one ether. Indeed, there have been claims that exposure to the boar sex pheromone, androstenone, may have a positive effect on different reproductive parameters in cows, such as earlier onset of cyclicity at puberty and better results from AI (Sokolov et al. 1995).

4.4 Discussion and Conclusion

Results from studies on bovine pheromones vary greatly. There are several possible explanations for this, such as differences in methods of sample collection, analysis and bioassay. First, signaling of estrus may involve several compounds with varying properties, making it difficult to get a true picture of the signal, especially if only one method of analysis is used and if relevant compounds are present only in minute amounts. Furthermore, different breeds could have different chemical profiles and it is also possible that there are individual components to the signal, even within species. Second, the environment in which animals are kept, feed and sexual experience might influence the result. Next, behavioral bioassays can be difficult to use because such methods are very sensitive to individual differences, both between subjects and between operators. Purely physiological measurements, such as hormonal fluctuations, follicular dynamics, and heart rate, may have an advantage in this aspect.

A challenge specific to the bovine species, compared to most other domestic species that have been studied, is that it is difficult to keep the animals in environments suitable to studying chemical communication. A lot of work has been done in this interesting field, but bovine pheromones remain elusive and some scientists doubt that they even exist. In the future, it would be of great interest to do a comparative study under controlled conditions of the compounds suggested to have pheromonal effects in cattle.

References

Aron C (1979) Mechanisms of control of the reproductive function by olfactory stimuli in female mammals. Physiol Rev 59:229–284

Beam SW, Butler WR (1999) Effects of energy balance on follicular development and first ovulation in postpartum dairy cows. J Reprod Fertil Suppl 54:411–424

Berardinelli J, Joshi P (2005a) Introduction of bulls at different days postpartum on resumption of ovarian cycling activity in primiparous beef cows. J Anim Sci 83:2106–2110

Berardinelli J, Joshi P, Tauck S (2007) Conception rates to artificial insemination in primiparous, suckled cows exposed to the biostimulatory effect of bulls before and during a gonadotropin-releasing hormone-based estrus synchronization protocol. J Anim Sci 85:848–852
Berardinelli JG, Joshi PS (2005b) Initiation of postpartum luteal function in primiparous restricted-suckled beef cows exposed to a bull or excretory products of bulls or cows. J Anim Sci 83:2495–2500
Butler WR, Smith RD (1989) Interrelationships between energy balance and postpartum reproductive function in dairy cattle. J Dairy Sci 72:767–783
Chenoweth PJ (1983) Reproduction management procedures in control of breeding. Anim Prod Aust 15:28–31
Custer EE, Berardinelli JG, Short RE, Wehrman M, Adair R (1990) Postpartum interval to estrus and patterns of LH and progesterone in first-calf suckled beef cows exposed to mature bulls. J Anim Sci 68:1370–1377
Dehnhard M, Claus R, Pfeiffer S, Schopper D (1991) Variation in estrus-related odors in the cow and its dependency on the ovary. Theriogenology 35:645–652
Dobson H, Walker SL, Morris MJ, Routly JE, Smith RF (2008) Why is it getting more difficult to successfully artificially inseminate dairy cows? Animal 2:1104–1111
Fernandez D, Berardinelli J, Short R, Adair R (1993) The time required for the presence of bulls to alter the interval from parturition to resumption of ovarian activity and reproductive performance in first-calf suckled beef cows. Theriogenology 39:411–419
Floyd LN, Lents CA, White FJ, Wettemann RP (2009) Effect of number of cows in estrus and confinement area on estrous behavior of beef cows. J Anim Sci 87:1998–2004
Forde N, Beltman ME, Lonergan P, Diskin M, Roche JF, Crowe MA (2011) Oestrous cycles in Bos taurus cattle. Anim Reprod Sci 124:163–169
Geary T, DeAvila D, Westberg H, Senger P, Reeves J (1991) Bulls show no preference for a heifer in estrus in preference tests. J Animal Sci 69:3999–4006
Hart G, Mead S, Regan W (1946) Stimulating the sex drive of bovine males in artificial insemination. Endocrinology 39:221–223
Hagemeyer P (2010) Pheromones in social mole-rats and implications for the study of mammalian chemical communication. Universität Duisburg-Essen, Fakultät für Biologie und Geografie. http://duepublico.uni-duisburg-essen.de/servlets/DocumentServlet?id=22783
Houpt KA, Rivera W, Glickstein L (1989) The flehmen response of bulls and cows. Theriogenology 32(3):343–350
Hradecky P (1986) Volatile fatty-acids in urine and vaginal secretions of cows during reproductive-cycle. J Chem Ecol 12:187–196
Hradecky P (1989) Possible induction by estrous cows of pheromone production in penmates. J Chem Ecol 15:1067–1076
Hurnik JF, King GJ, Robertson HA (1975) Estrous and related behaviour in postpartum Holstein cows. Appl Anim Ethol 2:55–68
Izard MK, Vandenbergh JG (1982a) The effects of bull urine on puberty and calving date in crossbred beef heifers. J Anim Sci 55:1160–1168
Izard MK, Vandenbergh JG (1982b) Priming pheromones from estrous cows increase synchronization of estrus in dairy heifers after PGF-2-alpha injection. J Reprod Fertil 66:189–196
Kiddy C, Mitchell D, Hawk H (1984) Estrus-related odors in body fluids of dairy cows. J Dairy Sci 67:388–391
Kiddy CA, Mitchell DS (1981) Estrus-related odors in cows: Time of occurrence. J Dairy Sci 64:267–271
Klemm W, Rivard G, Clement B (1994) Blood acetaldehyde fluctuates markedly during bovine estrous cycle. Anim Reprod Sci 35:9–26
Klemm WR, Hawkins GN, Delossantos E (1987) Identification of compounds in bovine cervico-vaginal mucus extracts that evoke male sexual-behaviour. Chem Senses 12:77–87
Landaeta-Hernández A, Giangreco M, Meléndez P, Bartolomé J, Bennet F, Rae D, Hernández J, Archbald L (2004) Effect of biostimulation on uterine involution, early ovarian activity and first postpartum estrous cycle in beef cows. Theriogenology 61:1521–1532

Larson C, Miller H, Goehring T (1994) Effect of postpartum bull exposure on calving interval of first-calf heifers bred by natural service. Can J Anim Sci 74:153–154

Lucy MC (2001) ADSA Foundation Scholar Award—Reproductive loss in high-producing dairy cattle: Where will it end? J Dairy Sci 84:1277–1293

Ma W, Clement B, Klemm W (1995) Cyclic changes in volatile constituents of bovine vaginal secretions. J Chem Ecol 21:1895–1906

McClintock MK (1971) Menstrual synchrony and suppression. Nature 229:244–245

McClintock MK (1978) Estrous synchrony and its mediation by airborne chemical communication (*Rattus norvegicus*). Horm Behav 10:264–276

McClintock MK (1983) Pheromonal regulation of the ovarian cycle: Enhancement, supression and synchrony. In: Vandenbergh JG (ed) Pheromones and reproduction in mammals, vol (Academic Press, New York, 1983). Academic, New York, pp 113–149

Miller V, Ungerfeld R (2008) Weekly bull exchange shortens postpartum anestrus in suckled beef cows. Theriogenology 69:913–917

Monje A, Alberio R, Schiersmann G, Chedrese J, Carou N, Callejas SS (1992) Male effect on the post-partum sexual activity of cows maintained on two nutritional levels. Anim Reprod Sci 29:145–156

Nagnan-Le Meillour P, Le Danvic C, Humblot P, Chemineau P, Briant C, Gérard O (2013) Method for stimulating the reproductive cpacity in a bull, and composition for stimulating the reproductive capacity in a bull. Patent INRA EP 255 2347 A1

Nishimura K, Utsumi K, Okano T, Iritani A (1991) Separation of mounting-inducing pheromones of vaginal mucus from estrual heifers. J Anim Sci 69:3343

Nordéus K, Båge R, Gustafsson H, Humblot P, Söderquist L (2012a) The influence of oestrous substances on cyclicity and oestrous behaviour in dairy heifers. Acta Vet Scand 54:26

Nordéus K, Båge R, Gustafsson H, Söderquist L (2012b) Changes in LH pulsatility profiles in dairy heifers during exposure to oestrous urine and vaginal mucus. Reprod Domest Anim 47:952–958

Nordéus K, Jergil E, Båge R, Lundeheim N, Hultén F, Söderquist L (2012c) Effects of pheromones on heart rate in bulls and heifers. Vet Rec 170:496–496

Nordéus K, Webster B, Söderquist L, Båge R, Glinwood R (2014) Cycle-characteristic odour of cow urine can be detected by the female face fly (*Musca autumnalis*). Reprod Domest Anim 49:903–908

Paleologou A (1977) The detection of oestrus cows for insemination by a new method based on bovine sex pheromones. J Instit Anim Technol 28:97–101

Presicce G, Brockett C, Cheng T, Foote R, Rivard G, Klemm W (1993) Behavioural responses of bulls kept under artificial breeding conditions to compounds presented for olfaction, taste or with topical nasal application. Appl Anim Behav Sci 37:273–284

Preti G (1984) Method for detecting bovine estrus by determining methyl heptanol concentrations in vaginal secretions. United States Patent (4467814)

Radford HM, Watson RH (1957) Influence of rams on ovarian activity and oestrus in Merino ewes in the spring and early summer. Aust J Agricul Res 8:460–470

Ramesh Kumar K, Archunan G, Jeyaraman R, Narasimhan S (2000) Chemical characterization of bovine urine with special reference to oestrus. Vet Res Comm 24:445–454

Rekwot P, Ogwu D, Oyedipe E (2000) Influence of bull biostimulation, season and parity on resumption of ovarian activity of zebu (*Bos indicus*) cattle following parturition. Anim Reprod Sci 63:1–11

Rivard G, Klemm W (1989) Two body fluids containing bovine estrous pheromone(s). Chem Senses 14:273–279

Royal MD, Darwash AO, Flint APE, Webb R, Woolliams JA, Lamming GE (2000) Declining fertility in dairy cattle: changes in traditional and endocrine parameters of fertility. Anim Sci 70:487–501

Sankar R, Archunan G (2004) Flehmen response in bull: role of vaginal mucus and other body fluids of bovine with special reference to estrus. Behav Proc 67:81–86

Sankar R, Archunan G (2008) Identification of putative pheromones in bovine (*Bos taurus*) faeces in relation to estrus detection. Anim Reprod Sci 103:149–153
Sankar R, Archunan G, Habara Y (2007) Detection of oestrous-related odour in bovine (*Bos taurus*) saliva: bioassay of identified compounds. Animal 1:1321–1327
Schank JC (2001) Do Norway rats (*Rattus norvegicus*) synchronize their estrous cycles? Physiol Behav 72:129–139
Schank JC (2004) Avoiding synchrony as a strategy of female mate choice. Nonlin Dynam Psychol Life Sci 8:147–176
Shinohara K, Morofushi M, Funabashi T, Kimura F (2001) Axillary pheromones modulate pulsatile LH secretion in humans. Neuroreport 12:893–895
Signoret JP (1991) Sexual pheromones in the domestic sheep - importance and limits in the regulation of reproductive physiology. J Steroid Biochem Mol Biol 39:639–645
Sokolov VE, Karavaeva EA, Belaya ZA, Zinkevich EP (1995) Interspecific action of boar sex pheromone: effects on oestrous cycle in cattle. Dokl Biol Sci 341:178–180
Stern K, McClintock MK (1998) Regulation of ovulation by human pheromones. Nature 392:177–179
Stumpf T, Wolfe M, Wolfe P, Day M, Kittok R, Kinder J (1992) Weight changes prepartum and presence of bulls postpartum interact to affect duration of postpartum anestrus in cows. J Anim Sci 70:3133–3137
Tauck S, Olsen J, Wilkinson J, Berardinelli J (2010a) Duration of daily bull exposure on resumption of ovulatory activity in postpartum, primiparous, suckled, beef cows. Anim Reprod Sci 118:13–18
Tauck SA, Berardinelli JG (2007) Putative urinary pheromone of bulls involved with breeding performance of primiparous beef cows in a progestin-based estrous synchronization protocol. J Anim Sci 85:1669–1674
Tauck SA, Olsen JR, Wilkinson JRC, Wedlake RJ, Davis KC, Berardinelli JG (2010b) Characteristics of temporal patterns of cortisol and luteinizing hormone in primiparous, postpartum, anovular, suckled, beef cows exposed acutely to bulls. Reprod Biol Endocrinol 8:1–8
Vanderlee S, Boot LM (1955) Spontaneous pseudopregnancy in mice. Acta Physiol Pharmacol Neerl 4:442–444
Weidong MA, Clement BA, Klemm WR (1997) Volatile compounds of bovine milk as related to the stage of the estrous cycle. J Dairy Sci 80:3227–3233
Whitten WK (1956) Modification of the oestrous cycle of the mouse by external stimuli associated withe the male. J Endocrinol 13:399–404
Zalesky DD, Day ML, Garcia-Winder M, Imakawa K, Kittok RJ, D'occhio MJ, Kinder JE (1984) Influence of exposure to bulls on resumption of estrous cycles following parturition in beef cows. J Anim Sci 59:1135–1139
Zhang J-X, Rao X-P, Sun L, Zhao C-H, Qin X-W (2007) Putative chemical signals about sex, individuality, and genetic background in the preputial gland and urine of the house mouse (*Mus musculus*). Chem Senses 32:293–303

Part II
Microbiomes and Chemical Signals

Chapter 5
Coding of Group Odor in the Subcaudal Gland Secretion of the European Badger *Meles meles*: Chemical Composition and Pouch Microbiota

Christina D. Buesching, H. Veronica Tinnesand, YungWa Sin, Frank Rosell, Terry Burke, and David W. Macdonald

5.1 Introduction

To convey discrete olfactory information to their conspecifics, many mammals have developed specialized scent glands as well as adaptive scent-marking behaviors (e.g., Brown and Macdonald 1985; Müller-Schwarze 2006). Diet (e.g., Ferkin et al. 1997) and parasites (e.g., Gangestad and Thornhill 1998), as well as the animal's sex (e.g., Kelliher 2007), age (e.g., Osada et al. 2003), endocrine status (e.g., Woodley and Baum 2003), and genetic makeup (Penn 2002) can affect primary gland products. Often, however, these primary secretions are metabolized by bacteria, and typically it is those secondary metabolites that generate the characteristic odor of the scent mark. The fermentation hypothesis (Albone 1984; Albone et al. 1974; Albone and Perry 1975; Gorman 1976; Archie and Theis 2011) predicts that odor profiles depend partly on the primary gland products excreted by the animal,

C.D. Buesching (✉) • D.W. Macdonald
Recanati-Kaplan Centre, Wildlife Conservation Research Unit, Department of Zoology, University of Oxford, Tubney House, Abingdon Road, Tubney OX13 5QL, UK
e-mail: christina.buesching@zoo.ox.ac.uk

H.V. Tinnesand • F. Rosell
Faculty of Arts and Sciences, Department of Environmental and Health Studies, Telemark University College, 3800 Bø, Telemark, Norway

Y. Sin
Recanati-Kaplan Centre, Wildlife Conservation Research Unit, Department of Zoology, University of Oxford, Tubney House, Abingdon Road, Tubney OX13 5QL, UK

NERC Biomolecular Analysis Facility – Sheffield, Department of Animal & Plant Sciences, University of Sheffield, Alfred Denny Building, Western Bank, Sheffield S10 2TN, UK

T. Burke
NERC Biomolecular Analysis Facility – Sheffield, Department of Animal & Plant Sciences, University of Sheffield, Alfred Denny Building, Western Bank, Sheffield S10 2TN, UK

B.A. Schulte et al. (eds.), *Chemical Signals in Vertebrates 13*,
DOI 10.1007/978-3-319-22026-0_5

partly on the composition of the bacterial flora metabolizing these primary gland products, and partly on the age of the scent mark (e.g., Buesching et al. 2002b; Goodwin et al. 2012; Zechman et al. 1984), e.g., the time the bacteria had to metabolize primary gland products. Whereas the management of these zoonotic microbiota is therefore paramount for mammals to ensure consistency and/or meaningful olfactory information content, the potential to vary odor profiles through fluctuating composition of the microbial communities ensures flexibility in body odor and scent marks.

One such example of essential flexibility in odor profiles is scent characteristics encoding group membership. In social mammals, shared group odors have been implicated in social acceptance (e.g., Hurst et al. 1993) and group cohesion (e.g., peccaries: Buyers 1985; badgers: Buesching et al. 2003). However, given that in the majority of species, offspring of one or both sexes disperse at least once in their lives, and in reality may change group affiliation several times even as adults (e.g., to avoid inbreeding: Greenwood 1980), olfactory advertisement of group membership must be reliable yet flexible. Thus, generation of shared group odors has long been suggested to rely predominantly on the assimilation of microbial communities amongst group members (Albone 1984). This has recently been shown to be the case in anal gland secretions of spotted hyenas *Crocuta crocuta* (Theis et al. 2012, 2013; for review of additional species see Wyatt 2010; Ezenwa and Williams 2014; James et al. 2013).

European badgers (*Meles meles*) are macrosmatic, and thus rely heavily on their sense of smell for intra-specific communication and recognition of others (Roper 2010). Their social organization is highly variable and ranges from solitary or pair-living to the formation of large social groups in the South of England (Johnson et al. 2000). In addition to other pathways of olfactory communication common amongst Mustelids, such as urine, feces, anal, and inter-digital glands (Macdonald 1985), badgers have evolved a subcaudal gland which is unique amongst the Carnivora (Macdonald 1985). It consists of several layers of apocrine and sebaceous cells, which secrete a margarine-like paste into a common lumen, the subcaudal pouch (Stübbe 1971), where it is stored until scent marking (Buesching et al. 2002c). The pouch opens at a 20–80 mm wide slit situated between the anus and the base of the tail. Both sexes possess this gland and use the secretion for scent marking by pressing the slit onto the substrate (Őstborn 1976; Buesching and Macdonald 2004) or conspecifics (Buesching et al. 2003). Cubs start to produce traces of subcaudal gland secretion when they are approximately 4 months old (Buesching et al. 2002c). Secretions are highly individual-specific, and encode information about sex, age, body condition, and reproductive status of the donor (Őstborn 1976; Gorman et al. 1984; Buesching et al. 2002a), but change with season and age of the scent mark (Buesching et al. 2002b). In addition, odor profiles are more similar amongst the members of a social group than they are between groups, indicating the existence of a shared group odor (Buesching et al. 2002a).

The subcaudal pouch has been shown to support a rich bacterial flora (Albone et al. 1978; Sin et al. 2012), and semiochemical analyses confirm that subcaudal gland secretions are of high chemical complexity, containing mainly medium- and long-chained carboxylic acids, which are usually of bacterial rather than mammalian origin (Albone 1984), water and protein (Gorman et al. 1984). A direct empirical link between pouch bacteria and scent profiles, however, remains to be established.

Here, we analyzed the bacterial pouch flora in relation to the chemical scent profile of adults belonging to six different social groups. The aim of this study was threefold: (1) to confirm that group membership is encoded in subcaudal gland secretions, (2) to investigate the overall relation between the bacterial communities and the chemical composition of the subcaudal-gland secretion, and (3) to investigate potential group differences in the bacterial communities of badger subcaudal gland secretions.

5.2 Material and Methods

5.2.1 Study Animals and Collection of Samples

Samples were collected in Wytham Woods, Oxfordshire, England (GPS reference: 51u469260N; 1u199190W; for a detailed description of the study site see Kruuk 1989; for details of the study population see Macdonald and Newman 2002), during two different trapping events: in spring (27th May–9th June 2010, $N=31$), and in summer (6th Sept.–18th Sept 2010, $N=35$). As part of an ongoing population study, and following the methodology described by Macdonald and Newman (2002), badgers were trapped overnight in cage traps baited with peanuts. At first capture (usually as cubs), all animals were tattooed with an individual number on the left inguinal region. Thus, after sedation with 0.2 ml ketamine hydrochloride/kg body weight (Thornton et al. 2005), all badgers could be identified individually and linked to their trapping history.

Subcaudal gland secretion (SGS) was scooped out of the subcaudal pouch using a rounded stainless steel spatula, and contact of the spatula with body parts other than the inside of the subcaudal pouch was strictly avoided to exclude bacteriological cross-contamination. Secretions were subdivided into two aliquots. Both were frozen immediately and stored at –20 °C until further analyses. To avoid contact with plasticizing agents, the aliquot for semiochemical analysis was stored first in a glass vial with Teflon lid, and the one for the bacteriological analyses was stored afterwards in a microcentrifuge tube. Between sampling different individuals, the spatula was wiped clean and sterilized three times with 90 % ethanol, which was flamed off to avoid contamination of the semiochemical sample.

All collected samples were used to investigate the correlation between microbiota and chemical composition. To investigate chemical and microbial differences between social groups, however, we used secretions collected from a total of six different social groups ($N_{CH}=5$, $N_{GAH}=5$, $N_{LS}=6$, $N_{MT}=6$, $N_{PO}=6$, $N_{RC}=5$) as established by biannual bait-marking surveys (following the methodology described by

Delahay et al. 2000), where we caught a minimum of five adult animals. To avoid pseudoreplication, only one (randomly chosen) sample from each individual was included ($N=33$, $N_{\mathrm{Males}}=14$, $N_{\mathrm{Females}}=19$).

5.2.2 *Chemical Analyses*

For each sample, 0.1 g of subcaudal gland secretion was extracted in 1.0 ml of dichloromethane (Pestanal® Grade, Sigma-Aldrich®, Oslo, Norway). Solutions were left at room temperature for 1 h and centrifuged at 3000 rpm through a 0.45 μm filter for 5 min. The resulting particle-free solution was transferred to a 1.5 ml GC-vial (Agilent®, Oslo, Norway).

Chemical composition was analyzed using a Hewlett-Packard (HP, Oslo, Norway) 6890 Series II gas chromatograph (GC) equipped with a nonpolar HP-5 MS 5 % phenyl-methyl-siloxane column (30.0 m long × 0.25 mm ID × 0.25 μm film thickness), connected to a HP 5973 Series mass spectrometer detector (MS) with a split/splitless inlet in splitless mode using helium as the carrier gas at a constant flow of 0.9 ml/min. Injection port temperature was set at 300 °C. The purge flow to split vent was 49.8 m/min at 1.00 min. An auto-injection system (Agilent 7683 Series Injector, Oslo, Norway) was used to inject 1.0 μl of the SGS solution into the GC-MS. The samples were cold-trapped at 40 °C on the column tip for 2 min and separated using a temperature program of 8 °C/min from 40° to 150 °C, then 6 °C/min from 150 °C to 200 °C, and finally 4 °C/min from 200° to 240 °C (holding for 15 min). A solvent delay of 5 min was set to prevent solvent damage to the detector. Samples were analyzed in random order, and after every five samples, a blank was run to ensure that there was no contamination left in the column. A mixture of unbranched alkanes between C_8 and C_{40} (Sigma-Aldrich, Oslo, Norway) was also run after every five samples to calculate Kováts Retention Indices (KRI), allowing standardization of retention times. The instrument was calibrated every morning to detect possible changes in sensitivity.

Compounds were matched between profiles by their retention times and mass spectra, and given an individual peak number. A tentative identification of the analytes was provided by cross-checking the best suggested matches from the Wiley 275 spectral library (Scientific Instrument Services Inc., Ringoes, USA) with the calculated KRI of the analytes. Identification of 12 compounds (hexanoic acid, octanoic acid, decanoic acid, dodecanoic acid, tetradecanoic acid methyl ester, tetradecanoic acid, hexadecanoic acid methyl ester, hexadecanoic acid, methyl linoleate, arachidic acid methyl ester (eicosanoic acid methyl ester), docosanoic acid methyl ester, tetracosanoic acid methyl ester; chosen to provide a good spread across the profile retention times) was confirmed through injection of commercial standards (Sigma-Aldrich, Oslo, Norway). Where the Wiley library did not provide a good match, analytes were given a name based on retention times and added to a new library.

5.2.3 Bacteriological Analyses

5.2.3.1 DNA Extraction

Following the methodology described in Sin et al. (2012), secretions were transferred to microcentrifuge tubes and re-suspended in 180 µl enzymatic lysis buffer [20 mM Tris–Cl (pH 8.0), 2 mM EDTA, 1.2 % Triton X-100] containing 20 mg/ml lysozyme. Following incubation at 37 °C for at least 1 h in a shaking incubator to lyse cell walls of Gram-positive bacteria before DNA purification, samples were incubated with proteinase K at 56 °C for at least 1 h, followed by the addition of 200 µl of ethanol. DNA isolation was continued by pipetting the mixture into the spin column according to the manufacturer's instructions for animal tissue samples using the DNeasy Blood & Tissue kit (Qiagen, Hilden, Germany).

5.2.3.2 T-RFLP Analysis

Internal fragments of 16S rRNA genes were amplified from the isolated DNA using universal bacterial primers 341f (5′-CCTACGGGAGGCAGCAG-3′; Muyzer et al. 1993) and 926r (5′-CCGTCAATTCMTTTRAGTTT-3′; Muyzer et al. 1995). Primer 341f was labeled at its 5′ end with the dye 6-carboxyfluorescein (6-FAM) and 926r labeled with hexachloro-6-carboxyfluorescein (HEX). PCR mixtures comprised 1× PCR buffer, 1 mM $MgCl_2$, each dNTP at a concentration of 0.2 mM, each primer at a concentration of 0.6 µM, as well as 0.5 U Taq polymerase (Bioline). Amplification was carried out in two separate 25 µL reaction mixtures under the following conditions: 95 °C for 3 min, amplified for 35 cycles of 95 °C/30s; 55 °C/60s; 72 °C/75 s, and a final extension at 72 °C for 10 min. For each sample, duplicate fluorescently labeled PCR products were pooled after PCR. PCR product sizes were confirmed by agarose gel electrophoresis using 100 bp DNA ladders as size markers and by staining with ethidium bromide. The amplified products (around 586 bp) were purified using the QIA quick gel extraction kit (Qiagen).

Fluorescently labeled PCR products were digested with MspI restriction enzymes (Applied Biosystems ABI) for 6 h at 37 °C, followed by 20 min at 80 °C for enzyme inactivation. Digested fragments were separated on the ABI 3730 Genetic Analyzer (ABI). Sizes of the fluorescently labeled terminal restriction fragments (TRFs) were determined by comparison with the GeneScan 500 ROX size standard (ABI). Before injection, 0.5 µL of the DNA sample was denatured in the presence of 9.5 µL Hi-Di formamide and 0.05 µL GS 500 ROX size standard (ABI) at 95 °C for 5 min. Injection was performed electrokinetically at 2 kV for 10s, and electrophoresis was run at 15 kV for 30 min. After electrophoresis, terminal restriction fragment length polymorphism (T-RFLP) electropherograms were imaged using the GENEMAPPER software (v. 3.7; ABI). The lengths of TRFs were determined by comparison with the internal standard, within the lower threshold at 50 bp and upper threshold at 500 bp. Only peaks with heights exceeding 50 fluorescence units were evaluated.

T-RFLP profiles were aligned using T-Align (Smith et al. 2005), which identified all fragments unequivocally with ±0.5 bp in all profiles generated, and determined the presence and absence of TRFs and their relative fluorescence in all samples. Log $(n+1)$ transformation was performed in order to normalize the data. The number of the operational taxonomic units (OTUs) and the bacterial identities of the OTUs that contributed to the TRFs were identified by cloning and sequencing of the 16S rRNA genes as detailed in Sin et al. (2012).

5.2.4 Statistical Analyses

All TRFs and chemical compounds that occurred in only one sample were excluded from all analyses.

5.2.4.1 Analyses of Scent Profiles

To account for variation in the total quantity of injected secretion, peak areas of each compound were standardized using the peak area of hexadecanoic acid, which was present in all scent profiles as one of the largest peaks. All data were square-root transformed to reduce the influence of the most abundant variables (Clarke and Warwick 2001), and a Bray–Curtis similarity index calculated between each possible sample pairing.

5.2.4.2 Analyses of Group Differences in Chemical Composition and TRFs

To investigate variation in chemical profiles and microbial communities among social groups, we first used principal coordinate (PCO) analysis based on the Bray–Curtis similarity index in order to visualize the patterns of variation among samples (Gower 1966). If the same animal was caught during both trapping events, only one sample, selected at random, was included in the analysis. Differences between social groups were then compared with a single factor PERMANOVA (Anderson 2001, McArdle and Anderson 2001) using 9999 permutations, applying both main and pairwise tests. PERMANOVA allows distance-based tests of significance for comparing a priori groupings. Significant ($P<0.05$) and marginally significant ($P<0.1$) differences between groups were investigated further using canonical analysis of principal coordinates (CAP: Anderson and Willis 2003), which obtains predictive models that search the multivariate data for the best discrimination between a priori groups. To present the reader with all results, we did not correct for multiple testing as suggested by Nakagawa (2004). The number of PCO axes, *m*, to use in the model and the predictive ability of the model to discriminate between the social groups was assessed by leave-one-out cross-validation (Anderson and Robinson 2003, Mardon et al. 2010). The software PRIMER V6.1.13 (Clarke and Gorley 2006) with the PERMANOVA+ V1.0.3 add-on package (Anderson et al. 2008) was used in all analyses.

5.2.4.3 Analyses of Correlation Between TRFs and Chemical Composition

To investigate any linkage between the TRFs and the chemical composition of the secretion, we performed the BEST procedure (Clarke et al. 2008). The rationale of BEST is to find the "best match" between the multivariate among-sample patterns of two data matrices, one being a fixed resemblance matrix (often biotic variables, or in our case the chemical compounds) and one an explanatory matrix (often environmental variables, in our case the TRF data: Clarke and Gorley 2006). The BVSTEP procedure in BEST selects the combination of explanatory variables that best explains the structure in the fixed matrix, whereas the global BEST test constitutes an overall significance test for the final subset of variables (Clarke et al. 2008).

To investigate correlations between specific TRFs and chemical compounds, we constructed a correlation matrix between all variables that occurred in > 10 % of the samples. All TRF-compound combinations with a correlation > 0.4 were tested for statistical significance using two-tailed tests. Correlation analyses were performed with the software packages JMP 10 (SAS Institute Inc., Cary, NC, USA) and SPSS Statistics 19 (SPSS Inc., New York, USA).

5.3 Results

5.3.1 Group Differences in Chemical Composition

When evaluating chemical variation among social groups including all compounds, an unconstrained 2D PCO explained 49.2 % of the variation in the data, and the third axis explained further 8.7 %, but there was no separation between badger social groups (Fig. 5.1). The main PERMANOVA comparing the 6 groups with ≥ 5 adults sampled was significant (pseudo $F=1.6692$, $P=0.0141$), and pairwise tests revealed significant differences between four pairs of groups (Table 5.1). The CAP analysis classified 33.3 % of the chemical profiles into the correct group using leave-one-out cross-validation and $m=16$ axes ($\delta_1^2=0.89814$, $P=0.028$, Fig. 5.2). The CAP plot, however, is a constrained plot, which views the data cloud through the filter of our hypothesis, and thus should be viewed in conjunction with the results from the cross-validation, which provide information the distinctiveness of each group, and how well the axes discriminate between each of them (Anderson

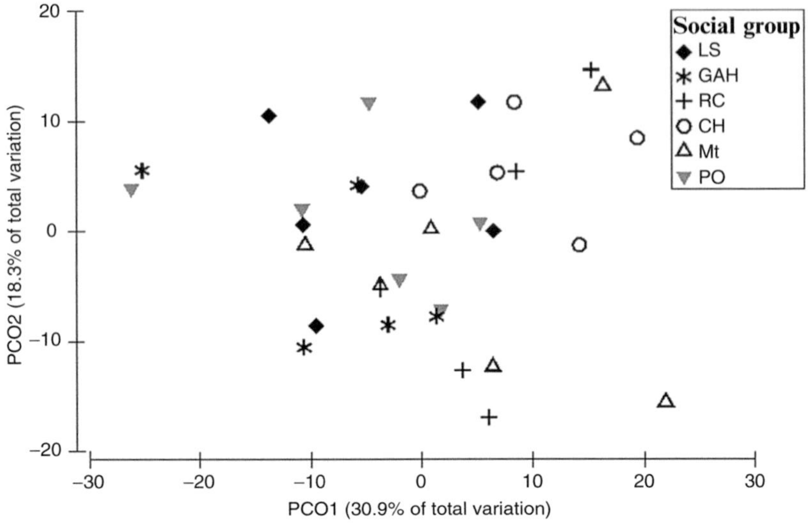

Fig. 5.1 Unconstrained 2D PCO including all chemical compounds from adults of six social groups with ≥5 SCG samples

Table 5.1 PERMANOVA results for pairwise comparisons between the chemical profiles and the microbiota of members of six different social groups (significant results highlighted in bold)

	Chemistry				Microbiota	
	All compounds included		Only compounds shared by all group members included		TRFs	
Group	*t*	P(perm)	*t*	P(perm)	*t*	P(perm)
LS vs. GAH	0.9084	0.5719	3.0404	**0.0013**	1.0487	0.3468
LS vs. RC	1.2738	0.1096	3.6784	**0.0035**	0.94561	0.4736
LS vs. CH	1.6147	**0.0281**	4.4595	**0.0025**	0.59353	0.6728
LS vs. Mt	1.3406	0.1041	3.5608	**0.0026**	1.052	0.313
LS vs. PO	0.9924	0.4065	3.4059	**0.0022**	0.64851	0.6137
GAH vs. RC	1.1914	0.1737	3.3558	**0.0018**	0.88187	0.6028
GAH vs. CH	1.5818	**0.0175**	4.2611	**0.0016**	0.7999	0.6341
GAH vs. Mt	1.1415	0.2353	4.5744	**0.0004**	1.6775	**0.0478**
GAH vs. PO	0.99478	0.4149	3.1995	**0.0008**	0.46955	0.8486
RC vs. CH	1.1533	0.1965	3.4795	**0.0094**	0.91271	0.5374
RC vs. Mt	0.91766	0.555	3.2032	**0.002**	1.5689	0.076
RC vs. PO	1.1016	0.2727	2.3542	**0.0036**	0.99276	0.4195
CH vs. Mt	1.1669	0.2369	3.9912	**0.0026**	0.96919	0.4436
CH vs. PO	1.441	**0.0535**	3.3926	**0.0021**	0.75341	0.5956
Mt vs. PO	0.93082	0.4903	3.0985	**0.0029**	1.239	0.1921

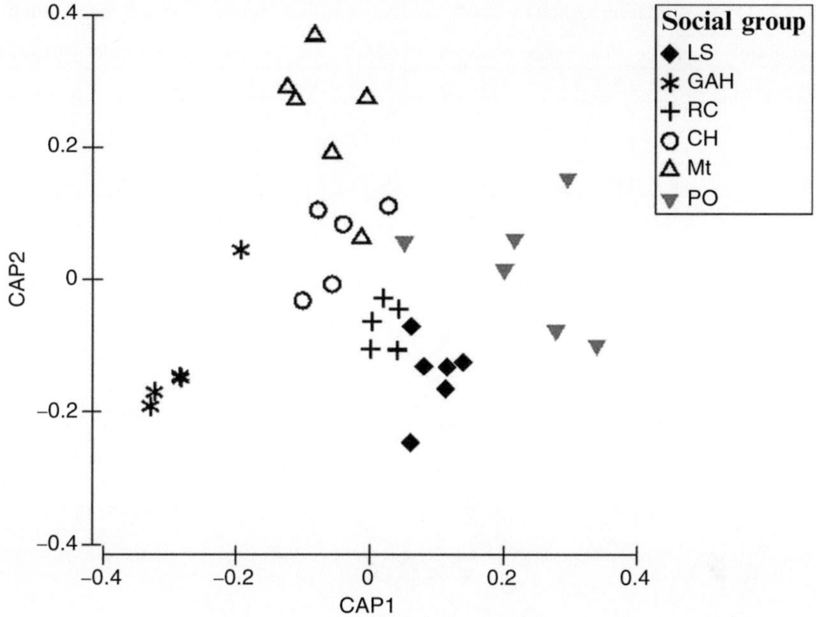

Fig. 5.2 CAP analysis including all chemical compounds from adults of six social groups with ≥5 SCG samples. Note that the figure only shows 2 CAP axes of the 15 axes generated in the model

et al. 2008). Although in our analyses the CAP plot suggested that semiochemical differences among social groups were clear, the cross-validation results showed that the predictive ability of the model was low. The cross-validation results, however, were still significant, as, with $n=6$ social groups, only approx. 16.7 % of samples would have been classified correctly by chance alone (Table 5.2).

However, if only compounds shared by all members in each group were included (*sensu* scent signature: Wyatt 2010), the unconstrained 2D PCO explained 54.8 % of the variation in the data (Fig. 5.3), and the third axis explained a further 18.1 %, with separations between social groups. The PERMANOVA comparing the six groups was highly significant (pseudo $F=12.39$, $P=0.0001$), as were all pairwise comparisons (Table 5.1). The CAP analysis classified 100 % of the chemical profiles into the correct group using $m=5$ axes ($\delta_1^2=0.98934$, $P=0.0001$, Fig. 5.4).

5.3.2 Group Differences in Microbiota

Comparing the TRFs to investigate variation in microbiota among groups, an unconstrained 2D PCO explained 82.3 % of the variation in the data (Fig. 5.5), and the third axis explained further 12.2 %, but there was no separation between social groups, and no significant differences in the TRFs of the 6 social groups were found (PERMANOVA: pseudo $F=1.007$, $P=0.4317$, Table 5.1).

Table 5.2 Cross-validation results for correct classification of samples for each social group

Social group	RC	PO	LS	GAH	CH	Mt	Total	% correctly identified
RC	0	1	1	1	2	0	5	0
PO	0	2	2	0	0	2	6	33.33
LS	1	1	3	0	1	0	6	50
GAH	1	0	2	1	0	1	5	20
CH	1	0	1	0	1	2	5	20
Mt	0	0	1	0	1	4	6	66.67

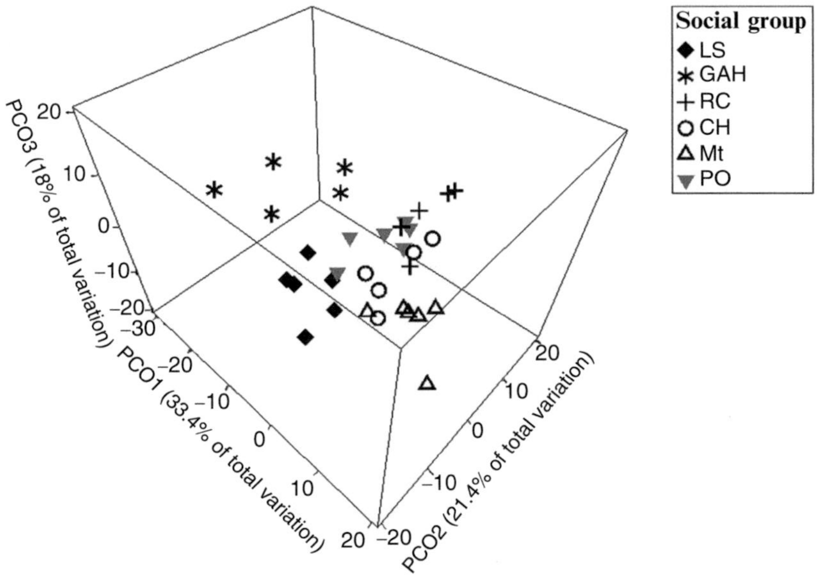

Fig. 5.3 Unconstrained 2D PCO including only chemical compounds which were shared by all adults belonging to the same social group (six social groups with ≥5 SCG samples)

5.3.3 *Correlation of Microbiota and Chemical Composition of Subcaudal-Gland Secretions*

In total, 50 TRFs and 125 different chemical compounds were found in the GC-MS profiles of 66 subcaudal-gland secretions. Using 9999 permutations, the BEST procedure showed an overall significant correlation between microbiota and chemical composition (Rho = 0.228, $P = 0.041$), with a subset of four TRFs (99.2, 138.9, 194.2, and 212.0) best explaining the structure in the chemical matrix (Table 5.3).

In the investigation of correlations between specific TRFs and chemical compounds, 5 TRFs and 23 chemical compounds showed a correlation > 0.4. All correlations were statistically significant ($P < 0.01$, Table 5.3).

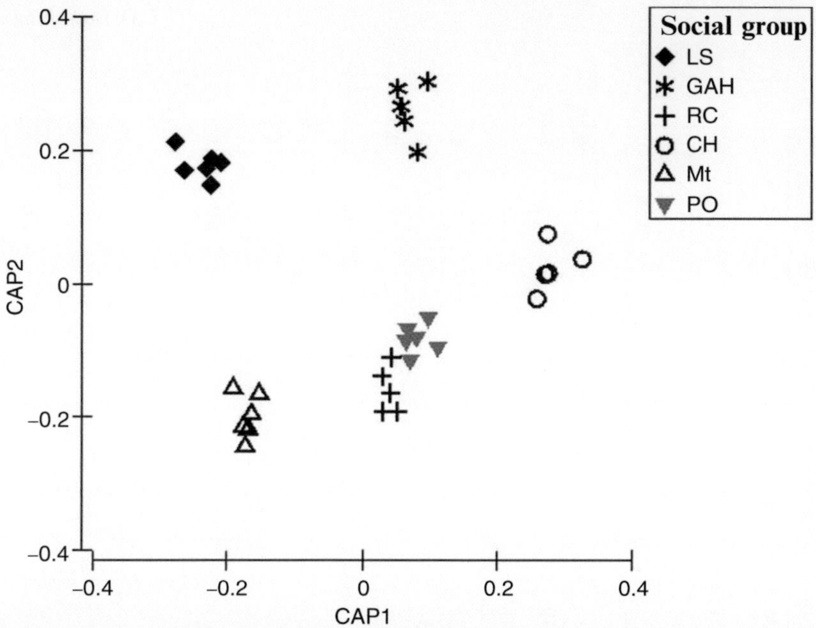

Fig. 5.4 CAP analysis including only chemical compounds which were shared by all adults belonging to the same social group (six social groups with ≥5 SCG samples). Note that the figure only shows 2 CAP axes of the 4 axes generated in the model

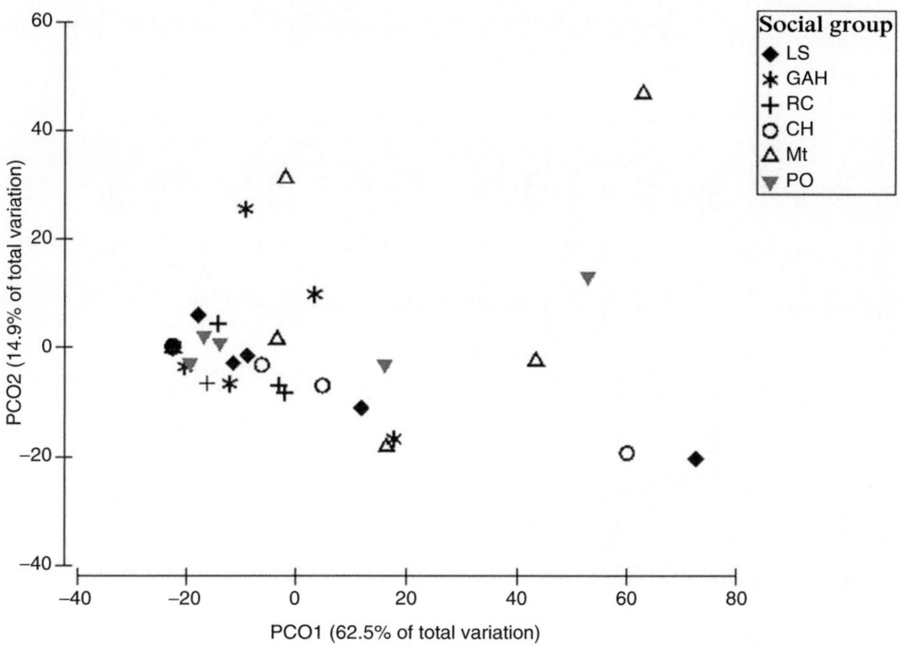

Fig. 5.5 Unconstrained 2D PCO comparing the TRFs from 6 social groups with ≥5 adults sampled

Table 5.3 Correlation of microbiota and chemical components in subcaudal gland secretions (BEST procedure using 9999 permutations)

Retention time	Compound	99.16	132.39	133.68	138.94	140.42	141.69	160.58	189.3	194.15	208.23	211.97
7.93	Benzaldehyde				**0.60**	**−0.36**						
9.30	Limonene				**0.56**							
10.92	2-Chloroethylbenzene										**0.38**	
10.97	Benzeneethanol	**0.37**			**0.30**							
12.08	Octanoic acid[a]						**0.31**				**0.36**	
14.60	Badger_SGS_14.6	**0.39**										
15.33	Badger_SGS_15.325											**0.54**
15.40	1,2,3,4-Tetrahydro-1,1, 1-trimethylnaphthalene					**−0.39**						**0.47**
15.45	Decanoic acid[a]	**0.30**					**0.31**					
15.76	Badger_SGS_15.756	**0.46**				**−0.34**				**−0.37**		**0.46**
16.03	Badger_SGS_16.032					**0.33**						
16.98	Badger_SGS_16.975											**0.33**
17.44	Badger_SGS_17.443	**0.49**										
17.77	Badger_SGS_17.77	**0.35**										
18.01	2-Methylundecanoic acid									**−0.34**		
18.17	Badger_SGS_18.168				**0.36**							
18.74	Dodecanoic acid[a]	**0.41**										
18.92	Badger_SGS_18.922	**0.35**										
19.67	Badger_SGS_19.672											**0.95**
20.55	Badger_SGS_20.548	**0.45**										
20.77	Badger_SGS_20.765	**0.55**										
20.98	Badger_SGS_20.981	**0.56**										
21.13	Badger_SGS_21.129											**0.34**

21.27	Badger_SGS_21.267				**0.35**							
21.32	Tetradecanoic acid, methyl ester[a]											**0.43**
21.56	Badger_SGS_21.556											**0.78**
21.76	Badger_SGS_21.760			**−0.33**								
21.82	Badger_SGS_21.821		**0.41**						**0.32**			
22.11	Tetradecanoic acid[a]	**0.46**										
22.47	Badger_SGS_22.466										**0.30**	
22.57	Badger_SGS_22.566											**0.46**
22.70	Badger_SGS_22.701					**0.31**						**−0.31**
23.00	Badger_SGS_23.000											**0.53**
23.05	Badger_SGS_23.054	**0.43**										
23.06	Badger_SGS_23.060	**0.35**										**0.41**
23.69	Badger_SGS_23.690				**0.50**							
26.02	Badger_SGS_26.021					**0.33**						**−0.31**
28.57	Linoleic acid											**0.58**
28.85	Linoleic acid, ethyl ester							**0.32**				
28.96	Oleic acid, ethyl ester							**0.40**				
29.03	Badger_SGS_29.025				**0.56**							
29.40	Octadecanoic acid, ethyl ester							**0.65**				
31.65	4-Octadecanolide											**0.34**

99.16 = Alphaproteobacteria/Betaproteobacteria; 132.39 Unknown OTU; 133.68 = Firmicutes/Alphaproteobacteria; 138.94 = Actinobacteria; 140.42 = Actinobacteria; 141.69 = Unknown OTU; 160.58 = Gammaproteobacteria/Betaproteobacteria/Actinobacteria; 189.3 = Firmicutes; 189.3 = Firmicutes; 194.15 = Unknown OTU; 208.23 = Bacteroidetes; 211.97 = Firmicutes

[a]Marks components validated through co-injection of standards

5.4 Discussion

Most mammals communicate a wide variety of information in their olfactory signals (Brown and Macdonald 1985; Müller-Schwarze 2006). Some species, such as the European badger, encode individual-specific information as well as group membership, seasonality, and age of the scent mark in a single scent mark (Buesching and Macdonald 2001). Whereas some of this information, such as individuality, is static and should thus remain stable over time, other signals are transient (e.g., information pertaining to reproductive receptivity) or need to be pliable (e.g., information pertaining to group membership) to accommodate changes in biology and behavior. As olfactory signals are partly dependent on primary gland products, but also on environmental factors (Wyatt 2010), precise management of the synergistic relationships between endo- and exogenous factors involved in olfactory signal creation is paramount for reliable communication.

The fermentation hypothesis predicts that mammalian scent profiles are heavily affected by bacteria metabolizing the primary gland products excreted by the individual (Albone et al. 1978, Albone and Perry 1975, Gorman 1976). In our population, TRF (Sin et al. 2012) as well as GC-MS (Gorman et al. 1984, Buesching et al. 2002a, b) profiles were indeed individual specific, indicating that many individual advertisement signals might in fact be affected by pouch bacteria (Buesching et al. 2003). Here, we found a significant correlation between the microbiota in the subcaudal pouch and the chemical composition of the subcaudal gland secretion. A subset of four TRFs (99.2, 138.9, 194.2, and 212.0) explained most of the effect. Sin et al. (2012) found TRFs 138.94 to belong to phylum Actinobacteria, and 97.6 and 211.97 to phylum Firmicutes (no phylum could be assigned from cloning and sequencing for TRF 194.2). Actinobacteria and Firmicutes include several well-known odor producers, and are also abundant in the paste of spotted hyenas (*Crocuta crocuta*: Theis et al. 2012, 2013). Several genera in these phyla produce a diverse array of short- and medium-chain fatty acids, which are prominent in the subcaudal gland secretion of badgers (Buesching et al. 2002a). Some Actinobacteria have been found to play a major role in the transformation of odorless steroids into odorous derivatives (e.g., Gower et al. 1986, Kohl et al. 2001). It is thus highly likely that bacteria belonging to these phyla will affect the chemical composition in the secretion, but further research using advanced 16SrRNA or metagenomic techniques is needed to confirm these associations.

In addition, our GC-MS results confirm the presence of shared group odors in this badger population. Nevertheless, while Buesching et al. (2002a) found clear group differences in samples collected from the same population 15 years prior to the present study also when all components detected by the GCMS were included in the analyses, in the present study group differences were obvious only if analyses were restricted to those components present in all members of the same social group. SGC profiles are chemically highly complex and have been reported to contain up to 58 of a possible ≥110 components, of which many occur only very rarely (Buesching et al. 2002a). By limiting our analyses to those compounds, which occurred at least in all members of one (or several) social group(s), we could increase statistical power considerably while simultaneously ensuring that biologically meaningful components were included. Although scent provision experiments have confirmed in a variety of species that individuals can discriminate between their own scent, scent of other members of their own group and scent from members of other social groups (e.g., badgers: Palphramand and White 2007; Bodin et al. 2006), few studies have tried to determine, what indicators animals utilize in this context, although learning of specific scent signatures appears important (Wyatt 2010). There is some evidence that it might be the case of recognizing components which are present also in their own scent/other members of their own group vs. components which are specific to a given different social group (Natynczuk and Macdonald 1994). Pairwise comparisons revealed significant differences between the groups, but the high number of axes necessary to separate all groups in the statistical analyses indicates that the coding of a specific group scent is complex and multidimensional. Although dispersal rates remain low in this population (Macdonald et al. 2008), and overall population density remained comparatively stable (Macdonald et al. 2009), individuals now move more frequently between social groups (i.e. excursions and visits to neighbors: Macdonald et al. 2008, Noonan et al. 2014) and extra-territorial matings appear to have increased from an estimated 42 % (Dugdale et al. 2007) to 48 % (Annavi et al. 2014). These changes appear to be associated with less pronounced group differences in odor profiles. Similar plasticity in group odors has also been reported in other species (e.g., elephants: Goodwin et al. 2012; meercats: Leclaire et al. 2014). Nevertheless, by restricting group-specific information to the relative amounts and ratios of a subset of components, other chemical compounds can be utilized effectively to encode other information, e.g., relating to individuality.

Interestingly, however, we found no group-specific differences in the microbiota, neither in the composition nor between any pair of social groups as determined by TRFs. Although this could indicate that other factors, such as genetic relationships (e.g., Todrank et al. 1998), are more important in creating the group odor in badgers, the sensitivity of the method applied in this study restricted the sensitivity of our analyses as the number of OTUs exceeded the numbers of TRFs, and a single TRF can comprise several OTUs, thus limiting resolution (Sin et al. 2012). In contrast, Theis et al. (2013) used a high-throughput sequencing approach to analyze the bacterial communities inhabiting the scent organs of spotted hyenas, which afforded a far more detailed analysis and allowed for the most comprehensive view of the bacterial communities inhabiting any specialized mammalian scent gland to date. As it is likely that the variation in metabolic activity is found at the species, subspecies, or even strain level in badgers, high-throughput sequencing can be expected to reveal more subtle differences in the microbial communities between social groups.

References

Albone E (1984) Mammalian semiochemistry: the investigation of chemical signals between vertebrates. Wiley, Chichester

Albone ES, Perry GC (1975) Anal sac secretion of the red fox, *Vulpes vulpes*; volatile fatty acids and diamines: implications for a fermentation hypothesis of chemical recognition. J Chem Ecol 2:101–111

Albone ES, Eglinton G, Walker JM, Ware GC (1974) The anal sac secretion of red fox (*Vulpes vulpes*); its chemistry and microbiology. A comparison with the anal sac secretion of the lion (*Panthera leo*). Life Sci 14:387–400

Albone ES, Gosden PE, Ware GC, Macdonald DW, Hough NG (1978) Bacterial action and chemical signalling in the red fox (*Vulpes vulpes*) and other mammals. In: Bullard RW (ed) Flavour chemistry of animal foods, vol 67, ACS Symposium Series. American Chemical Society, Washington, DC, pp 78–91

Anderson MJ (2001) A new method for non-parametric multivariate analysis of variance. Austral Ecol 26:32–46

Anderson MJ, Robinson J (2003) Generalized discriminant analyses based on distances. Aust N Zeal J Stat 45:301–318

Anderson MJ, Willis TJ (2003) Canonical analysis of principal coordinates: a useful method of constrained ordination for ecology. Ecology 84:511–525

Anderson MJ, Gorley RN, Clarke KR (2008) PERMANOVA+ for PRIMER: guide to software and statistical methods. PRIMER-E, Plymouth

Annavi G, Newman C, Dugdale HL, Buesching CD, Sin YW, Burke T, Macdonald DW (2014) Neighbouring-group composition and within-group relatedness drive extra-group paternity rate in the European badger (*Meles meles*). J Evol Biol 27:2191–2203

Archie EA, Theis KR (2011) Animal behaviour meets microbial ecology. Anim Behav 82:425–436

Bodin C, Benhamou S, Poulle ML (2006) What do European badgers (*Meles meles*) know about the spatial organisation of neighbouring groups? Behav Processes 72:84–90

Brown RE, Macdonald DW (eds) (1985) Social odours in mammals, vol Volume I and Volume II. Clarendon, Oxford

Buesching CD, Macdonald DW (2001) Scent-marking behaviour of the European badger (*Meles meles*): resource defence or individual advertisement. In: Marchlewska-Koj A, Lepri JL, Müller-Schwarze D (eds) Chemical signals in vertebrates 9. Kluwer Academic, New York, pp 321–327

Buesching CD, Macdonald DW (2004) Variations in object-marking activity and over-marking behaviour of European badgers (*Meles meles*) in the vicinity of their setts. Acta Theriol 49:235–246

Buesching CD, Waterhouse JP, Macdonald DW (2002a) Gas chromatographic analysis of the subcaudal gland secretion of the European badger (*Meles meles*) Part I: Chemical differences related to individual-specific parameters. J Chem Ecol 28:41–56

Buesching CD, Waterhouse JP, Macdonald DW (2002b) Gas chromatographic analysis of the subcaudal gland secretion of the European badger (*Meles meles*) Part II: Time-related variation in the individual-specific composition. J Chem Ecol 28:57–69

Buesching CD, Newman C, Macdonald DW (2002c) Variations in colour and volume of the subcaudal gland secretion of badgers (*Meles meles*) in relation to sex, season and individual-specific parameters. Z Säugetierk 67:1–10

Buesching CD, Stopka P, Macdonald DW (2003) The social function of allo-marking behaviour in the European badger (*Meles meles*). Behaviour 140:965–980

Buyers JA (1985) Olfaction related behaviour in peccaries. Z Tierpsychol 70:201–210

Clarke KR, Gorley RN (2006) PRIMER v6: user manual/tutorial. PRIMER-E, Plymouth

Clarke KR, Warwick RM (2001) Change in marine communities, 2nd edn. PRIMER-E, Plymouth

Clarke KR, Somerfield PJ, Gorley RN (2008) Testing of null hypotheses in exploratory community analyses: similarity profiles and biota-environment linkage. J Exp Mar Biol Ecol 366:56–69

Delahay RJ, Brown JA, Mallinson PJ, Spyvee PD, Handoll D, Rogers LM, Cheeseman CL (2000) The use of marked bait in studies of the territorial organization of the European badger (*Meles meles*). Mam Rev 30:73–87

Dugdale HL, Macdonald DW, Pope LC, Burke T (2007) Polygynandry, extra-group paternity and multiple-paternity litters in European badger (*Meles meles*) social groups. Mol Ecol 16:5294–5306

Ezenwa VO, Williams AE (2014) Microbes and animal olfactory communication: Where do we go from here? Bioessays 36:847–854

Ferkin MH, Sorokin ES, Johnston RE, Lee CJ (1997) Attractiveness of scents varies with protein content of the diet in meadow voles. Anim Behav 53:133–141

Gangestad SW, Thornhill R (1998) Menstrual cycle variation in women's preferences for the scent of symmetrical men. Proc Roy Soc B 265:927–933

Goodwin TE, Broederdorf LJ, Burkert BA, Hirwa IH, Mark DB, Waldrip ZJ, Kopper RA, Sutherland MV, Freeman EW, Hollister-Smith JA, Schulte BA (2012) Chemical signals of elephant musth: temporal aspects of microbially-mediated modifications. J Chem Ecol 38:81–87

Gorman ML (1976) A mechanism for individual recognition by odour in *Herpestes auropunctatus* (Carnivora: Viverridae). Anim Behav 24:141–145

Gorman ML, Kruuk H, Leitch A (1984) Social functions of the sub-caudal scent gland secretion of the European badger (*Meles meles*) (Carnivora: Mustelidae). J Zool Lond 203:549–559

Gower JC (1966) Some distance properties of latent root and vector methods used in multivariate analyses. Biometrika 53:325–338

Gower DB, Nixon A, Jackman PJH, Mallett AI (1986) Transformation of steroids by axillary coryneform bacteria. Int J Cosmet Sci 8:149–158

Greenwood PJ (1980) Mating systems, philopatry and dispersal in birds and mammals. Anim Behav 28:1140–1162

Hurst JL, Fang J, Barnard CJ (1993) The role of substrate odours in maintaining social tolerance between male house mice, *Mus musculus domesticus*. Anim Behav 45:997–1006

James AG, Austin CJ, Cox DS, Taylor D, Calvert R (2013) Microbiological and biochemical origins of human axillary odour. FEMS Microbiol Ecol 83:527–540

Johnson DD, Macdonald DW, Dickman AJ (2000) An analysis and review of models of the sociobiology of the Mustelidae. Mamm Rev 30:171–196

Kelliher KR (2007) The combined role of the main olfactory and vomeronasal systems in social communication in mammals. Horm Behav 52:561–570

Kohl JV, Atzmueller M, Fink B, Grammer K (2001) Human pheromones: Integrating neuroendocrinology and ethology. Neuro Endocrinol Lett 22:309–321

Kruuk H (1989) The social badger: ecology and behaviour of a group-living carnivore (*Meles meles*). Oxford University Press, Oxford

Leclaire S, Nielsen JF, Drea CM (2014) Bacterial communities in meerkat anal scent secretions vary with host sex, age, and group membership. Behav Ecol 25:996–1004

Macdonald DW (1985) The carnivores: order Carnivora. In: Brown M, Macdonald DW (eds) Social odours in mammals, vol 2. Clarendon, Oxford, pp 619–722

Macdonald DW, Newman C (2002) Badger (*Meles meles*) population dynamics in Oxfordshire, UK, Numbers, Density and Cohort life histories, and a possible role of climate change in population growth. J Zool 256:121–138

Macdonald DW, Newman C, Buesching CD, Johnson PJ (2008) Male-biased movement in a high-density population of the Eurasian badger (*Meles meles*). J Mammal 89:1077–1086

Macdonald DW, Newman C, Nouvellet PM, Buesching CD (2009) An analysis of Eurasian badger (*Meles meles*) population dynamics: Implications for regulatory mechanisms. J Mammal 90:1392–1403

Mardon J, Saunders SM, Anderson MJ, Cochoux C, Bonadonna F (2010) Species, gender, and identity: cracking petrels' sociochemical code. Chem Senses 35:309–321

McArdle BH, Anderson MJ (2001) Fitting multivariate models to community data: a comment on distance-based redundancy analysis. Ecology 82:290–297

Müller-Schwarze D (2006) Chemical ecology of vertebrates. Cambridge University Press, Cambridge

Muyzer G, De Waal EC, Uitterlinden AG (1993) Profiling of complex microbial populations by denaturing gradient gel electrophoresis analysis of polymerase chain reaction-amplified genes coding for 16S rRNA. Appl Environ Microbiol 59: 695–700

Muyzer G, Teske A, Wirsen CO, Jannasch HW (1995) Phylogenetic relationships ofThiomicrospira species and their identification in deep-sea hydrothermal vent samples by denaturing gradient gel electrophoresis of 16S rDNA fragments. Arc Microbiol 164: 165–172

Nakagawa S (2004) A farewell to Bonferroni: the problems of low statistical power and publication bias. Behav Ecol 15:1044–1045

Natynczuk SE, Macdonald DW (1994) Scent, sex, and the self-calibrating rat. J Chem Ecol 20: 1843–1857

Noonan MJ, Markham A, Newman C, Trigoni N, Buesching CD, Ellwood SA, Macdonald DW (2014) Climate and the individual: inter-annual variation in the autumnal activity of the European badger (*Meles meles*). PLoS One 9, e83156

Osada K, Yamazaki K, Curran M, Bard J, Smith BP, Beauchamp GK (2003) The scent of age. Proc Roy Soc Lond 270:929–933

Őstborn H (1976) Doftmarkering hos graveling. Zool Revy 38:103–112

Palphramand KL, White PC (2007) Badgers, *Meles meles*, discriminate between neighbour, alien and self scent. Anim Behav 74:429–436

Penn DJ (2002) The scent of genetic compatibility: sexual selection and the major histocompatibility complex. Ethology 108:1–21

Roper TJ (2010) Badger. Collins, London

Sin YW, Buesching CD, Burke T, Macdonald DW (2012) Molecular characterization of the microbial communities in the subcaudal gland secretion of the European badger (*Meles meles*). FEMS Microbiol Ecol 81:648–659

Smith CJ, Danilowicz BS, Clear AK, Costello FJ, Wilson B, Meijer WG (2005) T-Align, a web-based tool for comparison of multiple terminal restriction fragment length polymorphism profiles. FEMS Microbiol Ecol 54:375–380

Stübbe M (1971) Die analen Markierungsorgane des Dachses (*Meles meles*). Zool Garten NF 40:125–135

Theis KR, Schmidt TM, Holekamp KE (2012) Evidence for a bacterial mechanism for group-specific social odors among hyenas. Sci Rep 2:615

Theis KR, Venkataraman A, Dycus JA, Koonter KD, Schmitt-Matzen EN, Wagner AP, Holekamp KE, Schmidt TM (2013) Symbiotic bacteria appear to mediate hyena social odors. Proc Natl Acad Sci U S A 110:19832–19837

Thornton PD, Newman C, Johnson PJ, Buesching CD, Baker SE, Slater D, Johnson DDP, Macdonald DW (2005) Preliminary comparison of four anaesthetic regimes in badgers (*Meles meles*). Vet Anaesth Analg 32:40–47

Todrank J, Heth G, Johnston RE (1998) Kin recognition in golden hamsters: evidence for kinship odours. Anim Behav 55:377–386

Woodley SK, Baum MJ (2003) Effects of sex hormones and gender on attraction thresholds for volatile anal scent gland odors in ferrets. Horm Behav 44:110–118

Wyatt TD (2010) Pheromones and signature mixtures: defining species-wide signals and variable cues for identity in both invertebrates and vertebrates. J Comp Physiol 196:685–700

Zechman JM, Martin IG, Wellington JL, Beauchamp GK (1984) Perineal scent gland of wild and domestic cavies: bacterial activity and urine as sources of biologically significant odors. Physiol Behav 32:269–274

Chapter 6
The Role of Bacteria in Chemical Signals of Elephant Musth: Proximate Causes and Biochemical Pathways

Thomas E. Goodwin, Innocent H. Harelimana, Laura J. MacDonald, Daniel B. Mark, Aline Umuhire Juru, Qin Yin, James A. Engman, Randall A. Kopper, Cheryl F. Lichti, Samuel G. Mackintosh, James D. Shoemaker, Mark V. Sutherland, Alan J. Tackett, and Bruce A. Schulte

6.1 Introduction

6.1.1 Chemical Signals of Elephant Musth

Mammalian males of many species have long been noted for their use of secretions and excretions for conspecific communication. In 1871, Charles Darwin spoke to this topic: "The males are almost always the wooers; and they alone are

---

This chapter is dedicated to Dr. Eric S. Albone for his pioneering research on the role of bacteria in mammalian semiochemistry.

T.E. Goodwin (✉) • I.H. Harelimana • L.J. MacDonald • D.B. Mark
A.U. Juru • Q. Yin • R.A. Kopper
Department of Chemistry, Hendrix College, Conway, AR 72032, USA
e-mail: goodwin@hendrix.edu

J.A. Engman
Department of Biology, Henderson State University, Arkadelphia, AR 71999, USA

C.F. Lichti
Department of Pharmacology and Toxicology, University of Texas Medical Branch at Galveston, Galveston, TX 77555, USA

S.G. Mackintosh • A.J. Tackett
Department of Biochemistry and Molecular Biology, University of Arkansas for Medical Sciences (UAMS), Little Rock, AR 72205, USA

J.D. Shoemaker
E. A. Doisy Department of Biochemistry and Molecular Biology, Saint Louis University School of Medicine, St. Louis, MO 63104, USA

M.V. Sutherland
Department of Biology, Hendrix College, Conway, AR 72032, USA

B.A. Schulte
Department of Biology, Western Kentucky University, Bowling Green, KY 42101, USA

B.A. Schulte et al. (eds.), *Chemical Signals in Vertebrates 13*,
DOI 10.1007/978-3-319-22026-0_6

armed with special weapons for fighting with their rivals. They are provided, either exclusively or in a much higher degree than the females, with odoriferous glands. There is another and more peaceful kind of contest, in which the males endeavour to excite or allure the females by various charms. This may be affected by the powerful odours emitted by the males during the breeding-season; the odoriferous glands having been acquired through sexual selection" (Darwin 1981, p 313, 397; Kappeler 1998).

Darwin's characterizations are certainly relevant for male African (*Loxodonta africana*) and Asian (*Elephas maximus*) elephants in musth, a periodic state which is typically characterized by elevated serum androgens (specifically testosterone and dihydrotestosterone), heavy drainage from the temporal glands, urine dribbling, appetite reduction, increased activity to search for mates, intense competition for mating, and increased aggression (Eisenberg et al. 1971; Jainudeen et al. 1972; Poole and Moss 1981; Poole et al. 1984; Hall-Martin and van der Walt 1984; Poole 1987; Poole 1989; Rasmussen et al. 1996; Schulte and Rasmussen 1999; Ganswindt et al. 2005a, b; Brown et al. 2007; Rasmussen et al. 2008; Chelliah and Sukumar 2013; Ghosal et al. 2013). Unlike the rut in ungulates (e.g., moose and deer), musth among male Asian and African elephants in any given population is asynchronous.

African and Asian elephants have well-developed primary and secondary (vomeronasal) olfactory systems that are used to detect Darwin's "powerful odours" (chemical signals) from conspecifics (Rasmussen and Schulte 1998; Rasmussen et al. 2003). The number of confirmed mammalian chemical signals is relatively small (Albone 1984; Brown and Macdonald 1985; Schildknecht and Ubl 1986; Brennan and Keverne 2004; Burger 2005; Brennan and Zufall 2006; Sorensen and Hoye 2010; Charpentier et al. 2012; Apps 2013; Liberles 2014; Wyatt 2014). Therefore, it is rather startling that two insect pheromones have been identified as semiochemicals in Asian elephants: (1) *Z*-7-dodecen-1-yl acetate, a urinary signal of impending ovulation (Rasmussen et al. 1997), and (2) frontalin, a chemical signal of sexual state and social maturity in the temporal gland secretion (TGS) of mature males in musth (Rasmussen and Greenwood 2003; Greenwood et al. 2005).

Musth male TGS and urine emit distinctive odors which are easily detected from a great distance by conspecifics as well as humans (Poole 1987; Poole 1989; Hollister-Smith et al. 2008). Darwin (1981, p 279) described the odor of musth TGS as follows: "At this period the glands on the side of the face of the male elephant enlarge and emit a secretion having a strong musky odor." The pungent musth secretions and excretions, which are primarily TGS and urine, serve as vehicles for semiochemicals produced by physiological changes (Rasmussen and Perrin 1999; Schulte and Rasmussen 1999; Ganswindt et al. 2005a, b).

Urine is the solution by which mammals excrete not only excess water and electrolytes but also a myriad of metabolites, some of which function in chemical signaling (Albone 1984) and others in medical diagnosis (Ryan et al. 2011). Albone (1984, p 165) summarized this metabolic messaging as follows: "… urine necessarily conveys to the external world in its detailed composition much information concerning the internal physiological state of the animal concerned and thus provides the necessary basis for the evolution of specialized semiochemical systems."

Sexually dimorphic odors are significant factors in mammalian sexual selection (Blaustein 1981). Accordingly, African elephant musth urine trails evoke strong inspection by conspecifics (Poole 1987). It has been demonstrated that male African elephants can differentiate between the urine of female conspecifics in the luteal and follicular phases of estrus, as well as between musth and non-musth urine (Bagley et al. 2006; Hollister-Smith et al. 2008). The bark beetle pheromones frontalin and *endo*- and *exo*-brevicomin as well as their biosynthetic precursors (alken-2-ones; Vanderwel and Oehlschlager 1992; Keeling et al. 2013) have been identified in the urine of female African elephants, but their putative role in elephant chemical signaling has not yet been verified (Goodwin et al. 2006).

Rasmussen and Wittemyer (2002) reported much higher concentrations of volatile alkan-2-ones in the urine of wild African male elephants in musth when compared to non-musth urine. As this distinctive increase of alkan-2-ones has been demonstrated for both wild and captive male African and Asian elephant excretions and secretions, these authors proclaimed it as a "species-free" chemical signal of musth (Rasmussen and Wittemyer 2002).

Goodwin and colleagues used automated solid phase dynamic extraction (SPDE) and gas chromatography-mass spectrometry (GC-MS) (Goodwin et al. 2007, 2012; Goodwin and Schulte 2009) to find that the abundance of alkan-2-ones, alkan-2-ols, and several simple aromatic compounds is elevated in the headspace over musth urine of captive elephants. The production of these compounds continues in the urine exogenously due to bacterial fermentation. Several studies have shown that chemical changes occur in urine as it ages (Wellington et al. 1983; Saude and Sykes 2007; Mochalski et al. 2012; Kwak et al. 2013; Troccaz et al. 2013). Thus, we proposed that these simple compounds may serve as temporal signals from musth males to conspecifics.

The primary focus of the present report is on the alkan-2-ones and their corresponding alkan-2-ols. Specifically we discuss (1) the probable biosynthesis of these compounds via a secondary pathway for fatty acid metabolism, (2) the proximate cause for their increased abundance in musth urine, and (3) the role of bacteria in the increased abundance of these compounds exogenously in aged urine.

6.1.2 *Microorganisms in Mammalian Semiochemistry*

In his seminal book on mammalian semiochemistry, Albone (1984) includes a chapter entitled "Microorganisms in Mammalian Semiochemistry." In a subsection of this chapter (p 147–148), Albone discusses a "fermentation hypothesis" of chemical recognition which arose from his research with the red fox (*Vulpes vulpes*; Albone et al. 1974, 1977, 1978; Albone and Perry 1976), as well as from Gorman's research on the Indian mongoose (*Herpestes auropunctatus*; 1974; 1976). These studies propose that microorganisms associated with the anal sac/anal pocket metabolize organic substrates to produce a set of carboxylic acids that function for individual or group recognition. Simply put, the two tenets of the fermentation hypothesis can be expressed as follows.

- Symbiotic microorganisms in scent glands contribute to mammalian odor profiles
- Variations in mammalian host odor profiles mirror variations in their bacterial communities, and these odors have signaling relevance

The first of these tenets has long been known and amply demonstrated (cf. Leyden et al. 1981; Zechman et al. 1984; Voigt et al. 2005; Natsch et al. 2006; Penn et al. 2007; James et al. 2013), but in some cases it remains to be determined whether the specific metabolites that bacteria contribute are relevant in the context of signaling (Ezenwa and Williams 2014). The second tenet has only recently begun to be demonstrated (Zomer et al. 2009; Sin et al. 2012; Theis et al. 2013; Leclaire et al. 2014).

While the original fermentation hypothesis focused on mammalian scent glands, a broader definition has been proposed: "Although the fermentation hypothesis was developed for mammals that mark with scent glands, it might also accommodate mammals that mark with urine or faeces, or even nonmammalian species that use chemical communication. Therefore viewed broadly, the fermentation hypothesis has the potential to be an inclusive model for animal chemical recognition" (Archie and Theis 2011, p 428).

We propose herein that the fermentation hypothesis is relevant for exogenous, microbially mediated urinary chemical signals of elephant musth. There are many definitions of fermentation in biochemistry and microbiology (Li 2004; Willey et al. 2008); however, for the current context this simple one seems most appropriate: "… a process in which microorganisms produce chemical changes in organic substrates through the action of enzymes produced by these microorganisms (Li 2004, p 685)."

6.1.3 Hypotheses and Predictions

Our experiments were crafted with the following hypotheses in mind: (1) bacterial metabolism is responsible for the exogenous increase in abundance of the selected urinary compounds; (2) a higher abundance of the metabolites of interest would be observed in aged musth urine compared to aged non-musth urine; (3) the abundance of these metabolites could be reduced by centrifuging and filtering the urine to remove bacteria prior to aging; (4) increased bacterial metabolism would be restored by re-introduction of the centrifugation pellet or feces; (5) the bacteria would be killed by autoclaving the centrifugation pellet before reintroduction, thus lowering the abundance of the metabolites of interest; (6) the alkan-2-ones and alkan-2-ols are produced by fatty acid metabolism; therefore, (7) fatty acids would be found in the elephant urine, likely carried there bound to albumin (Kamijo et al. 2002; van der Vusse 2009).

6.2 Materials and Methods

6.2.1 Urine Collection

The elephants used in this research were examined routinely by veterinarians and were in good health at the times of urine collection (for more information about these elephants, see Goodwin et al. 2012, Table 6.1). Musth status was characterized by urine dribbling, heavy temporal gland secretion, and a pungent odor from excretions and secretions. A musth and non-musth urine sample from three African elephants and one Asian elephant were collected in clean, new glass containers by staff members at various facilities (see Acknowledgments), frozen soon after collection on dry ice or in a −70 or −80 °C freezer, and shipped on dry ice to Hendrix College. Samples were stored in a −70 °C freezer and thawed at approximately 37 °C immediately before analysis by SPDE/GC-MS). Bacterial contamination in human urine can alter the metabolic profile over time unless the urine is stored in an ultralow-temperature freezer (Saude and Sykes 2007).

6.2.2 SPDE/GC-MS Sample Preparation and Analysis

SPDE/GC-MS has been previously employed in a study of putative chemical signals in African elephant urine (Goodwin et al. 2006, 2007, 2012). SPDE features concentration of volatile organic chemicals by repetitive flow back and forth over an adsorbant polymer coating on the inside wall of a stainless steel syringe needle that is attached to a gastight syringe (Lipinski 2001). For a typical SPDE/GC-MS analysis of elephant urine, a 1.0 mL aliquot and a small Teflon®-coated stir bar were sealed in a 20 mL glass screw-top vial (autosamplerguys.com), which had been previously washed thrice with acetone and thrice with deionized water and was sterilized by autoclaving at 18 psi and 121 °C for 20 min.

The vials were sealed with a threaded, metallic septum cap (silicone/PTFE layered septum; autosamplerguys.com). Multiple samples were programmed to run automatically using the Combi PAL robot and associated SPDE hardware and software (Chromtech.com). The SPDE needle was internally coated with activated charcoal (Carboxen®)-polydimethylsiloxane (AC-PDMS). After incubating the stirred sample at 37 °C for 15 min, the headspace was extracted for 13 min (200 up-and-down 1 mL strokes of the syringe). Desorption of adsorbed analytes was at 250 °C in the GC inlet. GC-MS analyses were conducted using an Agilent 6890 N GC and 5973 N Mass Selective Detector. The capillary GC column was an Equity 1 (bonded; polydimethylsiloxane), 60 m × 0.32 mm ID, 1 μm film thickness (Supelco cat. No. 28058U). The GC oven was temperature programmed to hold for 2 min at 35 °C, followed by ramping to 180 °C at 3.75 °C/min where it was held for 5 min before ramping at 20 °C/min to a final temperature of 250 °C where it was held for 10 min. The mass spectrometer was programmed at 3.09 scans/s for a mass scan of 30–500 amu. The identities of all compounds reported herein from SPDE/GC-MS were verified by GC retention times and mass spectra when compared to commercial samples, as well as by comparison to the NIST mass spectral library.

6.2.3 Experimental Procedures to Assess Microbial Metabolism

To generate the data shown in Table 6.1, samples of musth and non-musth urine from three healthy African elephants were centrifuged (10–15,000 × *g* for 2 min) and filtered (0.22 μm sterilizing filter) to remove insolubles, including microbes. Glassware, plasticware, and stainless steel spatulas were washed with acetone and deionized water, and sterilized by autoclaving at 18 psi and 121 °C for 20 min. 1 mL aliquots of these centrifuged and filtered samples in SPDE vials were treated as follows in the bulleted list, sealed with a septum cap, and allowed to incubate at room temperature for 48 h before analyzing by SPDE/GC-MS. The abbreviations in parentheses refer to entries in Tables 6.1 and 6.2.

- Nothing was added to centrifuged and filtered musth and non-musth urine (CF)
- A non-musth centrifugation pellet was added to centrifuged and filtered musth and non-musth urine (CFNMP)
- A musth centrifugation pellet was added to centrifuged and filtered musth and non-musth urine (CFMP)
- An autoclaved musth centrifugation pellet was added to centrifuged and filtered musth urine (CFA)
- A stainless steel spatula was inserted into fresh, male African elephant feces; the visible feces was gently wiped off the spatula with a Kimwipe®; then the spatula tip was inserted into centrifuged and filtered musth urine and moved around for a few seconds to introduce fecal bacteria (CFF)

6.2.4 Culturing Bacteria from Elephant Urine

Bacteria were cultured by evenly spreading 0.1 mL of thawed urine over a tryptic soy agar plate using a sterile bent glass rod. The plate was then incubated 48 h at 37 °C. Sequential, selective streaking and culturing led to the isolation of several pure strains of bacteria.

6.2.5 Identifying Bacteria from Elephant Urine

Bacterial identification was based on sequencing the highly conserved 16S rRNA gene. Single colonies were placed in PCR tubes, and amplified using GoTaq(R) Hot Start PCR (Promega Corp., Madison, WI) using universal bacterial primers 8 F and 1492R, resulting in a ca. 1500 bp fragment. This was visualized by agarose gel electrophoresis, and purified with QIAquick PCR purification kits (Qiagen, Alameda, CA). The purified product was quantified using a Nanodrop 2000 spectrophotometer (Thermo Fisher Scientific, Inc., Waltham, MA) and adjusted to suggested concentrations. Samples were sequenced at the University of Arkansas for Medical Sciences (UAMS) Sequencing Core Facility using a 3130XL Genetic Analyzer (Applied Biosystems, Foster City, CA), using primers Fd1, 800 F, 1050 F, 800R, 1050R and

RP2. Sequences were aligned and edited using ChromasPro (Technelysium Pty Ltd, Brisbane). Identification was performed using searches for the closest matches in the NIH BLAST database (http://blast.ncbi.nlm.nih.gov/Blast.cgi).

6.2.6 *Urinary Fatty Acids*

Concentrations and identities of urinary fatty acids were determined by GC-MS as described by Shoemaker and Elliot (1991).

6.2.7 *Sequencing African Elephant Urinary Proteins*

Proteins were resolved by SDS-PAGE, visualized by colloidal Coomassie staining, excised, and treated in-gel with trypsin for protein identification (Smart et al. 2009). Peptides were subjected to tandem mass spectrometric analysis by nanospray ionization with a coupled nanoLC 2D reverse-phase liquid chromatography system (Eksigent) and Thermo LTQ mass spectrometer (Smart et al. 2009). Two proteins were identified by UniProt database searching with Mascot using the *Loxodonta africana* protein sequences (http://www.uniprot.org); The UniProt Consortium (2014) Activities at the Universal Protein Resource (UniProt). *Nucl Acids Res* 42(D1): D191–D198). A Schiff stain and a subsequent Coomassie blue stain were used to verify that one of the two proteins is highly glycosylated (Carlsson 1993).

6.2.8 *IACUC Approval*

All animal-related activities and research discussed herein were approved by the Institutional Animal Care and Use Committee (IACUC) or its equivalent at all participating institutions.

6.3 Results

6.3.1 *General Overview*

A series of alkan-2-ones, alkan-2-ols, and a few aromatic compounds have a higher concentration in the urine of musth male elephants than in urine from non-musth males (Rasmussen and Wittemyer 2002; Goodwin et al. 2012). After 48 h at the ambient temperature the abundance of these compounds in the expelled urine had increased, and this increase could be prevented by centrifuging and filtering to remove insolubles, including bacteria (Goodwin et al. 2012). In this section we compare relative abundances of these organic compounds in non-musth and musth African male elephant urine that has been aged for 48 h, centrifuged and filtered, and then a musth or non-musth centrifugation pellet or feces has been reintroduced. We also demonstrate that elephant urine contains fatty acids, the probable biochemical precursor of the alkan-2-ones and alkan-2-ols. Additionally, African elephant albumin, the probable carrier protein that ferries fatty acids into the urine, has been sequenced and shown to share a high sequence homology with Asian elephant albumin.

6.3.2 Relative Abundance of Selected Compounds in Aged African Elephant Urine

Following up on earlier findings that microorganisms are likely responsible for the exogenous increase in abundance of certain urinary ketones, alcohols, and aromatic compounds (Goodwin et al. 2012), a new set of experiments were designed to provide additional information regarding this phenomenon.

Tables 6.1 and 6.2 show compound abundances for aged centrifuged and filtered non-musth urine and musth urine (CF), and for centrifuged and filtered non-musth or musth urine with a non-musth or musth centrifugation pellet (CFNMP or CFMP, respectively) reintroduced before aging, as well as for centrifuged and filtered musth

Table 6.1 Comparison of selected compounds in non-musth and musth urine of three male African elephants under various conditions. The data are derived from mean peak areas from GC-MS total ion chromatograms normalized to the value for non-musth CF pentan-2-one, the compound of lowest abundance[a]

	Non-musth urine			Musth urine				
Compound	CF	CFNMP	CFMP	CF	CFA	CFF	CFNMP	CFMP
Pentan-2-one	1.00[b]	1.81[b]	3.96[b]	82.3	91.6[b]	**167**[b]	116	115
Hexan-2-one	–	–	1.63[b]	57.0	48.3[c]	92.7[c]	79.9	**181**
Hexan-2-ol	–	–	–	–	–	–	6.37[b]	–
Heptan-2-one	–	15.4	35.5[c]	27.0	20.1[b]	66.6	119	**679**
Heptan-2-ol	–	5.26[b]	–	–	–	8.27[b]	**27.2**[c]	10.6[c]
Octan-2-one	–	33.8	81.3	–	–	38.5	207	**808**
Octan-2-ol	–	13.5[b]	9.47[c]	–	–	21.5[b]	32.6[c]	**41.0**[c]
Acetophenone	4.67[c]	20.5	6.43[c]	49.9	45.3	**1620**	106	149
4-Methylphenol	3.34[c]	243	2290	763	1910	3640[c]	3500	**11,300**
Nonan-2-one	2.50[b]	25.6	54.3	52.0	15.2[b]	1260	4970	**12,300**
Nonan-2-ol	–	–	–	113	60.8	727[c]	**1600**[c]	194
Decan-2-one	–	–	9.57[b]	1.74[b]	–	104[c]	232	**1840**
Decan-2-ol	–	–	–	3.67[b]	–	90.5[b]	**648**[c]	3.63[b]
4-Ethylphenol	–	28.8[c]	168[c]	127	165	**1110**[c]	483	707
Undecan-2-one	–	4.73[b]	22.0[c]	–	–	352	2050	**4440**
Undecan-2-ol	–	–	–	14.5[b]	–	362[b]	**1300**[c]	28.6[b]

The maximum relative value for each compound is indicated in bold. All data were acquired 48 h after thawing. Unless otherwise noted, means are based on data from all three elephants. [CF=centrifuged and filtered; CFNMP=non-musth centrifugation pellet introduced; CFMP=musth centrifugation pellet introduced; CFA=autoclaved centrifugation pellet introduced; CFF=feces introduced]
[a]A dash indicates non-detectable or inadequate amount for accurate integration, [b]$N=1$, [c]$N=2$

urine with the following additions made before aging: autoclaved musth urine centrifugation pellet (CFA) or fresh African elephant fecal bacteria (CFF).

The results in Table 6.1 show the following:

- The compounds of maximum abundance in general were nonan-2-one and 4-methylphenol
- No compound exhibited its maximum abundance in either non-musth urine, the centrifuged and filtered musth urine, or the centrifuged and filtered musth urine to which the autoclaved centrifugation pellet was added (indeed, the largest number of undetected compounds was clearly for the CF non-musth samples)
- With the exception of pentan-2-one, the maximum abundance for the alkan-2-ones was evident in the CFMP samples
- The only compounds exhibiting maximum abundance in the CFNMP were four alkan-2-ols
- In all cases, the combined abundance of an alkan-2-one and the corresponding alkan-2-ol was highest in the CMFP samples
- For all musth compounds, the abundance was higher in the CFF samples than in the CF or the CFA samples

Table 6.2 Comparison of selected compounds in musth and non-musth urine of a male Asian elephant under various conditions. The data are derived from peak areas from GC-MS total ion chromatograms normalized to the value for non-musth CFNMP octan-2-one, the compound of lowest abundance[a]

	Non-musth			Musth				
Compound	CF	CFNMP	CFMP	CF	CFA	CFF	CFNMP	CFMP
Heptan-2-one	4.03	3.39	15.7	13.8	13.0	**37.9**	18.6	13.1
Heptan-2-ol	–	–	–	–	–	**3.93**	–	2.15
Octan-2-one	4.84	1.00	**46.2**	–	–	15.2	3.40	7.16
Octan-2-ol	–	–	16.9	–	–	–	–	–
Acetophenone	16.4	20.2	12.0	58.6	68.7	**264**	73.8	52.6
4-Methylphenol	–	–	**2890**	207	419	269	1050	1210
Nonan-2-one	–	–	53.7	3.23	2.43	**253**	140	241
Nonan-2-ol	–	–	11.3	2.44	5.67	2.43	43.2	**53.4**
1-Phenylpropan-2-one	–	–	–	123	0.81	**139**	22.1	71.5
Decan-2-one	–	–	**54.3**	–	–	–	4.40	–
Decan-2-ol	–	–	–	–	–	–	–	–
4-Ethylphenol	–	–	–	–	5.31	**19.9**	–	1.81
Undecan-2-one	–	–	–	–	-	15.8	**40.7**	24.5
Undecan-2-ol	–	–	–	–	–	–	3.72	**7.08**

[a]A dash indicates non-detectable or inadequate amount for accurate integration

The maximum relative value for each compound is indicated in bold. All data were acquired 48 h after thawing. [CF=centrifuged and filtered; CFNMP=non-musth centrifugation pellet introduced; CFMP=musth centrifugation pellet introduced; CFA=autoclaved centrifugation pellet introduced; CFF=feces introduced]

The results in Table 6.2 show the following:

- The compounds of maximum abundance were nonan-2-one and 4-methylphenol.
- Most compounds exhibited their maximum abundance in musth urine samples; however three compounds exhibited maximum abundance in the non-musth CFMP samples.
- The largest number of maximum compound abundances were in the musth CFF samples.

6.3.3 Relative Abundance of Medium- and Long-Chain Fatty Acids in Aged Elephant Urine

As will be detailed in the Discussion, we believe that the exogenously produced urinary alkan-2-ones are the result of bacterial metabolism of fatty acids, and that the alkan-2-ols are derived by reduction of the alkan-2-ones. The urine samples were analyzed on the day of thawing from −70 °C and were kept at the ambient temperature until the sixth day after thawing and analyzed once more. As shown in Table 6.3, 10 fatty acids were identified. The highest abundance for each acid is bolded.

6.3.4 African Elephant Urinary Proteins

Alignment of the*L. africana* (G3SMX8) and *E. maximus* (Q6B3Z0; Lazar et al. 2004) albumin sequences from the UniProt database showed sequence homology of over 99 % between the two species. A glycoprotein (a uromodulin; Devuyst et al. 2005) from African elephant urine was also sequenced (accession number G3SN28).

6.4 Discussion

6.4.1 Introduction

As previously stated, Rasmussen and Wittemyer (2002) found that a series of alkan-2-ones were present in higher concentration in the urine of wild African elephants in musth when compared to non-musth urine. Rasmussen and co-workers also found volatile ketones emanating from the temporal gland orifices (prior to drainage), temporal gland secretions (TGS), breath, and urine of captive and wild musth Asian elephants (Rasmussen et al. 1990; Perrin et al. 1996; Rasmussen and Perrin 1999). Similar sets of alkan-2-ones and the corresponding alcohols were present in higher abundances in musth urine of captive African elephants, and the

Table 6.3 Relative fatty acid abundances (mmol/mol creatinine as determined by GC-MS using an internal standard) from musth and non-musth urine of 3 African and 1 Asian elephants (#4)

Elephant	Hexanoic (D1) C-6	Hexanoic (D6)	Octanoic (D1) C-8	Octanoic (D6)	Capric (D1) C-10	Capric (D6)	Lauric (D1) C-12	Lauric (D6)	Myristic (D1) C-14	Myristic (D6)
#1 M	11.30	19.30	17.30	**25.70**	1.20	1.00	13.80	**52.50**	1.00	–
#1 NM	4.70	3.10	24.70	20.70	1.70	2.30	20.80	31.30	–	–
#2 M	7.00	11.40	11.30	5.80	2.00	2.50	29.70	25.10	10.10	–
#2 NM	5.80	7.90	12.60	13.40	1.60	1.90	25.60	28.50	3.80	–
#3 M	10.20	12.20	14.30	12.20	2.30	3.20	14.70	18.30	**19.30**	1.00
#3 NM	8.40	9.90	14.50	15.70	2.80	4.80	26.20	21.60	5.50	1.90
#4 M	5.00	4.90	3.80	4.60	1.70	1.60	34.10	43.60	7.40	–
#4 NM	28.00	**28.90**	16.00	23.00	**5.70**	4.70	18.60	11.40	9.10	5.40
Elephant	Palmitic (D1) C-16	Palmitic (D6)	Oleic (D1) *18:1 *cis*-9	Oleic (D6)	Linolenic (D1) *18:3 *cis,cis,cis*-6,9,12	Linolenic (D6)	Linoleic (D1) *18:2 *cis,cis*-9,12	Linoleic (D6)	Stearic (D1) C-18	Stearic (D6)
#1 M	6.80	7.70	11.10	24.60	–	–	7.90	4.80	10.80	11.10
#1 NM	8.30	7.30	12.80	22.60	–	–	9.30	5.40	15.20	11.70
#2 M	14.50	13.90	35.00	44.90	–	29.30	23.40	11.40	24.90	21.80
#2 NM	14.00	12.40	21.00	40.20	–	–	18.80	8.40	26.80	19.00
#3 M	16.10	**89.00**	21.40	**128.10**	**46.30**	40.50	22.60	**181.90**	22.60	**190.70**
#3 NM	36.70	46.60	40.50	117.90	–	–	92.60	72.50	77.10	101.40
#4 M	24.80	19.90	27.90	41.90	–	–	58.80	25.80	53.30	37.40
#4 NM	60.60	38.70	73.70	107.70	–	–	121.70	42.20	131.80	73.30

The data are normalized to the smallest number. (D1 = day of thawing; D6 = 6th day after thawing; C-X = number of carbon atoms; lipid numbers* are given for the unsaturated fatty acids: oleic, linolenic, and linoleic). The highest number for each acid is bolded abundances increased exogenously when the urine was aged at the ambient temperature (Goodwin et al. 2012).

Musth Asian males were also found to exhibit a large elevation of serum triglycerides and the lipase that catalyzes their hydrolysis to fatty acids and glycerol, leading to speculation that the alkan-2-ones are derived from fatty acids (Rasmussen 1999; Rasmussen and Perrin 1999). If this hypothesis is supported, then two questions naturally arise: (1) By what biochemical pathway are the alkan-2-ones synthesized from fatty acids? (2) Why are serum levels of fatty acids and urine levels of alkan-2-ones particularly elevated during musth? Both questions are addressed below, as is the role of bacteria in the continued urinary production of alkan-2-ones and the corresponding alkan-2-ols exogenously as the urine ages.

6.4.2 *Biosynthesis of Alkan-2-Ones from Fatty Acids During Musth*

Adrenalin (epinephrine) induces increased mobilization and lipolysis (hydrolysis) of adipose tissue triglycerides during exercise (Okuda 1975; Horowitz 2003). Analogously, malnutrition and untreated diabetes generate a need for energy derived from metabolism of stored triglycerides to replace energy normally supplied by carbohydrate metabolism (Miaskiewicz et al. 1989; Stryer 1995, p 612; Arner 1995). At such times some acetyl CoA, the usual end product of fatty acid metabolism, is diverted from the citric acid cycle to be used in biosynthesis ("ketogenesis") of the so-called ketone bodies (acetoacetate, acetone (propan-2-one), and β-hydroxybutyrate), which appear in the urine. Ketogenesis begins via condensation of two acetyl CoA subunits to produce acetoacetyl CoA, which is subsequently transformed into acetoacetate. β-Hydroxybutyrate and acetone are produced from acetoacetate by carbonyl reduction and decarboxylation, respectively (McClatchey 2002, p 526–527; Nelson and Cox 2008, p 666–667). Thus, metabolism by this secondary pathway can explain the increased production of propan-2-one by musth elephants.

Butan-2-one synthesis can occur in an analogous manner via condensation of acetyl CoA with propionyl CoA, a residue of catabolism of fatty acids with an odd number of carbon atoms (Stryer 1995, p 612); yet this type of process may not fully account for the biosynthesis of higher ketone homologs. The production of higher homologs of propan-2-one in plants via decarboxylation of β-keto acids was proposed long before the chemistry of "ketone body" formation was well understood (Dakin 1908). Much later, Rhodes et al. (1982, p 39) detected alkan-2-ones in the urine of diabetic rats and proposed their genesis from "decarboxylation of keto acids, in a manner similar to the well-known formation of acetone" (from acetoacetate). This was also demonstrated to be the likely pathway by which alkan-2-ones are generated by microorganisms from milk fats in cheese (McSweeney and Sousa 2000). Interestingly, some of propan-2-one's higher homologs (pentan-2-one, heptan-2-one, nonan-2-one, and several branched-chain, low-molecular-weight alkan-2-ones) have been detected as volatile metabolites in normal and diabetic human urine, possibly arising by the β-keto acid decarboxylation pathway (Zlatkis et al. 1973; Mills and Walker 2001).

Rasmussen and Perrin (1999) were the first to suggest that high levels of propan-2-one and a few other ketones in the breath of Asian elephants in musth were indicative of lipid catabolism. Rasmussen also proposed that decanoic acid might be converted into nonan-2-one, a compound identified in the temporal gland secretion of an Asian elephant in musth (Rasmussen et al. 1990; Rasmussen and Krishnamurthy 2000). In this and similar studies, nonan-2-one was invariably the third most abundant urinary alkan-2-one, with only propan-2-one and butan-2-one being present in higher concentrations (Rasmussen and Wittemyer 2002). Analogously, Novotny and co-workers (2007) found heptan-2-one to be a urinary chemical signal in female mice (*M. musculus domesticus*), and attributed its metabolic origin and that of similar compounds to fatty acid oxidation (see also Zhang et al. 2008).

Fatty acid metabolism ("the β-oxidation pathway") usually proceeds via a β-keto thioester which reacts with CoASH in a retro-Claisen condensation to provide one equivalent of acetyl CoA and a thioester which has been shortened by two carbons, and this process is repetitive (Scheme 6.1, primary pathway; Stryer 1995, p 605–611; Kunau et al. 1995; Laposata 1995; Bhaumik et al. 2005). However, when sufficient energy cannot be produced from carbohydrate metabolism (as with diabetes, malnutrition, or excessive exercise), unusually large quantities of stored triglycerides are hydrolyzed and released from adipocytes into the bloodstream as fatty acids. Elevated adrenaline and androgen levels also initiate mobilization of stored fatty acids into the bloodstream. When fatty acid concentrations overwhelm the ability of the normal catabolic production of acetyl CoA to keep up, a secondary pathway is activated. This process involves hydrolysis of the intermediary β-keto thioester to a β-keto acid which decarboxylates to produce an alkan-2-one with one less carbon (Scheme 6.1, secondary pathway). Because successive iterations of the primary pathway can produce a series of β-keto acids each two carbons shorter than its predecessor, a series of alkan-2-ones can result by the secondary pathway from metabolism of long-chain fatty acid precursors. The fatty acids or triglycerides can be of dietary origin, or may be synthesized in vivo from acetyl CoA (Stryer 1995, p 614–618; Laposata 1995).

There is good experimental evidence for the operation of this secondary pathway (Scheme 6.1) to produce alkan-2-ones (and thence reduction to the corresponding alkan-2-ols; Scheline 1973; Schulz and Dickschat 2007), notably in the catabolism of milk fatty acids by microorganisms to give a characteristic flavor and odor to certain cheeses (Lawrence and Hawke 1966; Kinsella and Hwang 1976a, b; McSweeney and Sousa 2000; Marilley and Casey 2004). Additionally, this pathway for biosynthesis of alkan-2-ones from fatty acids has been demonstrated to be operative in a wild tomato species (Fridman et al. 2005) and marine arctic bacteria (Dickschat et al. 2005).

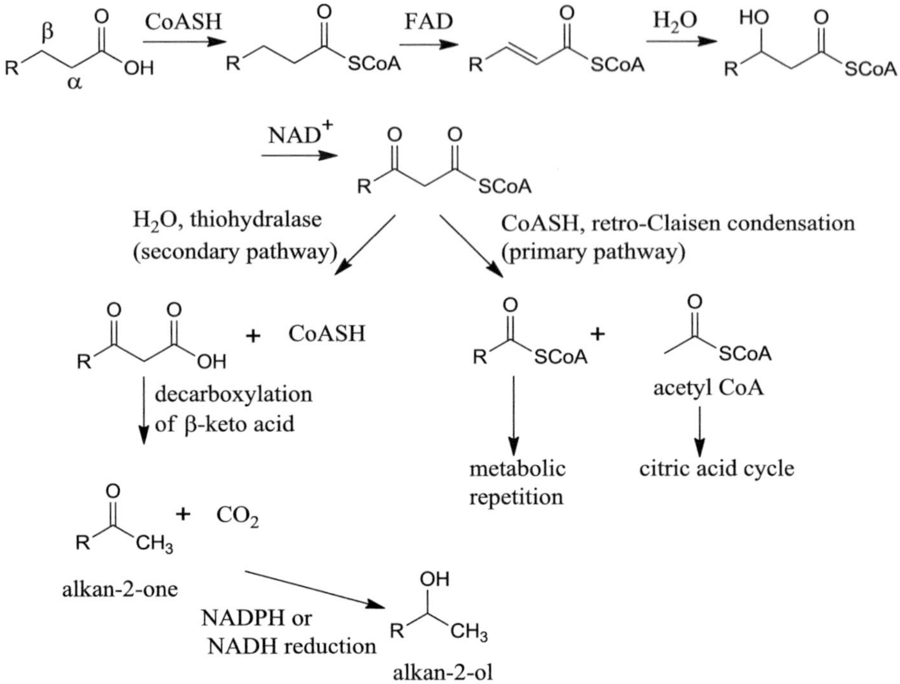

Scheme 6.1 Two pathways for fatty acid metabolism via β-oxidation

We suggest that the secondary biosynthetic pathway (Scheme 6.1) is followed for the production of alkan-2-ones and alkan-2-ols endogenously in elephants, and is followed exogenously as well in the urine via bacterial fermentation of fatty acids. It is likely followed in other mammals as well. For example, in the breeding season the urine of the male brown antechinus (*Antechinus stuartii*), a small Australian marsupial, contains a series of alkan-2-ones (hexan-2-one, octan-2-one, nonan-2-one, decan-2-one, and undecan-2-one) that are not present in the urine of females or castrated males (Toftegaard et al. 1999). The brown antechinus is one of only a few mammals (all in the Dasyuridae and Didelphidae families) in which the males die after one breeding season (i.e., they are semelparous). The cause of death is the deleterious effect of unabated high levels of cortisol and testosterone (Naylor et al. 2008).

6.4.3 Proximate Causes of Elevated Fatty Acid and Alkan-2-one Levels During Musth

A study on male moose (*Alces alces*) in rut provides experimental findings that are relevant to understanding certain aspects of elephant musth (Miquelle 1990). Mature male moose cease feeding completely (hypophagia) for about 2 weeks during their rutting period, and a similar appetite suppression is common for the rut in males of other ungulate species (Coblentz 1976). During the rut, male moose use urine marks to attract female conspecifics, and possibly to induce ovulation (Miquelle 1991). The peak timing of hypophagia coincides with maximum urine marking. A common proposal for elephants during musth is that decreased feeding and the resultant weight loss result from a high expenditure of energy as males search for and breed females, therefore having little time for foraging (Barnes 1982; Poole 1989). Miquelle (1990) found that during the rut, male moose spend approximately 45 % of their time standing inactive, and therefore have abundant free time to eat should they so desire. He proposed the following (p 150): "… hypophagia is a byproduct of a physiological process associated with scent-urination." There are good reasons to suggest that this physiological process is the increase in serum androgen concentrations and subsequent metabolism of an abundance of lipids as discussed below.

The androgen dependence of the chemical composition of male rodent urine during scent marking has been conclusively demonstrated (Roberts 2001). The increased mobilization of fatty acids into the bloodstream during moose rut and elephant musth is likely stimulated by the elevated level of androgens as well. This phenomenon has been proposed for elephants (Poole 1989; Rasmussen and Perrin 1999) and is well documented in humans (Kartin et al. 1944; Chioléro et al. 1997). Elevated androgen levels can lead to increased serum fatty acid concentrations through de novo lipogenesis from acetyl CoA, as well as from mobilization and lipolysis of stored triglycerides from adipose tissue (Swinnen and Verhoeven 1998; Saleh et al. 1999). Therefore it is reasonable to propose that male elephants decrease foraging and lose weight during musth not merely because they are busy pursuing females, but also due to elevated androgens and subsequent catabolism of triglycerides. In fact, Poole (1989, p 147) notes that captive Asian elephants have been reported to "lose weight during musth even when they are chained and given normal rations of food" (Deraniyagala 1955).

6.4.4 The Role of Bacteria in Chemical Signals of Elephant Musth

What we know:

- The ability of bacteria to produce alkan-2-ones and alkan-2-ols has been demonstrated previously (Dickschat et al. 2005; Schulz and Dickschat 2007), as well as in our preliminary studies when we inoculated urine with pure strains of bacteria (Goodwin, unpublished).
- There are bacteria in elephant urine (Sects. 6.2.4 and 6.2.5; Fig. 6.1).
- Musth male elephant urine has elevated levels of alkan-2-ones and alkan-2-ols compared to non-musth urine (Rasmussen and Wittemyer 2002; Goodwin et al. 2012; Tables 6.1 and 6.2).
- The abundance of these urinary ketones increases exogenously as the urine ages (Goodwin et al. 2012).

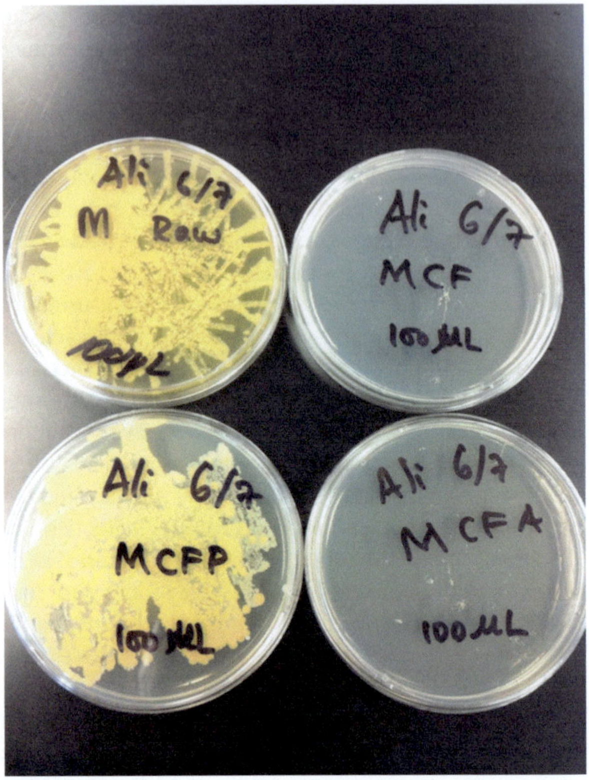

Fig. 6.1 Cultures of bacteria from musth urine. *Upper left*: Untreated musth urine; *lower left*: centrifuged and filtered, musth pellet reintroduced; *upper right*: centrifuged and filtered; *lower right*: centrifuged and filtered, autoclaved pellet reintroduced

- The exogenous production these compounds can be stopped by centrifuging (10–15,000 × *g* for 2 min) and filtering (0.22 μm sterilizing filter to remove insolubles, including microbes) (Tables 6.1 and 6.2; Fig. 6.1).
- We can restore the exogenous production of these compounds by reintroducing the centrifugation pellet (unless the pellet is autoclaved first) and by inoculation with feces.
- The ability of microbes to convert fatty acids to alkan-2-ones and alkan-2-ols has been demonstrated (Lawrence and Hawke 1966; Kinsella and Hwang 1976a).
- Elephant urine contains fatty acids (Table 6.3).
- Elephant urine contains albumin (Sects. 6.2.7 and 6.3.4).

6.5 Conclusion

In Sects. 6.4.2 and 6.4.3 we discussed proximate causes for higher levels of serum fatty acids during elephant musth, and how this could lead to the higher levels of alkan-2-ones and alkan-2-ols in musth urine. We also proposed that as the expelled urine aged, bacterial metabolism of urinary fatty acids leads to the observed increase in abundance of these ketones and alcohols. An obvious explanation for higher levels of alkan-2-ones and alkan-2-ols in musth urine versus non-musth urine would be that the concentrations of their fatty acid precursors are higher in musth urine, but that is not evident from the data in Table 6.3 (although a larger data set might be more revealing). There is yet another compelling explanation, as discussed below.

We propose that the fermentation hypothesis (Albone 1984) is operative in chemical signaling of elephant musth. Of particular interest for the present study is Albone's prescient statement that one function of microbial systems in mammalian scent production requires time modulation of fermentation product composition: "… as in signaling the occurrence of a particular physiological state in a mammal. The mammal is able to control the fermenting system by varying substrate composition and availability, and providing that microorganisms are present which have adapted to utilize the new substrate, the response in terms of signal production will be rapid" (Albone et al. 1977, p 38). Theis et al. (2013, p 19835) suggest a general "symbiotic hypothesis," the explanatory potential of which is "limited only by the capacity of hosts' social and physiological circumstances to alter the structure of their symbiotic microbial communities in ways that consequently affect hosts' odor profiles in signaling-relevant ways."

Thus elephants in musth may have a urinary microbial community that is somehow different from that when not in musth. This could explain why adding a musth urine centrifugation pellet (presumably containing bacteria) to centrifuged and filtered musth and non-musth urine generally results in higher abundances of alkan-2-ones and alkan-2-ols than does addition of a non-musth centrifugation pellet (Tables 6.1 and 6.2). It is also reasonable to propose that bacterial communities in elephant urine derive, at least in part, from the distal urethra, periurethral area, and/or the penile sheath (Spainea et al. 2000). Of interest in this regard is that of the two bacteria mentioned in Sect. 6.2.5, one matched most closely in the BLAST search to *Micrococcus xinjiangensis*, and the other to *Staphylococcus sp*. These are common types of skin bacteria (James et al. 2013).

So what directions might this field of research go in the future? Consider a recent review on microbially produced chemical signals in animals, in which Ezenwa and Williams (2014, p 3) stated the following": Mammals are the most-studied taxon in terms of microbial-based olfactory signaling, yet evidence from this group is still rather weak in two respects. First, only in a few cases have cause and effect relationships been established. Second, the specific microbes involved are typically unknown." These authors state, in specific reference to our elephant research (Goodwin et al. 2012), that "… captive individuals could be used to test for effects of microbial addition on receiver behavior." Thus behavioral bioassays could be devised in which responses to centrifuged and filtered elephant urine are compared to responses to untreated urine, and also to study whether responses change as the urine samples age. Additionally, as all of the selected ketones and alcohols are commercially available, mixtures of these could be bioassayed. In addition, the bacterial communities in elephant urine can be thoroughly surveyed as recently demonstrated with hyena secretions (Theis et al. 2013). In preliminary studies we have observed the ability of pure bacterial strains isolated from elephant urine to produce alkan-2-ones and alkan-2-ols when introduced to centrifuged and filtered elephant urine (Goodwin, unpublished), thus this avenue of research also deserves further attention.

The field of microbial chemical ecology is burgeoning as evidenced by a special issue of the *Journal of Chemical Ecology* (2013, 39:807–1054), yet none of the excellent articles contained therein focused on the role of microbes in mammalian chemical signaling. Many years ago Albone et al. (1977, p 35) wrote the following: "Although it has long been known that microorganisms are, at least partially, responsible for odor, such as axillary odor and halitosis, or breath odor, in man, only recently has it been suggested that microbially-derived odors might assume a chemical significance in mammals." Much has changed in the ensuing years as procedures, techniques, and instrumentation have improved dramatically, and as microbiologists, behavioral ecologists, chemists, and others have contributed new knowledge regarding the role of microbes in mammalian semiochemistry. We are glad that the majestic elephants and their tiny symbionts have a role to play in this fast-moving and fascinating field of research.

Acknowledgments Elephant urine samples were supplied by Disney's Animal Kingdom, Jacksonville Zoo, Miami Metro Zoo, and Riddle's Elephant and Wildlife Sanctuary. We are most grateful to the dedicated staff members at these facilities for their indispensable help. Preliminary studies related to elephant albumin sequencing were assisted by Rick Edmundson (UAMS) and Dustin Freyaldenhoven (Hendrix College), as well as Leighton Satterfield (Hendrix College) who also carried out the Schiff stain for carbohydrates in the uromodulin. Shelly Bradley (Hendrix College), Larry O'Kane (Agilent), and Ingo Christ (Autosampler Guys) provided much-needed expertise to keep the instruments running. Peggy Morrison and Matthew Windsor were patient and persistent in tracking down many references. We appreciate the assistance of Henderson State University undergraduate students in the Engman research group (Mark Castleberry, Shannon Fiser, Jonathan Shields, Lauren Story, and Charlotte Wetzlar) who helped by sequencing two strains of bacteria from elephant urine. We are grateful for valuable advice from Kevin Theis (Michigan State University), and from Scott and Heidi Riddle (Riddle's Elephant and Wildlife Sanctuary). We thank Hendrix College for financial support via a Distinguished Professor grant and the Julia Mobley Odyssey Professorship to T.E.G., as well as grants to students through the Odyssey Program. Additional funding in the early stages of this research was provided by the U.S. National Science Foundation (Award Nos. 02-17062, -17068, and -16862 to B.A.S., T.E.G. and the late L.E.L. Rasmussen, respectively). We would like to acknowledge the UAMS Proteomics Facility for mass spectrometric support. This work at UAMS was supported by National Institutes of Health grants P30GM103450, P20GM103429, and UL1TR000039.

References

Albone ES, Eglinton G, Walker JM, Ware GC (1974) The anal sac secretion of the red fox *Vulpes vulpes*; its chemistry and microbiology. A comparison with the anal sac secretion of the lion, *Panthera leo*. Life Sci 14:387–400

Albone ES, Perry GC (1976) Anal sac secretion of the red fox. Vulpes vulpes Volatile fatty acids and diamines Implications for a fermentative hypothesis of chemical recognition. J Chem Ecol 2:101–111

Albone ES, Gosden PE, Ware GC (1977) Bacteria as a source of chemical signals in mammals. In: Müller-Schwarze D, Mozell MM (eds) Chemical signals in vertebrates. Plenum, New York, pp 35–43

Albone ES, Gosden PE, Ware GC, Macdonald DW, Hough NG (1978) Bacterial action and chemical signalling in the red fox (*Vulpes vulpes*) and other mammals. In: Bullard RW (ed) Flavor chemistry of animal foods, ACS Symposium Series 67. American Chemical Society, Washington, DC, pp 78–91

Albone ES (1984) Mammalian semiochemistry: The investigation of chemical signals between mammals. Wiley-Interscience, New York

Apps PJ (2013) Are mammal olfactory signals hiding right under our noses? Naturwissenschaften 100:487–506

Archie EA, Theis KR (2011) Animal behaviour meets microbial ecology. Anim Behav 82:425–436

Arner P (1995) Impact of exercise on adipose tissue metabolism in humans. Int J Obes Relat Metab Disord 19:S18–S21

Bagley KR, Goodwin TE, Rasmussen LEL, Schulte BA (2006) Male African elephants, *Loxodonta africana*, can distinguish oestrus status via urinary signals. Anim Behav 71:1439–1445

Barnes RFW (1982) Mate searching behavior of elephant bulls in a semi-arid environment. Anim Behav 30:1217–1223

Bhaumik P, Koski MK, Glumoff T, Hiltunen JK, Wierenga RK (2005) Structural biology of the thioester-dependent degradation and synthesis of fatty acids. Curr Opin Struct Biol 15:621–628

Blaustein AR (1981) Sexual selection and mammalian olfaction. Am Nat 117:1006–1010

Brennan PA, Keverne EB (2004) Something in the air? New insights into mammalian pheromones. Curr Biol 14:R81–R89

Brennan PA, Zufall F (2006) Pheromonal communication in vertebrates. Nature 444:308–315

Brown RE, Macdonald DW (1985) Social odours in mammals. Clarendon, Oxford

Brown JL, Somerville M, Riddle HS, Keele M, Duer CK, Freeman EW (2007) Comparative endocrinology of testicular, adrenal and thyroid function in captive Asian and African elephant bulls. Gen Comp Endocrinol 151:153–162

Burger BV (2005) Mammalian semiochemicals. In: Schulz S (ed) The chemistry of pheromones and other semiochemicals II (Topics in current chemistry 240). Springer, Heidelberg, pp 231–278

Carlsson SR (1993) Isolation and characterization of glycoproteins. In: Fukuda M, Kobata A (eds) Glycobiology: a practical approach. Oxford University Press, New York, pp 14–16

Charpentier MJE, Barthes N, Proffit M, Bessière J-M, Grison C (2012) Critical thinking in the chemical ecology of mammalian communication: roadmap for future studies. Funct Ecol 26:769–774

Chelliah K, Sukumar R (2013) The role of tusks, musth, and body size in male-male competition among Asian elephants (*Elephas maximus*). Anim Behav 86:1207–1214

Chioléro R, Revelly J-P, Tappy L (1997) Energy metabolism in sepsis and injury. Nutrition 13:45–51

Coblentz BE (1976) Functions of scent-urination in ungulates with special reference to feral goats (*Capra hircus* L.). Am Nat 110:549–557

Dakin HD (1908) A synthesis of certain naturally occurring aliphatic ketones, with a suggestion of a possible mode of formation of these substances in the organism. J Biol Chem 4:221–225

Darwin C (1981) The descent of man and selection in relation to sex. Princeton University Press, Princeton, NJ, Photoreproduction of the 1871 edition published by J. Murray, London

Deraniyagala PEP (1955) Some extinct elephants, their relatives and the two living species. Ceylon Museums Publications, Government Press, Ceylon

Devuyst O, Dahan K, Pirson Y (2005) Tamm-Horsfall protein or uromodulin: New ideas about an old molecule. Nephrol Dial Transplant 20:1290–1294

Dickschat JS, Helmke E, Schulz S (2005) Volatile organic compounds from arctic bacteria of the *Cytophaga-Flavobacterium-Bacteroides* group: A retrobiosynthetic approach in chemotaxonomic investigations. Chem Biodivers 2:318–353

Eisenberg JF, McKay GM, Jainudeen MR (1971) Reproductive behavior of the Asiatic elephant (*Elephas maximus maximus* L.). Behaviour 38:193–225

Ezenwa VO, Williams AE (2014) Microbes and animal olfactory communication. BioEssays 36:847–854

Fridman E, Wang J, Iijima Y, Froehlich JE, Gang DR, Ohlrogge J, Pichersky E (2005) Metabolic, genomic, and biochemical analyses of glandular trichomes from the wild tomato species *Lycopersicon hirsutum* identify a key enzyme in the biosynthesis of methylketones. Plant Cell 17:1252–1267

Ganswindt A, Rasmussen HB, Heistermann M, Hodges JK (2005a) The sexually active states of free-ranging male African elephants (*Loxodonta africana*): defining musth and non-musth using endocrinology, physical signals, and behavior. Horm Behav 47:83–91

Ganswindt A, Heistermann M, Hodges JK (2005b) Physical, physiological, and behavioral correlates of musth in captive African elephants (*Loxodonta africana*). Physiol Biochem Zool 78:505–514

Ghosal R, Ganswindt A, Seshagiri PB, Sukumar R (2013) Endocrine correlates of musth in free-ranging Asian elephants (*Elephas maximus*) determined by non-invasive faecal steroid hormone metabolite measurements. PLoS One 8, e84787

Goodwin TE, Eggert MS, House SJ, Weddell ME, Schulte BA, Rasmussen LEL (2006) Insect pheromones and precursors in female African elephant urine. J Chem Ecol 32:1849–1853

Goodwin TE, Schulte BA (2009) Prospecting for mammalian chemical signals via solventless extraction techniques: an elephantine task. ChemoSense 11:9–15

Goodwin TE, Brown PA, Eggert MS, Evola MG, House SJ, Morshedi RJ, Weddell ME, Chen CJ, Jackson SR, Aubut Y, Eggert J, Schulte BA, Rasmussen LEL (2007) Use of automated solid phase dynamic extraction (SPDE)/GC-MS and novel macros in the search for African elephant pheromones. In: Hurst J, Beynon R, Wyatt T, Roberts C (eds) Chemical signals in vertebrates 11. Springer, New York, pp 25–35

Goodwin TE, Broederdorf LJ, Burkert BA, Hirwa IH, Mark DH, Waldrip ZJ, Kopper RA, Sutherland MV, Freeman EW, Hollister-Smith JA, Schulte BA (2012) Chemical signals of elephant musth: temporal aspects of microbially-mediated modifications. J Chem Ecol 38:81–87

Gorman ML, Nedwell DB, Smith RM (1974) An analysis of the contents of the anal scent pockets of *Herpestes auropunctatus*. J Zool 172:389–399

Gorman ML (1976) A mechanism for individual recognition by odour in *Herpestes auropunctatus* (Carnivora: Viverridae). Anim Behav 24:141–145

Greenwood DR, Comeskey D, Hunt MB, Rasmussen LEL (2005) Chemical communication: chirality in elephant pheromones. Nature 438:1097–1098

Hall-Martin AJ, van der Walt LA (1984) Plasma testosterone levels in relation to musth in the male African elephant. Koedoe 27:147–149

Hollister-Smith JA, Alberts SC, Rasmussen LEL (2008) Do male African elephants, *Loxodonta africana*, signal musth via urine dribbling? Anim Behav 76:1829–1841

Horowitz JF (2003) Fatty acid mobilization from adipose tissue during exercise. Trends Endocrinol Metab 14:386–392

Jainudeen MR, Katongole CB, Short RV (1972) Plasma testosterone levels in relation to musth and sexual activity in the male Asiatic elephant, *Elephas maximus*. J Reprod Fertil 29:99–103

James AG, Austin CJ, Cox DS, Taylor D, Calvert R (2013) Microbiological and biochemical origins of human axillary odour. FEMS Microbiol Ecol 83:527–540

Kamijo A, Kimura K, Sugaya T, Yamanouchi M, Hase H, Kaneko T, Hirata Y, Goto A, Fujita T, Omata M (2002) Urinary free fatty acids bound to albumin aggravate tubulointerstitial damage. Kidney Int 62:1628–1637

Kappeler PM (1998) To whom it may concern: the transmission and function of chemical signals in *Lemur catta*. Behav Ecol Sociobiol 42:411–421

Kartin BL, Man EB, Winkler AW, Peters JP (1944) Blood ketones and serum lipids in starvation and water deprivation. J Clin Invest 23:824–845

Keeling CI, Chiu CC, Aw T, Li M, Henderson H, Tittiger C, Weng H-B, Blomquist GJ, Bohlmann J (2013) Frontalin pheromone biosynthesis in the mountain pine beetle, *Dendroctonus ponderosae*, and the role of isoprenyl diphosphate synthases. Proc Natl Acad Sci U S A 110:18838–18843

Kinsella JE, Hwang D (1976a) Biosynthesis of flavors by *Penicillium roqueforti*. Biotechnol Bioeng 18:927–938

Kinsella JE, Hwang D (1976b) Enzymes of *Penicillium roqueforti* involved in the biosynthesis of cheese flavor. CRC Crit Rev Food Sci Nutr 8:191–228

Kunau W-H, Dommes V, Schulz H (1995) β-Oxidation of fatty acids in mitochondria, peroxisomes, and bacteria: a century of continued progress. Prog Lipid Res 34:267–342

Kwak J, Grigsby CC, Smith BR, Rizki MM, Preti G (2013) Changes in volatile compounds of human urine as it ages: Their interaction with water. J Chromatogr B 941:50–53

Laposata M (1995) Fatty acids: Biochemistry to clinical significance. Am J Clin Pathol 104: 172–179

Lawrence RC, Hawke JC (1966) The incorporation of [1-^{14}C]acetate into the methyl ketones that occur in steam-distillates of bovine milk fat. Biochem J 98:25–29

Lazar J, Rasmussen LEL, Greenwood DR, Bang I-S, Prestwich GD (2004) Elephant albumin: a multipurpose pheromone shuttle. Chem Biol 11:1093–1100

Leclaire S, Neilsen JF, Drea CM (2014) Bacterial communities in meerkat anal scent secretions vary with host sex, age, and group membership. Behav Ecol 25:996–1004

Leyden JJ, McGinley KJ, Hölzle E, Labows JN, Kligman AM (1981) The microbiology of the human axilla and its relationship to axillary odor. J Invest Dermatol 77:413–416

Li K-Y (2004) Fermentation: Principles and microorganisms. In: Hui YH, Meunier-Goddik L, Josephson J, Nip W-K, Stanfield PS (eds) Handbook of food and beverage fermentation technology, 2nd edn. CRC Press, Boca Raton, FL, p 685

Liberles S (2014) Mammalian pheromones. Annu Rev Physiol 76:151–175

Lipinski J (2001) Automated solid phase dynamic extraction: extraction of organics using a wall coated syringe needle. Fresenius J Anal Chem 369:57–62

Marilley L, Casey MG (2004) Flavour of cheese products: metabolic pathways, analytical tools and identification of producing strains. Int J Food Microbiol 90:139–159

McClatchey KD (2002) Clinical laboratory medicine, 2nd edn. Lippincott, Williams, and Wilkins, Philadelphia

McSweeney PLH, Sousa MJ (2000) Biochemical pathways for the production of flavour compounds in cheeses during ripening: a review. Lait 80:293–324

Miaskiewicz S, Levey GS, Owen O (1989) Case report: severe metabolic ketoacidosis induced by starvation and exercise. Am J Med Sci 297:178–180

Mills GA, Walker V (2001) Headspace solid-phase microextraction profiling of volatile compounds in urine: Application to metabolic investigations. J Chromatogr B 753:259–268

Miquelle DG (1990) Why don't male bull moose eat during the rut? Behav Ecol Sociobiol 27:145–151

Miquelle DG (1991) Are moose mice? The function of scent urination in moose. Am Nat 138:460–477

Mochalski P, Krapf K, Ager C, Wiesenhofer H, Agapiou A, Statheropoulos M, Fuchs D, Ellmerer E, Buszewski B, Amann A (2012) Temporal profiling of human urine VOCs and its potential role under the ruins of collapsed buildings. Toxicol Mech Methods 22:502–511

Natsch A, Derrer S, Flachsmann F, Schmid J (2006) A broad diversity of volatile acids, released by a bacterial aminoacylase from axilla secretions, as candidate molecules for the determination of human-body odor type. Chem Biodivers 3:1–19

Naylor R, Richardson SJ, McAllan BM (2008) Boom and bust: A review of the physiology of the marsupial genus *Antechinus*. J Comp Physiol B 178:545–562

Nelson DL, Cox MM (2008) Lehninger principles of biochemistry, 5th edn. W. H Freeman, New York

Novotny MV, Soini HA, Koyama S, Wiesler D, Bruce KE, Penn DJ (2007) Chemical identification of MHC-influenced volatile compounds in mouse urine. I: Quantitative proportions of major chemosignals. J Chem Ecol 33:417–434
Okuda H (1975) Mechanisms of actions of adrenaline and ACTH in fat mobilization. Pharmacol Biochem Behav 3:149–153
Penn DJ, Oberzaucher E, Grammer K, Fischer G, Soini HA, Wiesler D, Novotny MV, Dixon SJ, Xu Y, Brereton RG (2007) Individual and gender fingerprints in human body odour. J R Soc Interface 4:331–340
Perrin TE, Rasmussen LEL, Gunawardena R, Rasmussen RA (1996) A method for the collection, long-term storage, and bioassay of labile volatile chemosignals. J Chem Ecol 22:207–221
Poole JH (1987) Rutting behaviour in African elephants: The phenomenon of musth. Behaviour 102:283–316
Poole JH (1989) Announcing intent: the aggressive state of musth in African elephants. Anim Behav 37:140–152
Poole JH, Moss CJ (1981) Musth in the African elephant, *Loxodonta africana*. Nature 292:830–831
Poole JH, Kasman LH, Ramsay EC, Lasley BL (1984) Musth and urinary testosterone concentrations in the African elephant (*Loxodonta africana*). J Reprod Fertil 70:255–260
Rasmussen LEL, Hall-Martin AJ, Hess DL (1996) Chemical profiles of male African elephants, *Loxodonta Africana*: Physiological and ecological implications. J Mammal 77:422–439
Rasmussen LEL, Schulte BA (1998) Chemical signals in the reproduction of Asian (*Elephas maximus*) and African (*Loxodonta africana*) elephants. Anim Reprod Sci 53:19–34
Rasmussen LEL (1999) Evolution of chemical signals in the Asian elephant, *Elephas maximus*: Behavioural and ecological influences. J Biosci 24:241–251
Rasmussen LEL, Perrin TE (1999) Physiological correlates of musth: Lipid metabolites and chemical composition of exudates. Physiol Behav 67:539–549
Rasmussen LEL, Krishnamurthy V (2000) How chemical signals integrate Asian elephant society: The known and the unknown. Zoo Biol 19:405–423
Rasmussen LEL, Wittemyer G (2002) Chemosignaling of musth by individual wild African elephants, *(Loxodonta africana)*: Implications for conservation and management. Proc Biol Sci 269:853–860
Rasmussen LEL, Greenwood DR (2003) Frontalin: a chemical message of musth in Asian elephants (*Elephas maximus*). Chem Senses 28:433–446
Rasmussen LEL, Lazar J, Greenwood DR (2003) Olfactory adventures of elephantine pheromones. Biochem Soc Trans 31:137–141
Rasmussen LEL, Hess DL, Haight JD (1990) Chemical analysis of temporal gland secretions collected from an Asian bull elephant during a four-month musth episode. J Chem Ecol 16:2167–2181
Rasmussen LEL, Lee TD, Zhang A, Roelofs WL, Daves GD Jr (1997) Purification, identification, concentration and bioactivity of Z-7-dodecen-1-yl acetate: sex pheromone of the female Asian elephant, *Elephas maximus*. Chem Senses 22:417–437
Rasmussen HB, Ganswindt A, Douglas-Hamilton I, Vollrath F (2008) Endocrine and behavioral changes in male African elephants: linking hormone changes to sexual state and reproductive tactics. Horm Behav 54:539–548
Rhodes G, Holland ML, Wiesler D, Novotny M, Moore SA, Peterson RG, Felten DL (1982) Excretion of urinary volatile metabolites in response to alloxan induced diabetes of short duration in rats. J Chromatogr 228:33–42
Roberts SC (2001) Scent marking. In: Wolf JO, Sherman PW (eds) Rodent societies. University of Chicago Press, Chicago, pp 255–266
Ryan D, Robards K, Prenzler PD, Kendall M (2011) Recent and potential developments in the analysis of urine: A review. Anal Chim Acta 684:17–29
Saleh J, Sniderman AD, Cianflone K (1999) Regulation of plasma fatty acid metabolism. Clin Chem Acta 286:163–180
Saude EJ, Sykes BD (2007) Urine stability for metabolomic studies: Effects of preparation and storage. Metabolomics 3:19–27
Scheline RR (1973) Metabolism of foreign compounds by gastrointestinal microorganisms. Pharmcol Rev 25:451–523

Schildknecht H, Ubl J (1986) Ecochemistry of life: mammalian chemical signals. Interdiscip Sci Rev 11:153–157

Schulte BA, Rasmussen LEL (1999) Musth, sexual selection, testosterone and metabolites. In: Johnston RE, Müller-Schwarze D, Sorensen P (eds) Advances in chemical communication in vertebrates. Plenum, New York, pp 383–397

Schulz S, Dickschat J (2007) Bacterial volatiles: the smell of small organisms. Nat Prod Rep 24:814–842

Shoemaker JD, Elliot WH (1991) Automated screening of urine samples for carbohydrates, organic and amino acids after treatment with urease. J Chromatogr 562:125–138

Sin WS, Buesching CD, Burke T, Macdonald DW (2012) Molecular characterization of the microbial communities in the subcaudal gland secretion of the European badger (Meles meles). FEMS Microbiol Ecol 81:648–659

Smart SK, Mackintosh SG, Edmondson RD, Taverna SD, Tackett AJ (2009) Mapping the local protein interactome of the NuA3 histone acetyltransferase. Protein Sci 18:1987–1997

Sorensen PW, Hoye TH (2010) Pheromones in vertebrates. In: Mander LM, Liu H-W (eds) Comprehensive natural products chemistry II: Chemistry and biology. Elsevier, Oxford, pp 225–262

Spainea DM, Mamizukab EM, Cedenhoa AP, Srougia M (2000) Microbiologic aerobic studies on normal male urethra. Urology 56:207–210

Stryer L (1995) Biochemistry, 4th edn. W. H Freeman, New York

Swinnen JV, Verhoeven G (1998) Androgens and the control of lipid metabolism in human prostate cancer cells. J Steroid Biochem Mol Biol 65:191–198

Theis KR, Venkataraman A, Dycus JA, Koonter KD, Schmitt-Matzen EN, Wagner AP, Holekamp KE, Schmidt TM (2013) Symbiotic bacteria appear to mediate hyena social odors. Proc Natl Acad Sci U S A 110:19832–19837

Toftegaard CL, Moore C, Bradley AJ (1999) Chemical characterization of urinary pheromones in brown antechinus, *Antechinus stuartii*. J Chem Ecol 25:527–535

Troccaz M, Niclass Y, Anziani P, Starkenmann C (2013) The influence of thermal reaction and microbial transformation on the odour of human urine. Flavour Fragr 28:200–211

van der Vusse GJ (2009) Albumin as fatty acid transporter. Drug Metab Pharmacokinet 24:300–307

Vanderwel D, Oehlschlager AC (1992) Mechanism of brevicomin biosynthesis from (Z)-6-nonen-2-one in a bark beetle. J Am Chem Soc 114:5081–5086

Voigt CC, Caspers B, Speck S (2005) Bats, bacteria, and bat smell: Sex-specific diversity of microbes in a sexually selected scent organ. J Mammal 86:745–749

Wellington JL, Beauchamp GK, Wojciechowski-Metzler C (1983) Stability of chemical communicants in urine: individual identity and age of sample. J Chem Ecol 9:235–245

Willey JM, Sherwood LM, Woolverton CJ (2008) Prescott, Harley and Klein's microbiology, 7th edn. McGraw-Hill, New York, p 1066

Wyatt TD (2014) Pheromones and animal behavior, 2nd edn. Cambridge University Press, Cambridge

Zechman JM, Martin IG, Wellington JL, Beauchamp GK (1984) Perineal scent gland of wild and domestic cavies: Bacterial activity and urine as sources of biologically significant odors. Physiol Behav 32:269–274

Zhang J-X, Sun L, Zhang J-H, Feng Z-Y (2008) Sex- and gonad-affecting scent compounds and 3 male pheromones in the rat. Chem Senses 33:611–621

Zlatkis A, Bertsch W, Lichtenstein HA, Tishbee A, Shunbo F, Liebich HM, Coscia AM, Fleischer N (1973) Profile of volatile metabolites in urine by gas chromatography-mass spectrometry. Anal Chem 45:763–767

Zomer S, Dixon SJ, Xu Y, Jensen SP, Wang H, Lanyon CV, O'Donnell AG, Clare AS, Gosling LM, Penn DJ, Brereton RG (2009) Consensus multivariate methods in gas chromatography mass spectrometry and denaturing gradient gel electrophoresis: MHC-congenic and other strains of mice can be classified according to the profiles of volatiles and microflora in their scent-marks. Analyst 134:114–123

Chapter 7
Age-Related Variation in the Scent Pouch Bacterial Communities of Striped Hyenas (*Hyaena hyaena*)

Kevin R. Theis, Arvind Venkataraman, Aaron P. Wagner, Kay E. Holekamp, and Thomas M. Schmidt

7.1 Introduction

7.1.1 Fermentation Hypothesis for Animal Chemical Communication

Animals ubiquitously associate with microbes (Gilbert et al. 2012; McFall-Ngai et al. 2013), and one way these microbes can benefit their host animals is by contributing to their behavioral phenotypes (Archie and Theis 2011; Ezenwa et al. 2012). The fermentation hypothesis for animal chemical communication suggests that components of animals' chemical signals include metabolic products of their symbiotic microbes and that variation in these signals, both among and within animal species, is due in part to underlying variation in the structures of their odor-producing microbial communities (Albone et al. 1974; Gorman et al. 1974; Albone and Perry 1975; Gorman 1976; Albone 1984; Archie and Theis 2011). The hypothesis was originally proposed in the 1970s to explain the mechanistic origin of chemical signals among mammals that communicate with secretions from specialized

---

K.E. Holekamp and T.M. Schmidt are the co-senior authors.

K.R. Theis (✉) • K.E. Holekamp
Department of Zoology, Michigan State University, East Lansing, MI 48824, USA

BEACON Center for the Study of Evolution in Action, Michigan State University, East Lansing, MI 48824, USA
e-mail: theiskev@msu.edu

A. Venkataraman • T.M. Schmidt
Department of Internal Medicine, University of Michigan, Ann Arbor, MI 48109, USA

A.P. Wagner
BEACON Center for the Study of Evolution in Action, Michigan State University, East Lansing, MI 48824, USA

B.A. Schulte et al. (eds.), *Chemical Signals in Vertebrates 13*,
DOI 10.1007/978-3-319-22026-0_7

integumental scent glands, but, due in large part to difficulties inherent in accurately characterizing host-associated microbial communities, evaluations of this hypothesis have historically been ambiguous (Archie and Theis 2011). Specifically, the observed diversity and variation in scent gland bacterial communities could not account for the observed variation in scent gland odor profiles (Archie and Theis 2011). However, contemporary advances in molecular, culture-independent survey methods are affording researchers the capacity to more effectively characterize the structure of bacterial communities inhabiting mammalian scent glands, stimulating renewed interest in evaluating and further developing the fermentation hypothesis (Archie and Theis 2011; Douglas and Dobson 2013; Ezenwa and Williams 2014). As predicted by the hypothesis, it is becoming clear that adult mammalian scent glands are typically populated by odor-producing bacteria (Archie and Theis 2011), that the structures of scent gland bacterial communities can vary with chemically signaled host traits (Sin et al. 2012; Theis et al. 2012, 2013; Leclaire et al. 2014), and that the bacterial and odor profiles of scent gland secretions can covary (Theis et al. 2013). Therefore, these contemporary studies are beginning to provide empirical data in support of the fermentation hypothesis for chemical signaling among scent-marking mammals.

7.1.2 The Development of Scent Gland Bacterial Communities

A comprehensive understanding of any biological phenomenon, including the fermentation hypothesis, requires specific understanding of its phylogeny, function, mechanism, and ontogeny. The development of the scent gland bacterial communities of mammals remains poorly understood. Specifically, we do not yet know how these symbiotic communities are acquired and maintained nor whether these communities exhibit general patterns of development across mammalian clades.

Available data are limited but they suggest that mammals are not typically born with fully functional scent glands (Algard et al. 1966; Buesching et al. 2003; Theis et al. 2008; Sin et al. 2012). Instead, the morphology, physiology, and endocrinology of these glands appear to develop over time. For example, in Syrian hamsters (*Mesocricetus auratus*), flank scent organs increase in size and begin to produce sebum (i.e., the lipid-rich foundation of mammalian scent secretions) as an individual sexually matures (Algard et al. 1966). In European badgers (*Meles meles*) and spotted hyenas (*Crocuta crocuta*), similar patterns exist, as younger juveniles typically do not produce sebum in their subcaudal scent organs, older juveniles often produce appreciable volumes of sebum, and sexually mature individuals consistently produce substantial volumes of sebum in their scent organs (Matthews 1939; Buesching et al. 2003; Theis et al. 2008; Sin et al. 2012). This maturation process likely has important ramifications for the development of the scent gland bacterial communities of mammals (Theis et al. 2008; Sin et al. 2012).

The development of mammalian scent gland bacterial communities is probably a complex process that is both multifactorial and combinatorial. Potential host factors include the morphology, physiology, endocrinology, and immunology of developing scent organs, and potential bacterial influences include the metabolic phenotypes and competitive exclusivity of the organs' original colonizers, their current bacterial community members, and the environmental bacteria to which they are continuously being exposed. The possible sources of environmental bacteria include the physical environments the juvenile mammals inhabit, the conspecifics with whom they share space and interact (e.g., mothers, kin, social partners), and other areas of the juveniles' own bodies (e.g., intestine, general skin surface) (Palmer et al. 2007; Archie and Theis 2011; Capone et al. 2011; Funkhouser and Bordenstein 2013).

A general model for the development of scent gland bacterial communities, adapted from prior germane discussions of the fermentation hypothesis (Albone and Perry 1975; Theis et al. 2008; Sin et al. 2012), posits that as juvenile mammals' scent glands begin producing sebum, the environments these glands afford symbiotic microbes dramatically change, and they are increasingly colonized by bacteria to which they are exposed. These opportunistic bacteria capitalize on the new habitats and persist until bacterial types that typically inhabit mammals' scent glands come into contact with and subsequently colonize the glands. This process is facilitated by developing scent glands being exposed to bacteria that inhabit the respective scent glands of adult conspecifics, which can occur through juveniles and adults sharing the same spaces, coming into repeated physical contact, and/or juveniles overmarking the scent marks of adults (Buesching et al. 2003; Theis et al. 2008; Sin et al. 2012). Scent overmarking of adults' marks in particular could ensure that juveniles' scent glands are inoculated with and effectively colonized by the appropriate bacterial types (Albone and Perry 1975; Theis et al. 2008). Once the appropriate mature communities are established in scent glands, it is posited that their composition will become more stable and resistant to new colonizers (Sin et al. 2012), due to both host physiological and immunological influence and effective competitive exclusion by members of mature scent gland bacterial communities, as appears to be the case in other animal organs (McFall-Ngai et al. 2013). As such, the transmission of bacteria through scent overmarking among age classes is expected to be primarily unidirectional, from adults to juveniles. This general developmental model predicts that (1) the scent gland bacterial communities of juvenile and adult conspecifics should differ, (2) these communities should be more variable (i.e., exhibit greater dispersion) among juveniles than among adults, and (3) the structure of bacterial communities in the scent glands of adults should predict the composition of communities in the scent glands of juveniles.

Here we characterize age-related variation in the scent pouch bacterial communities of juvenile, young adult, and adult striped hyenas (*Hyaena hyaena*) and preliminarily interpret our findings in the context of this general developmental model.

7.1.3 Scent Pouch Bacterial Communities of Striped Hyenas

Striped hyenas are medium-sized carnivores found throughout North and East Africa and the Middle East (Holekamp and Kolowski 2009; Wagner 2013). Their social behavior is poorly understood but it appears that adults can be solitary, live in small family groups, establish monogamous pair bonds, or form polyandrous social groups comprised of one female and several males (Wagner et al. 2007, 2008; Holekamp and Kolowski 2009; Wagner 2013). Regardless, striped hyenas are largely behaviorally solitary, as they typically forage and feed alone, and direct interactions in other behavioral contexts appear to occur infrequently as well (Wagner et al. 2007, 2008; Holekamp and Kolowski 2009; Wagner 2013). Juvenile hyenas begin accompanying their mothers on foraging trips and can begin weaning as early as 6 months (Wagner 2013). Sexual maturity occurs at 2–3 years, with males and females being sexually size monomorphic. Striped hyenas have only a modest signaling repertoire, but they do appear to rely heavily on the odorous secretions of their subcaudal scent pouches for communication (Holekamp and Kolowski 2009; Wagner 2013). They extrude their scent pouches when greeting conspecifics, and they frequently scent mark when traveling, presumably to maintain spacing between competitors and potentially to facilitate reproductive advertisement (Wagner 2013). When scent marking (i.e., "pasting"), hyenas typically drag their extruded scent pouches across grass stalks, leaving behind secretions (i.e., "pastes") that can persist in the environment as active chemical signals for upward of a month (Mills 1990). The major volatile constituents of striped hyena pastes are carboxylic acids, aldehydes, lactones, and sulfur compounds (Buglass et al. 1990). We recently showed that the pastes of young adult and adult striped hyenas are densely populated with fermentative, odor-producing bacteria and that the bacterial and odor profiles of pastes covary, suggesting that variation in paste bacterial communities influences paste odor profiles (Theis et al. 2013). In the current study, we show that paste bacterial communities differ between juvenile and adult striped hyenas, that juveniles' bacterial communities are much more variable than those of adults, and that bacterial community data are consistent with the hypothesis that the paste bacterial communities of adult hyenas can serve as consistent sources of bacteria for juveniles' developing scent glands.

7.2 Materials and Methods

7.2.1 Sample Collection

Paste secretions were collected directly from the scent pouches of 39 anesthetized hyenas in the Laikipia (2001–2003) and Shompole (2007–2009) regions of Kenya. Each sampled hyena was anesthetized with Telazol (6.5 mg/kg) or a mixture of ketamine HCl (3.6 mg/kg) and medetomidine HCl (0.06 mg/kg), delivered from a blowpipe or air rifle after the hyena had been confined in a soft-catch foothold or cage trap (Wagner et al. 2007). The collected secretions were placed in sterile cryogenic vials, stored in liquid nitrogen, and transported to Michigan State University, where they remained frozen at −80 °C until analysis.

The sampled hyenas included juveniles ($N=8$; 6 months–1 year), young adults ($N=9$; 1–3 years), and adults ($N=22$; 3+ years). Age class assignments were based on known dates of birth or estimates from body measurements, weight, and tooth wear patterns (Wagner et al. 2008). Young adults were not yet reproductively active but were presumably on the verge of being so. Many of the samples, particularly those from young adults and adults, were included in a previous study evaluating differences between the bacterial and odor profiles of spotted and striped hyena pastes and assessing covariance between the bacterial and odor profiles of pastes in these two species (Theis et al. 2013). However, the current study incorporates new samples from six additional young striped hyenas (five juveniles, one young adult), enabling us to characterize and compare the bacterial communities in the scent pouches of juvenile, young adult, and adult hyenas, a subject not addressed in the previous study.

7.2.2 Sample Preparation and Next-Generation Sequencing

DNA was extracted from paste sample aliquots (0.07 ± 0.02 g) using MO BIO UltraClean fecal DNA kits. During this process, we did not observe any age-related variation in the consistency, texture, or color of striped hyena pastes—all fell within a beige to yellow spectrum. 16S rRNA genes in extractions were amplified using two broadly conserved, degenerate primers targeting the gene's V6–V4 variable regions (1046R: 5′-CGACRRCCATGCANCACCT-3′; 518F: 5′-CCAGCAG CYGCGGTAAN-3′). The PCR program included an initial dissociation step of 94 °C for 3 min, followed by 30 cycles of 94 °C for 30 s (denaturing), 60 °C for 45 s (annealing), and 72 °C for 1 min (extending) and a final extension step of 72 °C for 2 min. During amplification, unique molecular bar codes were incorporated into amplicons to permit multiplex sequencing of samples (Binladen et al. 2007). Samples were amplified in triplicate and subsequently pooled to minimize potential PCR bias, and non-template controls (all negative) were run for each bar-coded primer pair. Amplicons were purified using solid-phase reversible immobilization (AMPure XP beads) and quantified using an Invitrogen high-sensitivity Quant-iT dsDNA assay kit. Samples were then pooled at equimolar concentrations and sequenced using 454 GS FLX Titanium and GS Junior instruments at the Marine Biological Laboratory in Woods Hole and Michigan State University.

7.2.3 Sequence Analysis

The 454 run files were processed using mothur software, v 1.27.0 (Schloss et al. 2009), as described previously (Theis et al. 2013). Briefly, individual 16S rRNA gene sequences were discarded if they (1) contained any ambiguous base calls, (2) did not have perfect bar code and primer matches, (3) had a homopolymer run greater than eight bases, (4) had an average quality score of less than 35 using a sliding window of 50 bases, (5) contained fewer than 233 bases, or (6) were deemed chimeric by mothur's uchime tool (Schloss et al. 2011). All remaining sequences were binned into operational taxonomic units (OTUs) using mothur's average neighbor clustering algorithm and a 97 % sequence similarity cutoff. Taxonomic classification of OTUs was determined using the Ribosomal Database Project's classifier tool (trainset9) within mothur and an 80 % confidence threshold (Wang et al. 2007; Claesson et al. 2010).

Samples were subsampled to the depth of the least-represented sample (2173 sequences). The resultant data set contained 681 OTUs, 378 of which were singletons. The complete and subsampled data sets for this study, including detailed consensus taxonomy assignments, have been provided as supplementary material (Theis_etal_dataset1.xlsx). There was no correlation between the exact mass, to the nearest thousandth of a gram, of paste aliquots from which DNA was extracted and the number of OTUs recovered among any of the age classes or within the study overall (Spearman rank correlation; adult: $r_s = 0.17, p = 0.45$, young adult: $r_s = -0.23, p = 0.56$, juvenile: $r_s = -0.47, p = 0.25$, overall: $r_s = -0.13, p = 0.42$). With one exception (hyena HSYM112; Good's coverage = 92 %), the bacterial communities sampled from pastes had Good's coverage values exceeding 98 % and exhibited asymptotic rarefaction curves when plotted on equally scaled axes (unpublished results). These data indicate that, even with subsampling, thorough sample coverage was achieved in this study.

7.2.4 *Bacterial Community Analyses*

Similarities in composition and structure among sampled bacterial communities were characterized using Jaccard and Bray-Curtis similarity indices, respectively (Hammer 2011). Before calculating Bray-Curtis indices, OTU count data were $\log_{10}(x+1)$ transformed to temper the influences of highly prominent OTUs on analyses (Ramette 2007). That being said, analyses based on Bray-Curtis values yielded the same outcomes when we used the proportional abundances of OTUs in the complete data set and untransformed count data in the subsampled data set (unpublished results). As reported previously (Theis et al. 2013), there was no effect of sex on the composition or structure of paste bacterial communities among adult striped hyenas, nor was there any effect of sex among young adults or juveniles (unpublished results). Therefore, sex was not considered in the analyses presented here.

Similarity and dispersion among the paste bacterial communities of juvenile, young adult, and adult hyenas were visualized via nonmetric multidimensional scaling (nMDS) plots and statistically evaluated using analyses of similarity (ANOSIM) and permutational tests for homogeneity of multivariate dispersions (PERMDISP), respectively (Clarke 1993; McCune and Grace 2002; Anderson 2006). All ANOSIM and PERMDISP analyses were conducted using 9999 permutations.

Bonferroni corrections were not applied to permutational analyses of planned comparisons (Hammer 2011). All visual and statistical analyses were completed using PAST software, v 2.17 (Hammer et al. 2001; Hammer 2011), except for PERMDISP analyses which were completed using the PERMDISP2 executable (Anderson 2006) and heat map analyses which were generated using the Matrix2png tool (Pavlidis and Noble 2003).

7.2.5 Neutral Model Analysis

Transmission of bacteria from adult to juvenile scent pouches could potentially occur via adults and juveniles occupying shared spaces, coming into direct physical contact with one another, and/or juveniles overmarking the scent marks of adults. The neutral theory of community ecology is a convenient way to preliminarily test whether the composition of scent pouch bacterial communities among juveniles is consistent with transmission of these community members from adult pouches. This theory suggests that the composition of a local species pool (here juvenile scent pouches) is the product of random dispersal of species from a regional species pool (here adult scent pouches) and ecological drift (Hubbell 2005; Sloan et al. 2006). We implemented the theory by employing a previously described neutral model (Sloan et al. 2006; Morris et al. 2013; Venkataraman et al. 2015). In this model, the relative abundance of an OTU in adult scent pouch bacterial communities was used to calculate the probability of detecting that OTU in juvenile scent pouches due to dispersal and ecological drift. Ninety-five percent binomial proportion confidence intervals were constructed around these predicted probabilities. The empirically observed probability for each OTU was calculated as the number of juvenile scent pouches in which the OTU was detected out of the total number of juvenile scent pouches surveyed. In order to test the statistical significance of the neutral model, the root-mean-square error between the observed and modeled probabilities was calculated ($RMSE_{actual}$). Next, we randomized the observed and modeled probabilities 1000 times and calculated the RMSE for each randomization ($RMSE_{simulated}$). This set of 1000 simulated RMSEs provided a null distribution against which the $RMSE_{actual}$ could be compared. If the $RMSE_{actual}$ falls within the range of simulated RMSEs, then we conclude that the neutral model was unable to predict the observed juvenile scent pouch bacterial community composition ($p > 0.001$). Conversely, if the $RMSE_{actual}$ is lower than the minimum $RMSE_{simulated}$, we conclude that the neutral model cannot be rejected ($p < 0.001$) and proceed to interpret the model's findings. In this case, OTUs within the 95 % confidence intervals were viewed as being neutrally distributed in juvenile scent pouches and thus theoretically consistent with dispersal from adult pouches. OTUs that fell outside the upper and lower confidence intervals were deemed over- and under-represented, respectively, in juvenile scent pouches, given the abundances of these OTUs in adult pouches.

7.2.6 *IACUC Approval*

Our research, described in Animal Research Protocol IACUC 05/14-087-00, was approved most recently on April 29, 2014, by the Institutional Animal Care and Use Committee at Michigan State University, and complies with Kenyan Law.

7.3 Results and Discussion

7.3.1 *Taxonomic Characterization of Scent Pouch Bacterial Communities*

The majority of bacteria inhabiting striped hyena scent pouches could not be classified at the phylum level using a conservative 80 % confidence threshold (Table 7.1) (Wang et al. 2007; Claesson et al. 2010). However, as discussed previously (Theis et al. 2013), the closest characterized relatives of these bacteria are fermentative anaerobes in the phylum Firmicutes. Of the pouch bacteria that were confidently classified at the phylum level, Firmicutes were the most abundant (Table 7.1). Within this phylum, the genus *Peptoniphilus* was typically present at modest levels in pouches regardless of age class. *Anaerococcus* and *Lactobacillus* were highly abundant in the pouches of a few juveniles, as was *Propionibacterium* (phylum Actinobacteria). *Bacteroides* and *Porphyromonas* (both phylum Bacteroidetes) and *Fusobacterium* (phylum Fusobacteria) were highly abundant in the pouches of a few young adults. *Porphyromonas* was widely present, but not abundant, in the pouches of adults.

This apparent taxonomic variation among age classes could have importance for the fermentation hypothesis for striped hyena chemical communication because linear and branched short-chain fatty acids are primary volatile constituents of striped hyena pastes (Buglass et al. 1990; Theis et al. 2013), and members of these bacterial genera are known to typically produce these volatiles (Archie and Theis 2011; Theis et al. 2012). Although short-chain fatty acids are only one group of compounds potentially providing semiochemical information in mammalian scent gland secretions, they are widely hypothesized to do so (Burger 2005). In fact, the production of short-chain fatty acids by bacteria in mammalian scent glands figured prominently in Martyn Gorman's and Eric Albone's initial writings on the fermentation hypothesis (Albone et al. 1974; Gorman et al. 1974). Notably, symbiotic bacteria are the primary source of short-chain fatty acids in other animal organs as well, including the human large intestine and axillae (den Besten et al. 2013; Fredrich et al. 2013; James et al. 2013; Sharon et al. 2014; Troccaz et al. 2015). However, it is important to note that animal hosts, including striped hyenas, likely contribute to their own odor profiles indirectly by provisioning and modulating their odor-producing symbiotic microbial communities and directly by producing other odorants themselves.

Table 7.1 Phylum- and genus-level classification of bacteria inhabiting the scent pouches of juvenile, young adult, and adult striped hyenas

Bacterial phylum	Percentage of sequences in phylum (# of OTUs)	Genera recovered within phylum
Juvenile hyenas (*N*=8)		
Firmicutes	36.19±18.93 (103)	***Anaerococcus***, *Blautia*, *Clostridium sensu stricto*, *Clostridium XI*, *Clostridium* XIVb, *Faecalibacterium*, *Fastidiosipila*, *Finegoldia*, *Lachnospiracea incertae sedis*, ***Lactobacillus***, ***Peptoniphilus***, *Peptostreptococcus*, *Staphylococcus*
Actinobacteria	13.59±21.88 (11)	*Bifidobacterium*, *Corynebacterium*, ***Propionibacterium***
Proteobacteria	0.10±0.10 (13)	*Aquabacterium*, *Bradyrhizobium*, *Moraxella*, *Ralstonia*, *Sutterella*
Bacteroidetes	0.09±0.24 (10)	*Bacteroides*, *Parabacteroides*, *Prevotella*
Fusobacteria	0.01±0.03 (2)	*Fusobacterium*
Chloroflexi	0.01±0.02 (1)	….
Unclassified	50.02±32.50 (64)	….
Young adult hyenas (*N*=9)		
Firmicutes	23.89±8.72 (208)	*Allobaculum*, *Anaerococcus*, *Anaerovorax*, *Bacillus*, *Blautia*, *Clostridium sensu stricto*, *Clostridium XI*, *Clostridium XIVb*, *Clostridium XVIII*, *Coprobacillus*, *Faecalibacterium*, *Fastidiosipila*, *Finegoldia*, *Holdemania*, *Lachnospiracea incertae sedis*, *Lactobacillus*, *Moryella*, *Oscillibacter*, ***Peptoniphilus***, *Streptococcus*
Bacteroidetes	4.69±9.79 (60)	*Adhaeribacter*, *Alistipes*, ***Bacteroides***, *Hymenobacter*, *Mucilaginibacter*, *Odoribacter*, *Ohtaekwangia*, *Parabacteroides*, ***Porphyromonas***, *Prevotella*, *Rhodocytophaga*, *Segetibacter*
Fusobacteria	1.44±4.31 (8)	***Fusobacterium***
Proteobacteria	0.48±1.36 (51)	*Altererythrobacter*, *Aminobacter*, *Anaerobiospirillum*, *Bordetella*, *Desulfovibrio*, *Methylobacterium*, *Microvirga*, *Phenylobacterium*, *Pseudomonas*, *Ralstonia*, *Roseomonas*, *Skermanella*, *Sphingomonas*, *Steroidobacter*, *Sutterella*, *Telluria*
Actinobacteria	0.09±0.24 (15)	*Arthrobacter*, *Cryptosporangium*, *Geodermatophilus*, *Nocardioides*, *Propionibacterium*
TM7	0.02±0.05 (3)	TM7 *incertae sedis*
Acidobacteria	0.01±0.03 (2)	Gp4
Spirochaetes	<0.01±0.02 (1)	*Treponema*
Unclassified	69.37±21.59 (96)	….

(continued)

Table 7.1 (continued)

Bacterial phylum	Percentage of sequences in phylum (# of OTUs)	Genera recovered within phylum
Adult hyenas ($N=22$)		
Firmicutes	23.06 ± 8.35 (139)	*Anaerococcus*, *Clostridium sensu stricto*, *Faecalibacterium*, *Fastidiosipila*, ***Peptoniphilus***, *Staphylococcus*
Bacteroidetes	0.50 ± 1.25 (11)	***Porphyromonas***
Proteobacteria	0.01 ± 0.02 (4)	*Aquabacterium*, *Ralstonia*, *Sphingomonas*
Fusobacteria	<0.01 ± 0.01 (1)	....
Unclassified	76.42 ± 8.05 (110)	....

Taxonomic classifications of operational taxonomic units (OTUs) were determined using the classify.otu command in mothur and a conservative confidence threshold of 80 % (Wang et al. 2007; Claesson et al. 2009). Genera in bold type were notably widespread or abundant in the pastes of the respective age class. Percentage data are presented as means ± SD

Table 7.2 The diversity (i.e., richness and evenness) of scent pouch bacterial communities does not consistently vary among juvenile, young adult, and adult striped hyenas

Age group	Sample size	Number of OTUs	Chao1 richness	Shannon index	Simpson index $(1-D)$
Juveniles	8	44.3 ± 15.9	105.6 ± 41.8	1.74 ± 0.46	0.72 ± 0.12
Young adults	9	78.7 ± 76.5	197.1 ± 236.5	2.23 ± 0.60	0.79 ± 0.06
(w/o HSYM112)	(8)	(52.1 ± 15.5)	(115.0 ± 45.7)	(2.02 ± 0.13)	(0.77 ± 0.03)
Adults	22	47.4 ± 6.6	93.1 ± 32.0	1.99 ± 0.17	0.75 ± 0.07
Kruskal-Wallis test		$H=2.17$, $p=0.34$	$H=2.85$, $p=0.24$	$H=3.23$, $p=0.20$	$H=0.79$, $p=0.67$
(w/o HSYM112)		($H=1.18$, $p=0.55$)	($H=1.70$, $p=0.43$)	($H=2.78$, $p=0.25$)	($H=0.47$, $p=0.79$)

We provide diversity indices and statistical analyses for young adults with and without data from hyena HSYM112 included because this hyena's paste sample contained an abnormally high number (291) of operational taxonomic units (OTUs). Index data are presented as means ± SD

7.3.2 Age-Related Diversity in Scent Pouch Bacterial Communities

The alpha diversity (i.e., OTU richness and evenness) of scent pouch bacterial communities did not vary among juvenile, young adult, and adult striped hyenas (Table 7.2). Instead, alpha diversity was largely invariable. The only exception was the pouch bacterial community of the young adult hyena HSYM112, which contained 291 OTUs—no other sampled community contained more than 80 OTUs.

The beta diversity, or the composition (i.e., who is there) and structure (i.e., in what relative abundances), of scent pouch bacterial communities did vary with age (Fig. 7.1 and Table 7.3). The differences between juveniles and adults were robust

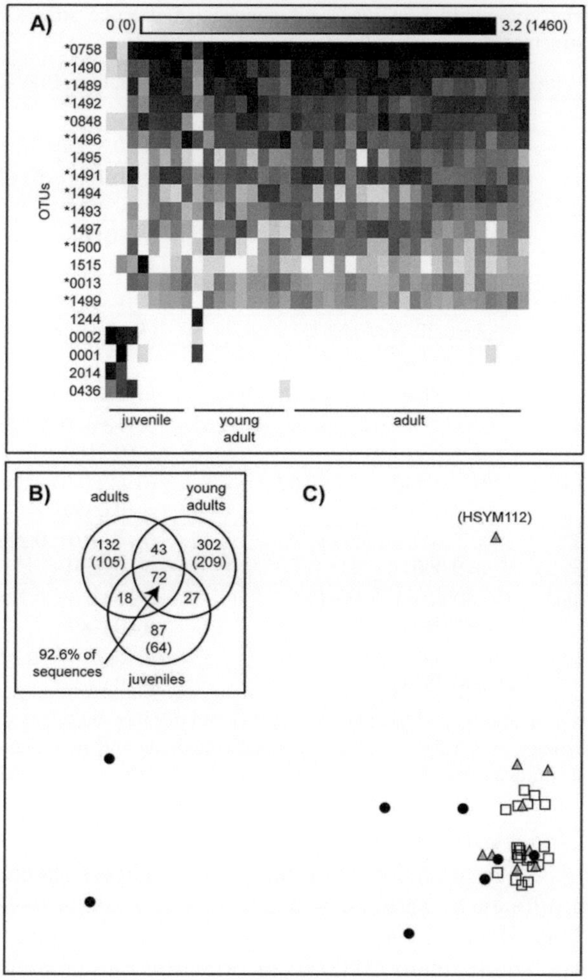

Fig. 7.1 Variation in scent pouch bacterial communities among juvenile, young adult, and adult striped hyenas. (**a**) Heat map illustrating variation in the relative abundances of prominent operational taxonomic units (OTUs) among age classes. Raw count data were log transformed before plotting. The scale bar reflects these transformed values, with corresponding raw count values provided in *parentheses*. The prominent OTUs include those whose minimum average abundance was 25 in any of the age classes, as well as those which were part of the core adult scent pouch bacterial community (indicated with an *asterisk*). (**b**) Venn diagram showing the degree to which paste OTUs were shared among age classes. For those OTUs which were unique to a specific age class, we have indicated the number of them that were singletons in *parentheses*. The vast majority of sequences in this study (92.6 %) clustered into OTUs that were shared among all age classes. (**c**) An nMDS plot illustrating variation in the structure (i.e., Bray-Curtis similarity index) of paste bacterial communities among age classes. The stress of the plot, seemingly reflecting strong similarities among the bacterial communities of older hyenas, was 0.1725. *Black circles*, *gray triangles*, and *white squares* indicate juveniles, young adults, and adults, respectively. We note the position of the community for hyena HSYM112, the only young adult whose scent pouch bacterial community was noticeably different from the typical bacterial community of adult striped hyenas

Table 7.3 Bacterial communities in the scent pouches of juvenile striped hyenas differ in composition and structure from those in the pouches of adults

	Composition (Jaccard index)	Structure (Bray-Curtis index)
ANOSIM: age		
Global effect	$R=0.34, p=0.0002$	$R=0.37, p=0.0001$
Juvenile vs. young adult	$R=0.06, p=0.1457$	$R=0.10, p=0.0633$
(w/o HSYM112)	($R=0.07, p=0.1318$)	($R=0.09, p=0.0817$)
Juvenile vs. adult	$R=0.57, p=0.0001$	$R=0.57, p=0.0001$
Young adult vs. adult	$R=0.18, p=0.0310$	$R=0.24, p=0.0118$
(w/o HSYM112)	($R=0.12, p=0.0981$)	($R=0.18, p=0.0353$)
PERMDISP: age		
Global effect	$p=0.0024$	$p=0.0035$
Juvenile vs. young adult	0.55 ± 0.03, 0.50 ± 0.04, $p=0.2641$	0.44 ± 0.07, 0.32 ± 0.06, $p=0.2180$
(w/o HSYM112)	(0.55 ± 0.03, 0.46 ± 0.02, $p=0.0176$)	(0.44 ± 0.07, 0.26 ± 0.02, $p=0.0350$)
Juvenile vs. adult	0.55 ± 0.03, 0.45 ± 0.01, $p=0.0001$	0.44 ± 0.07, 0.22 ± 0.01, $p=0.0001$
Young adult vs. adult	0.50 ± 0.04, 0.45 ± 0.01, $p=0.0497$	0.32 ± 0.06, 0.22 ± 0.01, $p=0.0008$
(w/o HSYM112)	(0.46 ± 0.02, 0.45 ± 0.01, $p=0.7028$)	(0.26 ± 0.02, 0.22 ± 0.01, $p=0.0426$)

We present results of analyses of similarity (ANOSIM) and distance-based tests for homogeneity of multivariate dispersions (PERMDISP) using spatial medians and permutations of the least-absolute-deviation residuals. For the latter, respective distances from spatial medians are reported as means ± SE

($R=0.6$; Table 7.3). We did not observe differences between juveniles and young adults, and the differences between young adults and adults were very modest ($R=0.2$; Table 7.3). Although there were many OTUs that were unique to juveniles, young adults, and adults, these OTUs were primarily rare community members, often singletons (Fig. 7.1b). Variation among age classes instead appeared to be driven primarily by the presence of a core scent pouch bacterial community in adults. Specifically, there were 12 OTUs that were present in all 22 adult scent pouches, and these OTUs accounted for 92.91 ± 2.13 % of the sequences obtained from adult samples (Fig. 7.1a). This is remarkable considering that these adult hyenas inhabited two regions of Kenya separated by 250 km. Notably, nine of the core adult scent pouch OTUs were also present in the pouches of all sampled young adults, and the remaining three were present in all but one young adult pouch. Among young adults, these OTUs accounted for 83.56 ± 27.52 % of analyzed sequences. However, among juveniles, only 3 of the core adult OTUs were present in all sampled scent pouches, and the 12 adult core OTUs accounted for only 63.28 ± 41.77 % of analyzed sequences. Although longitudinal data will ultimately be required for verification, these current data are consistent with the hypothesis that the acquisition of mature scent gland bacterial communities is a predictable ontogenetic process for mammals (Sin et al. 2012).

7.3.3 Age-Related Variation (i.e., Dispersion) in Scent Pouch Bacterial Communities

The scent pouch bacterial communities of juvenile striped hyenas not only differed from those of adults in composition and structure, they were also far more variable (Fig. 7.1a, c and Table 7.3). Excluding data from HSYM112, the aforementioned outlier, the scent pouch bacterial communities of juveniles were also more variable than those of young adults (Table 7.3). Similar patterns were found in the scent pouch bacterial communities of European badgers (Sin et al. 2012). Furthermore, these patterns among hyenas and badgers are comparable to observed patterns in the development of the human intestinal microbiome, as despite high variation among infant gut bacterial communities, they typically exhibit a more adult-like phenotype by 1–3 years of age (Palmer et al. 2007; Yatsunenko et al. 2012). As Palmer and colleagues (2007) explain, if a symbiotic microbial community is functionally important to its host, its development toward a stereotypical adult phenotype should be under strong selective pressure. The data on striped hyenas and European badgers suggest that their mature scent pouch bacterial communities represent highly specific, functionally integrated units.

7.3.4 Potential Transmission of Bacteria from Adult to Juvenile Scent Pouches

Whereas juvenile paste bacterial communities are variable, those of adults are highly specific and invariable (Fig. 7.1a, c and Table 7.3). As discussed above, a primary way by which this highly specific phenotype is hypothesized to develop among scent-marking mammals in general is via transmission of scent gland bacterial community members from adults to juveniles via shared space, direct physical contact, and/or scent overmarking (Buesching et al. 2003; Theis et al. 2008; Sin et al. 2012). We used the neutral community model to preliminarily test this hypothesis by specifically asking if the random dispersal of bacteria from adult scent pouches could theoretically explain the composition of bacterial communities in juvenile pouches. The neutral model was statistically significant ($p<0.001$; Fig. 7.2), indicating that we cannot currently reject the research hypothesis given the available data. Seventy-three percent of the 16S rRNA gene sequences recovered from juvenile pouches belonged to neutrally distributed OTUs. Importantly, 8 of the 12 OTUs that are part of the core scent pouch bacterial communities of adults were consistent with being neutrally distributed in juvenile from adult pouches. No OTUs were over-represented in the juvenile scent pouch and very few appeared to be under-represented, and even then only modestly so (Fig. 7.2). Our neutral model does include ecological drift explicitly as a neutral process in addition to dispersal, but we do not believe that ecological drift is having a pronounced effect here because scent pouch bacterial communities have very large population sizes (Orrock and

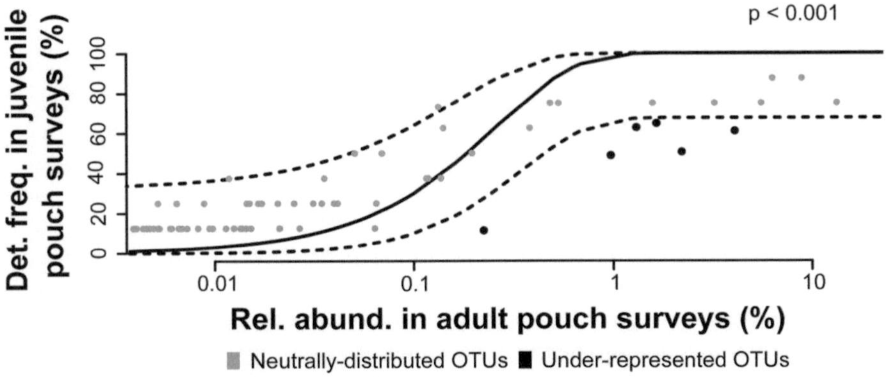

Fig. 7.2 Results of a neutral model analysis suggesting that the relative abundance of adult scent pouch bacterial community members predicts the membership of juvenile bacterial communities. Given that almost all of the operational taxonomic units (OTUs) shared between adult and juvenile scent pouches appear neutrally distributed, the model is consistent with the hypothesis that exposure to adult scent pouch bacterial communities, via shared space, physical contact, and/or scent overmarking, could be sufficient to largely explain the membership of juvenile scent pouch bacterial communities

Watling 2010). Moving forward, we must experimentally test, for example, through the use of genetically labeled symbionts (Cabello et al. 2005), that bacteria are indeed being horizontally transmitted from adult to juvenile scent pouches.

7.4 Conclusions and Future Directions

Like the scent pouches of young adult and adult striped hyenas, those of juveniles are primarily populated by fermentative bacteria from known odor-producing clades, suggesting that juveniles' symbiotic bacteria can influence their odor profiles. Nevertheless, the composition and structure of bacterial communities in the scent pouches of juveniles are different from, and more variable than, those of communities inhabiting the scent pouches of adults. Adult striped hyenas exhibit a core scent pouch bacterial community. These bacterial types are less widespread and abundant among juveniles. However, neutral modeling suggests that adult scent pouches could be a consistent source of bacterial types for the developing pouches of juveniles. These preliminary analyses are consistent with the hypothesis that striped hyena scent pouch bacterial communities converge on a species-typical, mature phenotype (Theis et al. 2008; Sin et al. 2012). Longitudinal sampling of individuals through ontogeny is required to verify that, regardless of their initial composition and structure, scent pouch bacterial communities do all culminate in what appears to be a highly stereotyped adult phenotype. Longitudinal sampling would also permit us to determine if there are predictable successional trajectories in the development of scent pouch bacterial communities in this species.

Acknowledgments We thank the Kenyan National Council for Science, Technology and Innovation; the Kenya Wildlife Service; the Loisaba, Mpala, and Kisima Ranches (Laikipia); and the Shompole and Olkirimatian Group Ranches for permissions and support of this research. This research was funded by National Science Foundation grants (IOS0920505, IOS1121475, DEB1353110) and the BEACON Center for the Study of Evolution in Action (DBI0939454). A.V. is supported by T32 HL007749.

References

Albone ES (1984) Mammalian semiochemistry. Wiley, New York

Albone ES, Perry GC (1975) Anal sac secretion of the red fox, *Vulpes vulpes*; volatile fatty acids and diamines: implications for a fermentation hypothesis of chemical recognition. J Chem Ecol 2:101–111

Albone ES, Eglinton G, Walker JM, Ware GC (1974) The anal sac secretion of the red fox (*Vulpes vulpes*); its chemistry and microbiology. A comparison with the anal sac secretion of the lion (*Panthera leo*). Life Sci 14:387–400

Algard FT, Dodge AH, Kirkman H (1966) Development of the flank organ (scent gland) of the Syrian hamster. 2. Postnatal development. Am J Anat 118:317–325

Anderson MJ (2006) Distance-based tests for homogeneity of multivariate dispersions. Biometrics 62:245–253

Archie EA, Theis KR (2011) Animal behaviour meets microbial ecology. Anim Behav 82:425–436

Binladen J, Gilbert MTP, Bollback JP, Panitz F, Bendixen C, Nielsen R, Willerslev E (2007) The use of coded PCR primers enables high-throughput sequencing of multiple homolog amplification products by 454 parallel sequencing. PLoS One 2:e197

Buesching CD, Stopka P, MacDonald DW (2003) The social function of allo-marking in the European badger (*Meles meles*). Behaviour 140:965–980

Buglass AJ, Darling FMC, Waterhouse JS (1990) Analysis of the anal sac secretion of the Hyaenidae. In: Macdonald DW, Muller-Schwarze D, Natynczuk SE (eds) Chemical signals in vertebrates 5. Kluwer, New York, pp 65–69

Burger BV (2005) Mammalian semiochemicals. Top Curr Chem 240:231–278

Cabello AE, Espejo RT, Romero J (2005) Tracing *Vibrio parahaemolyticus* in oysters (*Tiostrea chilensis*) using a Green Fluorescent Protein tag. J Exp Mar Biol Ecol 327:157–166

Capone KA, Dowd SE, Stamatas GN, Nikolovski J (2011) Diversity of the human skin microbiome early in life. J Invest Dermatol 131:2026–2032

Claesson MJ, O'Sullivan O, Wang Q, Nikkilä J, Marchesi JR, Smidt H, de Vos WM, Ross RP, O'Toole PW (2009) Comparative analysis of pyrosequencing and a phylogenetic microarray for exploring microbial community structures in the human distal intestine. PLoS One 4:e6669

Claesson MJ, Wang Q, O'Sullivan O, Greene-Diniz R, Cole JR, Ross RP, O'Toole PW (2010) Comparison of two next-generation sequencing technologies for resolving highly complex microbiota composition using tandem variable 16S rRNA gene regions. Nucleic Acids Res 38:e200

Clarke KR (1993) Non-parametric multivariate analysis of changes in community structure. Aust J Ecol 18:117–143

den Besten G, van Eunen K, Groen AK, Venema K, Reijngoud D-J, Bakker BM (2013) The role of short-chain fatty acids in the interplay between diet, gut microbiota, and host energy metabolism. J Lipid Res 54:2325–2340

Douglas A, Dobson A (2013) New synthesis: animal communication mediated by microbes: fact or fantasy? J Chem Ecol 39:1149

Ezenwa VO, Williams AE (2014) Microbes and animal olfactory communication: where do we go from here? Bioessays 36:847–854

Ezenwa VO, Gerardo NM, Inouye DW, Medina M, Xavier JB (2012) Animal behavior and the microbiome. Science 338:198–199

Fredrich E, Barzantny H, Brune I, Tauch A (2013) Daily battle against body odor: towards the activity of the axillary microbiota. Trends Microbiol 21:305–312

Funkhouser LJ, Bordenstein SR (2013) Mom knows best: the universality of maternal microbial transmission. PLoS Biol 11:e1001631

Gilbert SF, Sapp J, Tauber AI (2012) A symbiotic view of life: we have never been individuals. Q Rev Biol 87:325–341

Gorman ML (1976) A mechanism for individual recognition by odour in *Herpestes auropunctatus* (Carnivora: Viverridae). Anim Behav 24:141–145

Gorman ML, Nedwell DB, Smith RM (1974) An analysis of the contents of the anal scent pockets of *Herpestes auropunctatus* (Carnivora: Viverridae). J Zool 172:389–399

Hammer O (2011) PAST: PAleontological STatistics manual, 2.07 edn. Natural History Museum, Oslo, Norway

Hammer O, Harper DAT, Ryan PD (2001) PAST: PAleontological STatistics software package for education and data analysis. Palaeontol Electron 4:1–9

Holekamp KE, Kolowski JM (2009) Hyaenidae. In: Wilson D, Mittermeier R, Fonseca G (eds) Handbook of mammals of the world. Lynx Edicions, Madrid, pp 234–260

Hubbell SP (2005) Neutral theory in community ecology and the hypothesis of functional equivalence. Funct Ecol 19:166–172

James AG, Austin CJ, Cox DS, Taylor D, Calvert R (2013) Microbiological and biochemical origins of human axillary odour. FEMS Microbiol Ecol 83:527–540

Leclaire S, Nielsen JF, Drea CM (2014) Bacterial communities in meerkat anal scent secretions vary with host sex, age, and group membership. Behav Ecol 25:996–1004

Matthews LH (1939) Reproduction in the spotted hyaena, *Crocuta crocuta*. Philos Trans R Soc B 230:1–78

McCune B, Grace JB (2002) Analysis of ecological communities. MjM Software Design, Gleneden Beach, OR

McFall-Ngai M, Hadfield MG, Bosch TCG, Carey HV, Domazet-Lošo T, Douglas AE, Dubilier N, Eberl G, Fukami T, Gilbert SF, Hentschel U, King N, Kjelleberg S, Knoll AH, Kremer N, Mazmanian SK, Metcalf JL, Nealson K, Pierce NE, Rawls JF, Reid A, Ruby EG, Rumpho M, Sanders JG, Tautz D, Wernegreen JJ (2013) Animals in a bacterial world, a new imperative for the life sciences. Proc Natl Acad Sci U S A 110:3229–3236

Mills MGL (1990) Kalahari hyenas: comparative behavioral ecology of two species. Chapman & Hall, New York

Morris A, Beck JM, Schloss PD, Campbell TB, Crothers K, Curtis JL, Flores SC, Fontenot AP, Ghedin E, Huang L, Jablonski K, Kleerup E, Lynch SV, Sodergren E, Twigg H, Young VB, Bassis CM, Venkataraman A, Schmidt TM, Weinstock GM (2013) Comparison of the respiratory microbiome in healthy nonsmokers and smokers. Am J Resp Crit Care 187:1067–1075

Orrock JL, Watling JI (2010) Local community size mediates ecological drift and competition in metacommunities. Proc R Soc B Biol Sci 277:2185–2191

Palmer C, Bik EM, DiGiulio DB, Relman DA, Brown PO (2007) Development of the human infant intestinal microbiota. PLoS Biol 5:1556–1573

Pavlidis P, Noble WS (2003) Matrix2png: a utility for visualizing matrix data. Bioinformatics 19:295–296

Ramette A (2007) Multivariate analyses in microbial ecology. FEMS Microbiol Ecol 62:142–160

Schloss PD, Westcott SL, Ryabin T, Hall JR, Hartmann M, Hollister EB, Lesniewski RA, Oakley BB, Parks DH, Robinson CJ, Sahl JW, Stres B, Thallinger GG, Van Horn DJ, Weber CF (2009) Introducing mothur: open-source, platform-independent, community-supported software for describing and comparing microbial communities. Appl Environ Microbiol 75:7537–7541

Schloss PD, Gevers D, Westcott SL (2011) Reducing the effects of PCR amplification and sequencing artifacts on 16S rRNA-based studies. PLoS One 6:e27310

Sharon G, Garg N, Debelius J, Knight R, Dorrestein PC, Mazmanian SK (2014) Specialized metabolites from the microbiome in health and disease. Cell Metab 20:719–730

Sin YW, Buesching CD, Burke T, Macdonald DW (2012) Molecular characterization of the microbial communities in the subcaudal gland secretion of the European badger (*Meles meles*). FEMS Microbiol Ecol 81:648–659

Sloan WT, Lunn M, Woodcock S, Head IM, Nee S, Curtis TP (2006) Quantifying the roles of immigration and chance in shaping prokaryote community structure. Environ Microbiol 8:732–740

Theis KR, Heckla AL, Verge JR, Holekamp KE (2008) The ontogeny of pasting behavior in free-living spotted hyenas, *Crocuta crocuta*. In: Hurst JL, Beynon RJ, Roberts SC, Wyatt TD (eds) Chemical signals in vertebrates 11. Springer, New York, pp 179–188

Theis KR, Schmidt TM, Holekamp KE (2012) Evidence for a bacterial mechanism for group-specific social odors among hyenas. Sci Rep 2:615

Theis KR, Venkataraman A, Dycus JA, Koonter KD, Schmitt-Matzen EN, Wagner AP, Holekamp KE, Schmidt TM (2013) Symbiotic bacteria appear to mediate hyena social odors. Proc Natl Acad Sci U S A 110:19832–19837

Troccaz M, Gaia N, Beccucci S, Schrenzel J, Cayeux I, Starkenmann C, Lazarevic V (2015) Mapping axillary microbiota responsible for body odours using a culture-independent approach. Microbiome 3:3

Venkataraman A, Bassis CM, Beck JM, Young VB, Curtis JL, Huffnagle GB, Schmidt TM (2015) Application of a neutral community model to assess structuring of the human lung microbiome. mBio 6:e02284–14

Wagner AP (2013) *Hyaena hyaena*. In: Kingdon JS, Hoffmann M (eds) The mammals of Africa. Bloomsbury, London, pp 267–272

Wagner AP, Creel S, Frank LG, Kalinowski ST (2007) Patterns of relatedness and parentage in an asocial, polyandrous striped hyena population. Mol Ecol 16:4356–4369

Wagner AP, Frank LG, Creel S (2008) Spatial grouping in behaviourally solitary striped hyaenas, *Hyaena hyaena*. Anim Behav 75:1131–1142

Wang Q, Garrity GM, Tiedje JM, Cole JR (2007) Naive Bayesian classifier for rapid assignment of rRNA sequences into the new bacterial taxonomy. Appl Environ Microbiol 73:5261–5267

Yatsunenko T, Rey FE, Manary MJ, Trehan I, Dominguez-Bello MG, Contreras M, Magris M, Hidalgo G, Baldassano RN, Anokhin AP, Heath AC, Warner B, Reeder J, Kuczynski J, Caporaso JG, Lozupone CA, Lauber C, Clemente JC, Knights D, Knight R, Gordon JI (2012) Human gut microbiome viewed across age and geography. Nature 486:222–228

Chapter 8
Bacterial Communities Associated with Junco Preen Glands: Preliminary Ramifications for Chemical Signaling

Danielle J. Whittaker and Kevin R. Theis

8.1 Introduction

Renewed interest in the role of symbiotic bacteria in animal behavior—and particularly in producing chemical signals—has led to new insights and questions about the evolution of animal communication and host-symbiont coevolution (Archie and Theis 2011). The fermentation hypothesis for chemical recognition, first proposed in the 1970s (Albone et al. 1974; Gorman et al. 1974), suggests that symbiotic bacteria in mammalian scent glands produce volatile odorants that are used as recognition cues by the host animals and that variation in these bacterial communities contributes to variation in the animal scents. Support for this hypothesis has been demonstrated in several mammalian species, including spotted hyenas (Theis et al. 2012, 2013), meerkats (Leclaire et al. 2014), European badgers (Sin et al. 2012), North American porcupines (Roze et al. 2010), and greater sac-winged bats (Voigt et al. 2005). This hypothesis has rarely been applied outside of mammals and insects, although studies in birds and other taxa suggest that bacteria are present in glands related to chemical signaling (Ezenwa and Williams 2014). In this chapter, we explore whether symbiotic bacteria may play a role in manufacturing chemical signals in a songbird, the dark-eyed junco.

D.J. Whittaker (✉)
BEACON Center for the Study of Evolution in Action, Michigan State University, East Lansing, MI, USA
e-mail: djwhitta@msu.edu

K.R. Theis
BEACON Center for the Study of Evolution in Action, Michigan State University, East Lansing, MI, USA

Department of Zoology, Michigan State University, East Lansing, MI, USA

B.A. Schulte et al. (eds.), *Chemical Signals in Vertebrates 13*,
DOI 10.1007/978-3-319-22026-0_8

Chemical communication among birds, especially songbirds, is a long-neglected area of study due primarily to the persistent belief that most birds have little to no sense of smell. However, recent studies have revealed a potentially significant role for intraspecific chemical communication in avian behavior (Caro and Balthazart 2010). The best-studied source of volatile odors in birds is preen oil secreted by the uropygial or "preen" gland, the largest exocrine gland in most birds (Jacob and Ziswiler 1982). While preening, birds spread preen oil on their feathers, which protects them in a number of ways, including enhancing their insulative capacity, maintaining feather condition, and defense against pathogenic bacteria and fungi (Jacob and Ziswiler 1982; Moyer et al. 2003; Giraudeau et al. 2010). Symbiotic bacteria associated with the gland appear to provide some of these services, by producing antimicrobial compounds that help protect against feather-degrading bacteria and other pathogenic microbes (Shawkey et al. 2003; Martín-Vivaldi et al. 2009, 2010; Soler et al. 2010). Preen oil also emits volatile and semivolatile compounds that contribute to a bird's odor profile and thereby potentially function in communication (Mardon et al. 2010; Whittaker et al. 2010). These compounds vary among bird species (Soini et al. 2013), and within species, they can vary with individual identity (Mardon et al. 2010; Whittaker et al. 2010), sex (Soini et al. 2007; Whittaker et al. 2010), age (Shaw et al. 2011), and breeding condition (Whittaker et al. 2011b). In some bird species, they also reliably predict genetic and social reproductive success (Whittaker et al. 2013) and provide information about the relative quality of male rivals (Amo et al. 2012). Therefore, preen oil compounds may figure prominently in avian mate competition and choice. Importantly, many of these compounds, which include linear alcohols, methyl ketones, and carboxylic acids, are known to be end products of bacterial metabolism in other environments (Madigan et al. 2010; Agler et al. 2011), suggesting that the fermentation hypothesis for chemical recognition could explain the presence of these compounds in preen oil.

To our knowledge, no studies have yet tested whether bacteria play a role in avian chemical communication. However, several researchers have described relationships between preen glands and bacteria, including the effects of preen oil on feather-degrading bacteria, for example, in house finches (Shawkey et al. 2003), and beneficial bacteria found in breeding European hoopoe (Martín-Vivaldi et al. 2010) and wood hoopoe (Law-Brown and Meyers 2003) preen glands. Experimental work demonstrated that symbiotic bacteria in hoopoe uropygial glands produce preen oil volatile compounds with potent antimicrobial properties, though that study did not identify the symbiotic bacteria (Martín-Vivaldi et al. 2010). Shawkey and colleagues (2006) characterized the bacterial assemblages found on the plumage of several bird species and suggested that bacteria unique to the crested auklet could be a potential source for that species' distinctive tangerine odor, but did not explore potential links between auklet bacteria and odors (Hagelin et al. 2003; Shawkey et al. 2006). In this chapter, we bring together data on bacteria and preen oil volatile compounds in the context of chemical communication in a songbird.

One of the best characterized songbird chemical communication systems is that of the dark-eyed junco (*Junco hyemalis*), a widespread North American sparrow whose behavior, ecology, and physiology are well understood (Nolan et al. 2002).

Thirty-nine volatile and semivolatile compounds have been identified in junco preen oil, 17 of which varied seasonally or differed in relative concentration between the sexes and were selected for further research in subsequent studies: linear alcohols 1-decanol through 1-octadecanol, methyl ketones 2-undecanone through 2-pentadecanone, and carboxylic acids dodecanoic acid, tetradecanoic acid, and hexadecanoic acid (Soini et al. 2007; Whittaker et al. 2010). These 17 compounds vary with many aspects of junco biology, including population of origin, sex, hormone levels, and reproductive success (Soini et al. 2007; Whittaker et al. 2010, 2011b, 2013). Furthermore, juncos are able to detect and differentiate among preen oil odors from different individuals, sexes, and species (Whittaker et al. 2009, 2011a). These qualities make preen oil volatiles reliable candidate cues for mate assessment and choice, yet their mechanism of production remains unknown.

To begin evaluating whether the fermentation hypothesis for chemical recognition can account for variation in junco odor profiles, here we characterize the bacterial communities associated with the preen glands of breeding adult dark-eyed juncos and determine whether the taxa present would logically contribute to junco chemical signals. We compare diversity in bacterial community composition and structure among nesting pairs and between sexes. Finally, we consider the ramifications of our findings for future studies of chemical communication, mate choice, and kin recognition in birds.

8.2 Methods

We sampled the preen gland bacterial communities of wild adult Carolina dark-eyed juncos (*J. h. carolinensis*) at Mountain Lake Biological Station in Pembroke, VA. In this region of the Appalachian Mountains, juncos are primarily altitudinal migrants, moving down into the valleys during the winter and returning to higher elevations to breed. Males typically arrive on the breeding grounds in March, with females arriving about 2 weeks later (Nolan et al. 2002). Juncos are socially monogamous, typically forming pair bonds that are maintained throughout the breeding season (May to August). Juncos also demonstrate appreciable levels of extra-pair fertilization, with about 28 % of all offspring being sired by a male other than their mother's social mate (Ketterson et al. 1997; Gerlach et al. 2012). Juncos display biparental care, with females incubating the eggs and brooding the hatchlings, and both pair mates provisioning the young (Nolan et al. 2002).

Throughout May 2012, we identified junco nests during egg laying or incubation and monitored them through the nestling phase to fledging (day 11 or 12 post-hatching). On the morning of fledging day, we captured the adult female and male at the nest using mist nets. For this study, we captured 25 juncos from 13 nests (one nest did not have an attendant male). From each junco, we sampled the bacterial communities associated with the preen gland by rubbing the tip of the gland using a sterile cotton swab. This rubbing motion mimics the birds' own preen oil collection behavior when preening and is similar to that used to stimulate preen oil secretion for collection in capillary tubes (Whittaker et al. 2010). This collection method ensured that our samples included a small amount of preen oil and microbes from inside and outside the gland, which represent the mixture that birds collect on their bills in preparation for preening. We stored the samples at −80 °C until analysis.

We extracted DNA from the bacteria on swabs using MO BIO PowerSoil® DNA isolation kits (MO BIO Laboratories, Inc., Carlsbad, CA). We followed the manufacturer's recommended protocol, except we added an initial 10 min saturation step during which the swab bathed in bead solution within the bead tube, and we subsequently vigorously vortexed the bead tube for 1 min before removing the swab and proceeding to the step in which solution 1 is added. Each DNA extraction yielded a discernible band, consistent with 16S rDNA, on an agarose gel following PCR amplification of the 16S rRNA gene. Aliquots of the original DNA extractions were provided to the Michigan State University Research Technology Support Facility's Genomics Core, where the V4 region of the 16S gene was targeted for sequencing on the Illumina MiSeq platform. Sample preparation, sequencing, and preliminary quality filtering were completed using previously published protocols (Caporaso et al. 2011, 2012).

We processed the MiSeq run files using mothur software, v. 1.31.2 (Schloss et al. 2009; Kozich et al. 2013). Specifically, we removed all sequences that (1) contained any ambiguous base calls, (2) had homopolymer runs longer than eight bases, (3) did not start and end at our specific V4 primer positions when aligned to the Silva bacterial database, (4) were deemed chimeric by mothur's uchime tool, or (5) were classified as originating from mitochondria, chloroplasts, archaea, eukaryotes, or other nonbacterial sources using the Ribosomal Database Project's trainset9_032012 (Wang et al. 2007; Claesson et al. 2009). This process revealed that the preen sample of one male junco was not successfully sequenced, so data from this sample were discarded. Each of the 24 remaining samples were subsampled to a depth of 6000 sequences, and these sequences were binned into operational taxonomic units (OTUs) using mothur's average neighbor split-clustering algorithm and a 97 % sequence similarity cutoff. We then removed all singleton and doubleton OTUs from the data set and derived a consensus taxonomy for each of the remaining OTUs using a conservative 80 % confidence threshold (Claesson et al. 2009).

We generated a Clearcut cladogram, v 1.0.9, in mothur to illustrate the phylogenetic and taxonomic relationships among the prominent (i.e., widespread) OTUs in samples (Sheneman et al. 2006). We then used this information in conjunction with data on the typical production of volatiles by bacterial taxa, available through the mVOC database (Lemfack et al. 2014), to determine whether the prominent OTUs associated with junco preen glands are likely to manufacture any of the 17 volatile compounds of interest. Variation in the OTU profiles of samples among nests and between sexes was visualized via two-dimensional, principle coordinates analyses (PCoA) and statistically evaluated using nonparametric MANOVA with 10,000 permutations (Anderson 2001). We conducted these analyses using Dice and Bray-Curtis similarity indices, reflecting similarities in bacterial community membership and structure, respectively (Hammer 2011). Community membership addresses the shared presence or absence of OTUs, while community structure further considers similarities in their relative abundances. Prior to conducting community structure analyses, OTU abundance data were log10 $(x+1)$ transformed (Ramette 2007). All analyses were completed using PAST software, v 2.17 (Hammer et al. 2001; Hammer 2011).

8.3 Results and Discussion

Our preliminary bacterial survey suggests that junco preen glands, like mammalian scent glands, harbor diverse communities of symbiotic odor-producing bacteria. Figure 8.1 shows a cladogram of the most common OTUs found in our samples. Sixteen OTUs were unclassified, while 18 were assigned to 16 genera in the phyla Actinobacteria, Firmicutes, and Proteobacteria (classes Alphaproteobacteria, Betaproteobacteria, and Gammaproteobacteria). Several of these genera were previously found on the plumage of seabirds, chickens, and songbirds, including *Staphylococcus*, *Enterococcus*, *Pseudomonas*, and *Acinetobacter*; *Burkholderia* was found on the plumage of chickens and auklets (Shawkey et al. 2005, 2006). *Enterococcus* was also found in the preen glands and secretions of hoopoes and wood hoopoes (Martín-Vivaldi et al. 2009; Law-Brown and Meyers 2003). These findings suggest that these genera may be commonly found on birds, as they are on mammals.

Most of the bacterial genera associated with junco preen glands contain species that are known odor producers (Balkwill et al. 2006; Ezaki et al. 2006; Towner 2006; López del Castillo-Lozano et al. 2008; Blom et al. 2011; Latorre-Moratalla et al. 2011; Filipiak et al. 2012; Spraker et al. 2014). Most notably, *Burkholderia* and *Pseudomonas*, which can survive in a wide variety of habitats and utilize diverse nutrients, including oils (Haas and Défago 2005; Mahenthiralingam et al. 2005), can produce 9 of the 17 volatile compounds of interest (53 %) in juncos. Most of these nine volatile compounds are known to be produced by multiple species within each genus (e.g., nine species of *Burkholderia* and five species of *Pseudomonas* produce 2-tridecanone, Table 8.1). Thus, even without species-level identification in this study, the available data suggest that *Burkholderia* and *Pseudomonas* are strong candidates for the production of volatile compounds involved in junco chemical communication. We have begun efforts to successfully cultivate and metabolically characterize these bacteria from junco preen oil to directly test this hypothesis.

Not all of the identified genera contain odor producers, or at least they are not known to produce volatile compounds described in junco preen oil. However, while mammalian scent glands are specialized for a single purpose, secretions from avian preen glands perform diverse functions in addition to chemical signal production, including parasite defense, feather protection, and thermoregulation. Several genera associated with junco preen glands have documented antifungal (*Arthrobacter*, *Burkholderia*, *Pseudomonas*) or antibacterial (*Methylobacterium*, *Enterococcus*, *Pseudomonas*) effects (Fernando et al. 2005; Haas and Défago 2005; Green 2006; Jones and Keddie 2006; Soler et al. 2008; Groenhagen et al. 2013),

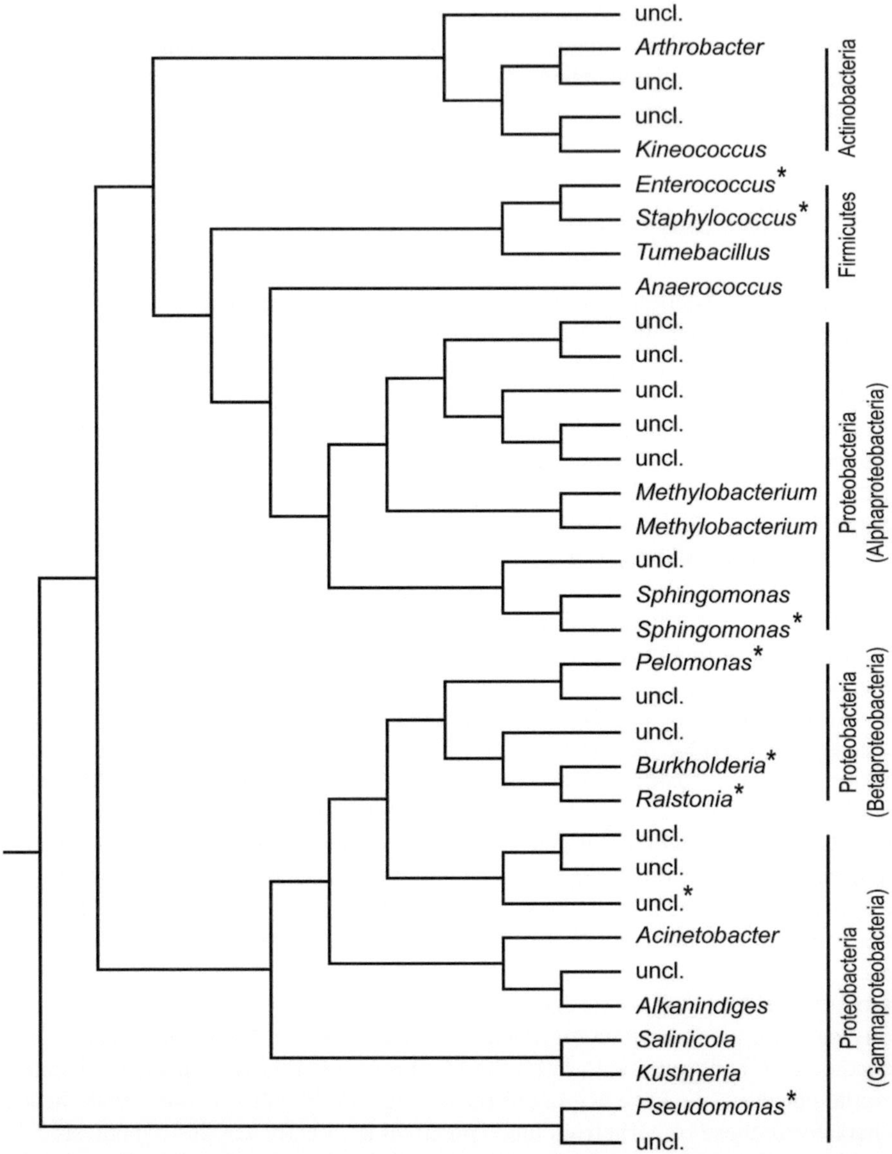

Fig. 8.1 Clearcut, relaxed neighbor-joining cladogram of the most common operational taxonomic units (OTUs) associated with junco preen glands. All OTUs listed were found in at least 50 % of birds sampled; those marked with *asterisks* were found in at least 75 % of the birds sampled. Taxonomic assignments were made using the Ribosomal Database Project's classifier tool in mothur, with an 80 % confidence threshold

Table 8.1 A count of how many species of *Burkholderia* and *Pseudomonas* have been found to emit junco volatile compounds

	# of species producing compound	
Volatile compound	*Burkholderia*	*Pseudomonas*
1-Decanol	0	4
1-Dodecanol	0	1
1-Heptadecanol	0	4
2-Undecanone	21	6
2-Dodecanone	1	0
2-Tridecanone	9	5
2-Pentadecanone	1	0
Dodecanoic acid	7	1
Tetradecanoic acid	5	3

and *Burkholderia* and *Pseudomonas* species are also used for biocontrol in agriculture and in bioremediation to clean up pollutants (Haas and Défago 2005; Mahenthiralingam et al. 2005). Therefore, these bacteria may be providing other valuable services to their junco hosts.

Although male and female juncos have significantly different volatile profiles (Whittaker et al. 2010), we did not find a consistent effect of sex on bacterial community composition or structure (NPMANOVA, $N=22$, df=1, Dice: $F=0.858$, $p=0.7713$, Bray-Curtis: $F=0.8024$, $p=0.8511$). Instead, paired males and females clustered together (Fig. 8.2). Similarly, in a study of captive zebra finches (*Taeniopygia guttata*), bacteria applied to the feathers of one zebra finch were found in the cloaca of its pair mate 24 h later, most likely transmitted via copulation (Kulkarni and Heeb 2007). Humans—and their pet dogs—living in the same household have significantly more similar skin microbiomes than those living in different households (Song et al. 2013), and individual human microbiomes quickly colonize the individual's surroundings (Lax et al. 2014). Since we collected our samples in May, near the beginning of the breeding season, our data suggest that adult birds that spend time in close physical proximity or have frequent contact may develop similar microbial communities in a fairly short period of time.

Previous studies have suggested that preen oil volatile compounds may be important for mate recognition and assessment, as they vary with species, sex, and individual identity (Soini et al. 2007; Mardon et al. 2010; Whittaker et al. 2010), and they predict reproductive success in juncos (Whittaker et al. 2013). These volatiles may also be important for recognizing kin (Leclaire et al. 2012) and one's home nest or burrow (Bonadonna and Bretagnolle 2002; Caspers and Krause 2010). Cross infection of symbiotic microbes between mates and between parents and offspring may contribute to the development of a recognizable, "signature" home scent (Archie and Theis 2011).

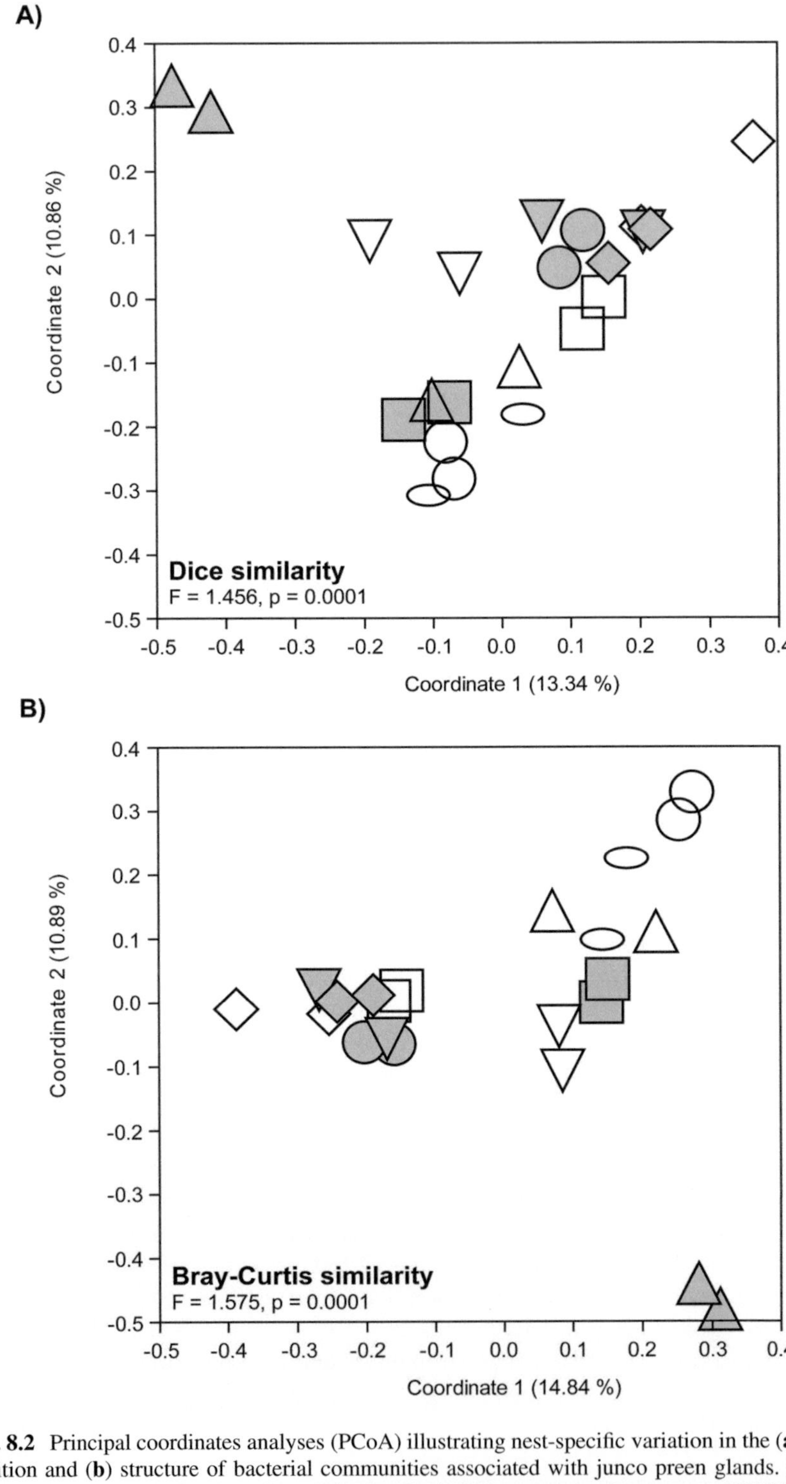

Fig. 8.2 Principal coordinates analyses (PCoA) illustrating nest-specific variation in the (**a**) composition and (**b**) structure of bacterial communities associated with junco preen glands. *Symbol shape* and *color* are indicative of nest identity. Statistical reports are for one-way NPMANOVA ($N=22$, df$=10$)

8.4 Conclusion

To our knowledge, this is the first study to examine avian microbiomes in the context of chemical signaling. Previous studies of symbiotic bacteria on birds' skin, feathers, or glands have focused on microbes that produce beneficial, antibacterial substances (e.g., Martín-Vivaldi et al. 2010) or on pathogenic microbes, especially feather-degrading bacteria (e.g., Shawkey et al. 2003; Saranathan and Burtt 2007; Saag et al. 2011). Our study suggests that the fermentation hypothesis for chemical recognition, originally formulated for mammals, may apply to a much broader range of taxa and opens new pathways for research. Future studies should evaluate covariance between preen gland bacterial communities and odors, the effect of manipulating bacterial communities on odor profiles, and the extent to which subsequent changes in odor profiles influence birds' behavioral responses to preen oil. They should also further elucidate the effect of social behavior on the transmission of preen gland bacterial communities and determine the extent to which preen gland bacterial communities differ from those associated with other avian organs.

Acknowledgments This work was supported by the BEACON Center for the Study of Evolution in Action (NSF Cooperative Agreement DBI-0939454). Fieldwork was supported by a Mountain Lake Biological Station Early Career Fellowship to DJW. All work was conducted in compliance with the Bloomington Institutional Animal Care and Use Committee guidelines (BIACUC protocol 09-037) and with permission from the US Department of Fish and Wildlife, the Virginia Department of Game and Inland Fisheries, and the US Forest Service. We thank Mountain Lake Biological Station, University of Virginia, as well as Mountain Lake Hotel and the Dolinger family for allowing research on their grounds. We thank Nicole Gerlach, Abby Kimmitt, Dustin Reichard, Samuel Slowinski, Sarah Wanamaker, and Joseph Welklin for assistance in the field, John Dover and Arvind Venkataraman for assistance with DNA extraction and sequence processing, and Ellen Ketterson for helpful discussion.

References

Agler MT, Wrenn BA, Zinder SH, Angenent LT (2011) Waste to bioproduct conversion with undefined mixed cultures: the carboxylate platform. Trends Biotechnol 29:70–78

Albone ES, Eglinton G, Walker JM, Ware GC (1974) The anal sac secretion of the red fox (*Vulpes vulpes*); its chemistry and microbiology. A comparison with the anal sac secretion of the lion (*Panthera leo*). Life Sci 14:387–400

Amo L, López-Rull I, Pagán I, Macías Garcia C (2012) Male quality and conspecific scent preferences in the house finch, *Carpodacus mexicanus*. Anim Behav 84:1483–1489

Anderson M (2001) A new method for non-parametric multivariate analysis of variance. Austral Ecol 26:32–46

Archie EA, Theis KR (2011) Animal behaviour meets microbial ecology. Anim Behav 82:425–436

Balkwill DL, Fredrickson JK, Romine MF (2006) *Sphingomonas* and related genera. In: Dworkin M, Falkow S, Rosenberg E, Schleifer K-H, Stackebrandt E (eds) The prokaryotes. Springer, New York, pp 605–629

Blom D, Fabbri C, Connor EC, Schiestl FP, Klauser DR, Boller T, Eberl L, Weisskopf L (2011) Production of plant growth modulating volatiles is widespread among rhizosphere bacteria and strongly depends on culture conditions. Environ Microbiol 13:3047–3058

Bonadonna F, Bretagnolle V (2002) Smelling home: a good solution for burrow-finding in nocturnal petrels? J Exp Biol 205:2519–2523

Caporaso JG, Lauber CL, Walters WA, Berg-Lyons D, Lozupone CA, Turnbaugh PJ, Fierer N, Knight R (2011) Global patterns of 16S rRNA diversity at a depth of millions of sequences per sample. Proc Natl Acad Sci U S A 108:4516–4522

Caporaso JG, Lauber CL, Walters WA, Berg-Lyons D, Huntley J, Fierer N, Owens SM, Betley J, Fraser L, Bauer M, Gormley N, Gilbert JA, Smith G, Knight R (2012) Ultra-high-throughput microbial community analysis on the Illumina HiSeq and MiSeq platforms. ISME J 6(8): 1621–1624

Caro SP, Balthazart J (2010) Pheromones in birds: myth or reality? J Comp Physiol A 196: 751–766

Caspers BA, Krause ET (2010) Odour-based natal nest recognition in the zebra finch (*Taeniopygia guttata*), a colony-breeding songbird. Biol Lett 7:184–186

Claesson MJ, O'Sullivan O, Wang Q, Nikkilä J, Marchesi JR, Smidt H, de Vos WM, Ross RP, O'Toole PW (2009) Comparative analysis of pyrosequencing and a phylogenetic microarray for exploring microbial community structures in the human distal intestine. PLoS One 4(8):e6669

Ezaki T, Li N, Kawamura Y (2006) The anaerobic gram-positive cocci. In: Dworkin M, Falkow S, Rosenberg E, Schleifer K-H, Stackebrandt E (eds) The prokaryotes. Springer, New York, pp 795–808

Ezenwa VO, Williams AE (2014) Microbes and animal olfactory communication: where do we go from here? Bioessays 36:847–854

Fernando WGD, Ramarathnam R, Krishnamoorthy AS, Savchuk SC (2005) Identification and use of potential bacterial organic antifungal volatiles in biocontrol. Soil Biol Biochem 37:955–964

Filipiak W, Sponring A, Baur MM, Filipiak A, Ager C, Wiesnhofer H, Nagl M, Troppmair J, Amann A (2012) Molecular analysis of volatile metabolites released specifically by *Staphylococcus aureus* and *Pseudomonas aeruginosa*. BMC Microbiol 12:113

Gerlach NM, McGlothlin JW, Parker PG, Ketterson ED (2012) Promiscuous mating produces offspring with higher lifetime fitness. Proc R Soc B 279:860–866

Giraudeau M, Duval C, Guillon N, Bretagnolle V, Gutierrez C, Heeb P (2010) Effects of access to preen gland secretions on mallard plumage. Naturwissenschaften 97:577–581

Gorman ML, Nedwell DB, Smith RM (1974) An analysis of the contents of the anal and scent pockets of *Herpestes auropunctatus* (Carnivora: Viverridae). J Zool 172:389–399

Green PN (2006) *Methylobacterium*. In: Dworkin M, Falkow S, Rosenberg E, Schleifer K-H, Stackebrandt E (eds) The prokaryotes. Springer, New York, pp 257–265

Groenhagen U, Baumgartner R, Bailly A, Gardiner A, Eberl L, Schultz S, Weisskopf L (2013) Production of bioactive volatiles by different *Burkholderia ambifaria* strains. J Chem Ecol 39:892–906

Haas D, Défago G (2005) Biological control of soil-borne pathogens by fluorescent pseudomonads. Nat Rev Microbiol 3:307–319

Hagelin JC, Jones IL, Rasmussen LEL (2003) A tangerine-scented social odour in a monogamous seabird. Proc R Soc B 270:1323–1329

Hammer O (2011) PAST: PAleontological STatistics manual, 2.07 edn. Natural History Museum, Oslo, Norway

Hammer O, Harper DAT, Ryan PD (2001) PAST: PAleontological STatistics software package for education and data analysis. Palaeontol Electron 4:1–9

Jacob JP, Ziswiler V (1982) The uropygial gland. In: Farner DS, King JR, Parkes KC (eds) Avian biology, vol 6. Academic, New York, pp 199–324

Jones D, Keddie RM (2006) The genus *Arthrobacter*. In: Dworkin M, Falkow S, Rosenberg E, Schleifer K-H, Stackebrandt E (eds) The prokaryotes. Springer, New York, pp 945–960

Ketterson ED, Parker PG, Raouf SA, Nolan V Jr, Ziegenfus C, Chandler CR (1997) The relative impact of extra-pair fertilizations on variation in male and female reproductive success in dark-eyed juncos (*Junco hyemalis*). Ornithol Monogr 1997:81–101

Kozich JJ, Westcott SL, Baxter NT, Highlander SK, Schloss PD (2013) Development of a dual-index sequencing strategy and curation pipeline for analyzing amplicon sequence data on the MiSeq Illumina sequencing platform. Appl Environ Microbiol 79:5112–5120

Kulkarni S, Heeb P (2007) Social and sexual behaviours aid transmission of bacteria in birds. Behav Proc 74:88–92

Latorre-Moratalla ML, Bosche-Fusté J, Bover-Cid S, Aymerich T, Vidal-Carou MC (2011) Contribution of enterococci to the volatile profile of slightly-fermented sausages. LWT Food Sci Technol 44:145–152

Law-Brown J, Meyers PR (2003) *Enterococcus phoeniculicola* sp. nov., a novel member of the enterococci isolated from the uropygial gland of the red-billed woodhoopoe, *Phoeniculus purpureus*. Int J Syst Evol Microbiol 53:683–685

Lax S, Smith DP, Hampton-Marcell J, Owens SM, Handley KM, Scott NM, Gibbons SM, Larsen P, Shogan BD, Weiss S, Metcalf JL, Urselll LK, Vázquez-Baeza Y, Van Treuren W, Hasan NA, Gibson MK, Colwell R, Dantas G, Knight R, Gilbert JA (2014) Longitudinal analysis of microbial interaction between humans and the indoor environment. Science 345:1048–1052

Leclaire S, Merkling T, Raynaud C, Mulard H, Bessière J-M, Lhuillier É, Hatch SA, Danchin É (2012) Semiochemical compounds of preen secretion reflect genetic make-up in a seabird species. Proc R Soc B 279:1185–1193

Leclaire S, Nielsen JF, Drea CM (2014) Bacterial communities in meerkat anal scent secretions vary with host sex, age, and group membership. Behav Ecol 25:996–1004

Lemfack MC, Nickel J, Dunkel M, Preissner R, Piechulla B (2014) mVOC: a database of microbial volatiles. Nucleic Acids Res 42(1):D744–D748

López del Castillo-Lozano M, Mansour S, Tâche R, Bonnarme P, Landaud S (2008) The effect of cysteine on production of volatile sulphur compounds by cheese-ripening bacteria. Int J Food Microbiol 122:321–327

Madigan MT, Martinko JM, Stahl DA, Clark DP (2010) Brock biology of microorganisms, 13th edn. Benjamin Cummings, San Francisco

Mahenthiralingam E, Urban TA, Goldberg JB (2005) The multifarious, multireplicon *Burkholderia cepacia* complex. Nat Rev Microbiol 3:144–156

Mardon J, Saunders SM, Anderson MJ, Couchoux C, Bonadonna F (2010) Species, gender, and identity: cracking petrels' sociochemical code. Chem Senses 35:309–321

Martín-Vivaldi M, Ruiz-Rodríguez M, Soler JJ, Peralta-Sánchez JM, Méndez M, Valdivia E, Martín-Platero AM, Martínez-Bueno M (2009) Seasonal, sexual and developmental differences in hoopoe *Upupa epops* preen gland morphology and secretions: evidence for a role of bacteria. J Avian Biol 40:191–205

Martín-Vivaldi M, Peña A, Peralta-Sánchez JM, Sánchez L, Ananou S, Ruiz-Rodríguez M, Soler JJ (2010) Antimicrobial chemicals in hoopoe preen secretions are produced by symbiotic bacteria. Proc R Soc B 277:123–130

Moyer BR, Rock AN, Clayton DH (2003) Experimental test of the importance of preen oil in rock doves (*Columba livia*). Auk 120:490–496

Nolan V Jr, Ketterson ED, Cristol DA, Rogers CM, Clotfelter ED, Titus R, Schoech SJ, Snajdr E (2002) Dark-eyed junco (*Junco hyemalis*), vol 716, The birds of North America. The Birds of North America, Inc., Philadelphia, PA

Ramette A (2007) Multivariate analyses in microbial ecology. FEMS Microbiol Ecol 62:142–160

Roze U, Leung KT, Nix E, Burton G, Chapman DM (2010) Microanatomy and bacterial flora of the perineal glands of the North American porcupine. Can J Zool 88:59–68

Saag P, Tilgar V, Mänd R, Kilgas P, Mägi M (2011) Plumage bacterial assemblages in a breeding wild passerine: relationships with ecological factors and body condition. Microbiol Ecol 61:740–749

Saranathan V, Burtt EH (2007) Sunlight on feathers inhibits feather-degrading bacteria. Wilson J Ornithol 119:239–245

Schloss PD, Westcott SL, Ryabin T, Hall JR, Hartmann M, Hollister EB, Lesniewski RA, Oakley BB, Parks DH, Robinson CJ, Sahl JW, Stres B, Thallinger GG, Van Horn DJ, Weber CF (2009) Introducing mothur: open-source, platform-independent, community-supported software for describing and comparing microbial communities. Appl Environ Microbiol 75:7537–7541

Shaw CL, Rutter JE, Austin AL, Garvin MC, Whelan RJ (2011) Volatile and semivolatile compounds in gray catbird uropygial secretions vary with age and between breeding and wintering grounds. J Chem Ecol 37:329–339

Shawkey MD, Pillai SR, Hill GE (2003) Chemical warfare? Effects of uropygial oil on feather-degrading bacteria. J Avian Biol 34:345–349

Shawkey MD, Mills KL, Dale C, Hill GE (2005) Microbial diversity of wild bird feathers revealed through culture-based and culture-independent techniques. Microbiol Ecol 50:40–47

Shawkey MD, Hussain J, Hagelin JC, Vollmer AC, Hill GE (2006) Use of culture-independent methods to compare bacterial assemblages on feathers of crested and least auklets (*Aethia cristatella* and *Aethia pusilla*) with those of passerines. Waterbirds 29:507–510

Sheneman L, Evans J, Foster JA (2006) Clearcut: a fast implementation of relaxed neighbor joining. Bioinformatics 22:2823–2824

Sin YW, Buesching CD, Burke T, MacDonald DW (2012) Molecular characterization of the microbial communities in the subcaudal gland secretion of the European badger (*Meles meles*). FEMS Microbiol Ecol 81:648–659

Soini HA, Schrock SE, Bruce KE, Wiesler D, Ketterson ED, Novotny MV (2007) Seasonal variation in volatile compound profiles of preen gland secretions of the dark-eyed junco (*Junco hyemalis*). J Chem Ecol 33:183–198

Soini HA, Whittaker DJ, Wiesler D, Ketterson ED, Novotny M (2013) Chemosignaling diversity in songbirds: chromatographic profiling of preen oil volatiles in different species. J Chromatogr A 1317:186–192

Soler JJ, Martín-Vivaldi M, Ruiz-Rodríguez M, Valdivia E, Martín-Platero AM, Martínez-Bueno M, Peralta-Sánchez JM, Méndez M (2008) Symbiotic association between hoopoes and antibiotic-producing bacteria that live in their uropygial gland. Funct Ecol 22:864–871

Soler JJ, Martín-Vivaldi M, Peralta-Sánchez JM, Ruiz-Rodríguez M (2010) Antibiotic-producing bacteria as a possible defence of birds against pathogenic microorganisms. Open Ornithol J 3:93–100

Song SJ, Lauber CL, Costello EK, Lozupone CA, Humphrey G, Berg-Lyons D, Caporaso JG, Knights D, Clemente JC, Nakielny S, Gordon JI, Fierer N, Knight R (2013) Cohabiting family members share microbiota with one another and with their dogs. eLife 2:e00458

Spraker JE, Jewell K, Roze LV, Scherf J, Ndagano D, Beaudry R, Linz JE, Allen C, Keller NP (2014) A volatile relationship: profiling an inter-kingdom dialogue between two plant pathogens, *Ralstonia solanacearum* and *Aspergillus flavus*. J Chem Ecol 40:502–513

Theis KR, Schmidt TM, Holekamp KE (2012) Evidence for a bacterial mechanism for group-specific social odors among hyenas. Sci Rep 2:615

Theis KR, Venkataraman A, Dycus JA, Koonter KD, Schmitt-Matzen EN, Wagner AP, Schmidt TM, Holekamp KE (2013) Symbiotic bacteria appear to mediate hyena social odors. Proc Natl Acad Sci U S A 110:19832–19837

Towner K (2006) The genus *Acinetobacter*. In: Dworkin M, Falkow S, Rosenberg E, Schleifer K-H, Stackebrandt E (eds) The prokaryotes. Springer, New York, pp 746–758

Voigt C, Caspers B, Speck S (2005) Bats, bacteria, and bat smell: sex-specific diversity of microbes in a sexually selected scent organ. J Mammal 86:745–749

Wang Q, Garrity GM, Tiedje JM, Cole JR (2007) Naive bayesian classifier for rapid assignment of rRNA sequences into the new bacterial taxonomy. Appl Environ Microbiol 73:5261–5267

Whittaker DJ, Reichard DG, Dapper AL, Ketterson ED (2009) Behavioral responses of nesting female dark-eyed juncos *Junco hyemalis* to hetero- and conspecific passerine preen oils. J Avian Biol 40:579–583

Whittaker DJ, Soini HA, Atwell JW, Hollars C, Novotny MV, Ketterson ED (2010) Songbird chemosignals: volatile compounds in preen gland secretions vary among individuals, sexes, and populations. Behav Ecol 21:608–614

Whittaker DJ, Richmond KM, Miller AK, Kiley R, Bergeon Burns C, Atwell JW, Ketterson ED (2011a) Intraspecific preen oil odor preferences in dark-eyed juncos (*Junco hyemalis*). Behav Ecol 22:1256–1263

Whittaker DJ, Soini HA, Gerlach NM, Posto AL, Novotny MV, Ketterson ED (2011b) Role of testosterone in stimulating seasonal changes in a potential avian chemosignal. J Chem Ecol 37:1349–1357

Whittaker DJ, Gerlach NM, Soini HA, Novotny MV (2013) Bird odour predicts reproductive success. Anim Behav 86:697–703

Part III
Intraspecific Signaling

Chapter 9
Age-Related Effects on Individual Discrimination Among Meadow Voles, *Microtus pennsylvanicus*

Michael H. Ferkin, Christian T. Vlautin, and Lyndsey M. Pierson

9.1 Introduction

For many terrestrial mammals, scent marks can indicate the senders or scent donor's presence in an area (Brown and Macdonald 1985) and its phenotype (Roberts 2007; Wyatt 2014; Ferkin 2015). Being able to discriminate between the scent marks of conspecifics would allow receivers to avoid or seek out an interaction with particular individuals (Thom and Hurst 2004). Discriminating between conspecifics could be as simple as an individual learning the identity of a familiar scent donor, habituating to its unique mark, and subsequently becoming more attentive to the scent marks of unfamiliar donors (Halpin 1986; Hurst and Beynon 2004; Johnston 2008; Kaur et al. 2014). For example, the amount of time golden hamsters (*Mesocricetus auratus*) spent sniffing the scent mark of an individual receiver decreased with each exposure to that receiver's scent marks. Later, when the hamsters were exposed to the scent mark of that same individual and the scent mark of a novel conspecific, they spent more time sniffing the scent mark of the novel conspecific. This finding suggests hamsters behave as if the scent marks of conspecifics were individually distinct (Johnston 1993, 2003; Todrank et al. 1998; Heth et al. 1999; Johnston and Bullock 2001; Mayeaux and Johnston 2002). Similar results were obtained in other studies using somewhat different methods in house mice, *Mus musculus* (Ferkin and Li 2005); Mongolian gerbils, *Meriones unguiculatus* (Halpin 1974, 1980); prairie voles, *Microtus ochrogaster* (Newman and Halpin 1988; Tang-Martinez and Bixler 2009); meadow voles, *M. pennsylvanicus* (Ferkin et al. 1999); ring-tailed lemurs, *Lemur catta* (Mertl 1975); chipmunks, *Tamias striatus* (Keevin et al. 1981);

M.H. Ferkin (✉) • C.T. Vlautin • L.M. Pierson
Department of Biological Sciences, The University of Memphis,
Ellington Hall, Memphis, TN 38152, USA
e-mail: mhferkin@memphis.edu

B.A. Schulte et al. (eds.), *Chemical Signals in Vertebrates 13*,
DOI 10.1007/978-3-319-22026-0_9

common marmosets, *Callithrix jacchus* (Smith 2006); Turkish hamsters, *Mesocricetus brandti* (Heth et al. 1999); and dwarf hamsters, *Phodopus campbelli* (Lai and Johnston 1994).

Studies that examined whether scent marks convey individually distinct information have focused on younger adult rodents as subjects. This may be problematic in that older adult rodents may not respond to scent marks in the same manner as would younger adult rodents. It has been well established that older adult rodents may have lowered olfactory sensitivity compared to younger adult rodents (Thompson and Spencer 1966; Gallagher and Rapp 1997; Wilson and Stevenson 2006), which may affect their ability to distinguish between scent marks of different senders. It is also possible that older rodents will have impairments that cause them to fail to learn, consolidate, or recollect the identity of conspecific scent donors (Doty and Kamath 2014).

Recent work has examined the effects of aging in meadow voles, *Microtus pennsylvanicus*. Meadow voles are weaned at 18–21 days of age, become sexually mature at 2 months of age, and can live up to 24 months in the wild (Tamarin et al. 1984; Keller 1985; Pugh, Ferkin, and Tamarin unpubl. data). We have discovered that aging affects scent marking and responses to scent marks among male and female meadow voles (Ferkin and Leonard 2008; Ferkin 2010). Male voles that were 10–13 months old spent more time self-grooming than did 2–9-month-old male voles when they were exposed to the odors of a sexually receptive female (Ferkin and Leonard 2008). Age of the female did not affect the amount of time female voles spent self-grooming when they were exposed to the odors of an adult male conspecific (Ferkin and Leonard 2008). Self-grooming increases the odor field around voles and enhances the attractiveness of the groomer's scent mark to opposite-sex conspecifics (Ferkin et al. 1996). In contrast to self-grooming, younger adult males deposited more scent marks but fewer over-marks in areas scent marked by a sexually receptive female than did older males (Ferkin 2010). Again, age did not alter the scent marking and over-marking responses of female voles (Ferkin 2010). Scent marking allows voles to indicate their phenotype and presence in an area (Ferkin 2011, 2015). Over-marking occurs when two or more scent marks overlap (Johnston et al. 1994, 1995; Ferkin 2011). As a whole, the results suggest that for meadow voles the effects of aging on self-grooming, scent marking, and over-marking differ between males and females (Ferkin and Leonard 2008; Ferkin 2010) and that each type of odor-related behavior may have its own vulnerability to the effects of aging. The observations also suggest that older males may have more mating opportunities than do younger males (Ferkin 1999). Such a speculation is not consistent with models that predict how aging affects the behavior of short-lived animals, such as rodents. The central prediction for short-lived animals is that mating success and fitness favor younger adult rodents over older adult rodents (Boonstra 1994). The converse is predicted for the effects of aging on reproduction long-lived species such as primates (Promislow 1991; Rose 1991).

The first aim of the present study was to test the hypothesis that aging affects the ability of meadow voles to discriminate between the scent marks of two different same-sex donors that share a similar phenotype. The second aim was to test the hypothesis that aging affects the ability of meadow voles to discriminate between the scent marks of an opposite sex and a same-sex scent donor. To test the first the hypothesis, we compared the amount of time that voles in four different age groups (2–3-month-old, 5–7-month-old, 10–12-month-old, and 15–18-month-old) spent investigating the scent marks of 5–7-month-old conspecifics during a habituation/test task. To test the second hypothesis, we compared the amount of time that voles in the four age groups spent investigating the scent mark of a male conspecific and that of a female conspecific in an odor preference test.

9.2 Materials and Methods

9.2.1 Animals

Meadow vole subjects were housed in the animal care facility at the University of Memphis, Memphis TN, USA. The colony consisted of meadow voles originally captured in Pennsylvania, Kentucky, and Ohio, USA. Every 18–24 months, the voles in the colony were mated with captured free-living voles. Meadow voles used in the present study were born and raised under long photoperiod (14:10 h, L:D, lights on at 0700 h CST). Female meadow voles are induced ovulators and do not undergo regular estrous cycles (Milligan 1982; Keller 1985). Female meadow voles reach sexual maturity between 1 and 2 months of age and males reach sexual maturity by 2 months of age (Nadeau 1985). In the present study, all male and female voles were sexually receptive; females were not currently pregnant or lactating. All voles used in this study were capable of reproducing and were not reproductively senescent (Ferkin unpubl. data).

All voles used in this study were weaned between 19 and 21 days of age, housed with littermates until 33–36 days of age, and thereafter housed singly in clear plastic cages ($18 \times 12.5 \times 10$ cm). Cages contained cotton nesting material, water, and food (Laboratory Rodent Diet #8640, Harlan Teklad, Madison, WI, USA).

9.2.2 Age Classes

Subjects were 2–3-month-old, 5–7-month-old, 10–12-month-old, and 15–18-month-old voles. Similar age groups have been used in other studies of odor-related behavior among meadow voles (Ferkin 1999, 2010; Ferkin and Leonard 2008). These four age groups roughly encompass the range of ages of most of the conspecifics that adult meadow voles will encounter in free-living populations (Sheridan and Tamarin 1988; Tamarin et al. 1984). The 2–3-month-old, 5–7-month-old, and 10–12-month-old voles encompass the adult cohorts found in free-living vole populations (Tamarin et al. 1984). The 2–3-month-old voles represent those individuals that have recently passed through puberty (Nadeau 1985); we consider these voles as being young adults (Ferkin 1999). The 5–7-month-old voles represent those individuals that are sexually mature, and in field populations are greatest in number (Tamarin et al. 1984). The 10–12-month-old voles represent those individuals that are sexually mature and have not yet reached senescence (Slade 1995). In field populations, these voles may have been born late in the previous breeding season, overwintered, and survived to mate again the following spring (Tamarin et al. 1984; Negus and Berger 1988; Sheridan and Tamarin 1988). The 15–18-month-old voles are also sexually mature, have not yet reached senescence (Slade 1995), and represent those voles that may have been born early in the previous breeding season, overwintered, and survived to mate again the following spring. Ten- to 18-month-old voles make up the smallest proportion of the population among free-living voles, approximately 5–10 % of the population (Tamarin et al. 1984; Sheridan and Tamarin 1988; Pugh et al. unpubl. data). In the present study, the 2–3-month-old voles were sexually receptive (Keller 1985), but sexually naïve. The 5–7-month-old voles, the 10–12-month-old voles, and the 15–18-month-old voles were sexually experienced (Keller 1985; Nadeau 1985), having sired or delivered a litter 3–8 weeks before being used in this study. Thus, all the voles used in this study were reproductively active, capable of mating and producing offspring.

9.2.3 Habituation Phase

In this study, we used a habituation/test task that had previously been used to test individual discrimination meadow voles (Ferkin et al. 1999), golden hamsters (Johnston 1993; Johnston et al. 1993), and house mice (Ferkin and Li 2005). The habituation/test task involves two phases with the habituation being the first phase and the test being second phase. All testing was conducted from April to August 2013 between 0830 and 1330 h (CST).

During the habituation phase, subjects were given four 5-min exposures to glass slides containing the single scent mark of an individual (donor A). The scent mark of donor A was placed on in the surface of a clean, clear glass, microscope slide (7.62 cm × 2.54 cm) by rubbing the slide gently across the anogenital region of the donor vole for 3–5 s. The scent mark was placed vertically in the center of the slide. The scent mark was 1.5–2.4 cm long and 0.5–1.0 cm wide. The glass slide was suspended by a clean wire hook and clasp 1 cm above the substrate in the subject vole's home cage, against the wall opposite the animal's nest, and always in the same location in the cage. We used a new glass slide and fresh anogenital area scent marks for each presentation during this phase. The inter-exposure interval between the four successive exposures during the habituation phase was 5 min. Thus, the habituation phase consisted of an initial 5-min exposure, a 5-min interval, a second 5-min exposure with a new slide containing the scent mark of the same donor as the first, another 5-min interval, and so forth. The habituation phase ended after the fourth 5-min presentation of the slide containing the scent marks of donor A to the subject. Subjects in all experiments were exposed to a unique combination of scent donors. The subjects and scent donors were unrelated and subjects were not familiar with donor A prior to the habituation phase.

We recorded the amount of time subjects spent licking or sniffing (nose within 1 cm) the scent mark of donor A during each of the four presentations during the habituation phase. We used a two-way repeated measures analysis of variance (factors: sex and habituation exposures) followed by a Holm-Sidák multiple comparison test to determine if male and female subjects spent significantly more time investigating the scent mark of donor A on the first exposure than it did on the fourth exposure. Significant differences were accepted at alpha = 0.05. A significant decrease in investigation time between the first and fourth exposure would indicate that the subject habituated to the scent mark of donor A (Johnston 1993; Johnston et al. 1993; Ferkin et al. 1999; Ferkin and Li 2005).

9.2.4 Test Phase

The test phase began 5 min after the completion of the habituation phase. During the test phase, subjects were presented with a slide containing the scent mark of donor A and a scent mark of another individual, donor B. The scent marks were placed on a new microscope slide that was divided vertically into three equal sections: two end sections and a middle section. Each section of the slide was 2.54 cm wide. One end section of the slide contained the scent mark from donor A. The other end section contained the scent mark from donor B. The placement of the scent marks from donor A and donor B on the left or right side of the slide was alternated. Each scent placed on the slide was approximately 24 cm long and 0.8 cm wide. The middle section contained no scent marks. The test slide was placed in the same location in the subject vole's home cage as were the slides used during the habituation phase.

We compared the amount of time that male and female subjects investigated the scent mark of donor A and that of donor B. We used a two-way repeated measures analysis of variance (factors: sex and scent donor (donor A vs. donor B)) followed by a Holm-Sidák multiple comparison test to determine if male and female subjects spent significantly more time investigating the scent mark of donor B than that of donor A. Significant differences in investigation time were accepted at $\alpha = 0.05$. We considered that the voles viewed the scent marks as being different if they spent significantly more time investigating the mark of donor B compared to donor A (Ferkin et al. 1999; Ferkin and Li 2005).

9.2.5 Stimulus Scent Marks

The anogenital area mark of a scent donor served as the stimulus during both phases of this experiment. Anogenital area marks are deposited by voles in runways and paths (Ferkin 2011) and are sexually discriminable (Ferkin et al. 1999), containing secretions from the vole's urogenital region glands and tissues, as well as its feces and urine (Ferkin and Johnston 1995).

9.2.5.1 Experiment 1

Subjects were selected from the four age groups: 2–3-month-old voles, 5–7-month-old voles, 10–12-month-old voles, and 15–18-month-old voles. We used 12 different males and 12 different females as subjects from each age group. Male subjects were exposed to the scent marks of two different female donors and female subjects were exposed to the scent marks of two different male donors. Eighteen female and 16 male 5–7-month-old voles provided the anogenital area scent marks and served as donor A or donor B. Male and female scent donors were not used as subjects in this experiment. We selected 5–7-month-old voles as scent donors because they are the most abundant age class of adults in free-living populations of meadow voles (Sheridan and Tamarin 1988; Tamarin et al. 1984; Pugh, Ferkin, Seamon, and Tamarin unpubl. data) and most likely to have their scent marks encountered by conspecifics.

9.2.5.2 Experiment 2

The methods for experiment 2 were the same as detailed in experiment 1 with these notable exceptions. Male subjects ($n=12$ per age group) were exposed to the scent marks of two different male donors and female subjects ($n=12$ per age group) were exposed to the scent marks of two different female donors. Fifteen female and 16 male 5–7-month-old voles provided the anogenital area scent marks. The subjects and scent donors used in this experiment were not used in experiment 1.

9.2.5.3 Experiment 3

Our previous work has shown that 2–3-month-old, 5–7-month-old, and 10–12-month-old male and female meadow voles can discriminate between the scent marks of male and female conspecifics, spending significantly more time investigating the anogenital area scent marks of opposite-sex conspecifics than those of same-sex conspecifics (Ferkin and Johnston 1995; Ferkin 1999, 2011). However, it is not known if 15–18-month-old voles can distinguish between the scent mark of a male and that of a female conspecific. Thus, we recorded the amount of time that 15–18-month-old male and female voles spent during a 5-min test investigating the anogenital area scent mark of male donor and that of a female donor. Seventeen ($n=9$ females and $n=8$ males) 5–7-month-old voles served as scent donors. The scent donors were unfamiliar and unrelated to the subjects. Subjects were 16 male and 16 female 15–18-month-old voles. The subjects and scent donors were not used in experiments 1 or 2.

The testing procedure was identical to the test phase used during experiments 1 and 2, except that here, the subject was presented with a microscope slide containing the scent mark of an unfamiliar male and the scent mark of an unfamiliar female donor. The test began when the slide was placed into the cage of the subject vole.

Each slide was used in only one trial and discarded. The placement of the scent marks from the male and female donor on the left or right side of the slide was alternated. We used paired *t*-tests to determine if statistically significant differences ($p<0.05$) existed in the amount of time voles spent investigating the scent mark of an unfamiliar male or that of an unfamiliar female donor.

9.3 Results

9.3.1 Experiment 1: Exposure to Scent Marks of Opposite-Sex Conspecifics

Across the four exposures during the habituation phase, 2–3-month-old voles ($F_{3, 30}=23.81$, $p<0.001$), 5–7-month-old voles ($F_{3, 30}=42.98$, $p<0.001$), and 10–12-month-old voles ($F_{3, 30}=58.43$, $p<0.001$) spent different amounts of time investigating the scent marks of donor A (Fig. 9.1a–c). Additionally, 2–3-month-old, 5–7-month-old, and 10–12-month-old male and female voles spent significantly more time investigating the mark of donor A on the first exposure than they did on the fourth exposure (Holm-Sidák multiple comparisons, $p<0.05$; Fig. 9.1a–c). In contrast, 15–18-month-old male and female voles spent similar amounts of time investigating the scent marks of donor A across the four exposures ($F_{3, 30}=0.82$, $p=0.49$; Fig. 9.1d).

During the test phase, 2–3-month-old ($F_{1, 20}=13.88$, $p<0.001$), 5–7-month-old ($F_{1,20}=22.41$, $p<0.001$), and 10–12-month-old male and female voles ($F_{1,20}=20.43$, $p<0.001$) spent more time investigating the scent mark of donor B, the novel individual, than that of donor A, the familiar individual (Fig. 9.1a–c). In contrast, 15–18-month-old male and female voles spent similar amounts of time investigating the scent mark of donor A and that of donor B ($F_{3, 30}=1.68$, $p=0.19$; Fig. 9.1d).

9.3.2 Experiment 2: Exposure to Scent Marks of Same-Sex Conspecifics

Across the four exposures during the habituation phase, 2–3-month-old voles ($F_{3, 30}=33.54$, $p<0.001$), 5–7-month-old voles ($F_{3, 30}=39.16$, $p<0.001$), and 10–12-month-old voles spent different amounts of time investigating the scent mark of donor A ($F_{3, 30}=58.43$, $p<0.001$; Fig. 9.2a–c). Additionally, 2–3-month-old, 5–7-month-old, and 10–12-month-old voles spent more time investigating the scent marks of donor A on the first exposure than they did on the fourth exposure (Holm-Sidák multiple comparisons, $p<0.05$; Fig. 9.2a–c). In contrast, 15–18-month-old male and female voles spent similar amounts of time investigating the scent marks of donor A across the four exposures ($F_{3, 30}=1.22$, $p=0.31$; Fig. 9.2d).

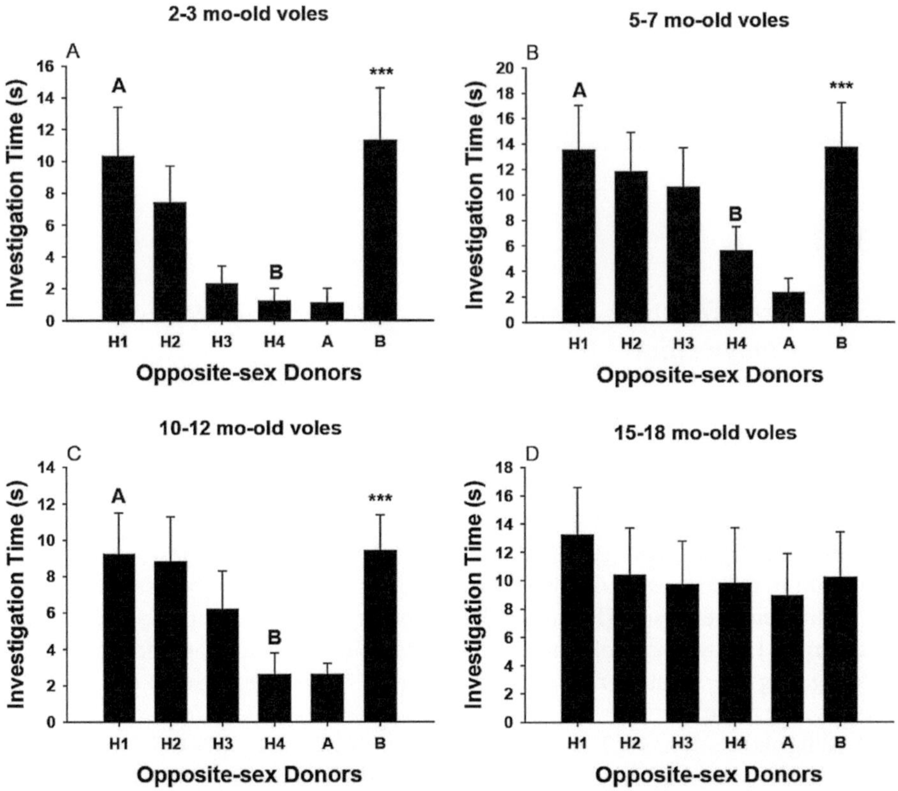

Fig. 9.1 (**a**–**d**) The amount of time (s) ± SEM spent investigating the scent mark of an opposite-sex conspecific(s) during the exposure phase and during the test phase of a habituation task by (**a**) 2–3-month-old, (**b**) 5–7-month-old, (**c**) 10–12-month-old, and (**d**) 15–18-month-old voles. Histograms capped with a different *letter* indicate significant differences ($p<0.001$) during the first and fourth exposure. *Asterisks* (***) denote significant differences ($p<0.001$) in the amount of time voles spent investigating the scent marks of the two donors during the test phase

During the test phase, 2–3-month-old voles ($F_{1,20}=13.88, p<0.001$), 5–7-month-old voles ($F_{1,\ 20}=22.41$, $p<0.001$), and 10–12-month-old voles ($F_{1,\ 30}=17.04$, $p<0.001$) spent more time investigating the scent mark of donor B (novel) than that of donor A (familiar) (Fig. 9.2a–c). However, 15–18-month-old voles spent similar amounts of time investigating the scent marks of donor A and donor B ($F_{1,20}=0.96$, $p=0.33$; Fig. 9.2d).

9.3.3 *Experiment 3: Discrimination of the Sexes*

Fifteen- to 18-month-old male ($t_{15}=2.81$, $p=0.01$) and 15–18-month-old female voles ($t_{15}=3.08$, $p=0.007$) spent significantly more time investigating the scent mark of an opposite-sex conspecific than that of a same-sex conspecific. Fifteen- to

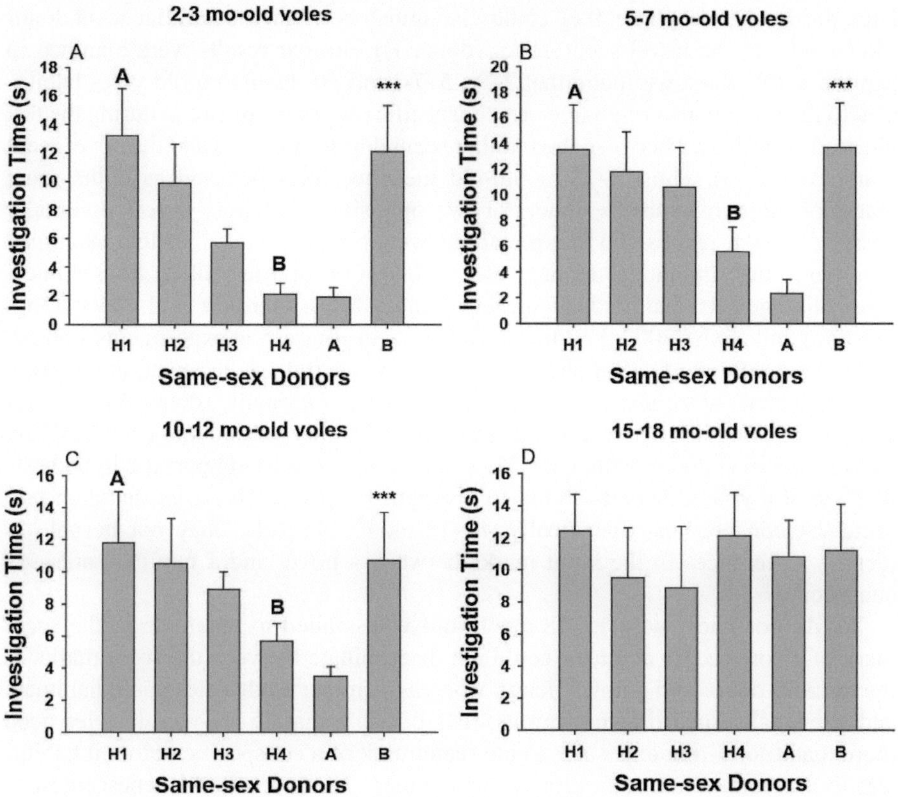

Fig. 9.2 (**a–d**) The amount of time (s) ± SEM spent investigating the scent mark of a same-sex conspecific(s) during the exposure phase and during the test phase of a habituation task by (**a**) 2–3-month-old, (**b**) 5–7-month-old, (**c**) 10–12-month-old, and (**d**) 15–18-month-old voles. Histograms capped with a different *letter* indicate significant differences ($p<0.001$) during the first and fourth exposure. *Asterisks* (***) denote significant differences ($p<0.001$) in the amount of time voles spent investigating the scent marks of the two donors during the test phase

18-month-old male voles spent 8.81 ± 0.55 s (mean ± SEM) and 5.58 ± 0.53 s investigating the scent marks of a female and that of a male conspecific, respectively. Fifteen- to 18-month-old female voles spent 7.72 ± 0.55 s and 4.39 ± 0.47 s investigating the scent marks of a male and that of a female conspecific, respectively.

9.4 Discussion

The first aim of the present study was to test the hypothesis that aging affects whether meadow voles can identify differences between the scent marks of two same-sex conspecifics. In experiment 1, we found that 2–3-, 5–7-, and 10–12-month-old voles habituated to the scent marks of an opposite-sex conspecific (donor A) and later, during the test phase, they could distinguish between the familiar scent donor (donor A) and the novel scent donor (donor B). Similar results were obtained in experiment 2, when we found that 2–3-, 5–7-, and 10–12-month-old voles habituated to the scent marks of a same-sex conspecific (donor A) and later, during the test phase, discriminated between the familiar scent donor (donor A) and the novel scent donor (donor B). Thus, 2–12-month-old meadow voles behaved as if the scent marks of the two same-sex donors or two opposite-sex donors were individually distinct. Similar results and interpretations were obtained in studies that examined individual discrimination among 3–10-month-old prairie voles, house mice, Mongolian gerbils, and golden hamsters (Halpin 1980; Johnston et al. 1993; Tang-Martinez and Bixler 2009; Ferkin and Li 2005). In contrast, in experiments 1 and 2, we discovered that 15–18-month-old meadow voles did not habituate to the scent mark of a same-sex conspecific or an opposite-sex conspecific (donor A) and later could not distinguish between scent mark of the familiar scent donor (donor A) and that of the novel donor (donor B). Thus, the present results support the hypothesis that age of a vole affects its ability to distinguish between the scent marks of two same-sex conspecifics. Specifically, 15–18-month-old voles may not be able to identify differences in the scent marks between a novel and a familiar same-sex conspecific.

We do not know why 15–18-month-old voles failed to habituate to the scent mark of a conspecific and later could not discriminate between the scent marks of the familiar donor and a novel donor, whereas younger adult voles could habituate and later make such discriminations. It is possible that 15–18-month voles need more than four 5-min exposures to the scent mark of a conspecific before it habituates to it. Shukitt-Hale et al. (2001) showed that 22–24-month-old, senescent male rats could habituate over an extended period to objects they encountered repeatedly in their arenas but were less likely to respond to novel objects placed in their arena. Somewhat similar results were found in a study showing that younger but not older adult house mice could habituate to familiar features and objects in their physical environment (Brennan et al. 1984). It is possible that in our study, 15–18-month-old voles like aged male rats have impairments in perception, processing, or memory, which prevent them from habituating to a familiar scent mark or object in their environment (Brennan et al. 1984; Matzel et al. 2011; Doty and Kamath 2014). It is also possible that 15–18-month-old voles cannot form individual representations of the scent marks of conspecifics, which can be formed by 3–13-month-old voles (i.e., Johnston and Bullock 2001).

Failure to discriminate between same-sex conspecifics may have profound effects on the behavior, survival, and fitness of 15–18-month-old voles. For example, 15–18-month-old male female and male voles may no longer be able to discriminate between residents and intruders, familiar and unfamiliar conspecifics, and kin and non-kin or may not be motivated to do so (Ferkin 2011, 2015). Thus, 15–18-month-old voles may be less likely to identify or respond to changes in their social environment, particularly with respect to their failure to identify differences among the scent marks of same-sex conspecifics. Such impairment in olfactory discrimination may also increase potentially costly interactions with same-sex competitors (Ferkin 2011). Aging studies on other rodents have reported that aged individuals may become more prone to risky or costly situations (Markowska et al. 1989; Matzel et al. 2011; Shao and Lee 2014).

The second aim of the study was to determine whether 15–18-month-old voles could distinguish between the scent mark of a male donor and that of a female donor. In experiment 3, 15–18-month-old male and female voles could discriminate between scent marks of a male conspecific and a female conspecific, spending more time investigating the scent mark of the opposite-sex donor to that of the same-sex donor. This finding is interesting in that previous work has shown that aging had a greater effect on the attractiveness of scent marks produced by male voles and their responses to the scent marks of a sexually receptive female conspecific than it did on female voles. Aging did not appear to affect the attractiveness of the scent marks of females to males or in the responses of females to the scent marks of male voles. Specifically, 10–13-month-old male voles produced scent marks that were more attractive to females compared to those produced by 2–9-month-old voles (Ferkin 1999). In addition, 10–12-month-old male but not female voles spent more time self-grooming when they encountered the odors of a sexually receptive, opposite-sex conspecific relative to the amount of time that 2–9-month-old voles spent self-grooming (Ferkin and Leonard 2008). In contrast, 2–3-month-old voles deposited more scent marks than older male voles in areas scent marked by a sexually receptive female conspecific. However, 5–7-month-old and 10–12-month-old male voles over-marked a higher proportion of the marks of females than did 2–3-month-old male voles (Ferkin 2010). From these observations, we can adduce that each type of odor-related behavior may have its own vulnerability to the effects of aging and that these effects differ between male voles and female voles.

Discriminating between familiar and unfamiliar conspecifics would provide benefits related to an individual's survival and fitness (Manning 1985; Kokko 1998; Beck and Powell 2000). Failure to do so may reduce the survival of 15–18-month-old voles if they encounter the scent marks of two same-sex conspecifics that are their competitors but increase the fitness of 15–18-month of voles if they encounter the scent marks of two opposite-sex conspecifics. Older, adult male voles also produce scent marks that are more attractive to those of younger males to females, and they behave in a manner that increases their ability to locate females than do younger male voles (Ferkin 1999, 2010, 2011; Ferkin and Leonard 2008). Indeed, older male voles may have more opportunities to mate than do younger male voles (Ferkin 1999, 2010; Ferkin and Leonard 2008) because they have lower residual reproductive value (i.e., Fischer et al. 2008). Similarly, older female voles may be more experienced mothers and they may provide more maternal care compared to younger female voles (McGuire and Novak 1984; Sabau and Ferkin 2013). As a whole this suggest that for 15–18-month-old voles, the ability to identify a mate may outweigh their inability to discriminate between two same-sex conspecifics for 15–18-month-old voles.

Our present findings add to a literature suggesting that the odor-related behaviors of meadow voles do not fit neatly into the predictions for short-lived animals (Ferkin 1999, 2010, 2011, 2015; Ferkin and Leonard 2008). It seems that each of the odor-related behaviors that have been studied, which includes self-grooming, over-marking, scent marking, and individual discrimination, has its own vulnerability to the effects of aging (Boonstra 1994). A more parsimonious model should incorporate the tactics that older and younger adult meadow voles use to communicate and respond to conspecifics (Ferkin 2011, 2015). Such models can be developed from existing models life history tradeoffs (i.e., Partridge and Endler 1987; Manning 1985; Kokko 1998) and modified to include the costs and benefits associated with discriminating among conspecifics for that species. Such life history characteristics that may influence age-related effects on individual discrimination should include the mating system and social system of the species in question, demography of the population, and the spatial and temporal distribution of conspecifics in that population (Tamarin et al. 1984; Sheridan and Tamarin 1988).

During most of the breeding season, meadow voles do not share nests with same-sex conspecifics. Female meadow voles are territorial, whereas males wander through large home ranges that encompass the territory of one of more females (Madison 1980). Thus, voles may have more interactions with opposite-sex rather than same-sex conspecifics. It is possible that 15–18-month-old voles have even fewer interactions with conspecifics, especially if they have reduced aerobic capacity, which could reduce their locomotor activity, running speed, and their ability to escape or avoid risky interactions (Ricklefs et al. 1996; Rezende et al. 2006; Boraty and Koteja 2009). In this case, 15–18-month-old voles may no longer need to distinguish between scent marks of same-sex competitors because they do not interact with them regularly. Thus, it is not a stretch to hypothesize that in species in which conspecifics have frequent and repeated social interactions with nearby conspecifics, such as prairie voles (McGuire et al. 1990), older individuals may be more likely to continue to encounter the scent marks of the same conspecifics. This would allow them to continue to discriminate between their scent marks and those of unfamiliar conspecific (Ferkin 2011, 2015).

Acknowledgments We thank Malle Carrasco, Dr. Bruce Schulte, and two anonymous reviewers for reading this manuscript and providing helpful suggestions. The research was supported by NIH grants HD 049525 and an ARRA supplement and NSF grant IOB 0444553 to MHF and funds from the Vice Provost of Research and the Department of Biological Sciences at the University of Memphis. This manuscript is dedicated to Dr. Robert E. Johnston, a pioneer and leader in the field of chemical communication in mammals.

References

Beck CW, Powell LA (2000) Evolution of female mate choice based on male age: are older males better mates? Evol Ecol Res 2:107–118

Boonstra R (1994) Population cycles in microtines: the senescence hypothesis. Evol Ecol 8:196–219

Boraty Z, Koteja P (2009) The association between body mass, metabolic rates and survival of bank voles. Funct Ecol 23:330–339

Brennan MJ, Allen D, Aleman D, Azmitia EC, Quartermain D (1984) Age differences in within-session habituation of exploratory behavior: effects of stimulus complexity. Behav Neural Biol 42:61–72

Brown RE, Macdonald DW (eds) (1985) Social odours in mammals. Oxford University Press, Oxford, UK

Doty RL, Kamath V (2014) The influences of age on olfaction: a review. Front Psychol 5:20. doi:10.3389/fpsyg.2014.00020

Ferkin MH (1999) Attractiveness of opposite-sex odor and responses to it vary with age and sex in meadow voles (*Microtus pennsylvanicus*). J Chem Ecol 4:757–769

Ferkin MH (2010) Age affects over-marking of opposite-sex scent marks in meadow voles. Ethology 116:24–31

Ferkin MH (2011) Odor-related behavior and cognition in meadow voles, *Microtus pennsylvanicus* (Arvicolidae, Rodentia). Folia Zool 60:262–276

Ferkin MH (2015) The response of rodents to scent marks: four broad hypotheses. Horm Behav 68:43–52

Ferkin MH, Johnston RE (1995) Meadow voles, *Microtus pennsylvanicus*, use multiple sources of scent for sex recognition. Anim Behav 49:37–44

Ferkin MH, Leonard ST (2008) Age of the subject and scent donor affects the amount of time that voles self-groom when they are exposed to odors of opposite-sex conspecifics. In: Hurst JL, Beynon RJ, Roberts SC, Wyatt TD (eds) Chemical signals in vertebrates 11. Springer, New York, pp 281–289

Ferkin MH, Li HZ (2005) A battery of olfactory-based screens for phenotyping the social and sexual behaviors of mice. Physiol Behav 85:489–499

Ferkin MH, Sorokin ES, Johnston RE (1996) Self-grooming as a sexually dimorphic communicative behaviour in meadow voles, *Microtus pennsylvanicus*. Anim Behav 51:801–810

Ferkin MH, Dunsavage J, Johnston RE (1999) Meadow voles can discriminate between the top and bottom scent of an over-mark and prefer the top scent. J Comp Psychol 113:43–51

Fischer K, Perlick J, Galetz T (2008) Residual reproductive value and male mating success: older males do better. Proc R Soc B 275:1517–1524

Gallagher M, Rapp PR (1997) The use of animal models to study the effects of aging on cognition. Annu Rev Psychol 48:339–370

Halpin ZT (1974) Individual differences in the biological odors of the Mongolian gerbil (*Meriones unguiculatus*). Behav Biol 11:253–259

Halpin ZT (1980) Individual odors and individual recognition: review and commentary. Biol Behav 5:233–248

Halpin ZT (1986) Individual odors among mammals: origins and functions. Adv Study Behav 16:39–70

Heth G, Todrank J, Johnston RE (1999) Similarity in the qualities of individual odors among kin and species in Turkish (*Mesocricetus brandti*) and golden (*Mesocricetus auratus*) hamsters. J Comp Psychol 113:321–326

Hurst JL, Beynon RJ (2004) Scent wars: the chemobiology of competitive signaling in mice. Bioassays 26:1288–1298

Johnston RE, Chiang G, Tung C (1994) The information in scent over-marks of golden hamsters. Anim Behav 48:323–330

Johnston RE, Munver R, Tung C (1995) Scent counter marks: selective memory for the top scent by golden hamsters. Anim Behav 49:1435–1442

Johnston RE (1993) Memory for individual scent in hamsters (*Mesocricetus auratus*) as assessed by habituation methods. J Comp Psychol 107:201–207

Johnston RE (2003) Chemical communication in rodents: from pheromones to individual recognition. J Mammal 84:1141–1162

Johnston RE (2008) Individual odors and social communication: individual recognition, kin recognition, and scent over-marking. Adv Study Behav 38:439–505

Johnston RE, Bullock T (2001) Individual recognition and the nature of individual representations in golden hamsters. Anim Behav 61:545–557

Johnston RE, Derzie A, Chiang G, Jernigan P, Lee HC (1993) Individual scent signatures in golden hamsters: evidence for specialization of function. Anim Behav 45:1061–1070
Kaur AW, Ackels T, Kuo T-H, Cichy A, Dey S, Hays C, Kateri M, Logan DW, Marton TF, Spehr M, Stowers L (2014) Murine pheromone proteins constitute a context-dependent combinatorial code governing multiple social behaviors. Cell 157:676–688
Keevin TM, Halpin ZT, McCurdy N (1981) Individual and sex-specific odors in male and female eastern chipmunks (*Tamias striatus*). Behav Biol 6:329–338
Keller BL (1985) Reproductive patterns. In: Tamarin RH (ed) Biology of new world *Microtus*, vol 8. Amer Soc Mammal Sp Publ, Lawrence, KS, pp 725–778
Kokko H (1998) Good genes, old age and life-history trade-offs. Evol Ecol 12:739–750
Lai S-C, Johnston RE (1994) The repertoire of individual odors in Djungarian hamsters, *Phodopus campbelli*. Ethology 96:117–126
Madison DM (1980) An integrated view of the social biology of *Microtus pennsylvanicus*. Biologist 62:20–33
Manning JT (1985) Choosy females and correlates of male age. J Theor Biol 116:349–354
Markowska AL, Stone WS, Ingram DK, Reynolds J, Gold PE, Conti LS, Pontecorvo MJ, Wenk GL, Olton DS (1989) Individual differences in aging: behavioral and neurobiological correlates. Neurobiol Aging 10:31–43
Matzel LD, Light KR, Wass C, Colas-Zelin D, Denman-Brice A, Waddel AC, Kolata S (2011) Longitudinal attentional engagement rescues mice from age-related cognitive declines and cognitive inflexibility. Learn Mem 18:345–356
Mayeaux DJ, Johnston RE (2002) Discrimination of individual odors by hamsters (*Mesocricetus auratus*) varies with the location of those odors. Anim Behav 64:269–281
McGuire BM, Novak MA (1984) A comparison of maternal behavior in the meadow vole (*Microtus pennsylvanicus*), prairie vole (*M. ochrogaster*), and pine vole (*M. pinetorum*). Anim Behav 32:1132–1141
McGuire B, Pizzuto T, Getz LL (1990) Potential for social interaction in a natural population of prairie voles (*Microtus ochrogaster*). Can J Zool 68:391–398
Mertl AS (1975) Discrimination of individuals by scent in a primate. Behav Biol 14:505–509
Milligan SR (1982) Induced ovulation in mammals. Oxf Rev Reprod Biol 4:1–46
Nadeau JH (1985) Ontogeny. In: Tamarin RH (ed) Biology of new world *Microtus*, vol 8. Amer Soc Mammal Sp Publ, Lawrence, KS, pp 254–285
Negus NC, Berger PJ (1988) Cohort analysis: environmental cues and diapause in microtine rodents. In: Boyce MS (ed) Evolution of life histories of mammals, theory and pattern. Yale University Press, New Haven, CT, pp 65–74
Newman KS, Halpin ZT (1988) Individual odours and mate recognition in the prairie vole, *Microtus ochrogaster*. Anim Behav 36:1779–1787
Partridge L, Endler JA (1987) Life history constraints on the effects of sexual selection. In: Bradbury JW, Andersson MB (eds) Sexual selection: testing the alternatives. Wiley, New York, pp 265–277
Promislow DEL (1991) Senescence in natural populations of mammals: a comparative study. Evolution 45:1869–1877
Rezende EL, Gomes FR, Malisch JL, Chappell MA, Garland T Jr (2006) Maximal oxygen consumption in relation to subordinate traits in lines of house mice selectively bred for high voluntary wheel running. J Appl Physiol 101:477–485
Ricklefs RE, Konarzewski M, Daan S (1996) The relationship between basal metabolic rate and daily energy expenditure in birds and mammals. Am Nat 147:1047–1071
Roberts SC (2007) Scent marking. In: Wolff JO, Sherman PW (eds) Evolution of life histories of mammals, theory and pattern. Chicago University Press, Chicago, IL, pp 255–267
Rose MR (1991) Evolutionary biology of aging. Oxford University Press, New York
Sabau RM, Ferkin MH (2013) Food restriction affects the mass of dams and pups and the amount of maternal behavior provided by female meadow voles, *Microtus pennsylvanicus*. J Mammal 94:1068–1076
Shao R, Lee TMC (2014) Aging and risk taking: toward an integration of cognitive, emotional, and neurobiological perspectives. Neurosci Neuroecon 3:47–62

Sheridan M, Tamarin RH (1988) Space use, longevity, and reproductive success in meadow voles. Behav Ecol Sociobiol 22:85–90

Shukitt-Hale B, Casadesus G, Cantu-Castelvetri I, Joseph JA (2001) Effect of age on object exploration, habituation, and response to spatial and nonspatial change. Behav Neurosci 115:1059–1064

Slade NA (1995) Failure to detect senescence in persistence of some grassland rodents. Ecology 76:863–870

Smith T (2006) Individual olfactory signatures in common marmosets (*Callithrix jacchus*). Am J Primatol 68:585–604

Tamarin RH, Reich LM, Moyer CA (1984) Meadow vole cycles within fences. Can J Zool 62:1796–1804

Tang-Martinez Z, Bixler A (2009) Individual discrimination by odors in sibling prairie voles (*Microtus ochrogaster*). J Chem Ecol 35:400–404

Thom MD, Hurst JL (2004) Individual recognition by scent. Ann Zool Fennici 41:765–787

Thompson RF, Spencer WA (1966) Habituation: a model phenomenon for the study of neuronal substrates of behavior. Psychol Rev 73:16–43

Todrank J, Heth G, Johnston RE (1998) Kin recognition in golden hamsters: evidence for kinship odours. Anim Behav 55:377–386

Wilson DA, Stevenson RJ (2006) Learning to smell: olfactory perception from neurobiology to behavior. The Johns Hopkins University Press, Baltimore, MD

Wyatt TD (2014) Pheromones and animal behavior. Chemical signals and signatures, 2nd edn. Cambridge University Press, Cambridge, UK

Chapter 10
Putative Pheromone of the Indian Crestless Porcupine, *Hystrix brachyura*

Biswatosh Ghosh, Mousumi Poddar-Sarkar, Sibdas Ray, and Ratan Lal Brahmachary

10.1 Introduction

Two of the three Indian porcupine species, *Hystrix indica* (Kerr 1792) and the smaller *Hystrix brachyura* (Linn. 1758) have become endangered in North and South Bengal. Even in the jungle, they usually emerge from their holes only after dusk (Sankhala 1993). In the last few decades the number of both *H. indica* and *H. brachyura* has been drastically reduced even in the rural areas, where they inhabit the rocky regions and make burrows under stones and rocks (Raha 2012).

Most of the behavioral studies of porcupines have focused on new world porcupines (Curtis and Kozicky 1944; Morris and Aarde 1985; Roze 1989, 2009; Ilse and Hellgren 2001). Although much is known about the reproductive behavior and odor communication of American porcupine, *Erethizion dorsatum* (Shadle 1946; Edwin and Shadle 1953; Brown and Macdonald 1985), we know little about odor communication, the sources of chemical cues, and their chemical composition about *H. indica* and *H. brachyura*. These endangered porcupines generally use urine and sometimes feces to mark their territory, for indicating their presence to potential mates, and possibly to defend against predators and conspecifics (Poddar-Sarkar et al. 2011; Raha et al. 2015). Thus, the goal of this present paper is to provide

B. Ghosh
Department of Zoology, Bidhannagar College, Salt Lake, Kolkata 700 064, India

M. Poddar-Sarkar (✉)
Department of Botany, University of Calcutta, Kolkata 700 019, India
e-mail: mousumipsarkar1@gmail.com

S. Ray
Department of Chemistry, University of Calcutta, Kolkata 700 009, India

R.L. Brahmachary
Biology Division, Indian Statistical Institute, Kolkata 700 108, India

B.A. Schulte et al. (eds.), *Chemical Signals in Vertebrates 13*,
DOI 10.1007/978-3-319-22026-0_10

preliminary observations of odor communication in *H. brachyura* by examining scent marking, the sources of chemical cues, and continuing the study of the chemical composition of these scents.

10.2 Materials and Methods

10.2.1 Observations of Behavior

We observed a total of five individual animals (three males and two females) and the behavior was video-taped between 2011 and 2014 across different seasons of the year with different sex combinations. Each observation was recorded for 2–3 h. All observations occurred during the day although porcupines are nocturnal, at the Gorchumuk Rehabilitation Center (GRC: 22°20.886′N and 88°04.379′E), near Kolkata under Zoo Authority of West Bengal where the animals are familiar with human beings (Fig 10.1). We were allowed to observe the porcupines shortly after dawn and shortly before dusk.

10.2.2 Chemical Analysis of Urine Samples

Urine deposited by the porcupines on the soil was aspirated into separate air-tight glass vials containing Hexane (Hx), Dichloromethane (DCM) and Acetonitrile (ACN). The urine samples were then brought into the laboratory under ice and stored at −20 °C for subsequent analyses.

Hx and DCM extract of the urine samples of both the sexes were subjected to two different Gas chromatography-mass spectrometry (GCMS) instruments attached with two different columns for identification of polar and non-polar low boiling volatiles with double verification. ACN extracts were utilized for analysis of comparatively less volatile, high boiling compounds with the aid of High Performance Liquid Chromatography (HPLC; Agilent, 1260 infinity series, USA). Solvents used for extraction were purchased from E. Merck, India (HPLC grade). The column temperature for Elite 5-MS attached in Perkin Elmer Clarus 600C GCMS was programmed at 50 °C for 2 min hold, then 80–280 °C with ramping at 5 °C/min and the column temperature for HP5-MS attached with Agilent GC-MS Agilent Technologies, GC-7890A, MS-5975C, USA was maintained at 70 °C for 1 min, then 70–190 °C with ramping at 5 °C/min, then 10 min hold, then again 190–260 °C with ramping at 4 °C/min with a final 5 min hold. In both the cases, flow of carrier gas was 1 ml/min. Compounds were identified by comparing the Relative Retention Time (RRT) and by considering the MS fragmentation database available by NIST and scrutinizing the MS fragmentation pattern.

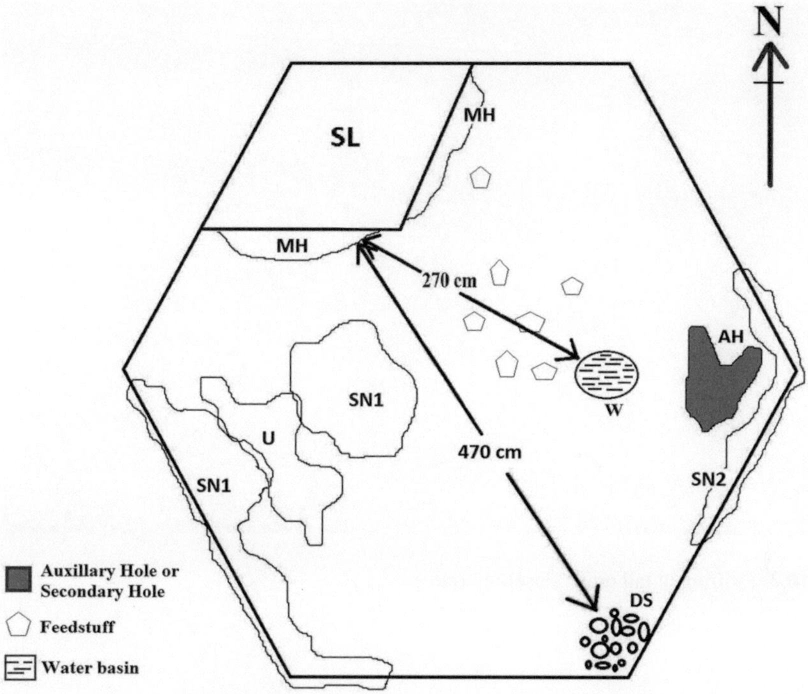

Fig. 10.1 Location of Dung heap Site (DS), site of Urination (U) and the area of interest for Sniffing (Sn1 and Sn2) with respect to the Main Hole (MH), Auxiliary Hole (AH) and Water basin (W) in a hexagonal enclosure. SL-Slab of stone above the main hole. (Figure not to scale)

For HPLC, we used ACN extracts of both male and female urine sample. HPLC system was attached with Zorbax SB-C18 column (4.6×150 mm, 3.5 μm) and equipped with Diode Array Detector. Gradient of two mobile phases: methanol (A) and water with 0.02 % aqueous H_3PO_4 (B) were set at: 25 % A+75 % B for 5 min→30 % A+70 % B for 10 min→45 % A+55 % B for 30 min→and 80 % A+20 % B for 45 min. The injection volume was 20 μl. The flow rate of the solvents was kept at 0.4 ml/min. Scanning of analytes at 275 nm provided a comprehensive output. In addition to the above procedure, peaks were also compared with the absorption spectra of the standards run under same condition (Sigma, USA; ChromaDex, USA).

Chloroform phase separated out by Bligh and Dyer's method (1959) was subjected to Thin Layer Chromatography (TLC) for fractionation of different putative fixative lipid classes present in male and female urine (Poddar-Sarkar 1996). The solvent we used for run was hexane: diethyl ether: acetic acid (90:10:1.5 v/v) (Stahl 1965). Spots were identified after spraying with 0.2 % 2,7-dichloro-fluorescein in ethanol (Christie 2003) after comparison with standards (Sigma, USA).

Fig. 10.2 Sniffing of tail quills (inset-tail quill)

10.3 Results and Discussion

10.3.1 Observations of Behavior

We discovered that *H. brachyura* deposited feces and urine at specific non-random locations of their enclosure. We observed males to sniff and lick the urine of the females and vice versa, suggesting that it may be a source of chemical cues. Porcupines also sniffed the quills of conspecifics (Fig. 10.2), a behavior observed by Raha (2012). It is interesting to note that just after urination male porcupines moistened both the quills of the lower hind part of their body and their tail quills with urine accumulated on the soil, after which the quills were shaken vigorously to scatter urine around the spot. Previous work has shown that porcupines use their hollow quills to carry water back to their burrow (Corbett 1953). Li et al. (1997) and Roze (2006) suggested that a specialized patch of skin on lower back, called rosette may be the possible source of (*R*)-δ-decalactone, the warning odor of New world porcupines. Our preliminary findings suggest that the specialized long-barbed quills may facilitate the dispersal of cresol, the putative pheromonal component of old world porcupine.

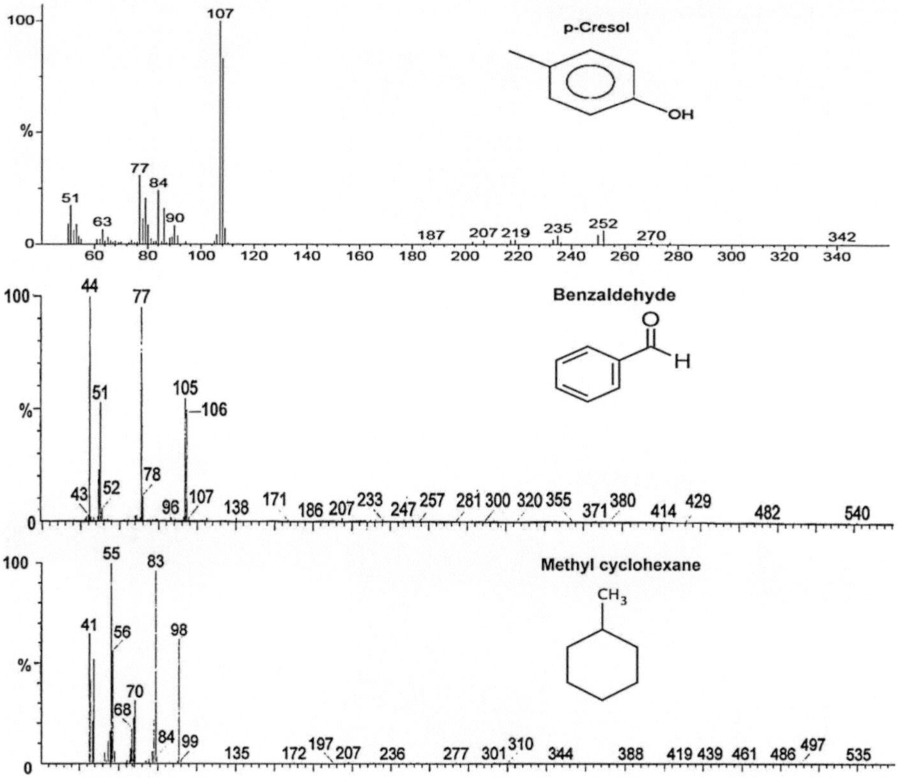

Fig. 10.3 Mass fragmentation pattern of cresol, benzaldehyde and methyl-cyclohexane present in urine samples

10.3.2 Chemical Composition

GCMS analysis of Hx extract of a female urine sample revealed five major peaks. Phenylacetylene (MW 102), Cresol (MW 108) and Phenol (MW 94) were convincingly identified. The molecular ion 102 and fragments 101, 76, 75, 50 suggest the presence of phenylacetylene. Phenol tautomerizes to the keto form resulting into cation radical 66, a ketone 28 and a very small peak 65. In the sample Hx of male urine the same procedure revealed cresol, benzaldehyde (MW 106) and methyl cyclohexane (MW 98) (Fig. 10.3). DCM extract revealed only one peak of cresol in GCMS. As isomeric compounds have specificity in pheromone signalling for many animals ranging from insects to mammals (Silverstein 1979; Mori 1996), we used HPLC to identify *p*- and *o*-cresol, separately (Fig. 10.4). We could not identify other compounds from the HPLC chromatogram. TLC analysis of the urine sample, however, revealed the presence of triglyceride (TAG), diglyceride (DAG). As a whole, our findings suggest a probable biogenetic pathway. This pathway begins

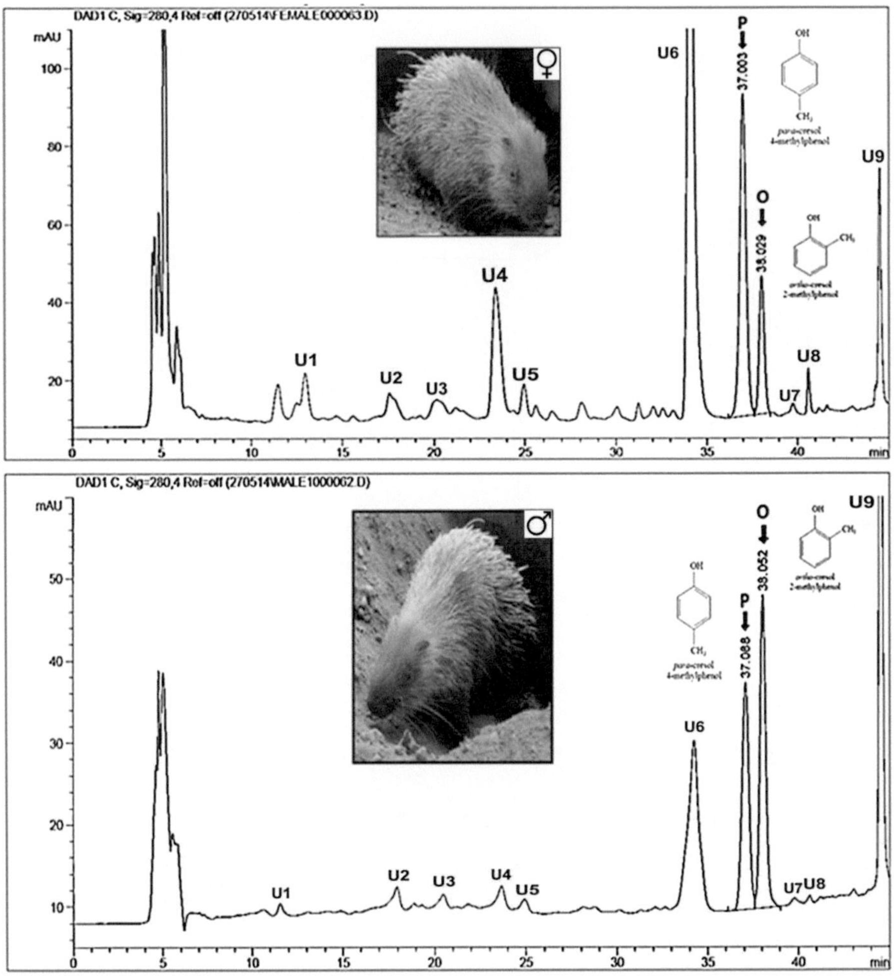

Fig. 10.4 HPLC of ACN extract of urine of female and male porcupine showing P=*p*-cresol and O=*o*-cresol. U1–U9 peaks were unidentified but present in different amount in chromatograms of both female and male

with cyclohexane (which, being very volatile, is unlikely to be detected), whose biochemical oxidation leads to phenol in one route, and subsequent methylation yields methyl cyclohexane in another route which further results (via toluene) into benzaldehyde and *o*- and *p*-cresol.

We found that the smell of urine of *H. brachyura* was similar to that of standard cresol. Būda et al. (2012) discussed the importance of *p*-cresol as a pheromone in the urine of mare but *p*-cresol occurs in both male and female porcupine. This substance has been found to be a sexual attractant in the urine of mares during estrus (Būda et al. 2012). *p*-cresol and *m*-cresol have been identified as major volatile components from the temporal gland secretion of African elephant *Loxodonta africana* (Adams et al. 1978; Wheeler et al. 1982). The primary component responsible for the aposematic behavior of new world porcupine was identified as (*R*)-δ-deca lactone (Li et al. 1997). Urine deposited on soil carries a smell to our nose even ~30 min later but pure cresol dropped on soil is not perceptible after ~5 min. Different classes of neutral lipid like TAG, DAG may be natural fixatives for volatile molecules found in the urine of porcupines, which is similar to that found in the urine of Bengal tigers (Brahmachary and Dutta 1981; Poddar-Sarkar 1996). These fixatives may increase the half-life of putative chemical cues (Poddar-Sarkar 1996).

Acknowledgements We acknowledge the financial help from DST-FIST programme, Government of India of Department of Botany, University of Calcutta for GCMS and HPLC facilities. We are thankful to our colleague Dr. Paramaa Raha for providing us with some important photographs. We are indebted to the PCCF, West Bengal and DFO, Howrah, Department of Forests, the Government of West Bengal, India for their kind cooperation. We are grateful to our field biologist friend Dr. Nilansu Das for his valuable comments and suggestions.

References

Adams J, Garcia III A, Foote CS (1978) Some chemical constituents of the secretion from the temporal gland of the African elephant (*Loxodonta africana*). J Chem Ecol 4:17–25

Bligh EG, Dyer WJ (1959) A rapid method of total lipid extraction and purification. Can J Biochem Physiol 37:911–917

Brahmachary RL, Dutta J (1981) On the pheromones of tigers—experiments and theory. Am Nat 118:561–567

Brown RE, Macdonald DW (1985) Social odours in mammals, vol 1. Clarendon Press, Oxford

Būda V, Mozūraitis R, Kutra J, Karlson AKB (2012) *P*-cresol: a sex pheromone component identified from the estrous urine of mares. J Chem Ecol 38:811–813

Christie WW (2003) Lipid analysis. In: Christie WW (ed) Isolation, separation, identification and structural analysis of lipids. The Oily Press, Bridgewater, England, pp 373–387

Corbett J (1953) Jungle lore. Oxford University Press, Oxford

Curtis JD, Kozicky EL (1944) Observations on the eastern porcupine. J Mammal 25:137–146

Edwin MA, Shadle AR (1953) Gross anatomy of the male reproductive system of the porcupine. J Mammal 34:210–220

Ilse L, Hellgren E (2001) Demographic and behavioural characteristics of North American porcupines (*Erethizon dorsatum*) in pinyon-juniper woodlands of Texas. Am Mid Nat 146:329–338

Li G, Roze U, Locke DC (1997) Warning odor of the porcupine (*Erethizon dorsatum*). J Chem Ecol 23:2737–2754

Mori K (1996) Molecular asymmetry and pheromone science. Biosci Biotechnol Biochem 60:1925–1932

Morris D, van Aarde RJ (1985) Sexual behaviour of the female porcupine, *Hystrix africaeaustralis*. Horm Behav 19:400–412

Poddar-Sarkar M (1996) The fixative lipid of tiger pheromone. J Lipid Mediat Cell Signal 15:89–101

Poddar-Sarkar M, Raha P, Chakroborty A, Bhar R, Brahmachary RL (2011) Ultrastructure and lipid chemistry of specialized epidermal structure of Indian porcupine and hedgehog. Acta Zool 92:134–140

Raha P (2012) Study on epidermal structure of some mammals with special reference to chemistry and ultrastructure. PhD Thesis, Jadavpur University, Kolkata
Raha P, Ghosh B, Bhar R, Das N, Poddar-Sarkar M (2015) Ultrastructure of trichome of old world porcupine of Indian subcontinent. Acta Zool 96:140–146
Roze U (1989) The North American porcupine. Smithsonian Institution Press, Washington, DC
Roze U (2006) Smart weapons: with an arsenal of quills and chemicals, the porcupine mounts one of nature's most robust defenses against predators. Natural History Magazine Featured Story 2006:49–53
Roze U (2009) The North American porcupine, 2nd edn. Cornell University Press, Ithaca
Sankhala K (1993) Return of the tiger. Lustre Press, New Delhi
Shadle AR (1946) Copulation in the porcupine. J Wildl Manage 10:159–160
Silverstein RM (1979) Enantiomeric composition and bioactivity of chiral semiochemicals in insects. In: Ritter FJ, Howse PE, Le Masne G (eds) Chemical ecology: odor communication in animals. Elsevier/North-Holland Biomedical Press, Amsterdam, pp 133–146
Stahl E (1965) Thin-layer chromatography—a laboratory handbook. Academic Press, New York, p 553
Wheeler JW, Rasmussen LE, Ayorinde F, Buss IO, Smuts GL (1982) Constituents of temporal gland secretion of the African elephant, *Loxodonta africana*. J Chem Ecol 8:821–835

Chapter 11
Chemical Cues, Hibernation and Reproduction in Female Short-Beaked Echidnas (*Tachyglossus aculeatus setosus*): Implications for Sexual Conflict

Rachel L. Harris, Elissa Z. Cameron, Noel W. Davies, and Stewart C. Nicol

11.1 Introduction

11.1.1 Seasonality of Reproductive Signals

Communication is vital for coordinating reproduction in many sexually-reproducing vertebrates (Andersson 1994; Bradbury and Vehrencamp 2011). In seasonal breeders, signals and cues related to reproduction (i.e. plumage, coloration, behavior, odor) tend to vary seasonally with reproductive state, thereby minimizing signaling costs (Gosling et al. 2000; Zala et al. 2004) and ensuring potential mates are not attracted at inappropriate times (Andersson 1994; Johansson and Jones 2007; Uhrig et al. 2012). Temporal changes in signaling activity and changes in composition may vary not only by season, but also show different patterns depending on the type of signal modality, since cost and information content may vary by signal type (Candolin 2003; Wyatt 2003). Here, we focus on olfactory (chemical) cues, the primary mode of communication in many mammals (Wyatt 2003).

Seasonality in mate-attracting signals and reproductive physiology might be particularly strong in animals which undergo periods of torpor or hibernation, as they have a reduced active period during which mate-locating, courtship and reproduction can occur. Along with depressing metabolic rate and body temperature (Geiser 2004, 2011), torpor and hibernation down-regulate ordinary behavior patterns

Electronic supplementary material: The online version of this chapter (doi:10.1007/978-3-319-22026-0_11) contains supplementary material, which is available to authorized users. Videos can also be accessed at http://link.springer.com/chapter/10.1007/978-3-319-22026-0_11.

R.L. Harris (✉) • E.Z. Cameron • S.C. Nicol
School of Biological Sciences, University of Tasmania, Hobart 7001, TAS, Australia
e-mail: rachel.harris@utas.edu.au

N.W. Davies
Central Science Laboratory, University of Tasmania, Hobart 7001, TAS, Australia

B.A. Schulte et al. (eds.), *Chemical Signals in Vertebrates 13*,
DOI 10.1007/978-3-319-22026-0_11

(Wimsatt 1969), endocrine system and reproductive organ function (Hudson and Wang 1979) and probably also inhibit signaling activity, since the size and activity of scent glands are under endocrine control (Ebling 1977). In most hibernating species, mating follows a period of euthermia which allows time for gonadal development, spermatogenesis and scent gland development (Dark 2005). However, winter torpor and mating activity are closely timed or even overlap in some vertebrates (Thomas et al. 1979; Boyles et al. 2006; Shine 2012), suggesting signal production may occur even when body temperature and metabolic rate are reduced (e.g. Parker and Mason 2009). However, garter snakes (LeMaster and Mason 2002) and some hibernating mammals (Exner et al. 2003; Blumstein et al. 2004) may overwinter in groups, so could use visual cues to locate potential mates once they have rewarmed. The effects of hibernation on mammalian chemical signals have not been directly investigated.

11.1.2 The Monotremes

Little is known about chemical signals in the egg-laying mammals (Order Monotremata), which comprise three genera including the platypus, short-beaked echidna and critically endangered long-beaked echidnas of New Guinea. Monotremes possess a unique mosaic of reptilian and mammalian reproductive characteristics (i.e. egg-laying, milk producing; Griffiths 1978) and diverged from the therian line approximately 161–271 million years ago (Phillips et al. 2009), hence they have been described as the ‘ultimate mammalian out-group’ (Nicol 2003). Reproduction in platypus and short-beaked echidnas is characterised by a distinct breeding season: after a short gestation period, females lay a single egg (usually two in platypus), and eggs hatch after ~10–11 days of incubation (Griffiths 1978; Morrow and Nicol 2012). After an initial stage where the neonate(s) remains with the mother and feeds on milk, both platypus and short-beaked echidnas leave their young in purpose-built ‘nursery burrows’ while the mother is away on foraging trips (Hawkins and Battaglia 2009; Morrow and Nicol 2012). In short-beaked echidnas, the egg is laid into a rudimentary pouch formed by hypertrophy of the mammary glands (Griffiths 1978; Morrow and Nicol 2012), and there is variation in the timing of nursery burrow use among different geographic subspecies (reviewed by Morrow and Nicol 2012; Nicol 2013).

11.1.3 Echidna Olfaction and Mating Behavior

The short-beaked echidna (*Tachyglossus aculeatus*; ‘echidna’) is the most common extant species of monotreme and is found throughout mainland Australia, Tasmania, several offshore islands and parts of New Guinea (Griffiths 1978; Augee 2008). Echidnas are sometimes described as ‘primitive’ (Nicol 2013), but their relatively large brain and complex olfactory structures suggest a large olfactory repertoire and significant brain capacity and memory (Ashwell 2013; Nicol 2013). In particular, anatomical and behavioral features suggest olfaction plays a key role in coordinating echidna reproduction and social behavior (reviewed by Harris 2014). Echidnas show a high degree of nocturnal activity (Nicol et al. 2004), use odor cues to locate food (Griffiths 1978), have large, well-developed olfactory sense organs (Home 1802; Broom 1895; Griffiths 1978; Schneider 2011) and a large olfactory bulb (Ashwell 2006; Nicol 2013). Both sexes have putative scent glands located in the cloaca and at the base of the spurs on the hind feet, although the spurs themselves are only retained in adult males (Harris et al. 2012). Scent-marking behavior (cloaca: Dobroruka 1960; Boisvert and Grisham 1988, spur: Morrow 2013) and seasonal variation in chemical composition (Harris et al. 2014) suggest these secretions are important for reproduction. In particular, echidna mating behavior appears strongly driven by male responses to female odor cues. Both sexes produce a strong odor during the mating season (Semon 1899; Nicol et al. 2004), and males are attracted to female scent (Rismiller 1992; Harris 2014). Rismiller (1992) suggested females may produce a 'pheromone' to attract males, although chemical analyses suggest changes in a suite of compounds are more important for signaling reproductive status to conspecifics (Harris et al. 2012, 2014). There is significant variation in mating behavior across geographic subspecies (Morrow et al. 2009) and therefore chemical cues may have different roles in coordinating reproduction across the echidna's distribution.

11.1.4 Tasmanian Echidna Mating Behavior and Odor

Tasmanian echidnas (*T. a. setosus*) are highly seasonal (Nicol and Andersen 2007; Nicol and Morrow 2012) and there is asynchrony in the timing of hibernation and reproduction between the sexes (Morrow and Nicol 2009; Nicol and Morrow 2012). Males and females hibernate for several months of the year, emerging prior to the winter breeding season (June–September). During hibernation, body temperature may fall to within 1 °C of substrate temperature (Nicol and Andersen 2002; Morrow et al. 2009). Both sexes are promiscuous (Morrow et al. 2009; Morrow and Nicol 2009) and females do not breed every year (Nicol and Morrow 2012), resulting in intense intra-male competition for access to reproductive females (Morrow et al. 2009). Males emerge from hibernation before females and are able to locate and mate with females that are still hibernating (Morrow and Nicol 2009). Hibernating females are usually completely concealed and use a variety of shelter sites (e.g. inside hollow logs, woodpiles, tussocks, or dug into soil; Nicol and Andersen 2002). Furthermore, echidnas at our study site do not use the same hibernaculum each year and often hibernate 0.5–1 m or more below the soil surface (unpubl. data), so female location cannot be predicted by males. In the absence of other cues, males seem to use odor to locate females (Harris et al. 2012), suggesting females produce an attractive odor while hibernating. Furthermore, non-reproductive females (sexually mature but not breeding in a given year) tend to hibernate through the mating season and emerge in mid-September (Nicol and Andersen 2002; Nicol and Morrow 2012), not usually disturbed by males (Harris 2014; Morrow 2013). Females are in better body condition (heavier) in reproductive than non-reproductive years (Nicol and Morrow 2012; Sprent et al. 2012) and only reproductive females seem to undergo estrous cycles (Morrow 2013). Although it might be expected that males could discriminate between females based on odor, reproductive females continue to attract males and mate during pregnancy (Morrow 2013; Harris 2014), suggesting males are unable to distinguish between pregnant and non-pregnant females.

To understand why reproductive female echidnas are attractive to males, even when hibernating or pregnant, we used a combination of organic chemical analyses, behavioral observations, urogenital smears (presence of spermatozoa confirmed mating) and external temperature loggers to investigate variations in female chemical cues during hibernation and the mating season in a free-living population. Female odor varies between hibernation and the mating season (Harris et al. 2014), but it is unclear whether odor composition and attractiveness changes during hibernation or after emergence, and what cues (e.g. temperature, photoperiod) might trigger such a change. No previous published studies have directly quantified changes in female chemical profiles during hibernation or pregnancy in a monotreme. We predicted that (1) female chemical profiles would change during hibernation; (2) reproductive and non-reproductive female profiles would differ, thereby allowing males to locate reproductive females; and (3) females would show changes in chemical profiles before, during and after fertilization.

11.2 Methods and Materials

11.2.1 Study Site and Animals

Fieldwork was conducted in the southern midlands of Tasmania (42° 28′ S, 142° 14′ E) between January 2010 and September 2012, encompassing three hibernating and mating seasons. Details of the site, fieldwork procedures and animal identification are described elsewhere (Nicol et al. 2011; Harris et al. 2012). A total of 35 female echidnas were sampled during the study period, including 18 sexually mature female echidnas which were fitted with RF tracking transmitters (RI-2C; Holohil, Canada) to allow serial sampling and monitoring. Ten females were monitored for consecutive years. Small external temperature loggers (i-Button, DS1992 L, Maxim Integrated Products Inc., Sunnyvale, California) were attached to transmitters and recorded temperature every hour (±0.5 °C). Loggers recorded events such as immergence and emergence from hibernation, periodic arousals, the entry of a male into a cold female hibernaculum, females entering nursery burrows and egg-laying (Nicol and Andersen 2002, 2006; Morrow and Nicol 2009). Radio-tracked animals were monitored weekly and sampled approximately monthly, and up to three times per week during the mating season (June–September), when reproductive females were found in mating aggregations (Morrow et al. 2009). Additional females ($n = 17$) were sampled opportunistically after being found walking around the property or in close proximity to radio-tracked males (e.g. in a mating aggregation); data from these females were included in our analyses where appropriate. Animals were captured by hand and samples (e.g. urogenital smears, odorant and blood samples) were collected while the animal was under light inhalation anesthesia. Body temperature (T_b) was measured by gently advancing a thermocouple probe approximately 3 cm into the cloaca. Animals were weighed (±0.01 kg) and returned to their place of capture after recovering from the anesthetic.

11.2.2 Odorant Sample Collection and Chemical Analyses

We followed previously described procedures for collecting and analyzing odorants (Harris et al. 2012). We concentrated our analyses on the volatile and nonvolatile components of female cloacal scent gland secretions: females scent-mark using their cloaca (Dobroruka 1960; Boisvert and Grisham 1988; Beard et al. 1992); free-ranging males show interest in female cloacal odor during the mating period (Harris 2014); and cloacal secretion composition varies between sexes and seasons (Harris et al. 2014). We collected cloacal swabs ($N = 209$ samples from 35 individuals) and cloacal wax secretions ($N = 113$, $n = 28$), all samples were stored in individual glass vials on ice while in the field (up to 8 h), then at −20 °C until analysis. Cloacal swabs were analyzed by thermal desorption followed by combined GC-MS using a Varian CP-3800 benchtop gas chromatograph (Varian, Palo Alto, CA, USA) and 1200L triple-quadrupole mass spectrometer (Varian, Palo Alto, CA, USA) or Bruker 300-MS TQ mass spectrometer (Bruker, Preston, Victoria, Australia). The Varian instrument was replaced by the Bruker instrument during the study, but all temperature programs and scanning conditions were held constant. Thermal desorption of cloacal swabs involves minimal interference with the sample (McLean et al. 2012) and was used primarily to describe changes in relatively 'volatile' compounds (b.p. <300 °C). Cloacal wax secretion samples were extracted in chloroform and derivatised to form TMS ethers, thereby improving peak separation of relatively 'non-volatile' scent components, then analyzed by GC-MS. All wax secretion analyses were performed using the same Varian GC and Bruker MS instruments described above. Relative abundance data were calculated for compounds occurring in >5 % of all samples analyzed (cloacal swabs $n = 68$ compounds; cloacal wax $n = 56$ compounds; Supplementary Table 11.1), using Varian instrument software (MS Data Review version 6.41). Components included short and long-chain fatty acids, sterols, methyl and ethyl esters of fatty acids, phenolics, terpenoids, aldehydes and sulphur and nitrogen-containing compounds (Supplementary Table 11.1; Harris et al. 2014).

11.2.3 Mating Activity and Estimating Timing of Pregnancy

Motion-activated cameras (Scoutguard SG550, HuntingCamOnline, Gadsden, SC, USA; Reconyx PC800, Holmen, WI, USA) were set up outside the entrances of female hibernacula prior to each mating season (Morrow and Nicol 2009). Females were selected for monitoring with cameras based on their relative body condition prior to entering hibernation and breeding activity in the previous year, if known. Cameras were used to record male behavior outside female hibernacula and their time of entry. Cameras were in continuous operation, collecting videos and still photos, and used an infra-red flash at night. Memory cards were downloaded every 3–7 days and batteries replaced fortnightly. Time stamps on recorded footage confirmed timing of events recorded by external temperature loggers. Non-reproductive females were not routinely monitored using cameras, but checked at least fortnightly for signs of male disturbance (e.g. hibernacula dug up, female moved location or in mating aggregation).

We used the presence and condition of spermatozoa collected from the female reproductive tract to confirm recent mating activity. Spermatozoa were recovered by collecting a urogenital smear, which was then stained and examined microscopically (Morrow and Nicol 2009), usually within 1–2 days of collection. Sperm collected immediately after mating have at least four curves or 'waves' along their length; after 4–5 days sperm have fewer curves and begin to fragment (Morrow and Nicol 2009). Date of egg-laying was estimated from temperature logger data: egg-laying coincides with a distinct trough or peak in otherwise stable temperatures recorded while female echidnas were in nursery burrows (Morrow and Nicol 2012). Gestation in Tasmanian echidnas is approximately 21 euthermic days (Nicol and Morrow 2012), so the date of fertilization, and therefore timing of pregnancy, could be estimated once date of egg-laying was known. This estimate was supported by the presence of fresh sperm in the female's reproductive tract within 2 days of the estimated date of fertilization.

11.2.4 Data Selection and Statistical Analyses

Females were classified as 'reproductive' or 'non-reproductive' in a given year, depending on their body condition and observed mating activity (Nicol and Morrow 2012). Twelve females (10 with radio-transmitters, 2 opportunistically) were monitored for multiple years, giving a total of 18 reproductive and 9 non-reproductive years of odorant sample data. Since females show significant variation in physiological condition between years, temperature and behavioral data collected from the same individuals in different years were treated as if they were different animals (i.e. statistically independent; Sprent et al. 2012). Within years, we categorized animal activity at the time of sampling according to individual behavior and physiology, e.g. in mating aggregation with males or alone, euthermic (active, $T_b \approx 32$ °C) or hibernating (slow movement, T_b well below normal 32 °C; Sprent et al. 2012). The maximum number of samples collected from a single individual was 26 cloacal swabs and 20 wax secretions. We included additional temperature and camera data collected prior to the study period (2008–2009; Morrow 2013) in our analyses where available.

Female chemical profiles were compared using mixed model permutational multivariate analysis of variance (PERMANOVA; Anderson 2001). Cloacal swab and wax secretion samples were analyzed as two separate datasets. Chemical relative abundances for each dataset were square-root transformed and converted into Bray-Curtis matrices prior to the distance-based PERMANOVA analyses. The overall detectable composition of cloacal swab samples varies significantly between mass spectrometers (but has minimal effect on main factors such as sex or seasonal differences, see Harris et al. 2014). We included 'Year' (fixed) and 'Animal ID' (random) in all tests, to allow for multiple samples from individuals, sometimes over multiple years. We have not directly tested the effect of mass spectrometer due to sample size restrictions, but this effect should be approximated by 'Year', since most samples collected in a given year were also analyzed by the same instrument. We excluded interactions involving Year or Animal ID as the data were too sparse to test them reliably (Anderson et al. 2008). PERMANOVA *P*-values were obtained using 9999 permutations of residuals under a reduced model with Type I (sequential) sums of squares (Anderson et al. 2008) using Primer version 6 (Clark and Warwick 2001) with the PERMANOVA add-on package (Anderson et al. 2008). Significance was set at $\alpha < 0.05$ for all statistical tests. Reported values are means with standard deviations unless specified.

First, we compared body condition, timing of hibernation, mating activity and chemical profiles in reproductive and non-reproductive females (i.e. *between* years). Specifically, we tested for differences in chemical profiles between reproductive and non-reproductive females prior to and during stages of hibernation, including during the mating season when males locate hibernating reproductive females. Body mass at the time of sampling was converted to a percentage of the long-term mean (mass%, calculated using at least 3 years of adult body mass data) as an indicator of body condition (Nicol and Morrow 2012; Sprent et al. 2012). Females may re-enter hibernation after fertilization (Morrow and Nicol 2009; Nicol and Morrow 2012) and samples collected from hibernating, pregnant females were excluded from analyses.

Next, we compared chemical profiles and mating activity with the timing of hibernation and pregnancy only in reproductive females (i.e. *within* years). Differences in chemical profiles between reproductive females which were hibernating or euthermic and in mating aggregations or alone during the mating season were compared using PERMANOVA. This analysis also includes data from opportunistically sampled females found in mating aggregations ($n = 15$). Differences in female chemical profiles during pregnancy were compared by categorizing odorant samples according to gestation stage, as inferred from external temperature logger data, mating activity and presence/absence of spermatozoa: 'fertilization' (±2 days); 'mid-gestation' (3–11 days post-fertilization); 'late gestation'

(12–20 days post-fertilization, or just prior to entry into a nursery burrow, whichever occurred first); and 'nursery' (female in a plugged nursery burrow, egg either in pouch or detected via ultrasound in utero; for methods see Morrow 2013). Some females also showed mating activity and subsequently re-entered hibernation up to several weeks prior to fertilization (range: 5–41 days, $n=5$); samples from these females were categorized as 'pre-fertilization.'

11.3 Results

11.3.1 Inter-annual Differences in Female Body Condition, Hibernation and Chemical Profiles

Reproductive females were in better body condition (mass%) than non-reproductive females prior to entering hibernation (January–March; reproductive females mean ± standard deviation: 116.7 ± 9.6 %, $n=17$; non-reproductive females: 104.6 ± 9.2 %, $n=8$; two-tailed t-test for unequal variances, $t_{1,14}=3.04$, $P<0.01$) and during the breeding season (June–September; reproductive females: 104.4 ± 7.2 %, $n=19$; non-reproductive females: 95.7 ± 9.1 %, $n=7$; two-tailed t-test for unequal variances, $t_{1,9}=2.27$, $P<0.05$). There was no difference in the mean date that females entered hibernation (range: 17 February–16 April, $N=30$; reproductive females: 19 March ± 11.1 days, $n=20$; non-reproductive females: 15 March ± 16.8 days, $n=10$; t-test for unequal variances, $t_{1,13}=0.59$, $P>0.5$; Fig. 11.1). There was no detectable difference in chemical profiles between reproductive and non-reproductive females prior to or during hibernation, including during the mating season when males entered hibernacula of reproductive females (Table 11.1).

11.3.2 Mating Activity and Emergence from Hibernation

All camera traps set up over reproductive females in the 2010–2012 breeding seasons recorded at least one male entering the female's hibernaculum ($n=6$). Cameras also recorded males outside a reproductive female's hibernaculum, but not entering, on several occasions in the weeks prior to a male actually entering her hibernaculum (echidna 2957, Fig. 11.1a). Digging activity was not observed outside the hibernacula of non-reproductive females, and males were never recorded entering hibernacula of non-reproductive females, except for one female where cameras recorded individual males outside on several occasions, although they did not enter the female's hibernation chamber (echidna 6865, 2010 and 2011; Fig. 11.1b).

Male entry into female hibernacula coincided with a distinct, abrupt increase in temperature recorded by the loggers attached to the females' transmitters (Fig. 11.1 inset d) clearly distinguishable from a normal periodic arousal (Fig. 11.1 inset c).

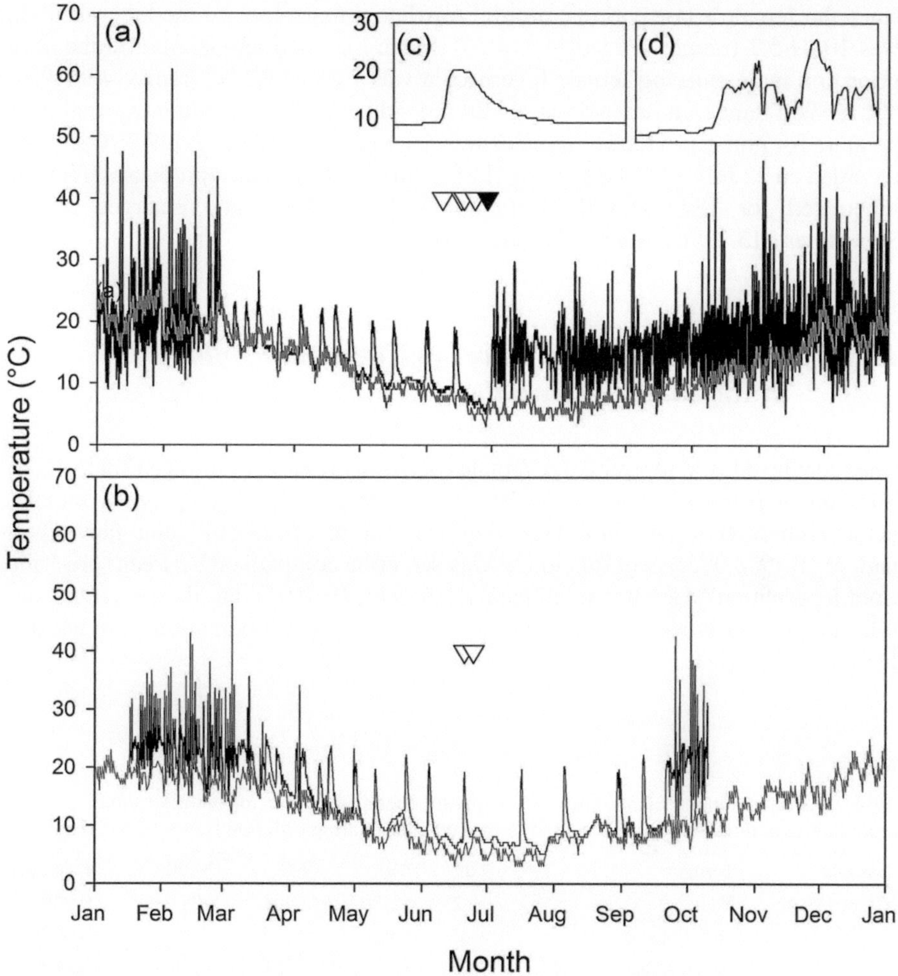

Fig. 11.1 Temperature records from external temperature loggers attached to (**a**) reproductive female (echidna 2957) in 2012 and (**b**) non-reproductive female (echidna 6865) in 2011. Symbols: *inverted open triangle* males recorded by cameras outside female hibernacula but did not enter; *inverted filled triangle* males entered the female's hibernaculum (reproductive female only). *Grey lines*: soil temperature (Bureau of Meteorology data). *Inset*: detail showing difference in temperature changes during a normal periodic arousal (**c**) and when male enters female hibernacula (**d**) over a 7-day period

These temperature patterns were used to calculate when a male entered female hibernacula for 12 additional reproductive females with complete temperature data and observed mating activity, but which were not closely monitored with cameras (2008–2012). Reproductive females hibernated for 108 ± 17.5 days (range: 70–154 days, n = 20) before being disturbed by a male entering their hibernaculum (mean date: 3 July ± 16.1 days (range: 4 June–31 July, n = 20). The mean number of days since the female's last periodic arousal to when a male entered the hibernaculum was 10.1 ± 5.2 (range: 1–25 days, $n = 20$). Mean recorded temperature in the 48 h prior to a male entering female hibernacula was 7.25 ± 1.22 °C (range: 4.96–9.38 °C; $N = 960$ temperature readings, $n = 20$ individuals). The minimum external temperature for non-reproductive females was 5.1 ± 1.79 °C (range: 2.1–9.2 °C, $n = 10$) recorded on 22 July ± 5.9 days (range: 11–31 July, $n = 10$). Non-reproductive females hibernated for 184 ± 24.0 days (range: 137–217 days) and emerged on 15 September ± 13.7 days (range: 31 August–10 October, $n = 10$).

11.3.3 Intra-annual Differences in Chemical Profiles in Reproductive Females

Chemical profiles in reproductive females varied by hibernation stage (Table 11.1). Subsequent pair-wise tests showed profiles changed significantly when females entered hibernation (i.e. 'pre-hibernation' was different from 'mid' and 'late' hibernation; all $P < 0.05$, except for cloacal wax secretion comparison between 'pre-' and 'mid-hibernation' were not significantly different, $P = 0.06$), but there were no significant changes *during* hibernation (i.e. 'mid' and 'late' hibernation were not different; all $P > 0.1$).

Table 11.1 Results of PERMANOVAs comparing chemical profiles of odorants collected from reproductive and non-reproductive adult female echidnas during different stages of hibernation

Sample	Term	df	SS	MS	Pseudo-*F*	*P* (perm)
Cloaca swab	Year	2	12,802	6401	5.05	**0.0001**
	Hibernation[a]	2	6365	3182	2.89	**0.0006**
	BreedYear	1	2252	2252	1.21	0.2879
	Animal ID (BreedYear)	20	27,432	1372	1.31	**0.0207**
	Hibernation[a] BreedYear	2	2016	1008	0.96	0.4874
	Residuals	70	73,555	1051	–	–
Cloaca wax secretion	Year	2	424	212	2.20	**0.0081**
	Hibernation[a]	2	559	279	4.19	**0.0002**
	BreedYear	1	75.9	76	0.56	0.8485
	Animal ID (BreedYear)	11	1347	122	2.15	**0.0001**
	Hibernation[a] BreedYear	2	135	67	1.18	0.2957
	Residuals	17	967	57	–	–

'BreedYear' indicates whether the female was reproductive or non-reproductive in a given calendar year. Significant results in bold type. Sample sizes: cloacal swab ($n = 98$), cloacal wax ($n = 36$)
[a]Odorants categorized as: 'pre-hibernation' (active, euthermic, $T_b \approx 32$ °C, prior to entering hibernation, usually January–March), 'mid-hibernation' (in deep hibernation for approximately 2 months, slow response to touch, low T_b, usually April–May), and 'late hibernation' (in deep hibernation during the mating season, slow response to touch, low T_b, June–July)

Table 11.2 Results of PERMANOVAs comparing chemical profiles of odorants collected from reproductive female echidnas which were alone or in mating aggregations ('Mating'), and torpid or euthermic ('Temperature')

Sample	Term	df	SS	MS	Pseudo-*F*	*P* (perm)
Cloaca swab	Year	2	21,997	10,999	6.07	**0.0001**
	Mating	1	1160	1160	1.00	0.4211
	Temperature	1	3882	3882	3.59	**0.0013**
	Animal ID	25	37,535	1501	1.48	**0.0035**
	Mating* temperature	1	1420	1420	1.40	0.1870
	Residuals	58	58,764	1013	–	–
Cloaca wax secretion	Year	2	3711	1856	11.43	**0.0001**
	Mating	1	235	235	1.92	0.0641
	Temperature	1	290	290	2.53	**0.0217**
	Animal ID	20	3322	166	1.60	**0.0062**
	Mating* temperature	1	72	72	0.69	0.6821
	Residuals	44	4577	104	–	–

Significant results in bold type. Sample sizes: cloacal swab ($n=89$), cloacal wax ($n=70$)

A total of 86 mating aggregations, comprising one female in close proximity to one or more males (max: 4), were found over the three mating seasons between 2010 and 2012 (mean date: 30 July ± 20.1 days; range: 12 June–6 September). The female was accessible in 69 mating aggregations, and females were torpid ($T_b < 28$ °C) in 15 (22 %) of these aggregations. Chemical profiles varied significantly between torpid and euthermic females during the mating season, but there were no differences between females which were alone or in mating aggregations (Table 11.2).

11.3.4 Chemical Profiles and Mating Activity During Pregnancy

Fertilization occurred on 20 July ± 15.7 days (range: 26 June–18 August, $n=16$); egg-laying occurred on 11 August ± 14.9 days (range: 17 July–7 September, $n=16$). All reproductive females monitored after fertilization were found in mating aggregations during pregnancy ($N=16$, $n=11$ individuals; Fig. 11.2). One female (echidna 4815 in 2011) was found accompanied by two males just 2 days prior to egg-laying. Nine females mated up to 14 days after estimated fertilization (mating was confirmed by presence of fresh sperm in the females' reproductive tracts). There were no significant changes in volatile or nonvolatile scent components between females at the time of fertilization compared with females which showed mating activity several weeks prior to fertilization, nor during different stages of pregnancy or at egg-laying (Table 11.3).

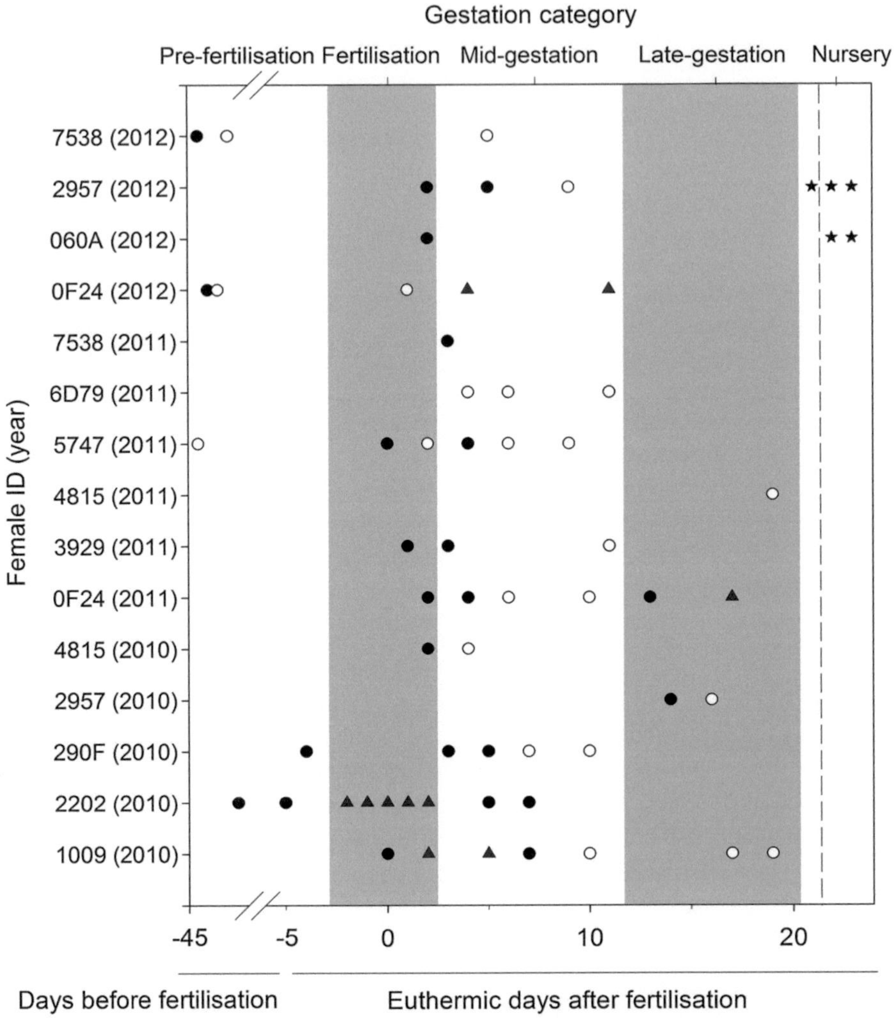

Fig. 11.2 Observed mating activity relative to reproductive state (prior to fertilization, during pregnancy and during egg incubation in nursery burrow) in female short-beaked echidnas. *Shaded areas* indicate gestation stage categories used for PERMANOVA analyses. Symbols: *filled circle* mating aggregation, sperm recovered from female reproductive tract; *open circle* mating aggregation, old sperm recovered; *filled triangle* mating aggregation, female not accessible; *star* male digging activity outside nursery burrows (data from Harris and Nicol 2014). *Dashed line*: mean date of egg-laying (21.5 days after fertilization; data from Morrow and Nicol 2012; Nicol and Morrow 2012). Note break in *x*-axis scale as females sampled more than 5 days prior to the time of fertilization re-entered hibernation after mating

Table 11.3 Results of PERMANOVAs comparing chemical profiles of odorants collected from reproductive female echidnas during different stages of gestation, ranging from prior to fertilization through to pregnancy and egg-laying (total five categories, details in main text), and torpid or euthermic (temperature)

Sample	Term	df	SS	MS	Pseudo-*F*	*P* (perm)
Cloaca swab	Year	2	14,245	7122	4.65	**0.0001**
	Gestation	4	4659	1165	1.03	0.4185
	Animal ID	10	17,221	1722	1.66	**0.0033**
	Residuals	45	46,641	1037	–	–
Cloaca wax secretion	Year	2	2471	1236	7.35	**0.0001**
	Gestation	4	708	177	1.40	0.1103
	Animal ID	10	1899	190	1.68	**0.0066**
	Residuals	34	3852	113	–	–

Significant results in bold type. Sample sizes: cloacal swab ($n=62$), cloacal wax ($n=51$)

11.4 Discussion

11.4.1 Hibernation, Reproductive State and Odor Cues

During hibernation, body temperature and metabolic rate decreases and non-essential body functions shut down (Hudson and Wang 1979; Dark 2005; Geiser 2011). We have previously found that chemical signal production in echidnas also appears to shut down during hibernation, since both sexes produce less cloacal wax secretion during hibernation (Harris et al. 2014), odor profiles in females vary depending on T_b and secretion composition does not change after entering hibernation (this study). We previously suggested females may produce a pheromone or attractive odor cue during late hibernation (Harris et al. 2012), but our present results do not support this hypothesis. The duration of hibernation varied widely between individual reproductive females (range: 70–154 days), as did the date that males entered reproductive female hibernacula (range: 3 June–31 July). These observations suggest photoperiod, prolonged exposure to low temperatures and metabolic changes during hibernation do not cause predictable changes in female signal composition which attracts males. Our results contrast with those studies on amphibians and reptiles, which describe qualitative and quantitative increases in female pheromone production in response to prolonged exposure to low temperatures during winter torpor (reviewed by Parker and Mason 2009). This difference is probably a reflection of fundamental differences between hibernation in endotherms and low temperature dormancy in ectotherms, but there are no similar published mammalian studies for comparison. In echidnas, odor production could continue during hibernation (but not necessarily change in composition), through, for example, microbial activity (Alberts 1992). The time at which males enter female hibernacula might therefore be driven by when female scent accumulates to a threshold level of intensity or attractiveness which males can detect (e.g. Schöne and Tengö 1981).

Although some mammals do show overlap between hibernation and reproduction (e.g. torpor during pregnancy; see Geiser and Brigham 2012), very few show overlap between mating and hibernation. In some vespertilionid bats, males mate with torpid females (Boyles et al. 2006), but possible links between olfactory cues and hibernation have not specifically been investigated. Consequently, the broader influence of hibernation on chemical signals, and their potential role in mating activity during hibernation, remains unclear.

In many vertebrates, odor advertises female body condition and males prefer scent from heavier females or those more likely to be receptive in the near future (e.g. reptiles, Cooper and Pèrez-Mellado 2002; LeMaster and Mason 2002; Shine et al. 2003; Bryant et al. 2011; mammals, Swaisgood et al. 2002; Ferkin et al. 2004; Fernández-Vargas et al. 2008). We observed several differences between female echidnas of different reproductive states, consistent with previous studies. In breeding years, females were in better body condition (Nicol and Morrow 2012; Sprent et al. 2012) and males entered their hibernacula (Morrow and Nicol 2009; Nicol and Morrow 2012), while non-breeding females were generally not disturbed by males and hibernated throughout the winter breeding season, emerging in spring (Nicol and Andersen 2002; Nicol and Morrow 2012). Although reproductive females probably have developed follicles prior to entering hibernation (Morrow 2013), we found no detectable differences in odor profiles between reproductive and non-reproductive females, and have no evidence that reproductive females scent mark their hibernacula prior to entering hibernation. Alternatively, males could target reproductive females by first using sex-specific odor differences to locate hibernating females (Harris et al. 2014), then variations in signal *intensity* to assess reproductive status, not measured here. For example, reproductive females could have greater fat reserves and hence more fat-derived scent components in their cloacal secretions (e.g. sterols, fatty acids; Supplementary Table 11.1), thereby contributing to a more intense odor signal and facilitating discrimination by males. Field data indicate echidnas are capable of such discrimination, as males do not simply enter every female hibernaculum they find: males were recorded outside the hibernacula of non-reproductive females (this study; Morrow 2013); and outside the hibernaculum of a reproductive female several times in the weeks prior to a male actually entering. Signal intensity (or absolute amount) of overall chemical cues or specific pheromones are important for conspecific response to female odor in many insects (e.g. Carlson et al. 1984; Carrière and McNeil 1990; Iyengar et al. 2001; reviewed by Johansson and Jones 2007), although there are few examples in vertebrates (e.g. Kopena et al. 2011; Apps 2013). Male-female interactions prior to entering hibernation could also be important for assessing female readiness to breed: males could use individual-specific odor cues (this study; Harris et al. 2014) to later recognize and target hibernating females in reproductive condition. Similarly, male 13-lined ground squirrels *Spermophilus tridecemlineatus* assess female reproductive status and prioritize later courtship behavior towards estrous females, potentially requiring significant cognitive ability (Schwagmeyer 1995; Schwagmeyer et al. 1998).

Females continued to attract males and mate during pregnancy, consistent with Morrow (2013), but we found no detectable changes in female odor profiles prior to or at the time of fertilization, or during pregnancy. Fertility in female echidnas, although apparently not advertised in odor, has several important influences on male and female mating behavior. Females of several species show extended sexual receptivity during pregnancy, including primates (Heistermann et al. 2001; Engelhardt et al. 2007; Fürtbauer et al. 2011), rodents (Huck et al. 1982; Jeppsson 1986; Exner et al. 2003) and carnivores (Packer and Pusey 1983). Since fertilization is not possible, polyandrous mating during non-fertile stages probably serves functions other than reproduction (Fürtbauer et al. 2011), or perhaps male ability to assess female reproductive status is imperfect. Many of the possible explanations for female promiscuity (reviewed by Wolff and Macdonald 2004), such as sperm competition or guarding against male infertility, do not account for multi-male mating during pregnancy when paternity has already been determined. Although it has previously been suggested that pregnant female echidnas may 'trade-up' and mate with a male of higher genetic quality (Morrow and Nicol 2009), females do not abort their pregnancy despite mating with additional males (Morrow 2013). Males provide no parental care, and females do not gain benefits such as territories or nuptial gifts from mating (Morrow 2013). One explanation for mating during pregnancy in echidnas is to confuse paternity as a counter-strategy to possible male infanticide behavior (Harris and Nicol 2014). Males of many species use mating history as a proxy for likelihood of paternity and are therefore less likely to kill offspring belonging to females that the male has previously mated with (Hrdy 1979; Ebensperger 1998; Fürtbauer et al. 2011). We suggest female echidnas could benefit from multi-male mating by confusing paternity among males (Ebensperger 1998; Wolff and Macdonald 2004), thereby decreasing the risk of infanticide to her offspring. Infanticide is feasible in echidnas, since females experience a high degree of male harassment throughout the mating period, and rates of offspring mortality during the egg-incubation stage are high (Morrow and Nicol 2012). Males enter and cause significant damage to nursery burrows (Harris and Nicol 2014), which would disrupt the stable thermal environment essential for offspring survival and development (Morrow and Nicol 2012), exposing the egg or young to cold external temperatures (Harris and Nicol 2014).

If chemical signals reliably signal reproductive physiology (Johansson and Jones 2007), why then do female echidnas show no changes in chemical profiles before, during or after fertilization and into late pregnancy, as seen in other species (e.g. Jemiolo et al. 1987; Chase-Crawford and Drea 2015)? It is possible that changes in female profiles were not detected because of a lack of statistical power, but we detected consistent differences in volatile and nonvolatile secretion components associated with year, individual identity and T_b. Alternatively, cues of female fertility may be present in ratios of specific combinations of compounds (Apps 2013), or in spur secretions, which were not investigated here. Direct, causal links between reproductive hormones and odor cues are difficult to demonstrate, but estrogen increases scent-marking behaviour (*Oryctolagus cuniculus*, Hudson et al. 1990) and female attractiveness (*Microtus pensylvanicus*, Ferkin et al. 2004; *Thamnophis sirtalis parietalis*, Mendonça and Crews 2001; Mason and Parker 2010), even inducing female sex pheromone production in male conspecifics (*Thamnophis sirtalis parietalis*, Parker and Mason 2012). Conversely, progesterone decreases scent-marking behavior and attractiveness (Johnston 1980; Molteno et al. 1998). In echidnas, estrogen levels are below standard detection thresholds (Dean 2000), but progesterone levels become elevated during pregnancy (Nicol et al. 2005; Morrow 2013), and Tasmanian echidnas appear to be induced ovulators (Morrow and Nicol 2009). Since males are attracted to females regardless of reproductive state, if there are odor changes associated with fertilization or pregnancy, male echidnas may also have difficulty detecting them.

11.4.2 Odor Cues and Sexual Conflict

Reproduction in Tasmanian echidnas is characterized by roving male searching behavior and intra-sexual competition for access to widely dispersed females in reproductive condition (Morrow et al. 2009; Nicol and Morrow 2012; Morrow 2013), consistent with scramble competition (Ims 1988; Murphy 1998). Males cannot monopolize females during the mating season (Morrow 2013), so male fitness is primarily driven by sensory and behavioral traits which maximize ability to locate females (Trivers 1972; Murphy 1998; Lane et al. 2009). Sexual selection for mate-locating ability will be especially strong when females are sparsely (spatially and temporally) distributed and difficult to detect (Ims 1988; Endler 1992; Kokko and Rankin 2006), and this appears to be the case in our study population, where male echidnas must locate sparsely-distributed, reproductive females buried within their hibernacula. Intense intra-sexual competition has driven male Tasmanian echidnas to emerge from hibernation as early as physiologically possible, resulting in males being ready to breed before females have emerged from hibernation (Morrow 2013). However, searching for mates can be a costly activity (Kokko and Wong 2007) and male echidnas spend a significant amount of time and energy searching for females during the mating season (Morrow 2013); weight loss is maximal during this period (Nicol and Morrow 2012). Males can minimize search costs and increase reproductive fitness by being sensitive to variations in female-specific chemical cues, allowing them to quickly locate females and even target those in reproductive condition (Svensson 1996; Thomas 2011). Echidna males are strongly attracted to female odor (Rismiller 1992; Harris 2014) and use scent to locate females which have not yet emerged from hibernation. Odor differences between euthermic and hibernating female echidnas (this study) raise the possibility that males can distinguish between these females, and even seek out hibernating females as they may be less able to resist forced copulations (e.g. Shine et al. 2005), and because hibernating females are less likely to be pregnant (Morrow 2013; Harris 2014), but see Nicol and Morrow (2012). A clear extension of the present study would incorporate results from robust behavioral trials and confirm whether males detect and use the information present in female odor cues in their mate-searching behavior (Harris 2014).

Female echidnas experience a high degree of male harassment throughout the mating period, beginning prior to her final emergence from hibernation and continuing throughout pregnancy (this study; Morrow 2013; Harris 2014). There were no differences in chemical profiles between euthermic females which were alone or with males during the mating season, suggesting females do not control when males are attracted to them. Disturbance during hibernation would lead to reduced energy savings from hibernation, and continued harassment prior to entering a nursery burrow would also disrupt normal foraging behavior, leading to decreased ability to conserve energy and potentially reducing capacity for lactation and long-term reproductive output or survival. Male disturbance of nursery burrows could also negatively affect pregnancy and egg or offspring survival. Females may be constrained in their ability to avoid males during the mating season, as physiological processes involved with producing an attractive female-specific odor may also be related to pregnancy or lactation, or simply due to sex differences, which cannot simply be 'switched off'. However, the costs for females associated with male harassment and mating multiply may be outweighed by several benefits. For example, a high degree of individual variation in sensory sensitivity is expected in populations with intense selection for ability to locate mates (Ronald et al. 2012). In some species, females which produce small amounts of pheromone may gain reproductive benefits by indirectly selecting for those high quality males which are best able to locate her quickly and precisely (Phelan 1992; Svensson 1996; Wiley and Poston 1996). In echidnas, males which enter female hibernacula may be of particularly high quality, since they have emerged from hibernation early, shortening the period of energy saving, and may be olfactorily sensitive, making them capable of locating hibernating females (Harris 2014). Furthermore, the costs associated with mating may be outweighed by the potential fitness benefits of confusing paternity and reducing infanticide risk (Harris 2014).

In situations of sexual conflict, selection favors recipients skilled at discriminating genuine signals from those that are not (Seyfarth and Cheney 2003). Our research indicates male and female Tasmanian echidnas use chemical signals to increase their reproductive success in different ways, often at the expense of the other sex. Males may pay 'opportunity costs' by wasting time and energy locating, courting, mating with and guarding pregnant females, which decreases mating opportunities with other females (Shine 2012; Morrow 2013). Therefore, male echidnas should benefit from detecting female reproductive status (pregnancy) via olfactory cues, but this can simultaneously decrease female reproductive success by reducing their potential number of mates (Thomas 2011). These processes can result in an evolutionary arms race, with increasing subtlety in female signals and a corresponding increase in male sensory ability (Stumpf et al. 2011). In echidnas, this arms race between signalers and receivers reflects differences in parental investment, and results from a complex interplay between chemical signals, behavior, and ecological and other selective pressures.

Acknowledgments We thank Gemma Morrow for her assistance in the field and allowing us to use some of her data, and the McShane family for allowing us access to their property. We are grateful for comments from two anonymous reviewers and for statistical advice from Barbara Holland and Leon Barmuta. Research was carried out under University of Tasmania Animal Ethics Approvals A100426, A12320, and the Tasmanian Department of Primary Industries, Parks, Water and the Environment Permits FA10045, FA11033, FA12069. Research was supported by the Holsworth Wildlife Research Fund, the MA Ingram Trust, and the National Geographic Committee for Research and Exploration.

References

Alberts AC (1992) Constraints on the design of chemical communication systems in terrestrial vertebrates. Am Nat 139:S62–S89

Anderson MJ (2001) A new method for non-parametric multivariate analysis of variance. Austral Ecol 26:32–46

Anderson MJ, Gorley RN, Clarke KR (2008) PERMANOVA+ for PRIMER: guide to software and statistical methods. Primer-E Ltd, Plymouth

Andersson M (1994) Sexual selection. Princeton University Press, Princeton

Apps PJ (2013) Are mammalian olfactory signals hiding right under our noses? Naturwissenschaften 100:487–506

Ashwell KWS (2006) Chemoarchitecture of the monotreme olfactory bulb. Brain Behav Evol 67:69–84

Ashwell KWS (2013) Reflections: monotreme neurobiology in context. In: Ashwell KWS (ed) Neurobiology of the monotremes: brain evolution in our distant mammalian cousins. CSIRO Publishing, Collingwood, pp 285–298

Augee ML (2008) Short-beaked echidna. In: Van Dyck S, Strahan R (eds) The mammals of Australia. Reed New Holland, Sydney, pp 37–39

Beard LA, Grigg GC, Augee ML (1992) Reproduction by echidnas in a cold climate. In: Augee ML (ed) Platypus and echidnas. Royal Zoological Society NSW, Mosman, pp 93–100

Blumstein DT, Im S, Nicodemus A, Zugmeyer C (2004) Yellow-bellied marmots (*Marmota flaviventris*) hibernate socially. J Mammal 85:25–29

Boisvert M, Grisham J (1988) Reproduction of the short-nosed echidna, *Tachyglossus aculeatus*, at the Oklahoma City Zoo. Int Zoo Yearb 27:103–108

Boyles JG, Dunbar MB, Whitaker JO (2006) Activity following arousal in winter in North American vespertilionid bats. Mamm Rev 36:267–280

Bradbury JW, Vehrencamp SL (2011) Principles of animal communication, 2nd edn. Sinauer Associates, Inc., Sunderland

Broom R (1895) On the organ of Jacobson in the Monotremata. J Anat Physiol 30:70–80

Bryant GL, Bateman PW, Fleming PA (2011) Tantalising tongues: male carpet pythons use chemoreception to differentiate among females. Aust J Zool 59:42–48

Candolin U (2003) The use of multiple cues in mate choice. Biol Rev 78:575–595

Carlson DA, Nelson DR, Langley PA, Coates TW, Davis TL, Leegwater-Van Der Linden ME (1984) Contact sex pheromone in the tsetse fly *Glossina pallidipes* (Austen): identification and synthesis. J Chem Ecol 10:429–450

Carrière Y, McNeil JN (1990) Effect of age and mating status on the mating behaviour and pheromone titre in alfalfa blotch leafminer females *Agromyza frontella* (Diptera: Agromyzidae). J Insect Physiol 36:457–461

Chase-Crawford J, Drea C (2015) Baby on board: olfactory cues indicate pregnancy and fetal sex in a non-human primate. Biol Lett 11:20140831

Clark K, Warwick R (2001) Change in marine communities: an approach to statistical analysis and interpretation, 2nd edn. Primer-E Ltd, Plymouth

Cooper WE, Pèrez-Mellado V (2002) Pheromonal discriminations of sex, reproductive condition, and species by the lacertid lizard *Podarcis hispanica*. J Exp Zool 292:523–527

Dark J (2005) Annual lipid cycles in hibernators: integration of physiology and behavior. Annu Rev Nutr 25:469–497

Dean KM (2000) Aspects of the reproductive biology of the echidna, *Tachyglossus aculeatus*. PhD Thesis, University of Melbourne, Melbourne

Dobroruka VLJ (1960) Einige Beobachtungen an Ameisenigeln, *Echidna aculeata* Shaw (1792). Z Tierpsychole 17:178–181

Ebensperger LA (1998) Strategies and counterstrategies to infanticide in mammals. Biol Rev Camb Philos Soc 73:321–346

Ebling FJ (1977) Hormonal control of mammalian skin glands. In: Müller-Schwarze D, Mozell MM (eds) Chemical signals in vertebrates 1. Plenum, New York, pp 17–39

Endler JA (1992) Signals, signal conditions, and the direction of evolution. Am Nat 139: S125–S153

Engelhardt A, Hodges JK, Heistermann M (2007) Post-conception mating in wild long-tailed macaques (*Macaca fascicularis*): characterization, endocrine correlates and functional significance. Horm Behav 51:3–10

Exner C, Wehrend A, Hospes R, Einspanier A, Hoffmann B, Heldmaier G (2003) Hormonal and behavioural changes during the mating season and pregnancy in Alpine marmots (*Marmots marmota*). Reproduction 126:775–782

Ferkin MH, Lee DN, Leonard ST (2004) The reproductive state of female voles affects their scent marking behavior and the responses of male conspecifics to such marks. Ethology 110:257–272

Fernández-Vargas M, Tang-Martínez Z, Phelps SM (2008) Olfactory responses of neotropical short-tailed singing mice, *Scotinomys teguina*, to odors of the mid-ventral sebaceous gland: discrimination of conspecifics, gender, and female reproductive condition. J Chem Ecol 34:429–437

Fürtbauer I, Heistermann M, Schulke O, Ostner J (2011) Concealed fertility and extended female sexuality in a non-human primate (*Macaca assamensis*). PLoS One 6:e23105

Geiser F (2004) Metabolic rate and body temperature reduction during hibernation and daily torpor. Annu Rev Physiol 66:239–274

Geiser F (2011) Hibernation: endotherms, Encyclopedia of life sciences. Wiley, Chichester. doi:10.1002/9780470015902.a0003215.pub2, http://www.els.net

Geiser F, Brigham RM (2012) The other functions of torpor. In: Ruf T, Bieber C, Arnold W (eds) Living in a seasonal world: thermoregulatory and metabolic adaptations. Springer, Vienna, pp 109–122

Gosling LM, Roberts SC, Thornton EA, Andrew MJ (2000) Life history costs of olfactory status signalling in mice. Behav Ecol Sociobiol 48:328–332

Griffiths M (1978) The biology of the monotremes. Academic, London

Harris RL (2014) Chemical communication for reproduction in the Tasmanian short-beaked echidna, *Tachyglossus aculeatus setosus*. PhD thesis, School of Biological Sciences, University of Tasmania, Hobart

Harris RL, Nicol SC (2014) Observations of breeding behaviour and possible infanticide in a wild population of Tasmanian echidnas (*Tachyglossus aculeatus setosus*). Aust Mamm 36:108–112

Harris RL, Davies NW, Nicol SC (2012) Chemical composition of odorous secretions in the Tasmanian short-beaked echidna (*Tachyglossus aculeatus setosus*). Chem Senses 37:819–836

Harris RL, Holland BR, Cameron EZ, Davies NW, Nicol SC (2014) Chemical signals in the echidna: differences between seasons, sexes, individuals and gland types. J Zool 293:171–180

Hawkins M, Battaglia A (2009) Breeding behaviour of the platypus (*Ornithorhynchus anatinus*) in captivity. Aust J Zool 57:283–293

Heistermann M, Ziegler T, van Schaik CP, Launhardt K, Winkler P, Hodges JK (2001) Loss of oestrus, concealed ovulation and paternity confusion in free-ranging Hanuman langurs. Proc R Soc B Biol Sci 268:2445–2451

Home E (1802) Description of the anatomy of the *Ornithorhynchus hystrix*. Philos Trans R Soc B 92:348–364

Hrdy SB (1979) Infanticide among animals: a review, classification, and examination of the implications for the reproductive strategies of females. Ethol Sociobiol 1:13–40

Huck UW, Soltis RL, Coopersmith CB (1982) Infanticide in male laboratory mice: effects of social status, prior sexual experience, and basis for discrimination between related and unrelated young. Anim Behav 30:1158–1165

Hudson JW, Wang LCH (1979) Hibernation: endocrinologic aspects. Annu Rev Physiol 41:287–303
Hudson R, González-Mariscal G, Beyer C (1990) Chin marking behavior, sexual receptivity, and pheromone emission in steroid-treated, ovariectomized rabbits. Horm Behav 24:1–13
Ims RA (1988) The potential for sexual selection in males: effect of sex ratio and spatiotemporal distribution of receptive females. Evol Ecol 2:338–352
Iyengar VK, Rossini C, Eisner T (2001) Precopulatory assessment of male quality in an arctiid moth (*Utetheisa ornatrix*): hydroxydanaidal is the only criterion of choice. Behav Ecol Sociobiol 49:283–288
Jeppsson B (1986) Mating by pregnant water voles (*Arvicola terrestris*): a strategy to counter infanticide by males? Behav Ecol Sociobiol 19:293–296
Jemiolo B, Andreolini F, Wiesler D, Novotny M (1987) Variations in mouse (*Mus musculus*) urinary volatiles during different periods of pregancy and lactation. J Chem Ecol 13:1941–1956
Johansson BG, Jones TM (2007) The role of chemical communication in mate choice. Biol Rev 82:265–289
Johnston RE (1980) Responses of male hamsters to odors of females in different reproductive states. J Comp Physiol Psychol 94:894–904
Kokko H, Rankin DJ (2006) Lonely hearts or sex in the city? Density-dependent effects in mating systems. Philos Trans R Soc B 361:319–334
Kokko H, Wong BBM (2007) What determines sex roles in mate searching? Evolution 61:1162–1175
Kopena R, Martín J, López P, Herczeg G (2011) Vitamin E supplementation increases the attractiveness of males' scent for female European green lizards. PLoS One 6:e19410
Lane JE, Boutin S, Gunn MR, Coltman DW (2009) Sexually selected behaviour: red squirrel males search for reproductive success. J Anim Ecol 78:296–304
LeMaster MP, Mason RT (2002) Variation in a female sexual attractiveness pheromone controls male mate choice in garter snakes. J Chem Ecol 28:1269–1285
Mason RT, Parker MR (2010) Social behavior and pheromonal communication in reptiles. J Comp Physiol A 196:729–749
McLean S, Davies N, Wiggins N (2012) Scent chemicals of the brushtail possum, *Trichosurus vulpecula*. J Chem Ecol 38:1–22
Mendonça MT, Crews D (2001) Control of attractivity and receptivity in female red-sided garter snakes. Horm Behav 40:43–50
Molteno AJ, Sliwa A, Richardson PRK (1998) The role of scent marking in a free-ranging, female black-footed cat (*Felis nigripes*). J Zool 245:35–41
Morrow GE (2013) Optimising reproduction in the Tasmanian echidna *Tachyglossus aculeatus setosus*: the influence of an obligatory hibernation period and intense sexual conflict. PhD thesis, University of Tasmania, Hobart
Morrow G, Nicol SC (2009) Cool sex? Hibernation and reproduction overlap in the echidna. PLoS One 4:e6070
Morrow GE, Nicol SC (2012) Maternal care in the Tasmanian echidna (*Tachyglossus aculeatus setosus*). Aust J Zool 60:289–298
Morrow G, Andersen NA, Nicol S (2009) Reproductive strategies of the short-beaked echidna—a review with new data from a long-term study on the Tasmanian subspecies (*Tachyglossus aculeatus setosus*). Aust J Zool 57:275–282
Murphy CG (1998) Interaction-independent sexual selection and the mechanisms of sexual selection. Evolution 52:8–18
Nicol S (2003) Monotreme biology. Comp Biochem Phys A 136:795–798
Nicol SC (2013) Behaviour and ecology of monotremes. In: Ashwell KWS (ed) Neurobiology of the monotremes: brain evolution in our distant mammalian cousins. CSIRO Publishing, Collingwood, pp 17–30
Nicol SC, Andersen NA (2002) The timing of hibernation in Tasmanian echidnas: why do they do it when they do? Comp Biochem Phys B 131:603–611

Nicol SC, Andersen NA (2006) Body temperature as an indicator of egg-laying in the echidna, *Tachyglossus aculeatus*. J Therm Biol 31:483–490
Nicol SC, Andersen NA (2007) The life history of an egg-laying mammal, the echidna (*Tachyglossus aculeatus*). Ecoscience 14:275–285
Nicol SC, Morrow G (2012) Sex and seasonality: reproduction in the echidna (*Tachylossus aculeatus*). In: Ruf T, Bieber C, Arnold W, Millesi E (eds) Living in a seasonal world: thermoregulatory and metabolic adaptations. Springer, Heidelberg, pp 143–153
Nicol SC, Vedel-Smith C, Andersen NA (2004) Behaviour, body temperature, and hibernation in Tasmanian echidnas (*Tachyglossus aculeatus*). In: Barnes BM, Carey HV (eds) Life in the cold: evolution, mechanisms, adaptation and application Twelfth International Hibernation Symposium. University of Alaska Fairbanks, Fairbanks, pp 377–388
Nicol SC, Andersen NA, Jones SM (2005) Seasonal variations in reproductive hormones in free-ranging echidnas (*Tachyglossus aculeatus*): interaction between reproduction and hibernation. Gen Comp Endocrinol 144:204–210
Nicol SC, Vanpé C, Sprent JA, Morrow G, Andersen NA (2011) Spatial ecology of a ubiquitous Australian anteater, the short-beaked echidna (*Tachyglossus aculeatus*). J Mammal 92:101–110
Packer C, Pusey AE (1983) Adaptations of female lions to infanticide by incoming males. Am Nat 121:716–728
Parker MR, Mason RT (2009) Low temperature dormancy affects the quantity and quality of the female sexual attractiveness pheromone in red-sided garter snakes. J Chem Ecol 35:1234–1241
Parker MR, Mason RT (2012) How to make a sexy snake: estrogen activation of female sex pheromone in male red-sided garter snakes. J Exp Biol 215:723–730
Phelan PL (1992) Evolution of sex pheromones and the role of asymmetric tracking. In: Roitberg BD, Isman MB (eds) Insect chemical ecology: an evolutionary approach. Chapman & Hall, New York, pp 265–314
Phillips MJ, Bennett TH, Lee MSY (2009) Molecules, morphology, and ecology indicate a recent, amphibious ancestry for echidnas. Proc Natl Acad Sci USA 106:17089–17094
Rismiller PD (1992) Field observations on Kangaroo Island echidnas (*Tachyglossus aculeatus multiaculeatus*) during the breeding season. In: Augee ML (ed) Platypus and echidnas. The Royal Zoological Society of New South Wales, Mosman, pp 101–105
Ronald KL, Fernandez-Juricic E, Lucas JR (2012) Taking the sensory approach: how individual differences in sensory perception can influence mate choice. Anim Behav 84:1283–1294
Schneider NY (2011) The development of the olfactory organs in newly hatched monotremes and neonate marsupials. J Anat 219:229–242
Schöne H, Tengö J (1981) Competition of males, courtship behaviour and chemical communication in the digger wasp *Bembix rostrata* (Hymenoptera, Sphecidae). Behaviour 77:44–66
Schwagmeyer PL (1995) Searching today for tomorrow's mates. Anim Behav 50:759–767
Schwagmeyer PL, Parker GA, Mock DW (1998) Information asymmetries among males: implications for fertilization success in the thirteen-lined ground squirrel. Proc R Soc B Biol Sci 265:1861–1865
Semon R (1899) In the Australian bush and on the coast of the Coral Sea: being the experiences and observations of a naturalist in Australia, New Guinea and the Moluccas. Macmillan, London
Seyfarth RM, Cheney DL (2003) Signalers and receivers in animal communication. Annu Rev Psychol 54:145–173
Shine R (2012) Sex at the snake den: lust, deception, and conflict in the mating system of red-sided garter snakes. Adv Stud Behav 44:1–51
Shine R, Phillips B, Waye H, LeMaster M, Mason RT (2003) The lexicon of love: what cues cause size-assortative courtship by male garter snakes? Behav Ecol Sociobiol 53:234–237
Shine R, Wall M, Langkilde T, Mason RT (2005) Battle of the sexes: forcibly inseminating male garter snakes target courtship to more vulnerable females. Anim Behav 70:1133–1140

Sprent J, Jones SM, Nicol SC (2012) Does leptin signal adiposity in the egg-laying mammal, *Tachyglossus aculeatus*? Gen Comp Endocrinol 178:372–379

Stumpf RM, Martinez-Mota R, Milich KM, Righini N, Shattuck MR (2011) Sexual conflict in primates. Evol Anthropol 20:62–75

Svensson M (1996) Sexual selection in moths: the role of chemical communication. Biol Rev Camb Philos Soc 71:113–135

Swaisgood RR, Lindburg DG, Zhang H (2002) Discrimination of oestrous status in giant pandas (*Ailuropoda melanoleuca*) via chemical cues in urine. J Zool 257:381–386

Thomas ML (2011) Detection of female mating status using chemical signals and cues. Biol Rev 86:1–14

Thomas DW, Fenton MB, Barclay RM (1979) Social behavior of the little brown bat, *Myotis lucifugus* I. Mating behavior. Behav Ecol Sociobiol 6:129–136

Trivers RL (1972) Parental investment and sexual selection. In: Campbell BG (ed) Sexual selection and the descent of man 1871–1971. Aldine Publishing Company, Chicago, pp 136–179

Uhrig EJ, Lutterschmidt DI, Mason RT, LeMaster MP (2012) Pheromonal mediation of intraseasonal declines in the attractivity of female red-sided garter snakes, *Thamnophis sirtalis parietalis*. J Chem Ecol 38:71–80

Wiley RH, Poston J (1996) Perspective: indirect mate choice, competition for mates, and coevolution of the sexes. Evolution 50:1371–1381

Wimsatt WA (1969) Some interrelations of reproduction and hibernation in mammals. In: Woolhouse HW (ed) Dormancy and survival, symposia of the Society for Experimental Biology, vol 23. Cambridge University Press, London, pp 511–549

Wolff JO, Macdonald DW (2004) Promiscuous females protect their offspring. Trends Ecol Evol 19:127–134

Wyatt TD (2003) Pheromones and animal behaviour: communication by smell and taste. Cambridge University Press, Cambridge

Zala SM, Potts WK, Penn DJ (2004) Scent-marking displays provide honest signals of health and infection. Behav Ecol 15:338–344

Chapter 12
Chemical Analyses Reveal Family-Specific Nest Odor Profiles in Zebra Finches (*Taeniopygia guttata*): A Pilot Study

Sarah Kohlwey, E. Tobias Krause, Markus C. Baier, Caroline Müller, and Barbara A. Caspers

12.1 Introduction

Birds need to find their own nest, and, if they live in colonies, they also need to distinguish their own nest from conspecific nests, to lay and incubate their eggs, to feed the chicks, or to receive shelter in unsecure circumstances. Studies investigating the factors important in nest recognition led to the assumption that in many cases nest sites are recognized and not particularly the nest itself. Whereas visual nest site recognition might be a sufficient mechanism for diurnal species (Hughes et al. 1995), as nests are immobile, visual cues might be insufficient for species that return to their nest in the absence of light (Bonadonna et al. 2003). The first evidence for olfactory nest recognition has been provided by Minguez (1997), who found that British storm petrel chicks (*Hydrobates pelagicus*) are able to find their nest using olfactory cues. British storm petrels belong to the order of Procellariiformes, which are known to have a well-developed sense of smell (Bang and Cobb 1968; Bonadonna et al. 2003; Nevitt and Bonadonna 2005). The ability to recognize their own nest based on olfactory cues has also been demonstrated for other

S. Kohlwey • B.A. Caspers (✉)
Department of Animal Behaviour, Bielefeld University, Morgenbreede 45, 33615 Bielefeld, Germany
e-mail: barbara.caspers@uni-bielefeld.de

E.T. Krause
Department of Animal Behaviour, Bielefeld University, Morgenbreede 45, 33615 Bielefeld, Germany

Institute of Animal Welfare and Animal Husbandry, Friedrich-Loeffler-Institut, Dörnbergstr. 25-27, 29223 Celle, Germany

M.C. Baier • C. Müller
Department of Chemical Ecology, Bielefeld University, Universitätsstraße 25, 33615 Bielefeld, Germany

B.A. Schulte et al. (eds.), *Chemical Signals in Vertebrates 13*,
DOI 10.1007/978-3-319-22026-0_12

Procellariiformes species, such as the diving petrel (*Pelecanoides urinatrix*) and the South-Georgian diving petrel (*P. georgicus*), both of which return in the absence of light to their nest (Bonadonna et al. 2003).

Recent studies on a laboratory population of zebra finches (*Taeniopygia guttata*) have shown that olfactory-based nest recognition is not restricted to nocturnal sea birds but also occurs in diurnal songbirds (Caspers and Krause 2011; Krause and Caspers 2012). The fledglings of the colony-breeding zebra finch and adult females show a preference for their home nest odor (Caspers and Krause 2011; Krause and Caspers 2012) and avoid the odor of a foreign nest. The ability to recognize their own nest odor and to distinguish from other conspecific nest odors (Caspers and Krause 2011; Krause and Caspers, 2012; Krause et al. 2012) suggests that family-specific nest odor profiles should exist.

Whereas olfactory nest recognition in nature might also be based on different nesting material and different food sources used (Petit et al. 2002; Gwinner and Berger 2008; Mennerat 2008; Mennerat et al. 2009; Gwinner 2013), laboratory studies, in which all individuals experienced the same food and nesting material, led to the assumption that olfactory nest recognition is based on family-specific mixtures of individual body odors (Caspers and Krause 2011). Based on the hypothesis that different families should have distinguishable family-specific nest odors, we performed a pilot study in which we analyzed the volatiles of nest odors of different families.

12.2 Methods

12.2.1 Breeding Conditions

Domesticated zebra finches (*Taeniopygia guttata*) were allowed to breed pairwise in two-compartment cages (80×40×30 cm) at Bielefeld University from April to November 2012. This study was part of a larger experiment, in which the influence of relatedness on nest odor recognition was tested. Therefore, pairs were assigned to create families which were completely unrelated or families with higher levels of relatedness (Fig. 12.1). A wooden nest box (15×15×15 cm) was attached to each cage, and coconut fibers were provided as nest material. All breeding pairs were housed in the same room under a 14 h:10 h dark/light rhythm. All birds received the identical diet, i.e., a seed mix ad libitum, protein-rich egg-food and germinated seeds, which were provided three times a week. All birds had ad libitum access to water.

12.2.2 Chemical Analysis of Nest Odors

The aim of this study was to test volatiles of zebra finch nests. Therefore nest samples were collected from four pairs (Fig. 12.1) that successfully bred and raised their chicks. Nest samples were collected at the mean brood age of 10 days. At this age,

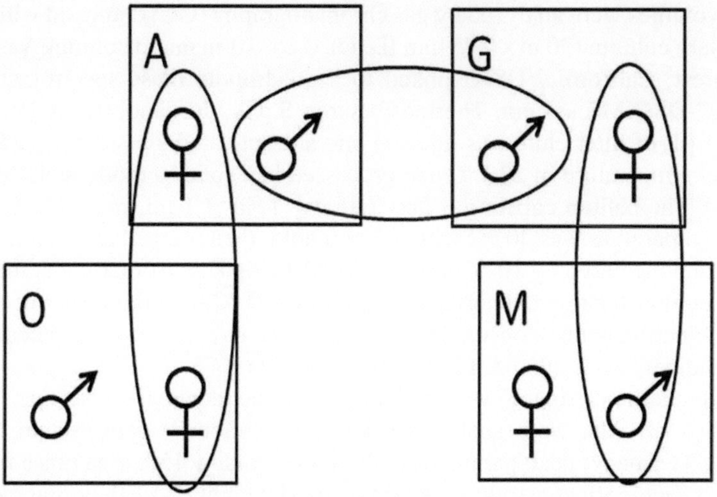

Fig. 12.1 Relationship of the four zebra finch breeding pairs used as nest sample donors for chemical analysis. The letters represent the pair ID. Siblings are encircled, i.e., female from breeding pair A and female from breeding pair O are sisters

all offspring are still in the nest and thus rely fully on parental care. Moreover, at this developmental stage of the offspring, mothers have been shown to identify their own nest odors (Krause and Caspers 2012). For the subsequent chemical analysis, nest samples were collected by cutting approximately 2 g of nest material out of each nest. Using these samples we aimed to compare the volatile organic compounds (VOCs) of the nest odors of different families. Furthermore, to test for intra-nest variability, we collected a second sample for each nest at the same day, but from a different site within the same nest. Additionally, at each day of sample collection, unused coconut fibers were taken as a control, to be able to subtract the nest material-specific compounds from further analysis later.

VOCs of each sample were collected using headspace analyses in a closed-loop stripping apparatus (CLSA; Lorbeer et al. 1984; Krause et al. 2014). Therefore, samples of each nest were enclosed in 150 ml Erlenmeyer flasks. The flasks were closed using rubber plugs, and each plug was perforated by two 14-gauge needles connected to a pumping system. Each vacuum pump (DC12/08FK, Fürgut GmbH, Tannheim, Germany) created a continuously circulating gas flow (1.3 L/min) over the respective sample's headspace. Via de-airing cannula, the withdrawn air was conducted through a 1.5-mg charcoal sorbent filter (Brechbuehler AG, Schlieren, Switzerland). Flasks containing unused nest material, i.e., coconut fibers, were sampled with each run as background controls. After 25 h of extraction at 37 °C, the components that had been adsorbed by the sorbent filter were eluted with 80 μL of dichloromethane (Carl Roth GmbH and Co. KG, Karlsruhe, Germany) containing 0.2 mM 1-bromodecane (Sigma-Aldrich, St. Louis, MO, USA) as an internal standard and stored at −80 °C until chemical analysis.

Nest volatiles were analyzed by gas chromatography (GC) equipped with a VF-5 ms capillary column (30 m × 0.25 mm ID, DF 0.25, 10-m guard column, Varian Inc., Lake Forest, California, USA) linked to a quadrupole mass spectrometer (MS; Focus GC-DSQ MS system, Thermo Electron S.p.A., Rodano, Italy). Per trapped sample, 1 μL of filter elute was injected into a deactivated glass wool-packed liner at an inlet temperature of 220 °C and processed in a split 10 mode with 20 ml/min split flow. The helium carrier gas flow rate was held at 1 mL/min. The initial GC column temperature was 40 °C with a 2 min hold, then ramped at 4 °C/min to 180 °C, and then increased by 10 °C/min to 240 °C, being kept isothermal at the end for 3 min. The transfer line temperature was set to 270 °C, and the mass spectra were taken in electron ionization mode at 70 eV from *m/z* 45–500. To determine the Kovats indices, an *n*-alkane mixture (ranging from C8 to C20, Sigma-Aldrich-Fluka, Steinheim, Germany) was analyzed under the identical conditions. The processed GC-MS data were analyzed using the Xcalibur™ data system (Thermo Scientific, Germany; peak parameters—baseline window 150, area noise factor 80, peak noise factor 80; constrain peak width—peak height (%) 1.0, tailing factor 5.0) and the area of each peak was noted. Peaks were defined as substances to be discriminated based on their retention time and the similarity of the mass spectra. Substances were putatively identified using a combination of Kovats indices and the NIST library. The relative proportion of each peak to the sum of all peaks was used for further analysis.

12.2.3 Statistical Analysis

For further statistical analysis, we computed a similarity matrix based on all peaks that comprised at least 0.1 % of the overall area (Caspers et al. 2011) of the chromatogram and of those that were present in at least two of all samples. Pairwise similarities of odor profiles were calculated using the Bray-Curtis similarity index (Clarke and Warwick 1994) on the transformed data (presence/absence as well as (log $(x+1)$)). We hypothesized that each nest has its own family-specific odor profile that enables individuals to distinguish between different nest odors. To statistically determine whether the samples of the same nest were more similar to each other than the samples of different nests (i.e., repeatability of nest odor composition), a multivariate nonparametric analysis of similarities (ANOSIM) was used (PRIMER-E Ltd., Plymouth, United Kingdom). All peaks found in the VOC profiles of the coconut fiber controls were excluded from the statistical analyses.

12.3 Results

Eight different nest samples originating from four families were analyzed by GC-MS. After excluding the coconut fiber-specific substances, 113 compounds were found. Two example chromatograms of nest samples and of a control are

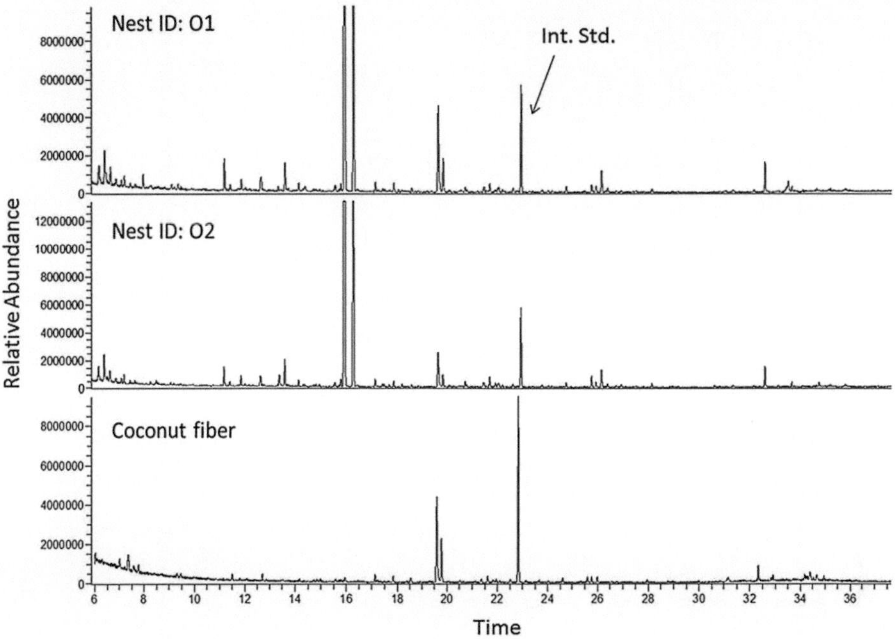

Fig. 12.2 Chromatograms of two samples of one nest and a control containing coconut fibers only. Int. Std. – internal standard

shown in Fig. 12.2 and the relative amounts of the 15 major substances of all nest samples are shown in Table 12.1.

The comparison of nest odor profile similarity within and between the nest odor samples shows that the similarity of volatiles within one nest was higher than among different nests (presence/absence transformed data, ANOSIM, factor: nest ID, Global $R=0.438$, $p=0.048$; Fig. 12.3). The same groups were statistically determined when using the log $(x+1)$ transformed data (ANOSIM, factor: nest ID, Global $R=0.625$; $p=0.038$). Although all samples originated from closely related families, they show family- or nest-specific differences in their volatile composition, but not in the existence of family-specific substances.

12.4 Discussion

The chemical analysis of nest volatiles supports the existence of family-specific nest odor profiles. The nest odor profiles did not differ in single components, i.e., we could not find specific substances present in one nest only. Nest-specific odors were characterized by differences in the composition of compounds.

Since this study was performed in the laboratory, in which the birds had identical environmental conditions, differences in nest odor profiles have to be influenced by

Table 12.1 Relative amounts of the 15 major substances of nests of *Taeniopygia guttata*

Putative substances	Kovats index	RT	G1	G2	M1	M2	A1	A2	O1	O2
Unidentified		**6.33**	0.94	2.58	2.54	12.33	0	0.03	1.06	0.8
2,3-Butanediol (CAS# 513-85-9)		**6.55**	1.08	4.91	9.3	8.26	0.4	1.7	1.76	1.57
Hexanal (CAS# 66-25-1)	**803**	**6.77**	1.23	4.92	2.48	0.56	0.58	3.73	0.72	0.47
Unidentified	**811**	**7**	0.56	1.37	0.77	0.95	0.5	0.71	0.25	0
Unidentified	**824**	**7.34**	0.,68	2.5	1.38	1.66	0.93	1.31	0.41	0.33
1,4-Dimethylbenzene (CAS# 106-42-3)	**871**	**8.65**	0.47	0	2.45	0	0	0	0	0
(E)-2-Heptenal (CAS# 18829-55-5)	**960**	**11.27**	1	2.84	1.95	0.13	0	4.06	1.24	0.81
1-Octen-3-ol (CAS# 3391-86-4)	**981**	**11.92**	0.31	0.2	0.47	1.23	0	0.84	0.42	0.44
Octanal (CAS# 124-13-0)	**1006**	**12.68**	0.51	1.01	2.05	0.55	0.18	0.51	0.56	0.45
Unidentified	**1054**	**14.17**	0.41	1.4	0.95	0.98	0.47	0.79	0.27	0.21
Unidentified	**1099**	**15.55**	0.38	1.23	0.76	0.66	0.21	0.43	0.25	0
Nonanal (CAS# 124-90-6)	**1107**	**15.79**	1.62	2	2.07	0.44	0	1.08	0.27	0.28
4-Methyl-1-propan-2-ylbicyclo[3.1.0] hexan-3-one (3-Thujanone) (CAS# 1125-12-8)	**1112**	**15.94**	17.76	0.2	0	0	62.22	33.12	54.23	59.35
(1S,4R,5R)-4-Methyl-1-propan-2-ylbicyclo[3.1.0]hexan-3-one (α-Thujone) (CAS# 546-80-5)	**1124**	**16.28**	5.94	0	0	0	16.03	10.23	18.47	19.27
Unidentified	**1886**	**33.81**	0.43	0	12.87	0.34	0	0	0	0

Substances were putatively identified using the Kovats indices and the NIST library. The relative proportion of each specific substance is given in relation to the sum of all compounds. RT is the retention time and G, M, A, and O symbolize the nest names. Two samples of each family were analyzed

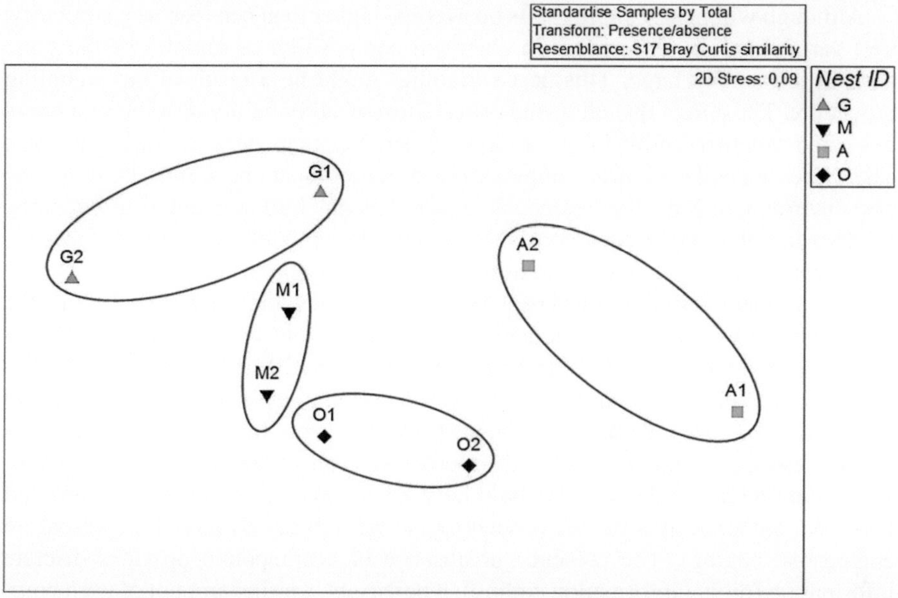

Fig. 12.3 Nest odor similarities of nests of *Taeniopygia guttata*. Each symbol represents one nest odor sample. The axes are dimensionless. The closer the symbols appear on the plot, the more similar the two odor samples are. Each symbol type and letter encodes one family and the number behind the letter indicates the sample number (one and two) of each family nest. The odor profiles from the same nest are on average more similar (symbolized by the circles) than between two different nest samples (ANOSIM, Global $R=0.438$, $p=0.048$)

different individuals and their respective secretions and excretions present in the nest. Nest odors are probably mixtures of self-produced substances in the form of preen oil secretion, feces or saliva, or products of different microbe communities present on the bird itself or in the nest.

Some of the putative substances found here (Table 12.1) have also been found on feathers or in preen oil secretion of other bird species (Douglas et al. 2001; Hagelin et al. 2003; Martín-Vivaldi et al. 2010; for review see Campagna et al.2012). For example, aldehydes, such as hexanal, octanal, and heptanal, have been found on feathers in crested auklets (Hagelin et al. 2003). One interesting compound, thujone, was found in most of our nest odor samples in high concentrations. Thujone is a common floral compound and can be found, for example, in different species of *Artemisia* (Rezaeinodehi and Khangholi 2008). It has been described that this compound has anthelmintic (Albert-Puleo 1978) and antimicrobial (Dorman and Deans 2000) functions, potentially influencing nest microecology and hygiene. Another substance, 2,3-butanediol, is a well-known product of microbial metabolism (Syu 2001). To gain a more detailed knowledge about the origin of substances important in nest recognition and their potential ecological relevance (Charpentier et al. 2012), further studies are needed.

Although within nest similarity is on average higher than between nest similarity, nest variability, i.e., differences in chemical composition of samples of the same nest, is reasonably large. This nest variability might be a result of our sampling procedure. The aim of this pilot study was to test whether family-specific nest odors exist and whether within nest variation is smaller than between nest variation. Therefore, we collected nest samples from different locations within the nest. The nest environment is not homogenously covered by any bird material. Consequently, the two samples might have consisted of different amounts of feathers, feces, or other relevant material, leading to differences in the group of compounds.

In a number of studies, it has been demonstrated now that zebra finches are able to distinguish between their own and a foreign nest exclusively based on olfactory cues (Caspers and Krause 2011; Krause and Caspers 2012; Krause et al. 2012). Thereby, the family-specific odor is learned around or shortly after hatching (Caspers et al. 2013). Family-specific nest odors in zebra finches are not the result of the presence and absence of family-specific substances, i.e., the presence of substances in one nest only (which would have been equal to digital-like coding), but more the outcome of different concentrations of substances (which represent an analog-like coding). The presence or absence of a compound provides discrete information (digital information coding), whereas the varying amount of each compound allows continuous information coding (or analog information coding) (Bradbury and Vehrencamp 2011). The ability of fledglings to recognize their nest based on a continuous signal may underline the sensitivity of the sensory modality and the importance of chemical cues for intraspecific communication (Caspers and Krause 2013).

Acknowledgments The study was part of the Master's thesis of SK. ETK was funded by the Volkswagen Foundation (85994). BAC is funded by a Freigeist Fellowship from the Volkswagen Foundation.

References

Albert-Puleo M (1978) Mythobotany, pharmacology, and chemistry of thujone-containing plants and derivatives. Econ Bot 32:65–74

Bang BG, Cobb S (1968) Size of olfactory bulb in 108 species of birds. Auk 85:55–61

Bonadonna F, Cunningham GB, Jouventin P, Hesters F, Nevitt GA (2003) Evidence for nest-odour recognition in two species of diving petrel. J Exp Biol 206:3719–3722

Bradbury JW, Vehrencamp SL (2011) Principles of animal communication. Sinauer Associates, Sunderland, MA

Campagna S, Mardon J, Celerier A, Bonadonna F (2012) Potential semiochemical molecules from birds: a practical and comprehensive compilation of the last 20 years studies. Chem Senses 37:3–25

Caspers BA, Krause ET (2011) Odour-based natal nest recognition in the zebra finch (*Taeniopygia guttata*), a colony-breeding songbird. Biol Lett 7:184–186

Caspers BA, Krause ET (2013) Intraspecific olfactory communication in zebra finches (*Taeniopygia guttata*)—potential information apart from visual and acoustic cues. In: East ML, Dehnhard M (eds) Chemical signals in vertebrates 12. Springer Verlag, Berlin, pp 341–351

Caspers BA, Schroeder FC, Franke S, Voigt CC (2011) Scents of adolescence: the maturation of the olfactory phenotype in a free-ranging mammal. PLoS ONE 6, e21162

Caspers BA, Hoffman JI, Kohlmeier P, Krüger O, Krause ET (2013) Olfactory imprinting as a mechanism for nest odour recognition in zebra finches. Anim Behav 86:85–90

Charpentier MJE, Barthes N, Proffit M, Bessière JM, Grison C (2012) Critical thinking in the chemical ecology of mammalian communication: roadmap for future studies. Funct Ecol 26:769–777

Clarke KR, Warwick RM (1994) Similarity-based testing for community pattern: the two-way layout with no replication. Mar Biol 118:167–176

Dorman HJD, Deans SG (2000) Antimicrobial agents from plants: antibacterial activity of plant volatile oils. J Appl Microbiol 88:308–316

Douglas H, Jones T, Conner W (2001) Heteropteran chemical repellents identified in the citrus odor of a seabird (crested auklet: *Aethia cristatella*): evolutionary convergence in chemical ecology. Naturwissenschaften 88:330–332

Gwinner H (2013) Male European starlings use odorous herbs as nest material to attract females and benefit nestlings. In: East ML, Dehnhard M (eds) Chemical signals in vertebrates 12. Springer Verlag, Berlin, pp 353–362

Gwinner H, Berger S (2008) Starling males select green nest material by olfaction using experience-independent and experience-dependent cues. Anim Behav 75:971–997

Hagelin JC, Jones IL, Rasmussen LEL (2003) A tangerine-scented social odour in a monogamous seabird. Proc R Soc Lond B 270:1323–1329

Hughes BO, Petherick JC, Brown MF, Waddington D (1995) Visual recognition of key nest site stimuli by laying hens in cages. Appl Anim Behav Sci 42:271–281

Krause ET, Caspers BA (2012) Are olfactory cues involved in nest recognition in two social species of estrildid finches? PLoS ONE 7, e36615

Krause ET, Krüger O, Kohlmeier P, Caspers BA (2012) Olfactory kin recognition in a songbird. Biol Lett 8:327–329

Krause ET, Brummel C, Kohlwey S, Baier MC, Müller C, Bonadonna F, Caspers BA (2014) Differences in olfactory species recognition in the females of two Australian songbird species. Behav Ecol Sociobiol 68:1819–1827

Lorbeer E, Mayr M, Hausmann B, Kratzl K (1984) Zur Identifizierung flüchtiger Substanzen aus biologischem Material mit Hilfe der CLSA (Closed loop stripping apparatus). Monatsh Chem 115:1107–1112

Martín-Vivaldi M, Peña A, Peralta-Sánchez JM, Sánchez L, Ananou S, Ruiz-Rodríguez M, Soler JJ (2010) Antimicrobial chemicals in hoopoe preen secretions are produced by symbiotic bacteria. Proc R Soc Lond B 277:123–130

Mennerat A (2008) Blue tits (*Cyanistes caeruleus*) respond to an experimental change in the aromatic plant odour composition of their nest. Behav Processes 79:189–191

Mennerat A, Perret P, Lambrechts MM (2009) Local individual preferences for nest materials in a passerine bird. PLoS ONE 4, e5104

Minguez E (1997) Olfactory nest recognition by British storm-petrel chicks. Anim Behav 53:701–770

Nevitt GA, Bonadonna F (2005) Sensitivity to dimethyl sulphide suggests a mechanism for olfactory navigation by seabirds. Biol Lett 1:303–305

Petit C, Hossaert-McKey M, Perret P, Blondel J, Lambrechts MM (2002) Blue tits use selected plants and olfaction to maintain an aromatic environment for nestlings. Ecol Lett 5:585–589

Rezaeinodehi A, Khangholi S (2008) Chemical composition of the essential oil of *Artemisia absinthium* growing wild in Iran. Pak J Biol Sci 11:946–949

Syu MJ (2001) Biological production of 2,3-butanediol. Appl Microbiol Biotechnol 55:10–18

Chapter 13
Exploring the Use of Olfactory Cues in a Nonsocial Context in Zebra Finches (*Taeniopygia guttata*)

E. Tobias Krause, Johanna Kabbert, and Barbara A. Caspers

13.1 Introduction

The use of chemical signals in bird communication received increasing attention over the last years (Roper 1999; Hagelin and Jones 2007; Caro and Balthazart 2010; Caspers and Krause 2013; Caro et al. 2015). Several avian taxa have been found to make use of olfactory cues in numerous contexts (Table 13.1). Even songbirds have been shown to be able to use their sense of smell, although they have been long thought to be anosmic due to their relative small olfactory bulbs (Bang and Cobb 1968) and the lack of obvious odor-guided behavior. However, several studies revealed that songbirds, including the zebra finch, possess numerous olfactory receptor genes (Steiger et al. 2008, 2009; Warren et al. 2010). The zebra finch is one of the most used avian laboratory model organisms (Zann 1996; Griffith and Buchanan 2010). As all songbirds, zebra finches are mainly visual and acoustically dominated (Zann 1996), but apart from these sensory modalities they make also use of olfactory cues in several social contexts. Fledglings as well as mothers can distinguish between their own and a conspecifics nest based on olfactory cues (Caspers and Krause 2011; Krause and Caspers 2012; Kohlwey et al. 2015; Fig. 13.1a).

E.T. Krause (✉)
Department of Animal Behaviour, Bielefeld University, Morgenbreede 45, 33615 Bielefeld, Germany

Institute of Animal Welfare and Animal Husbandry, Friedrich-Loeffler-Institut, Dörnbergstr. 25-27, 29223 Celle, Germany
e-mail: tobias.krause@fli.bund.de

J. Kabbert • B.A. Caspers
Department of Animal Behaviour, Bielefeld University, Morgenbreede 45, 33615 Bielefeld, Germany

B.A. Schulte et al. (eds.), *Chemical Signals in Vertebrates 13*,
DOI 10.1007/978-3-319-22026-0_13

Table 13.1 Some examples when olfactory cues are used in nonsocial context by birds

Nonsocial context	Species	Reference
(a) Navigation	Pigeons	Papi et al. (1974), Gagliardo (2013)
	Antarctic prions	Nevitt and Bonadonna (2005)
	Cory's shearwaters	Gagliardo et al. (2013)
	Catbird	Holland et al. (2009)
(b) Foraging	Blue tit	Mennerat et al. (2005)
	Great tit	Amo et al. (2013)
	Several procellariiformes	Nevitt et al. (1995)
	Penguins	Wright et al. (2011)
(c) Response to predators cues	Blue tit	Amo et al (2008)
	House finch	Roth et al. (2008)
	Chicken	Zidar and Løvlie (2012)
(d) Nest material selection	Blue tit	Petit et al. (2002) Mennerat (2008) Mennerat et al. (2009)
	European starlings	Gwinner and Berger (2008) Gwinner (2013)

Fig. 13.1 (**a**) Fledgling zebra finch at the entrance of its natal nest, zebra finches fledge at an age of around 19 days post-hatching; (**b**) a zebra finch brood shortly after hatching, one of the chicks is raising the head for begging; and (**c**) interspecific interaction between a diamond firetail (front) and a zebra finch

Zebra finch chicks learn around hatching (Fig. 13.1b) the scent of their family (Caspers et al. 2013) and can use odors to discriminate kin from non-kin (Krause et al. 2012). Odor also seems to play a role at an interspecific level as zebra finches have different olfactory fingerprints than individuals from a sympatric close-related species and can use those to distinguish between the scent of conspecifics and heterospecifics (Krause et al. 2014; Fig. 13.1c). Our studies so far focused on the use of chemical cues for social communication. Here we explore the use of olfactory cues in zebra finches (*Taeniopygia guttata*) in a nonsocial context without any semiochemical cues.

Here we put our emphasis on the use of olfaction for foraging. Zebra finches are granivores that usually forage on the ground (Zann 1996). In experimental contexts, food can be used as a good motivational stimulus to engage birds in learning and/or exploration tasks (Bischof et al. 2006; Boogert et al. 2008; Krause and Naguib 2011; Brust et al. 2014). A previous study explored the role of olfactory cues in a foraging-related context. Kelly and Marples (2004) suggested that neophobia to a familiar food was not triggered when additionally a new artificial odor was presented. Neophobia was induced when the food was dyed in a new color or when the new color of the food was accompanied by the new artificial odor. We wanted to investigate whether zebra finches can use the scent of their known food alone to locate the food patch.

13.2 Methods

We used 23 adult zebra finch females from the domesticated lab stock at Bielefeld University (Forstmeier et al. 2007; Hoffman et al. 2014). The birds were kept in single-sex groups of 3–4 birds in cages (30×40×83 cm). The birds had *ad libitum* access to standard seed food and water and received a mixture of germinated seeds and egg food (CéDé Premium Eggfood) once a week. The floor of each cage was cleaned twice a week. All birds were housed in the experimental room with a light–dark cycle of 14:10 h. No other birds were housed in this room.

13.2.1 Olfactory Foraging Experiment

The test arena consisted of four identical choice arms and one start box (Fig. 13.2a). It was built out of gray plastic (PVC) and was covered during the test by transparent Plexiglas, to avoid birds from escaping. In each of the four choice arms, a wooden divider was located (Fig. 13.2b). The wooden dividers were varnished with white paint. Behind this divider the odor samples or controls were located, in a way that the birds needed to hop around the divider to see the sample behind. Each divider had small holes (Fig. 13.2c) with a slope of 45°. This way we guaranteed that an air

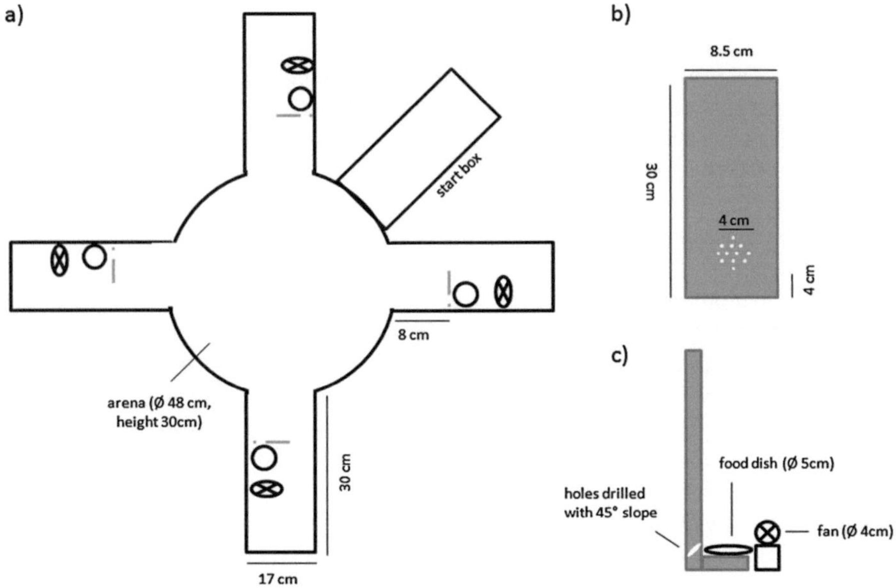

Fig. 13.2 Experimental test arena. (**a**) Test arena from a top view perspective. The arena consisted of four arms where food could potentially be hidden behind dividers. The birds are released to the test arena from a start box. (**b**) Each divider has small holes that allowed an airstream to pass through, created by a fan. (**c**) Holes in the dividers had a slope of about 45° to prevent birds from looking through the holes

stream could pass through the barrier, whereas birds cannot look through. Behind the dividers a white plastic food dish for odor samples or controls and a small fan were placed (Fig. 13.2c). In the experiments, in three of the four arms, the food dish remained empty (control) and in only one arm food was placed in the respective dish to provide a food-related odor. The fans ran in all four arms. A mixture of germinated seeds and protein-rich egg food was used as the food sample. This food represents high-quality nutrition to zebra finches (e.g., Krause and Naguib 2011). We used the odor of the bird's familiar food only, as we wanted to know whether in their daily life, food-related cues might be potentially relevant. Despite this, germinated seeds are a natural food source (Zann 1996) and zebra finches in natural population also occasionally take in insects (Zann and Straw 1984). After each trial the entire test arena was uncovered from the Plexiglas to allow an air exchange with fresh air. After the habituation period and after each experimental trial, the arena and the plastic food dishes were cleaned with 70 % ethanol and water. All experiments were observed using four wireless cameras (dnt, QuattSecure Profiset).

Individuals were habituated to the arena before the tests began. Therefore, a group of 3–4 birds (i.e., all birds from their home cage) was housed inside the arena for 48 h. During this habituation period, water was provided in the central area of the test arena and food was offered in each of the four arms visible to the birds beside the dividers. The start box was not accessible to birds during the habituation phase. Birds were deprived for food in their home cages for about 3 h (mean, 213 min ± 45 min S.D.) prior to the experiment, to ensure that they participate in the behavioral experiments (Krause and Naguib 2011). Each bird was tested individually at four subsequent days and on four trials per day. At the beginning of each trial, birds were placed in the start box and were allowed to habituate there for three minutes. Thereafter, the start box was opened and the birds were individually released to the test arena. During the experiments a fan in each arm created an air stream directed to the center of the arena. Only one of the four air streams transported the smell of food. The location of the food rewarded arm was randomly altered in each trial. Each trial lasted up to 15 min. If a bird located the food in a trial, it was allowed to feed for 10 s before the trial ended. The four trials each day were conducted in a row.

We recorded in each trial and measured on a daily average basis: (1) the latency to leave the start box, (2) the first choice (whether it was correct (=1) or not (=0), the chance level was 0.25), (3) the time to enter the rewarded arm, (4) the time to find the food (i.e., looking behind the barrier in the rewarded arm), and (5) the number of errors (i.e., entering wrong arms). The first choice has been shown to be a valid measure in olfactory choice experiments (e.g., Bonadonna et al. 2006; Amo et al. 2012; Krause et al. 2014).

13.2.2 Statistical Analysis

All measures were analyzed as a mean for each respective experimental day (1–4). We analyzed the average first choices with one-sample t-tests, testing whether the bird's performance differed significantly from the expected probability that the choice was correct (1/4, i.e., 0.25) by chance. Performance significantly better than chance level would indicate that they have used olfactory cues to find the food. We further tested whether the bird's performance differed between days using a Friedman test. With this test we examined whether their performance improved over time, i.e., whether the bird, for example, learned to use olfactory cues to search the food in the tests. The average first choices were analyzed for correlation across the four experimental days using Spearman rank correlations; however due to multiple comparisons the p-values of all correlation have been Bonferroni corrected. The average time to leave the start boxes, to enter the correct arm, and to find the food and the number of errors were analyzed using a Friedman test, to check whether the performance of the birds improved over time or not. All tests were calculated using SPSS 22.

13.3 Results

13.3.1 *First Choice*

The mean first choice at all days of the birds was not significantly different from chance level (0.25) at any of the 4 days (all one-sample t-tests, $t > -0.42$, $p > 0.68$; Fig. 13.3a). Birds did not show any improvement over the subsequent 4 days (Friedman test, df = 3, $X^2 = 4.05$, $p = 0.26$; Fig. 13.3a). However, the mean first choice ratio between day 3 and day 4 correlated significantly (Fig. 13.3b; Table 13.2).

13.3.2 *Latency for Leaving the Start Box*

Birds in general left the start box relatively fast, but latency for leaving the start box reduced significantly over days (medians: day 1, 3.3 s; day 2, 2.5 s; day 3, 1.3 s; day 4, 1.5 s; Friedman test; $N = 23$, df = 3, $X^2 = 21.24$, $p = 0.00009$).

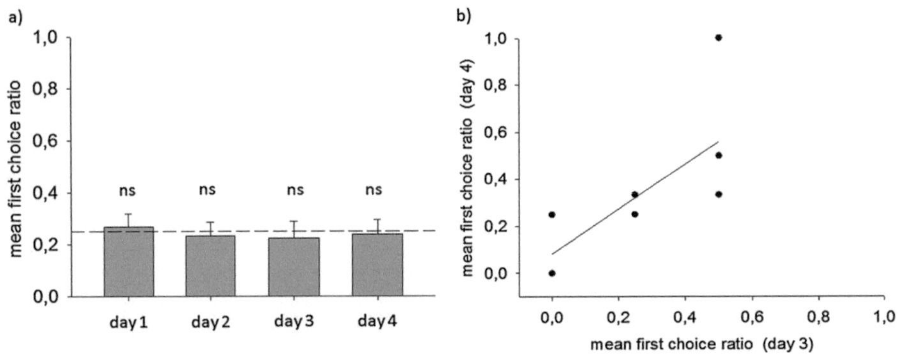

Fig. 13.3 (**a**) Birds' first choice in the test arena shown as means (± SE) for all four test days. The dashed line indicates the chance level, i.e., here 0.25. (**b**) Significant correlation between the mean first choices at days 3 and 4 of testing in the arena. Some data point overlap

Table 13.2 The Spearman correlations of the average first choices of all four testing days are shown. The p-values in all cases have been Bonferroni corrected as indicated by *p#*-values

	Mean first choice ratio			
	Day 1	Day 2	Day 3	Day 4
Day 1	–	$r = -0.18$ $p_{\#} = 1$	$r = 0.01$ $p_{\#} = 1$	$r = -0.35$ $p_{\#} = 1$
Day 2		–	$r = 0.12$ $p_{\#} = 1$	$r = 0.09$ $p_{\#} = 1$
Day 3			–	$r = 0.87$ $p_{\#} = 0.0001$

13.3.3 Time to Enter Rewarded Arm

The time to enter the correct arm did not differ significantly between days, i.e., birds did not become faster (medians: day 1, 620 s; day 2, 552 s; day 3, 687 s; day 4, 631 s; Friedman test $N=23$, df$=3$, $X^2=1.57$, $p=0.67$).

13.3.4 Time to Find the Food

The time to find the food did not differ between days and the birds did not become faster in finding the food over the 4 days (Friedman test, $N=23$, df$=3$, $X^2=1.35$, $p=0.14$; Fig. 13.4).

13.3.5 Number of Errors

The mean number of errors did not differ between days (Friedman test, $N=23$, df$=3$, $X^2=3.93$, $p=0.27$; Fig. 13.5). The error rate, i.e., number of errors per time in the test, did not differ across test days (median errors per min, day 1, 0.20; day 2, 0.29; day 3, 0.05; day 4, 0.11; Friedman $N=23$, df$=3$, $X^2=1.09$, $p=0.78$).

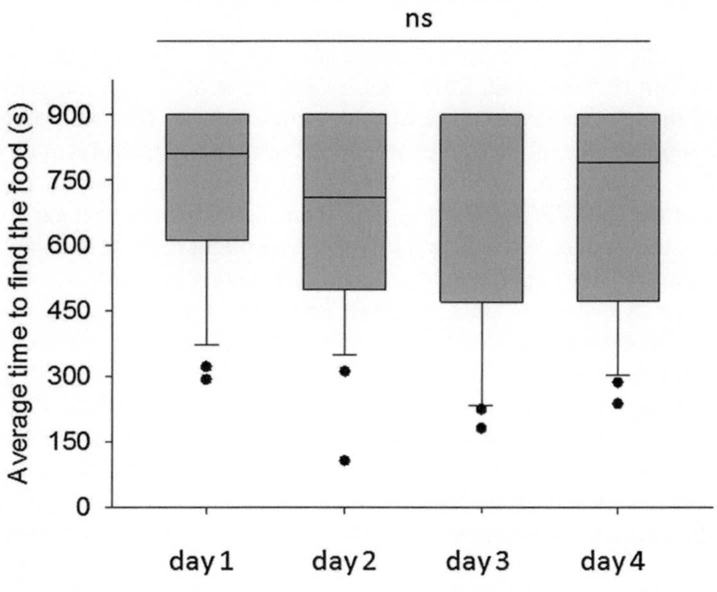

Fig. 13.4 The median time to find the food did not significantly improve over days. Birds showed no training effect, with respect to this parameter. Shown are box plots with median and the quartiles. The dots represent outliers, i.e., data points that lie outside the 10th and 90th percentiles

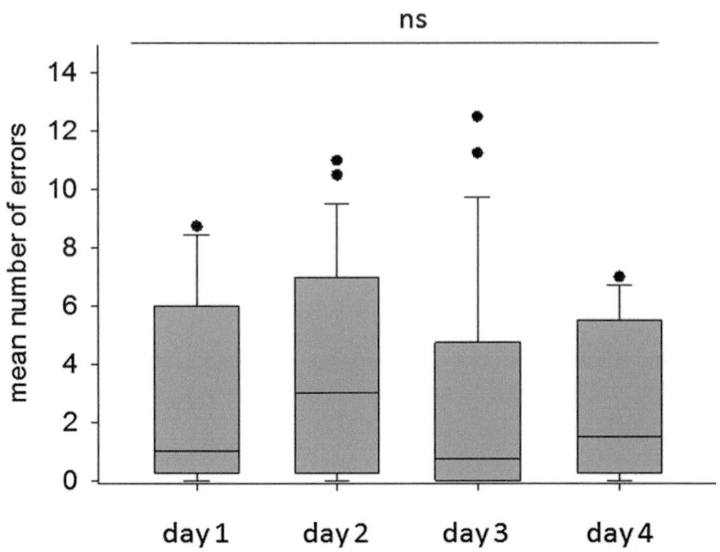

Fig. 13.5 The median number of errors did not significantly decrease over days, indicating that no training effect occurs. Shown are box plots with median and the quartiles. The dots represent outliers, i.e., data points that lie outside the 10th and 90th percentiles

13.4 Discussion

Although zebra finches are known to be able to use olfactory cues for social communication (Krause et al. 2012, 2014; Caspers et al. 2013), our present data shows that they do not seem to use olfactory cues in every context. Zebra finches found the food at none of the four experimental days at a higher probability than at chance level. In addition, the time to find the food and the number of errors until ending up at the food did not improve over time. Zebra finches become faster over the different test days in leaving the start box which might indicate that they become more familiar with the experimental setup and the testing procedures.

This suggests that zebra finches do not primarily rely on olfactory cues while foraging and that the use of olfaction is context dependent in this species. However, their first choices were correlated at the last two days of testing probably giving at least a hint that some initiated an association between the odor of the food and the respective location. In this nonsocial foraging context, investigated here, zebra finches might not primarily rely on chemical cues, which may highlight the context specificity of chemical communication in birds and songbirds in particular. Although other birds, including passerines (e.g., Mennerat et al. 2005) and non-passerines (e.g., Nevitt et al. 1995, Wright et al. 2011), have been shown to use chemical cues for foraging, it always needs to be considered that different species are faced with different natural environments and different selective pressures. Zebra finches forage mainly on seeds, which can probably be easily located by vision. The context specificity has also been found to be important in social contexts, where adult zebra finch females react differently to their own nest odor depending on age of their chicks and the stimulus odor that is presented simultaneously (Krause and Caspers 2012).

It might also be possible that zebra finches do not learn this task. Kelly and Marples (2004) found a similar pattern although they aimed to initiate aversive reactions. However, in other context, such as spatial learning, it is possible to train zebra finches to learn the association between location and food (e.g., Krause and Naguib 2011; Mayer et al. 2010). An alternative explanation might be that the food-related odor used in our apparatus was not strong enough to be located by the zebra finches or that turbulences in the air streams inhibit the birds to find the food in our arena. It would be interesting to test whether zebra finches can find food when the stimuli are reinforced by artificial odors (e.g., Würdinger 1990). As we wanted to investigate whether food-related odors may play a role in the normal housing conditions, we used a food source our zebra finches are quite familiar with.

It could also be possible that zebra finches do not use food-related odors directly, but odors that are linked to the foraging sites as it is known from Procellariiformes and great tits. Procellariiformes, for example, use dimethyl sulfide (DMS) as a foraging cue (Nevitt et al. 1995). DMS is produced by phytoplankton and is an indicator for productive areas. Great tits use chemical cues of infested plants to locate potential insect prey items (Amo et al. 2013). Although this is a fascinating idea, it has so far only been demonstrated in non-herbivorous species, in which the prey is feeding on plants and the plants release olfactory cues. It seems rather unlikely that a similar mechanism may be present in herbivore species, but we cannot rule out this possibility and it might be interesting to explore this idea in future studies.

Whether zebra finches have not perceived or just did not react to the food-related odors cannot fully be answered here. However, it seems extremely unlikely that they cannot perceive the odors as they have been shown to be able to smell in several other experimental studies, but in other mainly social context (Caspers and Krause 2011; Krause et al. 2012; Krause and Caspers 2012; Caspers et al. 2013; Krause et al. 2014). The fact that the outcomes of the third and fourth trial are highly correlated might lead to the conclusion that longer training and testing procedures are needed to train zebra finches to locate food on the basis of smell. This needs to be tested in future experiments. At the moment we can summarize that olfactory cues do not seem to be of major importance for foraging in zebra finches. Taken together, we have shown that the use of chemical cues may be context dependent in zebra finches. In a nonsocial context, such as individual foraging, olfactory cues do seem not to play a primarily role. However, based on previous work, it is well documented that in social context olfactory cues provide an important source of information to these birds (Caspers et al. 2013; Krause et al. 2012, 2014).

Acknowledgments ETK was funded by the Volkswagen Foundation (85994) and BAC is a Freigeist Fellow of the Volkswagen Foundation. The experiments were part of a student research module of JK. We are grateful to the constructive and helpful comments of two anonymous reviewers.

References

Amo L, Galván I, Tomás G, Sanz JJ (2008) Predator odour recognition and avoidance in a songbird. Funct Ecol 22:289–293

Amo L, Avilés JM, Parejo D, Peña A, Rodríguez J, Tomás G (2012) Sex recognition by odour and variation in the uropygial gland secretion in starlings. J Anim Ecol 81:605–613

Amo L, Jansen JJ, van Dam NM, Dicke M, Visser ME (2013) Birds exploit herbivore-induced plant volatiles to locate herbivorous prey. Ecol Lett 16:1348–1355

Bang BG, Cobb S (1968) Size of olfactory bulb in 108 species of birds. Auk 85:55–61

Bischof HJ, Lieshoff C, Watanabe S (2006) Spatial memory and hippocampal function in a non-foodstoring songbird, the zebra finch (Taeniopygia guttata). Rev Neurosci 17:43–52

Bonadonna F, Caro S, Jouventin P, Nevitt GA (2006) Evidence that blue petrel, *Halobaena caerulea*, fledglings can detect and orient to dimethyl sulphide. J Exp Biol 209:2165–2169

Boogert NJ, Giraldeau LA, Lefebvre L (2008) Song complexity correlates with learning ability in zebra finch males. Anim Behav 76:1735–1741

Brust V, Krüger O, Naguib M, Krause ET (2014) Lifelong consequences of early nutritional conditions on learning performance in zebra finches (*Taeniopygia guttata*). Behav Process 103: 320–326

Caspers BA, Krause ET (2011) Odour-based natal nest recognition in the zebra finch (*Taeniopygia guttata*), a colony-breeding songbird. Biol Lett 7:184–186

Caspers BA, Krause ET (2013) Intraspecific olfactory communication in zebra finches (*Taeniopygia guttata*) – Potential information apart from visual and acoustic cues. In: East ML, Dehnhard M (eds) Chemical signals in vertebrates 12. Springer Verlag, Berlin, pp 341–351

Caspers BA, Hoffman JI, Kohlmeier P, Krüger O, Krause ET (2013) Olfactory imprinting as a mechanism for nest odour recognition in zebra finches. Anim Behav 86:85–90

Caro SP, Balthazart J (2010) Pheromones in birds: myth or reality? J Comp Physiol A 196: 751–766

Caro SP, Balthazart J, Bonadonna F (2015) The perfume of reproduction in birds: Chemosignaling in avian social life. Horm Behav 68:25–42

Forstmeier W, Segelbacher G, Mueller JC, Kempenaers B (2007) Genetic variation and differentiation in captive and wild zebra finches (*Taeniopygia guttata*). Mol Ecol 16:4039–4050

Gagliardo A (2013) Forty years of olfactory navigation in birds. J Exp Biol 216:2165–2171

Gagliardo A, Bried J, Lambardi P, Luschi P, Wikelski M, Bonadonna F (2013) Oceanic navigation in Cory's shearwaters: evidence for a crucial role of olfactory cues for homing after displacement. J Exp Biol 216:2798–2805

Griffith SC, Buchanan KL (2010) The zebra finch: the ultimate Australian supermodel. Emu 110:v–xii

Gwinner H, Berger S (2008) Starling males select green nest material by olfaction using experience-independent and experience-dependent cues. Anim Behav 75:971–976

Gwinner H (2013) Male European starlings use odorous herbs as nest material to attract females and benefit nestlings. In: East ML, Dehnhard M (eds) Chemical Signals in Vertebrates 12. Springer Verlag, Berlin, pp 353–362

Hagelin JC, Jones IL (2007) Bird odors and other chemical substances: a defense mechanism or overlooked mode of intraspecific communication? Auk 124:741–761

Hoffman JI, Krause ET, Lehmann K, Krüger O (2014) MC1R genotype and plumage colouration in the zebra finch (*Taeniopygia guttata*): Population structure generates artefactual associations. PLoS One 9, e86519

Holland RA, Thorup K, Gagliardo A, Bisson IA, Knecht E, Mizrahi D, Wikelski M (2009) Testing the role of sensory systems in the migratory heading of a songbird. J Exp Biol 212: 4065–4071

Kelly DJ, Marples NM (2004) The effects of novel odour and colour cues on food acceptance by the zebra finch, *Taeniopygia guttata*. Anim Behav 68:1049–1054

Kohlwey S, Krause ET, Baier MC, Müller C, Caspers BA (2015) Chemical analysis reveal family specific nest odour profiles in Zebra finches (*Taeniopygia guttata*): A pilot study. In: Schulte BA, Ferkin MH, Goodwin TE (eds) Chemical Signals in Vertebrates 13. Springer Verlag, Berlin, pp

Krause ET, Naguib M (2011) Compensatory growth affects exploratory behaviour in zebra finches, *Taeniopygia guttata*. Anim Behav 81:1295–1300

Krause ET, Krüger O, Kohlmeier P, Caspers BA (2012) Olfactory kin recognition in a songbird. Biol Lett 8:327–329

Krause ET, Caspers BA (2012) Are olfactory cues involved in nest recognition in two social species of estrildid finches? PLoS One 7, e36615

Krause ET, Brummel C, Kohlwey S, Baier MC, Müller C, Bonadonna F, Caspers BA (2014) Differences in olfactory species recognition in the females of two Australian songbird species. Behav Ecol Sociobiol 68:1819–1827

Mayer U, Watanabe S, Bischof HJ (2010) Hippocampal activation of immediate early genes Zenk and c-Fos in zebra finches (*Taeniopygia guttata*) during learning and recall of a spatial memory task. Neurobiol Learn Mem 93:322–329

Mennerat A, Bonadonna F, Perret P, Lambrechts MM (2005) Olfactory conditioning experiments in a food-searching passerine bird in semi-natural conditions. Behav Processes 70:264–270

Mennerat A (2008) Blue tits (*Cyanistes caeruleus*) respond to an experimental change in the aromatic plant odour composition of their nest. Behav Process 79:189–191

Mennerat A, Perret P, Bourgault P, Blondel J, Gimenez O, Thomas DW, Heeb P, Lambrechts MM (2009) Aromatic plants in nests of blue tits: positive effects on nestlings. Anim Behav 77: 569–574

Nevitt GA, Veit RR, Kareiva P (1995) Dimethyl sulphide as a foraging cue for Antarctic procellariiform seabirds. Nature 376:680–682

Nevitt GA, Bonadonna F (2005) Sensitivity to dimethyl sulphide suggests a mechanism for olfactory navigation by seabirds. Biol Lett 1:303–305

Papi F, Ioalé P, Fiaschi V, Benvenuti S, Baldaccini NE (1974) Olfactory navigation of pigeons: the effect of treatment with odorous air currents. J Comp Physiol 94:187–193

Petit C, Hossaert-McKey M, Perret P, Blondel J, Lambrechts MM (2002) Blue tits use selected plants and olfaction to maintain an aromatic environment for nestlings. Ecol Lett 5:585–589

Roth TC II, Cox JG, Lima SL (2008) Can foraging birds assess predation risk by scent? Anim Behav 76:2021–2027

Roper TJ (1999) Olfaction in birds. Adv Stud Behav 28:247–332

Steiger SS, Fidler AE, Valcu M, Kempenaers B (2008) Avian olfactory receptor gene repertoires: evidence for a well-developed sense of smell in birds? Proc R Soc Lond B 275:2309–2317

Steiger SS, Kuryshev VY, Stensmyr MC, Kempenaers B, Mueller JC (2009) A comparison of reptilian and avian olfactory receptor gene repertoires: Species-specific expansion of group gamma genes in birds. BMC Genomics 10:446

Warren WC, Clayton DF, Ellegren H, Arnold AP, Hillier LW, Künstner A, Searle S, White S, Vilella AJ, Fairley S et al (2010) The genome of a songbird. Nature 464:757–762

Wright KL, Pichegru L, Ryan PG (2011) Penguins are attracted to dimethyl sulphide at sea. J Exp Biol 214:2509–2511

Würdinger I (1990) Die Reaktionen von Zebrafinken (*Taeniopygia guttata*) auf Düfte—eine Pilotstudie. Vogelwarte 35:359–367

Zidar J, Løvlie H (2012) Scent of the enemy: behavioural responses to predator faecal odour in the fowl. Anim Behav 84:547–554

Zann RA, Straw B (1984) Feeding ecology and breeding of zebra finches in farmland in northern Victoria. Aust Wildl Res 11:533–552

Zann RA (1996) The Zebra Finch—a synthesis of field and laboratory studies. Oxford University Press, Oxford

Chapter 14
Variation in Urinary Amino Acids in the Mozambique Tilapia: A Potential Signal of Dominance or Individuality?

Olesya Kutsyna, Zélia Velez, Adelino V.M. Canário, Tina Keller-Costa, and Peter C. Hubbard

14.1 Introduction

Chemical communication is generally held to be of vital importance in fishes, especially during reproduction; however, given the wide taxonomic and biological diversity within this group, the chemical identification of the pheromones involved is restricted to few species (Stacey and Sorensen 2009). We have recently identified two pregnanetriol 3-glucuronates in the urine of male Mozambique tilapia (*Oreochromis mossambicus* Peters 1852; hereafter "tilapia") which induce an endocrine response in conspecific females (Huertas et al. 2014; Keller-Costa et al. 2014). Moreover, the urinary concentration of these steroids glucuronates is positively correlated with the social rank (i.e., dominance index) of the donor. However, male urine contains other odorants, the identity and function of which remain unclear.

Male tilapia use urine to signal to both potential mates and rival males (Almeida et al. 2005; Miranda et al. 2005; Barata et al. 2007, 2008). In addition, the urinary bladder of dominant male tilapia is larger than that of subordinate males, or females. Thus, male tilapia can store more urine for release in the appropriate social context (Miranda et al. 2005; Keller-Costa et al. 2012). In addition to the endocrine effect on females (Huertas et al. 2014; Keller-Costa et al. 2014), males also use urine to signal dominance to rival males and to reduce aggression in territorial disputes (Barata et al. 2007; Keller-Costa et al. 2012). However, attempts to reproduce these effects with the steroid glucuronates alone have met with limited success (Keller-Costa 2014). It seems, therefore, that this urinary pheromone consists of a complex mixture of different odorants.

O. Kutsyna • Z. Velez • A.V.M. Canário • T. Keller-Costa (✉) • P.C. Hubbard
Centro de Ciências do Mar, Universidade do Algarve, Campus de Gambelas, 8005-139 Faro, Portugal
e-mail: tkeller@ualg.pt

B.A. Schulte et al. (eds.), *Chemical Signals in Vertebrates 13*,
DOI 10.1007/978-3-319-22026-0_14

What could these odorants be? One possibility is amino acids. It is well known that fish, in general, have high olfactory sensitivity to amino acids (Hara 1994; Kasumyan 2004). Furthermore, fish urine often contains amino acids at surprisingly high concentrations (up to 40 mM; see Ng et al. 1996; Sato and Suzuki 2001; Pane et al. 2005), and amino acids—or their derivatives—have pheromonal roles in some species (Kawabata 1993; Yambe et al. 2006). The current study, therefore, assessed whether urinary amino acids play a role in chemical communication in the tilapia. Specifically, we tested (1) whether amino acids are present in urine from male tilapia, (2) if so, what is their contribution to the olfactory potency of the urine, and (3) can their concentrations be correlated with the social rank of the donor?

14.2 Materials and Methods

14.2.1 Experimental Animals and Urine Collection

Fish care and experimentation complied with the guidelines of the European Union Council (86/609/EU) and Portuguese legislation for the use of laboratory animals under a "group-1" license issued by the Veterinary General Directorate of the Ministry of Agriculture, Rural Development and Fisheries of Portugal. Sexually mature tilapia were raised in captivity from a brood-stock maintained at the University of Algarve (Faro, Portugal). Males and females were kept together in large 500 l stock tanks until used in experiments. All animals were maintained in recirculated and bio-filtered freshwater at 27 °C, with sand substrate and at a 12 h light/12 h dark photoperiod; they were fed once a day with commercial cichlid food pellets. Individual urine samples from males ($N=22$) of different social rank were taken from social groups consisting of five males and five females of similar standard length (SL in mm) and body weight (BW in g; coefficient of variation of BW less than 5 %) that were maintained for 9 days in 200 l tanks (93×55×50 cm). Males were tagged (T-Bar anchor FD94, Floy Tag Inc., Seattle, WA, USA) and systematic focal observation of their behavior carried out daily. The frequency of submissive displays during agonistic interactions, or absence of dark coloration without social interaction, and dominant behaviors such as aggression (biting, chasing, lateral displays, circling, or mouth-to-mouth fights), nest digging, courtship towards females, or dark coloration without social interaction was recorded over 5 min for each male. A dominance index (DI) ranging from zero to one was calculated for every male each day as the sum of all dominant behaviors and subsequent division by the sum of all dominant and subordinate behaviors. Accordingly, after 5 days of observation, the mean DI was calculated for each male (Keller-Costa et al. 2012, 2014). Males with $DI \geq 0.5$ were defined as "dominant" ($N=9$) whereas males with $DI < 0.5$ were defined as "subordinate" ($N=13$).

Urine was collected from each male by applying gentle pressure to the abdomen immediately anterior to and above the genital papilla, and collecting the resultant stream of urine in a plastic tube. Individual samples were immediately frozen (−20 °C) until subsequent treatment. In addition, a pool of urine was produced from males ($N=7$) that were each kept in a 200 l aquarium together with four or five females; each male contributed equally to this pool.

To remove the potent steroid glucuronates, both pooled urine and individual samples were C18-solid phase extracted prior to measurement of amino acids or use in electro-olfactogram (EOG) recording (see below).

14.2.2 Measurement of Amino Acids

Individual urine samples were diluted in distilled water (1:2 v/v) and passed through C18 solid phase extraction cartridges (tC18 Sep-Pak®, 1 cc, 100 mg, Waters). The aqueous fractions (C18-SPE flow-through) of the urine samples were derivatized with *N,O*-bis(trimethylsilyl)trifluoroacetamide (BSTFA; Supelco, Sigma-Aldrich, Portugal) plus trimethylchlorosilane (TMCS; Merck, Sigma-Aldrich, Portugal) following the procedure described by Deng et al. (2005). Briefly, samples (300 μl) and the internal standard (L-norvaline; Sigma-Aldrich, Portugal), 5 μg ml^{-1}) were placed into a 1.5 ml screw-cap vial and lyophilized. When completely dry, 100 μl acetonitrile and 100 μl BSTFA+TMCS (98:2, v/v) were added. The reaction was performed under microwave irradiation at a power of 750 W for 60s. After cooling to room temperature, 1 μl of each sample was injected into the gas chromatography-mass spectrometry (GC-MS) machine. Each sample was assayed in triplicate and the arithmetic mean used in subsequent analysis.

The GC-MS system employed was a Varian 450-GC, 240 MS ion trap detector (Emilio de Azevedo Campos, SA, Portugal). The capillary column used for chromatography was a 30 m × 0.25 mm I.D.; 0.25 μm film thickness (Bruker, BR-5MS, Bruker, Portugal). Helium (1 ml min^{-1}) was used as carrier gas. The initial oven temperature was set at 60 °C during 5 min followed by an increase to 80 °C at 20 °C min^{-1} and then to 150 °C at 5 °C min^{-1} followed by an increase to 280 °C at 15 °C min^{-1} and this final temperature was maintained for 5 min. The injection port temperature was set at 250 °C and that of the transfer line and the trap at 220 °C and 250 °C respectively. Splitless mode was selected; the mass-selective detector was operated in electron impact (EI) mode at 70 eV of activation energy. To confirm the mass fragment of the derivatives, data were obtained in the full scan mode from m/z 70 to 250.

To construct the calibration curves, amino acids (L-glutamic acid, L-arginine, and L-phenylalanine) were prepared at various concentrations (5, 12.5, 25, 37.5, and 50 μg ml^{-1}) and treated with the same derivatization and GC-MS analytical procedure as for the samples. The calibration curves were obtained by plotting the peak area ratio between the derivatives of amino acids and that of L-norvaline (internal standard). To measure the limits of quantification and detection, standards were serially diluted and processed according to the procedure described above.

14.2.3 Recording the Electro-olfactogram (EOG)

The electro-olfactogram is a D.C. field potential measured in the water (or mucus layer, in the case of terrestrial vertebrates) immediately above the olfactory epithelium, and is extensively used to assess olfactory sensitivity (Scott and Scott-Johnson 2002). The EOG was recorded from tilapia as previously described (Frade et al. 2002). Briefly, tilapia of both sexes (26–164 g) were anesthetised by immersion in water containing 200 mg l^{-1} MS222 (ethyl-3-aminobenzoate methanesulfonate salt; Sigma-Aldrich) and 400 mg l^{-1} $NaHCO_3$. After voluntary movement had stopped, the fish were injected intramuscularly with 3 mg kg^{-1} gallamine triethiodide (in 0.9 % NaCl), and maintained in a padded support with aerated water (containing 100 mg l^{-1} MS222) pumped through the gills *via* a tube placed in the mouth. The olfactory epithelium was exposed by cutting around the nostril, and the end of glass tube with a constant flow of freshwater (approximately 6 ml min^{-1}) was placed in the nasal cavity. Stimulus solutions were introduced into this flow *via* a computer-controlled solenoid valve (4 s). Borosilicate glass micropipettes filled with 4 % agar in 0.9 % NaCl were placed, one (recording electrode) near the center of the raphe, and the other (reference electrode) lightly in contact with the skin of the head. The DC voltage signal was amplified (Neurolog NL102, Digitimer Ltd, Welwyn Garden City, UK) and digitized (Digidata 1330A; Axon Instruments, Inc., now Molecular Devices, Sunnyvale, CA, USA) and stored on a PC running "Axoscope" software (Axon Instruments).

14.2.4 Data Treatment and Statistical Analysis

All odorants were presented as a 4 s pulse in order of increasing concentration with at least 1 min between stimuli. The EOG amplitude was measured (in mV) from the baseline to the peak of the initial downward deflection of the trace. This was blank-subtracted (blank water—the same water used to dilute the stimuli—given as a stimulus) and normalized to the response to 10^{-5} M L-serine, similarly blank-subtracted. For the urine, amino acid mixture, and individual amino acids, linear regression was applied to a plot of normalized EOG amplitude plus 1.5 against log(dilution), using only concentrations giving responses significantly greater than blanks, and the calculated thresholds of detection (intercept on the x axis; Hubbard et al. 2011) and compared by Student's t test for paired data (urine and amino acid mixture) or one-way ANOVA (individual amino acids). The exponent α was calculated as the slope of linear regression to a plot of log(normalized EOG) amplitude against log[amino acid] and compared by Student's t test for paired data (aqueous urine fraction and amino acid mixture) or one-way ANOVA (individual amino acids). The "factor γ" (number of log units increase in concentration to induce one log unit increase in response) is $1/\alpha$ (Byrd and Caprio 1982). Thresholds, normalized EOG amplitudes and α-values were correlated using the Pearson Product Moment. Urinary amino acid concentrations were compared between dominant and subordinate males by the Mann–Whitney rank sum test (as data were not always normally distributed). In all cases, a P value of less than 0.05 was taken as statistically significant. Data are shown as mean ± S.E.M., unless otherwise stated. Multivariate analyses were performed using the software package Canoco for Windows 4.5. Male individuals ($N=22$) were introduced as "samples." Their urinary amino acid concentrations [L-arginine], [L-glutamate], [L-phenylalanine] were introduced as descriptors of each "sample." Male size (i.e., standard length) and social status (i.e., dominance index) were introduced as independent variables. Amino acid concentrations and male size were normalized by dividing each individual value by the sum of all values of a given variable; hence scaling was between 0 and 1 for all variables. First, detrended correspondence analysis (DCA) was run to decide between linear or unimodal modes of data distribution. Subsequently, principal components analysis (PCA) was selected as an appropriate, unconstrained linear ordination method fitting well the extent of variability within the dataset. Amino acid concentrations were square root transformed to prevent a few high values from unduly influencing the ordination. PCA analyses were performed either focussing the scaling on "inter-sample distances" or on "dependent variables correlations" (Lepš and Šmilauer 2003). Results were plotted in a PCA ordination triplot. Redundancy analysis (RDA) and Monte-Carlo-Permutation tests (999 permutations) were used to determine whether the independent variables "dominance index" and "male size" significantly influence the overall variation of the dataset.

14.3 Results

14.3.1 Amino Acid Concentrations in a Pool of Urine from Dominant Males

The aqueous fraction of pooled urine from reproductively active dominant male tilapia contained 1.95 mM L-arginine, 0.71 mM L-glutamic acid, and 0.53 mM L-phenylalanine; no other proteinogenic amino acids were detected A mixture of these three amino acids, at these concentrations, was made in distilled water. This mixture was then aliquoted and frozen (−20 °C) until use for comparison with the aqueous fraction of the urine pool (see below).

14.3.2 Olfactory Potency of Aqueous Urine Fraction and Artificial Amino Acid Mixture

Both the aqueous fraction of male urine and the artificial mixture of amino acids evoked large amplitude, concentration-dependent, normal "fish-type" EOGs in tilapia (Fig. 14.1a). The olfactory potency of the aqueous fraction of the urine pool

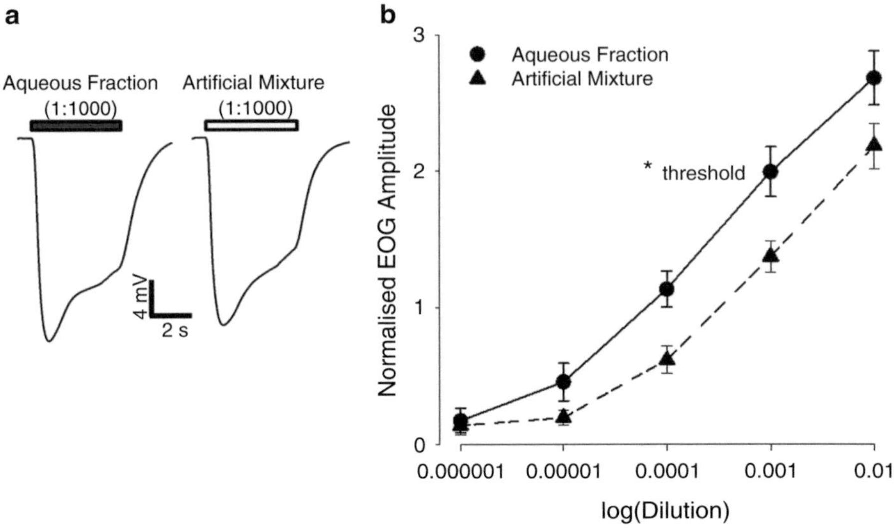

Fig. 14.1 Olfactory sensitivity to the aqueous fraction of male urine, and an artificial mixture of amino acids (L-arginine, L-glutamate, and L-phenylalanine) based on their concentrations in that fraction, in the tilapia. (**a**) Typical EOG traces in response to the aqueous fraction of pooled urine from dominant males (*dark grey horizontal bar*) and the artificial mixture of amino acids (*empty horizontal bar*). Both EOGs were recorded from the same fish; a downward deflection of the trace is negative. (**b**) Semi-logarithmic plot of normalized EOG amplitude against dilution of the aqueous fraction of pooled male urine (*circles*) from pooled male urine and artificial amino acid mixture (*triangles*). Note that most, but not all, olfactory potency of the aqueous fraction of urine can be explained by these amino acids. Data are shown as mean ± SEM ($N=8$). The threshold of detection of the aqueous urine fraction ($1{:}10^{6.35\pm0.26}$) was lower than that of the artificial mixture ($1{:}10^{5.67\pm0.17}$), Students *t*-test for paired data, $^*P<0.05$

was slightly, but significantly, greater than that of the artificial amino acid mixture (Fig. 14.1b). The threshold of detection of the aqueous urine fraction was $1{:}10^{6.35\pm0.26}$ whereas that of the artificial mixture was $1{:}10^{5.67\pm0.17}$ ($P<0.05$). The α values (slope of the log-log plot), however, were similar (aqueous fraction 0.31 ± 0.04; mixture 0.36 ± 0.03, $P=0.11$). At the highest concentration used (1:100), the artificial mixture could account for about 80 % of the olfactory potency of the aqueous fraction of the urine pool.

14.3.3 Olfactory Sensitivity to Amino Acids

Tilapia had olfactory sensitivity to all twenty proteinogenic amino acids (Fig. 14.2), giving quasi-linear or exponential concentration-response curves when plotted semi-logarithmically (for example, see Fig. 14.3). The most potent were the sulfur-containing amino acids L-methionine and L-cysteine with thresholds of detection of

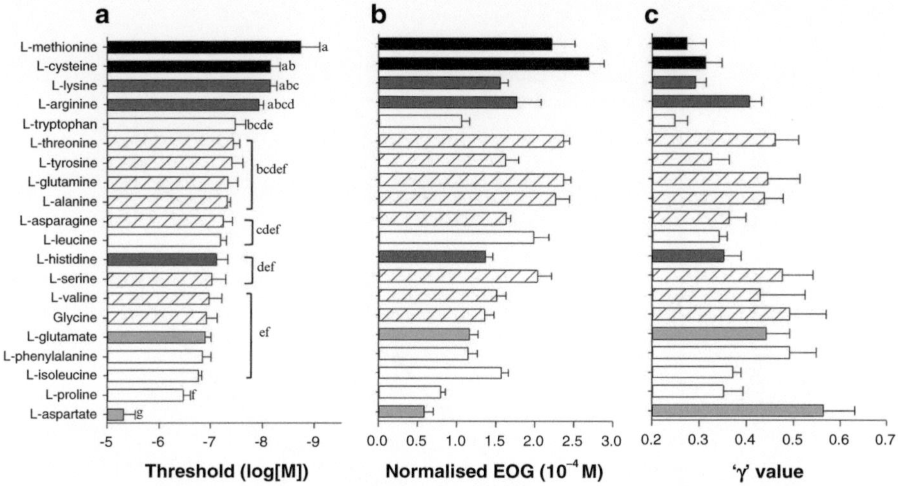

Fig. 14.2 Olfactory sensitivity to the twenty proteinogenic amino acids in the tilapia. (**a**) Calculated thresholds of detection of amino acids, (**b**) normalized amplitudes (at 10^{-4} M), and (**c**) γ values calculated from individual concentration–response curves. Different shades of *grey*/patterns of the bars refer to different chemical properties of the side-chains of the amino acids: sulfur-containing (*black*), positively charged (*dark grey*), negatively charged (*light grey*), hydrophobic (empty), polar/neutral (coarse). Data are shown as mean + SEM ($N = 6$–8). Different letters above bars indicate significant differences between the calculated thresholds of detection of amino acids (one way ANOVA followed by the Tukey post hoc method, $F_{19,110} = 13.915$, $P < 0.001$)

$10^{-8.73 \pm 0.36}$ M and $10^{-8.14 \pm 0.19}$ M, respectively. These were followed by the two amino acids with basic side-chains, L-lysine and L-arginine (thresholds $10^{-8.14 \pm 0.13}$ M and $10^{-7.92 \pm 0.10}$ M, respectively). Thus, of the three amino acids found in significant concentrations in male urine, the most abundant amino acid in male urine, L-arginine, was also the most potent. This group was followed by the rest of the amino acids with thresholds around 10^{-7} M, including L-glutamate ($10^{-6.89 \pm 0.11}$ M) and L- phenylalanine ($10^{-6.84 \pm 0.17}$ M), acidic and aromatic amino acids, respectively, also found in significant quantities in male urine. The other acidic amino acid, L-aspartate, was clearly the least potent amino acid tested, with a threshold of $10^{-5.32 \pm 0.22}$ M. Generally, those amino acids with the lower thresholds also gave the higher amplitude EOGs (correlation coefficient −0.671, $P = 0.0012$), although there were some exceptions (e.g., L-tryptophan and L-serine). On the other hand, the exponent α values were positively correlated with the threshold (coefficient 0.668, $P = 0.0012$), but not amplitude ($P = 0.434$). Thus, those amino acids with the lowest thresholds also tended to have the widest dynamic range, at the expense of sensitivity to changes and/or differences in concentrations. The concentration-response curves for the three urinary amino acids, L-arginine, L-glutamate, and L-phenylalanine, are shown in Fig. 14.3.

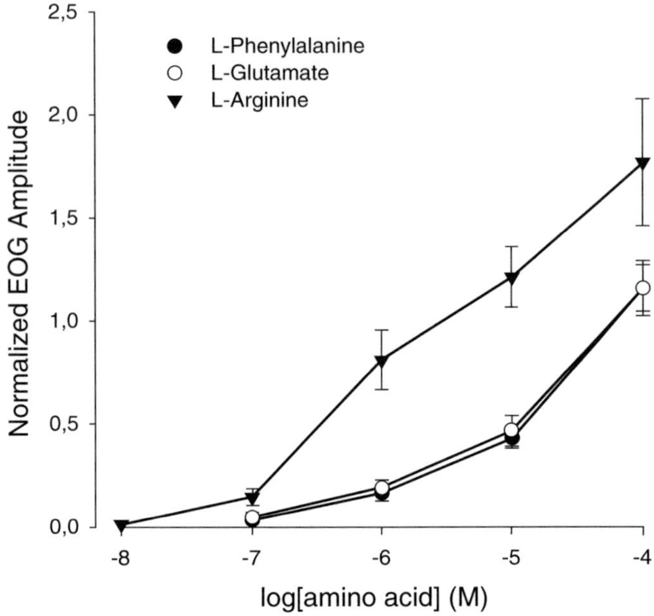

Fig. 14.3 Olfactory sensitivity to urinary amino acids in tilapia. Semi-logarithmic plots of normalized EOG amplitude against amino acid concentration for the three amino acidsL L-phenylalanine (*black circles*), L-glutamate (*empty circles*), and L-arginine (*black inverted triangles*) found in male tilapia urine. Data are shown as mean ± SEM (N=6–8)

14.3.4 Variation of Urinary Amino Acid Concentrations with Social Status

The chemical profiles of the aqueous urine fraction varied between male individuals (Fig. 14.4). The concentrations of L-arginine, but not L-glutamate, were significantly greater in dominant (DI ≥ 0.5) than subordinate males (DI < 0.5); whereas subordinate males had higher concentrations of L-phenylalanine than dominant males (Fig. 14.5a–c). However, there was no correlation between urinary [L-arginine] and [L-glutamate] (P=0.540); two individuals with the highest urinary L-arginine levels, for example, are different from those two with the highest L-glutamate levels. A Principal components (PCA) ordination triplot focussing on "inter-amino-acid correlations" (Fig. 14.5d) was created to visualize correlations between [L-ARGININE], [L-glutamate], and [L-phenylalanine] concentrations in the urine samples of 22 tilapia males. The horizontal axis accounted for 56.8 % of the variation in the data, whereas the vertical axis accounted for 26.7 %; hence more than 83 % of the overall variation in the dataset is explained in this ordination diagram. Redundancy analysis (RDA) and Monte-Carlo-Permutations tests revealed that dominance index (F=5.547; P=0.004) significantly influenced variation in urinary amino acid

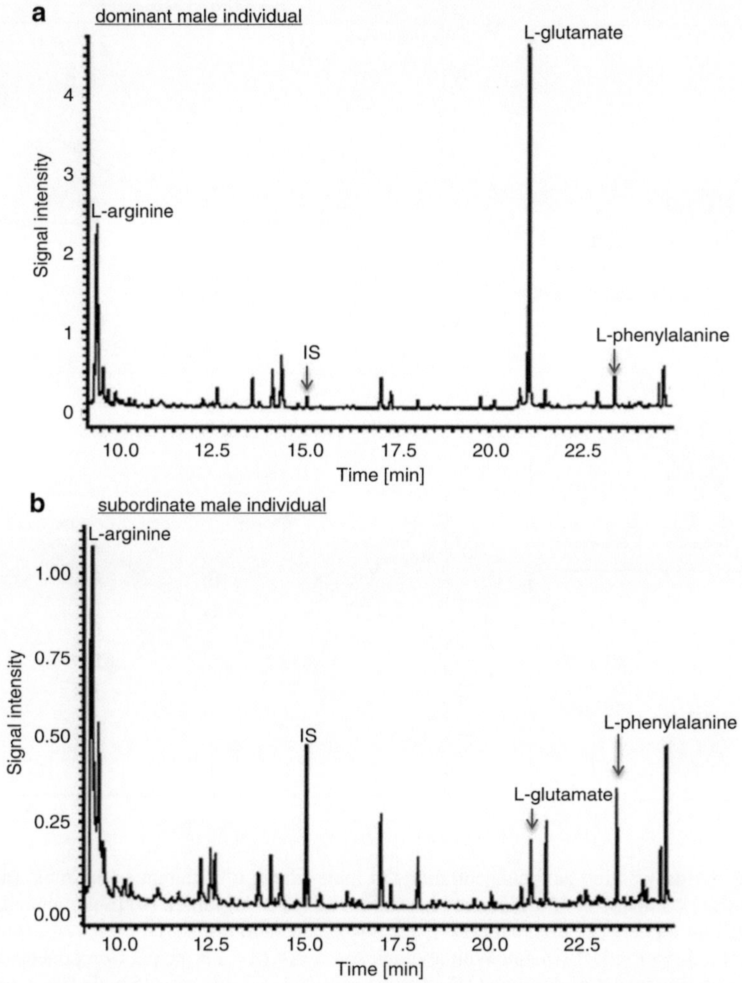

Fig. 14.4 Chemical profiles of the aqueous urine fraction of male Mozambique tilapia. Example (partial) GC-MS chromatograms of the aqueous urine fraction from a dominant (**a**) and subordinate (**b**) male individual. IS—internal standard L-norvaline

concentrations (as indicated also by the long arrow), whereas male size did not ($F=2.671$; $P=0.057$; short arrow). The arrows representing L-arginine concentration and dominance index in the diagram highlight the positive correlation between these two variables as they point in the same direction. To group the tilapia male individuals of different social status in respect of their urinary amino acid concentrations, a second PCA ordination triplot was created, focussing on "inter-male distances" (Fig. 14.5e); males with high amino acid concentrations are grouped further apart from the intercept. Euclidean distances between subordinate males were in general much shorter and hence they clustered together, while dominants are more spread out following their greater variations in urinary amino acids.

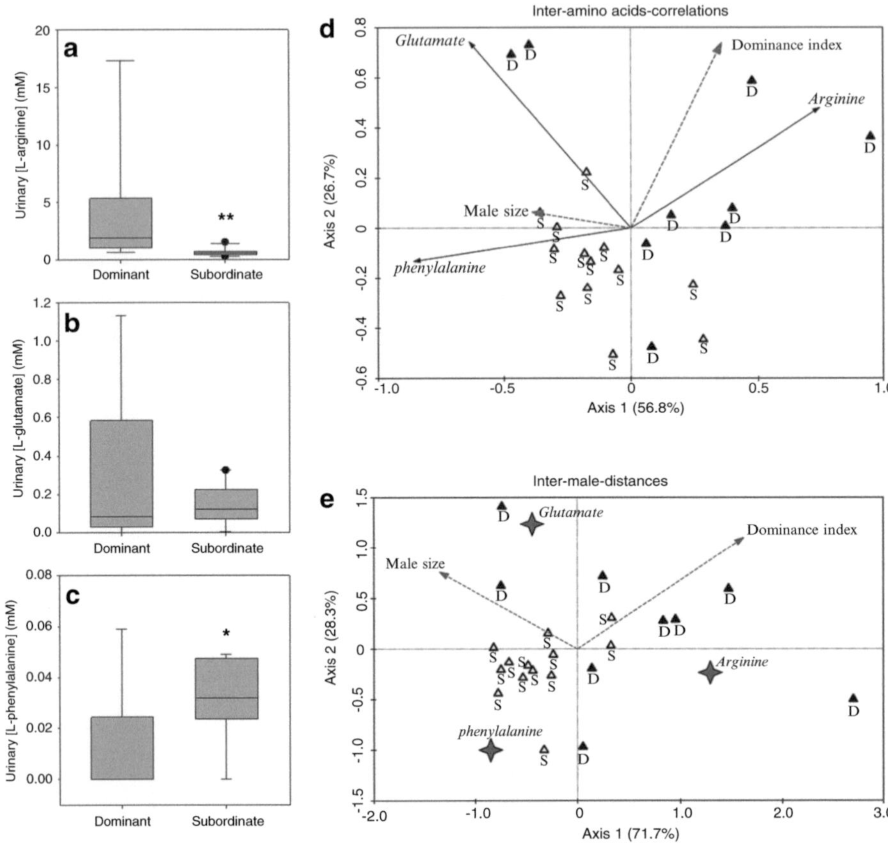

Fig. 14.5 Urinary amino acid concentrations in male tilapia of different social rank. (**a–c**) Box plots showing comparison of urinary concentrations of (**a**) [L-arginine], (**b**) [L-glutamate], and (**c**) [L-phenylalanine] between dominant (DI $\geq$ 0.5; N=9) and subordinate (DI < 0.5; N=13) male tilapia; * P<0.05, ** P<0.01 (Mann–Whitney rank sum test). (**d**, **e**) Principal Components Analysis (PCA) ordination triplot of dominant (D - *black triangles*) and subordinate male (S - *empty triangles*) individuals (N=22; "samples"), the respective urinary amino acid concentrations (dependent variables), and the respective dominance index and body size (independent variables; *dashed grey arrows*). *Arrows* pointing in the same direction approximate properties that share positive correlation. Pairs of properties whose arrows diverge at $\geq$ 90° possess no or negative (linear) correlation. (**d**) Focus on "dependent variables (i.e., amino acid concentrations) correlations": [L-arginine] concentration and social status were predicted to be positively correlated with each other. Male size and dominance index were not correlated as a result of the experimental design: males in each social group were size-matched. (**e**) Focus on "inter-sample (i.e., male) distances": subordinate male individuals cluster together while dominant males are more scattered and spread out based on their larger variation in urinary amino-acid concentrations. *Large stars* represent centroid positions of these amino acids

14.4 Discussion

The current study has shown that urine from dominant male Mozambique tilapia has substantial concentrations (up to 17 mM) of the amino acids L-arginine, L-glutamate, and L-phenylalanine, and that the concentrations of L-arginine correlate positively with the social status of the donor. Conversely, urinary L-phenylalanine concentrations were significantly higher in subordinate than dominant males, whereas L-glutamate concentrations were more varied in dominant males than subordinates. In common with many other species (see reviews; Hara 1994; Kasumyan 2004), the tilapia has high olfactory sensitivity to amino acids, and those present in the urine constitute a large proportion of the olfactory activity of the aqueous urine fraction. Together, this suggests that urinary amino acids play a role in chemical communication in this species.

We have previously shown that 5β-pregnane-3α,17α,20β-triol-3α-glucuronate, and its epimer 5β-pregnane-3α,17α,20α-triol-3α-glucuronate, act as urinary pheromones in the tilapia, at least in a reproductive context (Keller-Costa et al. 2014). However, these steroids alone do not mimic the aggression-modulating effect of untreated urine in male–male interactions (Keller-Costa et al. 2012, 2014). Given that urinary [L-arginine] and, possibly, [L-glutamate] are positively correlated with dominance index (DI), whereas L-phenylalanine is present at lower concentrations in dominant males, and that the tilapia has olfactory sensitivity to all three, it is possible that these amino acids play a role in chemical communication and potentially signal social dominance when mixed with steroids or other compound found in the urine.

Nevertheless, the nature of the correlation of urinary concentration of the steroid glucuronates with dominance index is different from that of the urinary amino acids; several dominant males had lower urinary [L-arginine] and/or [L-glutamate], sometimes even as low as those of subordinate males, whereas the correlation of DI with the urinary steroid glucuronates is tighter and much more linear (Keller-Costa et al. 2014). Furthermore, the variance in urinary amino acid concentrations was greater in dominant males than in subordinates. Despite the correlation of both [L-arginine] and [L-glutamate] with dominance index, there was no correlation *between* the two amino acids. Therefore, the presence of these amino acids in the urine is unlikely simply due to dietary excess and certainly not due to different feed (all tilapia received the same). Another possible explanation is that the amino acids are responsible for providing an individual "signature mixture" to the urinary dominance pheromone (Wyatt 2010). Such that, male tilapia may identify previous rivals via urinary olfactory cues in a way that is used by lobsters (Karavanich and Atema 1998; Johnson and Atema 2005) and mice (see Hurst and Beynon 2004, for review).

It is possible, too, that the identified amino acids and other unidentified odorant(s) present in the aqueous fraction of male urine are involved in mate-choice by females (Barata et al. unpublished). For example, female Nile tilapia can discriminate between high-protein-fed and low-protein-fed males by olfaction. They spend more time in the vicinity of holding water from well-fed males than conditioned water from males fed with low-protein diet (Giaquinto et al. 2010). Male nutritional state may be a good indicator of the male's health and genetic quality. It is conceivable that these females differentiated between the males on the basis of their secretory nitrogenous end products. MHC-related and/or diet-derived odorants from males are important in mate-choice by female sticklebacks (Milinski et al. 2005; Rafferty and Boughman 2006; Ward et al. 2007), whereas polar conspecific-derived odorants (possibly amino acids) synergize the effects of prostaglandin pheromones and act as species-specific markers in cyprinids (Levesque et al. 2011; Lim and Sorensen 2011). Territorial and non-territorial fathead minnow males also differ in their urinary profiles: trimethylamine, a hydrophilic nitrogenous excretory end product of teleost urine, has been suggested as potential chemical signals of social status in this species (Martinovic-Weigelt et al. 2012).

14.5 Olfactory Sensitivity to Amino Acids

The olfactory sensitivity of the tilapia to amino acids was similar to many other teleosts, with thresholds of detection within the range 10^{-9} to 10^{-5} M (Kasumyan 2004). The sulfur containing amino acids, L-methionine and L-cysteine, were the most potent whereas L-proline and L-aspartate were least potent. This is fairly consistent across different species and seems largely independent of phylogeny, diet or habitat (for example, see; Byrd and Caprio 1982; Velez et al. 2005; Hubbard et al. 2011; Yacoob et al. 2004; Silver 1982). Furthermore, there was a negative correlation between threshold and EOG amplitude (at 10^{-4} M)—those amino acids with low detection thresholds evoked larger amplitude EOGs—but there were several outliers (e.g., L-tryptophan and L-serine). Also, those amino acids with low detection thresholds tended to have lower α values (again, with several exceptions; L-phenylalanine and L-threonine, for example); high olfactory sensitivity is usually accompanied by a wide dynamic range. This would appear to be at the cost of sensitivity to *changes* in concentration; L-aspartate had the highest α value (i.e., largest change in response for a given concentration change) and highest threshold of detection. Nevertheless, olfactory sensitivity to the three urinary amino acids is sufficient to detect them in urine diluted down to about 1:100,000. Furthermore, although sensitivity to these three was unexceptional compared to other amino acids, and other species, it is curious that the order of sensitivity followed the order of relative abundance in urine. In the natural environment, however, other amino acids from a variety of different sources may complicate conspecific signaling with amino acids; salmonids, for example, use amino acids present in the water to identify natal streams (Shoji et al. 2000; Yamamoto et al. 2010). More information on both environmental concentrations and sources of individual amino acids, and the olfactory receptor mechanisms involved are needed to understand olfactory sensitivity to amino acids in tilapia.

14.6 Urinary Amino Acids and Social Status

There are differences between teleost fishes and land animals in nitrogen excretion. Teleosts, for example, are unable to synthesize urea *via* the Krebs ornithine–citrulline–arginine cycle (main route of uorogenesis in mammals). Normally, the chief end product in teleosts is amonina, mostly excreted *via* the gills (Marshall and Grosell 2006). Hence, only a small fraction of the total nitrogen excreted by teleosts appears in the urine (Foster and Goldstein 1969). Nevertheless, considerable urinary L-arginine and L-glutamate concentrations were found in tilapia but they varied independently of each other; both varied more widely in dominant than subordinate males. How, then, are the urinary levels of L-arginine and L-glutamate regulated? If the presence of amino acids in the urine is not simply due to dietary excess, then this suggests the existence of distinct renal mechanisms either for the active extrusion of L-arginine and L-glutamate into the urine, or the prevention of their tubular reabsorption, in dominant males. Furthermore, these renal mechanisms must be somehow influenced by social status; significantly, in this context, the kidneys of male tilapia are larger than those of females (Keller-Costa et al. 2012).

L-arginine and L-phenylalanine are among the ten essential amino acids in fishes required for normal growth and metabolism (the others are: L-histidine, L- isoleucine, L-leucine, L-lysine, L-methionine, L-threonine, L-tryptophan, and L-valine (Wilson and Halver 1986). Tilapia require about 30 to 40 % of their diet to be protein, four percent of which needs to be L-arginine. This raises the question: why would a male excrete an essential compound (i.e., L-arginine) in such large quantities (up to 17 mM)? All urine donor males in this study appeared healthy and active, so we do not believe that the increased amino acid concentrations in the urine are due to illness or compromised renal function. It is possible that high rates of amino acid loss *via* the urine can only be withstood by males in good condition and/or with access to sufficient food, and thereby signaling this information to conspecifics.

Why are L-arginine, L-glutamate, and L-phenylalanine, in particular, present at high concentrations in urine? It is significant that the three urinary amino acids have side chains with markedly different chemical properties (basic, acidic and aromatic, respectively). It is likely, therefore, that they bind to different olfactory receptors, and are consequently perceived as different by the nervous system (Caprio and Byrd 1984; Friedrich and Korsching 1997). If so, this would have clear implications for recognition of "individual signatures."

Acknowledgements T.K.-C. acknowledges a doctoral fellowship (grant number: SFRH/BD/46192/2008) and Z.V. acknowledges o postdoctoral fellowship (grant number: SFRH/BPD/63417/2009), both from the Foundation for Science and Technology of Portugal.

References

Almeida OG, Miranda A, Frade P, Hubbard PC, Barata EN, Canário AVM (2005) Urine as a social signal in the Mozambique tilapia (*Oreochromis mossambicus*). Chem Senses 30:i309–i310

Barata EN, Hubbard PC, Almeida OG, Miranda A, Canário AVM (2007) Male urine signals social rank in Mozambique tilapia (*Oreochromis mossambicus*, Peters 1852). BMC Biol 5:54

Barata EN, Fine JM, Hubbard PC, Almeida OG, Frade P, Sorensen PW, Canário AVM (2008) A sterol-like odorant in the urine of Mozambique tilapia males likely signals social dominance to females. J Chem Ecol 34:438–449

Byrd RP Jr, Caprio J (1982) Comparison of olfactory receptor (EOG) and bulbar (EEG) responses to amino acids in the catfish, *Ictalurus punctatus*. Brain Res 249:73–80

Caprio J, Byrd RP Jr (1984) Electrophysiological evidence for acidic, basic, and neutral amino acid olfactory receptor sites in the catfish. J Gen Physiol 84:403–422

Deng C, Yin X, Zhang L, Zhang X (2005) Development of microwave-assisted derivatization followed by gas chromatography/mass spectrometry for fast determination of amino acids in neonatal blood samples. Rapid Commun Mass Spectrom 19:2227–2234

Foster RP, Goldstein L (1969) Formation of excretory products. In: Hoar WS, Randall DJ (eds) Fish Physiology Volume 1: excretion, ionic regulation and metabolism, vol 1. Academic, New York, pp 313–350

Frade P, Hubbard PC, Barata EN, Canário AVM (2002) Olfactory sensitivity of the Mozambique tilapia to conspecific odors. J Fish Biol 61:1239–1254

Friedrich RW, Korsching SI (1997) Combinatorial and chemotopic odorant coding in the zebrafish olfactory bulb visualized by optical imaging. Neuron 18:737–752

Giaquinto PC, Berbert CMD, Delicio HC (2010) Female preferences based on male nutritional chemical traits. Behav Ecol Sociobiol 64:1029–1035

Hara TJ (1994) The diversity of chemical stimulation in fish olfaction and gustation. Rev Fish Biol Fisher 4:1–35

Hubbard PC, Barata EN, Ozório ROA, Valente LMP, Canário AVM (2011) Olfactory sensitivity to amino acids in the blackspot seabream (*Pagellus bogaraveo*): a comparison between olfactory receptor recording techniques in seawater. J Comp Physiol A 197:839–849

Huertas M, Almeida OG, Canário AVM, Hubbard PC (2014) Tilapia male urinary pheromone stimulates the female reproductive axis. Gen Comp Endocrinol 196:106–111

Hurst JL, Beynon RJ (2004) Scent wars: the chemobiology of competitive signalling in mice. Bioessays 26:1288–1298

Johnson ME, Atema J (2005) The olfactory pathway for individual recognition in the American lobster *Homarus americanus*. J Exp Biol 208:2865–2872

Karavanich C, Atema J (1998) Individual recognition and memory in lobster dominance. Anim Behav 56:1553–1560

Kasumyan AO (2004) The olfactory system in fish: structure, function, and role in behavior. J Ichthyol 44(Suppl 2):S180–S223

Kawabata K (1993) Induction of sexual behavior in male fish (*Rhodeus ocellatus ocellatus*) by amino acids. Amino Acids 5:323–327

Keller-Costa T (2014) Chemical identification of dominance pheromones in Mozambique tilapia males. Doctoral thesis, Univseristy of Evora, Evora

Keller-Costa T, Lopes OS, Almeida O, Hubbard PC, Iacovella A, Lima M, Barata EN, Canário AVM (2012) Muscular hypertrophy of urinary bladders in dominant tilapia facilitates the control of aggression through urinary signals. Behaviour 149:953–975

Keller-Costa T, Hubbard PC, Paetz C, Nakamura Y, da Silva JP, Rato A, Barata EN, Schneider B, Canario AVM (2014) Identity of a tilapia pheromone released by dominant males that primes females for reproduction. Curr Biol 24:2130–2135

Lepš J, Šmilauer P (2003) Multivariate analysis of ecological data using CANOCO. Cambridge University Press, Cambridge

Levesque HM, Scaffidi D, Polkinghorne CN, Sorensen PW (2011) A multi-component species identifying pheromone in the goldfish. J Chem Ecol 37:219–227

Lim H, Sorensen PW (2011) Polar metabolites synergize the activity of prostaglandin F_{2a} in a species-specific hormonal sex pheromone released by ovulated common carp. J Chem Ecol 37:695–704

Marshall SW, Grosell M (2006) Ion transport, osmoregulation and acid-base balance. In: Evans DH, Claiborne JB (eds) The physiology of fishes, 3rd edn. CRC Press, New York, pp 200–2001

Martinovic-Weigelt D, Ekman DR, Villeneuve DL, James CM, Teng Q, Collette TW, Ankley GT (2012) Fishy aroma of social status: urinary chemo-signalling of territoriality in male fathead minnows (*Pimephales promelas*). PLoS ONE 7(11), e46579

Milinski M, Griffiths S, Wegner KM, Reusch TBH, Haas-Assenbaum A, Boehm T (2005) Mate choice decisions of stickleback females predictably modified by MHC peptide ligands. Proc Natl Acad Sci U S A 102:4414–4418

Miranda A, Almeida OG, Hubbard PC, Barata EN, Canário AVM (2005) Olfactory discrimination of female reproductive status by male tilapia (*Oreochromis mossambicus*). J Exp Biol 208:2037–2043

Ng WK, Hung SSO, Herold MA (1996) Poor utilization of dietary free amino acids by white sturgeon. Fish Physiol Biochem 15:131–142

Pane EF, Bucking C, Patel M, Wood CM (2005) Renal function in the freshwater rainbow trout (*Oncorhynchus mykiss*) following acute and prolonged exposure to waterborne nickel. Aquat Toxicol 72:119–133

Rafferty NE, Boughman JW (2006) Olfactory mate recognition in a sympatric species pair of three-spined sticklebacks. Behav Ecol 17:965–970

Sato K, Suzuki N (2001) Whole-cell response characteristics of ciliated and microvillous olfactory receptor neurons to amino acids, pheromone candidates and urine in rainbow trout. Chem Senses 26(9):1145–1156

Scott JW, Scott-Johnson PE (2002) The electoolfactogram: a review of its history and uses. Microsc Res Tech 58:152–160

Shoji T, Ueda H, Ohgami T, Sakamoto T, Katsuragi Y, Yamauchi K, Kurihara K (2000) Amino acids dissolved in stream water as possible home stream odorants for masu salmon. Chem Senses 25:533–540

Silver WL (1982) Electrophysiological responses from the peripheral olfactory system of the American eel, *Anguilla rostrata*. J Comp Physiol A 148:379–388

Stacey N, Sorensen P (2009) Hormonal pheromones in fish. In: Pfaff DW, Arnold AP, Etgen AM, Fahrbach SE, Rubin RT (eds) Hormones, brain and behavior, vol 1, 2nd edn. Academic Press, San Diego, pp 639–681

Velez Z, Hubbard PC, Barata EN, Canário AVM (2005) Evidence for functional asymmetry in the olfactory system of the Senegalese sole (*Solea sensgalensis*). Physiol Biochem Zool 78: 756–765

Ward AJW, Webster MM, Hart PJB (2007) Social recognition in wild fish populations. Proc R Soc B 274:1071–1077

Wilson RP, Halver JE (1986) Protein and amino acid requirements of fishes. Annu Rev Nutr 6:225–244

Wyatt TD (2010) Pheromones and signature mixtures: defining species-wide signals and variable cues for identity in both invertebrates and vertebrates. J Comp Physiol A 196:685–700

Yacoob SY, Browman HI, Jensen PA (2004) Electroencephalogram recordings from the olfactory bulb of juvenile (0 year) Atlantic cod in response to amino acids. J Fish Biol 65:1657–1664

Yamamoto Y, Hino H, Ueda H (2010) Olfactory imprinting of amino acids in lacustrine sockeye salmon. PLoS ONE 5(1), e8633

Yambe H, Kitamura S, Kamio M, Yamada M, Matsunaga S, Fusetani N, Yamazaki F (2006) L-Kynurenine, an amino acid identified as a sex pheromone in the urine of ovulated female masu salmon. Proc Natl Acad Sci U S A 103:15370–15374

Chapter 15
A Low Phytoestrogen Diet Reduces the Proceptivity But Not the Attractivity of Meadow Voles (*Microtus pennsylvanicus*)

Lyndsey Pierson, Matthew Hetherington, and Michael H. Ferkin

15.1 Introduction

Phytoestrogens are nonsteroidal compounds produced by plants that have the ability to mimic or antagonize the effects of some steroid hormones. Phytoestrogens are frequently described as weakly "estrogen-like" compounds, as they are capable of binding to estrogen receptors and may induce estrogenic action (Whitten and Naftolin 1992; Patisaul et al. 2003; Luine et al. 2006) or antiestrogenic action (Santell et al. 1997). For instance, some studies have shown that phytoestrogens such as the soy-derived isoflavones can affect behavior, mood, and memory in rats, in some cases stimulating behavior, whereas in others inhibiting behavior (Santell et al. 1997; Lephart et al. 2004). More importantly, because of their estrogenic and antiestrogenic effects, and their affinity for estrogen receptors, consumption of phytoestrogens like isoflavones may also affect an animal's reproductive physiology, endocrine milieu, and behaviors associated with reproduction.

The literature is replete with studies using a variety of fish, birds, rodents, ungulates, and primates as subjects to understand the effects of phytoestrogens on reproductive physiology (Wasserman et al. 2013; Brown et al. 2014). Despite this rich source of information, studies of the effects of dietary phytoestrogens on the reproductive physiology of vertebrates provide mixed results. Some studies have shown that phytoestrogens exert a stimulatory effect or estrogenic effect in vertebrates (Musey et al. 1995; Lephart et al. 2002; Song et al. 2006; Zanoli et al. 2009; Zavatti et al. 2009). Other studies have reported inhibitory effects or antiestrogenic effects (Newsome and Kitts 1980; Adams 1995; Cline et al. 1996; Opalka et al. 2006; Tsao et al. 2006; Corbitt et al. 2007) or no effect of dietary phytoestrogens on reproductive

L. Pierson (✉) • M. Hetherington • M.H. Ferkin
Department of Biological Sciences, The University of Memphis,
Ellington Hall, Memphis, TN 38152, USA
e-mail: lmperson@memphis.edu

B.A. Schulte et al. (eds.), *Chemical Signals in Vertebrates 13*,
DOI 10.1007/978-3-319-22026-0_15

physiology of a variety of vertebrates (Anthony et al. 1996; Foth and Cline 1998; Stevenson et al. 2011; Brown et al. 2014). Still other studies have discovered that dietary phytoestrogens have mixed effects such that one element of a subject's reproductive physiology is stimulated, whereas another element is unaffected or inhibited (Howell et al. 1980; Kiparissis et al. 2003; Lephart et al. 2005). Potential explanations to account for such variable results have been attributed to differences in species, methods, type of phytoestrogen used and amount, the aspect of reproductive physiology in question, and age and condition of the subjects (Whitten et al. 1995; Pierson and Ferkin 2014). For example, the effects of isoflavones on sexual behavior vary depending on an animal's initial physiological estrogen levels (Brzezinski and Debi 1999; Lephart et al. 2002; Zanoli et al. 2009), the length and timing of phytoestrogen exposure (Whitten et al. 1995; Lephart et al. 2002), and whether or not it is an omnivore or herbivore (Pierson and Ferkin 2014).

The impact of phytoestrogens on sexual behavior has focused primarily on female mice and rats (Mills and Childs 1998; Wuttke et al. 2002; Lephart et al. 2005). Again, the results of these studies are mixed and it remains unclear if phytoestrogens cause estrogenic or antiestrogenic effects on sexual behavior (Pierson and Ferkin 2014). For example, ovariectomized rats treated with estradiol or the phytoestrogen ferutinin showed increased sexual receptivity when compared to control ovariectomized animals (Zanoli et al. 2009). However, when ovariectomized female rats were treated with estradiol and ferutinin together, it did not increase their interest in male rats (Zanoli et al. 2009). Patisaul et al. (2003) found that phytoestrogens were not sufficient to induce ovariectomized rats to increase their interest in males or likelihood of mating. In another study, however, female rats treated with estradiol and ferutinin showed increases in their sexual receptivity (Zavatti et al. 2009). Male rats treated with phytoestrogens had alterations in the sexually dimorphic areas of the brain (Lephart et al. 2002, 2005), as well as having heavier prostate glands (Tou et al 1998).

The mixed results about the effects of phytoestrogens on sexual behavior may be due to the fact that rats and mice are omnivores and may be rather limited in the amount of phytoestrogens that they consume. It is possible the impact of phytoestrogens on the sexual behavior of rodents may become clearer if we examine its effects on herbivores such as voles. For example, meadow voles (*Microtus pennsylvanicus*) inhabit ephemeral grasslands that differ in plant species density and patchiness seasonally and annually (Getz 1985; Bergeron and Jodoin 1987; Bergeron et al. 1990). Thus, meadow voles may live in fields containing plants such as clover and alfalfa that have high-phytoestrogen content, or they may live in fields that contain plants that have low-phytoestrogen content, such as hay (Batzli 1985; Getz 1985; Cos et al. 2003), which may vary within and between seasons (Eldridge and Kwolek 1983). More importantly, however, male and female voles differ in their space use and the sizes of the areas they inhabit. Male meadow voles live in large, overlapping home ranges that encompass the territories of one or more females (Madison 1980). Thus, males have access to a variety of grasses and forbs that differ in their phytoestrogen content (Batzli 1985; Getz 1985). Conversely, female meadow voles live in small territories (Madison 1980) and may be more limited to foraging on the plants in their territory, which may differ among females and also differ in their phytoestrogen content (Eldridge and Kwolek 1983; Bergeron et al. 1990), which may affect the vole's phenotype and subsequent sexual behaviors.

Two components of sexual behavior are attractivity and proceptivity (Beach 1976). For meadow voles, like many other terrestrial mammals, an individual's attractivity depends on whether its odors and scent marks attract opposite-sex conspecifics (Pierce et al. 2005). Traditionally, the behaviors displayed by females to show interest in male conspecifics are referred to as proceptive behaviors. These behaviors include ear wiggling in female rats and the preference of many female rodents for the odors produced by sexually receptive conspecific males (Beach 1976; Johnston 1979; Brown and Macdonald 1985). However, Pierce et al. (2005) and Hobbs and Ferkin (2012) have included comparable behaviors by males to show interest in female voles. Proceptive-like behaviors for males may include the amount of time they spend investigating scent mark of a female conspecific (Pierce et al. 2005). Together, attractivity and proceptivity establish communication between potential mates and allow them to coordinate behaviors that facilitate or inhibit mating (Beach 1976; Stopka and Macdonald 1998).

Studies have shown that phytoestrogens may affect the circulating gonadal hormone titers of mammals (Cameron 2003; Lephart et al. 2004; Gore 2008) and high titers of gonadal steroids are necessary for meadow voles to produce scent marks that are attractive to opposite-sex conspecifics and maintain preferences for their odors (Ferkin and Zucker 1991; Ferkin and Gorman 1992; Ferkin et al. 1992). In the present study, we hypothesize that phytoestrogens may also affect attractivity and proceptivity in meadow voles because they may interact with estrogen receptors and affect the action of estrogen (i.e., Luine et al. 2006; Pierson and Ferkin 2014). Thus, if phytoestrogens exert an estrogenic effect, we predict that female voles fed a diet low in phytoestrogen content will no longer produce scent marks that are attractive to males and they will no longer spend more time investigating the scent marks of a male conspecific than that of a female conspecific. We predicted the converse for the attractivity and proceptivity of female voles fed a diet containing a moderate amount of phytoestrogens and for male voles fed either a diet containing a moderate amount of phytoestrogen or a diet containing a low amount of phytoestrogens.

15.2 Methods

15.2.1 Animals

The meadow voles used in this study were fourth–sixth generation captive animals, born and raised in a room that was maintained between 23 and 25 °C and under long photoperiod (14:10 h light:dark, lights on at 07:00 h, CST). This photoperiod simulates the day length during the breeding season in free-living meadow voles (Ferkin and Zucker 1991; Ferkin and Gorman 1992). All of the female meadow voles used in the study were between 70 and 120 days of age, similar in body weight to one another (within 3–5 g), and sexually naïve. Female meadow voles are induced ovulators (Milligan 1982) and do not undergo estrous cycles (Keller 1985). Therefore, all female voles in the present study were considered to be in similar reproductive condition. Male meadow voles were between 120 and 150 days of age, similar in body weight to one another (within 3–5 g), and sexually experienced, having previously sired a litter when they were 70–80 days of age. Males and females had continuous access to standard laboratory chow (Harlan Teklad Rodent Diet, #8640, Madison, WI) prior to their inclusion in this study.

15.2.2 Treatment Groups

All voles were raised from birth on a standard rodent chow diet (Harlan Teklad Rodent Diet, #8640) until they were placed into one of two diet groups. One group of males and females continued to be fed the standard rodent chow diet, which contained 350–650 mg/kg soy-derived phytoestrogens (PE); we considered this to group to be the control (control-PE group). The other group of males and females were switched to a relatively low phytoestrogen diet (Harlan Teklad Rodent Diet 2020X, Madison, WI, USA), which typically contains 20 mg/kg phytoestrogens or less; we considered this to be the low phytoestrogen group (low-PE group). The voles in the low-PE group were fed this diet for 30 days prior to testing.

15.2.3 Attractivity

15.2.3.1 Scent Donors and Subjects

We used 36 female and 36 male meadow voles as scent donors. Donors were 12 females and 12 males in the low-PE group and 24 males and 24 females in the control-PE group. We used 24 different control-PE female and 24 different control-PE male voles as subjects. The low-PE voles and the control-PE voles were used as a subject once and as a scent donor once in random order. Subjects were not tested with their own scent marks. We did not use more than two individuals from the same litter as subjects or scent donors to eliminate the potential for litter effects. Subjects and their scent donors were unfamiliar and unrelated to one another.

15.2.3.2 Testing Procedure

All attractivity tests took place within 24 h after the scent donors were fed the low-PE diet for 30 days. During the attractivity test, subjects underwent a single 5-min test in which we recorded the amount of time in seconds that male and female control-PE subjects spent investigating the anogenital area scent mark of an opposite-sex, low-PE conspecific versus that of an opposite-sex, control-PE conspecific.

Thus, male subjects were tested with marks of female scent donors and female subjects were tested with the marks of male scent donors. There were 12 control-PE male and 12 control-PE female subjects per group. Each subject was exposed to the scent marks of a unique pair of opposite-sex scent donors. We also used 12 control-PE males, 12 control-PE females, 12 low-PE males, and 12 low-PE females as scent donors.

The scent mark of each of the two donors were placed on a clean glass microscope slide (2.5 × 7.6 cm), which was divided in three equal sections, each 2.5 cm long. The middle section contained no stimulus odor and each end section contained a scent from a corresponding scent donor. For example, for female subjects, one end section of the slide contained a scent mark from a low-PE male donor, while the other end section of the slide contained a scent mark from the control male donor. The methods of the attractivity test follow those detailed in Pierce et al. (2005). Briefly, to obtain the scent marks, the anogenital area of a scent donor was rubbed for approximately 5 s against the left or right side of a clean slide. The position of the scent marks of the two different scent donors was alternated for each vole subject. After both scent marks were placed on the slide, the slide was suspended on a clip and hook apparatus 1 cm above the substrate in the subject's home cage, against the wall opposite the subject's nest. During the 5-min attractivity test, an observer continuously recorded the amount of time (seconds) that each subject licked or sniffed (the subject's nose comes within approximately 1–2 cm) each section of the slide. The observer was blind to the test condition. The test began when the slide was placed into the cage of the subject (Pierce et al. 2005).

We used anogenital area scent marks as the odor stimulus. The anogenital area may contain components of scent from multiple sources such as urine, feces, sex organs, and sebaceous glands and are responded to similarly by conspecifics (Ferkin and Johnston 1995); the anogenital area scent marks are deposited by voles in their runways (Ferkin et al. 2001). Each slide was used only once and discarded. Immediately prior to each test, fresh scent marks from the anogenital area were obtained from each scent donor. The experimenter wore disposable latex gloves to minimize human scent transfer while handling all slides.

15.2.4 Proceptivity

15.2.4.1 Scent Donors and Subjects

All proceptivity tests were conducted within 24 h after the low-PE subjects were fed their diet for 30 days. We followed the methods for proceptivity testing developed by Pierce and collaborators (2005). Briefly, each subject was exposed to a glass slide that contained the scent mark of a control-PE opposite-sex conspecific and that of a control-PE same-sex conspecific. Subjects were 12 low-PE females, 12 low-PE males, 12 control males and 12 control females. Scent donors were 24 voles fed a control-PE diet since birth; these voles were not used in the attractivity test, were not currently pregnant or lactating, and were between 70 and 150 days of age.

15.2.4.2 Testing Procedure

During the 5-min proceptivity test, we continuously recorded the amount of time (seconds) that each subject licked or sniffed each scent mark and the clean section of the slide. The test began when the slide was placed into the cage of the subject (Pierce et al. 2005). Voles were considered to display proceptive behavior if they spent significantly more time (separate paired *t*-tests, $p<0.05$) investigating the odors of the opposite-sex conspecific than those of the same-sex conspecific (Pierce et al. 2005, 2007).

15.3 Results

15.3.1 Attractivity

Male subjects on the control phytoestrogen diet spent similar amounts of time ($t_{22}=0.134$, $p=0.895$) investigating the scent mark of the control female and that of the low-PE female (Fig. 15.1a). Likewise, female subjects on the control phytoestrogen diet spent similar amounts of time ($t_{22}=0.157$, $p=0.877$) investigating the scent mark of the control male scent and that of a low-PE male (Fig. 15.1a). Male and female subjects on the control diet spent similar amounts of time investigating the scent marks of two opposite-sex conspecifics fed the control diet (male subjects: $t_{22}=0.536$, $p=0.597$, female subjects: $t_{22}=0.284$, $p=0.779$, Fig. 15.1b).

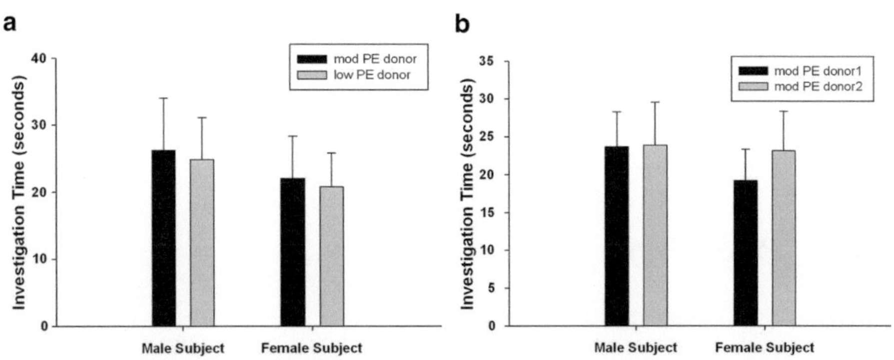

Fig. 15.1 (**a**) The mean amount of time (s) ± sem that male and female subjects spent investigating the scent mark of an opposite-sex conspecific fed a diet with a moderate content of phytoestrogens and that of an opposite-sex conspecific fed a diet with a low content of phytoestrogens. There were no statistical differences in investigation times ($p>0.5$). (**b**) The mean amount of time (s) ± sem that male and female subjects spent investigating the scent mark of an opposite-sex conspecific fed a diet with a moderate content of phytoestrogens and that of an opposite-sex conspecific fed a diet with a moderate content of phytoestrogens. There were no statistical differences in investigation times ($p>0.5$)

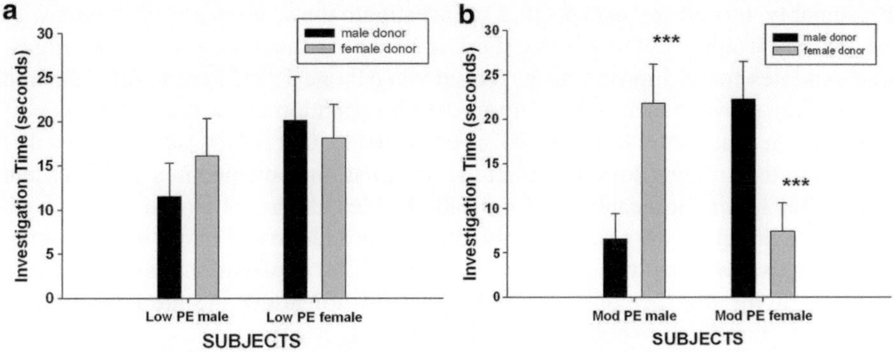

Fig. 15.2 (**a**) The mean amount of time (s) ± sem that male and female subjects fed a diet with low phytoestrogen content spent investigating the scent mark of an opposite-sex conspecific fed a diet with a moderate content of phytoestrogens and that of a same-sex conspecific fed a diet with a moderate content of phytoestrogens. There were no statistical differences in investigation times ($p>0.5$). (**b**) The mean amount of time (s) ± sem that male and female subjects fed a diet with moderate phytoestrogen content spent investigating the scent mark of an opposite-sex conspecific fed a diet with a moderate content of phytoestrogens and that of a same-sex conspecific fed a diet with a moderate content of phytoestrogens. *Asterisks* indicate statistical differences in investigation times ($p<0.001$)

15.3.2 Proceptivity

Low-PE males ($t_{22}=0.807$, $p=0.428$) and low-PE females ($t_{22}=0.323$, $p=0.75$) spent similar amounts of time investigating the scents of a same-sex conspecific and that of an opposite-sex conspecific (Fig. 15.2a). In contrast, control females ($t_{22}=8.670$, $p<0.001$) and control males ($t_{22}=10.28$, $p<0.001$) spent more time investigating the scent mark of an opposite-sex, control vole than that of a same-sex, control vole (Fig. 15.2b).

15.4 Discussion

We found that the scent marks deposited by low-PE voles and control voles were similar in their attractiveness to opposite-sex conspecifics. Attractivity is an important aspect of sexual behavior as it allows for the establishment of olfactory communication between potential mates (Beach 1976; Pierce et al. 2005). For many species of terrestrial mammals, individuals can detect differences in the attractiveness of the scent marks of opposite-sex conspecifics (Brown and Macdonald 1985). The differences in the attractiveness of marks from these scent donors reflect differences in some aspect of the donor's phenotype or genotype (Rich and Hurst 1999; Roberts 2007; Wyatt 2014). Conspecifics that investigate these scent marks can use this information to discriminate between donors.

Presumably, this allows conspecifics to investigate the scent marks of donors with whom they would want to interact (Ferkin 2011). As these scent marks are digestive exudates found in urine, feces, or saliva (Albone 1984; Ferkin and Johnston 1995), they provide accurate and up-to-date information about the donor's condition (Ferkin et al. 1997; Roberts 2007; Ferkin 2011). Our results suggest that a diet low in phytoestrogens does not affect the information contained in a donor's scent mark. This finding indicates that for meadow voles, dietary phytoestrogens have a negligible effect on the tissues involved in the production of odors that are sexually attractive to opposite-sex conspecifics. A similar explanation may be adduced to account for the fact that phytoestrogens do not affect the attractiveness of Siamese fighting fish to conspecifics (Brown et al. 2014).

It appears that in voles, phytoestrogens were not exerting an endocrine effect on the sensitivity of these odor-producing tissues as do gonadal steroids. Several studies on meadow voles have demonstrated that high titers of gonadal steroids are necessary for individuals to be attracted to the scent marks of opposite-sex conspecifics (Ferkin and Zucker 1991; Ferkin et al. 1992; Ferkin and Johnston 1993; Leonard and Ferkin 1999). This is not a unique observation; in other rodents, such as mice, gonadal steroids are vital to odor communication or signaling of physical and hierarchal status (Johnson and Phoenix 1976; McGraw et al. 2006; Ferkin 2011; Asaba et al. 2014). Our results suggest that a diet low in phytoestrogen content was sufficient to inhibit male and female meadow voles from showing preferences for the scent marks of opposite-sex conspecifics. Zavatti et al. (2009) ovariectomized Sprague-Dawley rats and found that they no longer displayed partner preferences for male over female conspecifics. However, treating ovariectomized rats with ferutinin, a phytoestrogen, or estradiol was sufficient to reinstate partner preferences in these animals. In contrast, Patisaul et al. (2003) found that a soy dietary supplement containing isoflavones induced ovariectomized Longs Evans rats to display lower levels of proceptive behavior in the form of reduced hopping and darting. In addition, when the phytoestrogen ferutinin or the phytoestrogens daidzein and genistein were administered simultaneously with estradiol, gonadectomized rats became less proceptive towards opposite-sex conspecifics (Patisaul et al. 2003; Zavatti et al. 2006; Zanoli et al. 2009). Other studies on gonadectomized rats have reported no increase in proceptivity after receiving phytoestrogens (Henry and Witt 2002). Examples also exist for other species: soy-derived isoflavones inhibited testosterone secretion in adult Bilgoraj ganders (*Anser anser*) (Opalka et al. 2006), and decreased the rate of reproductive readiness in male dark-eyed juncos (*Junco hyemalis*) (Corbitt et al. 2007).

Our results show that voles reared on a diet with moderate amounts of phytoestrogens and switched to a diet of with negligible to low amounts of phytoestrogens no longer showed an interest in opposite-sex conspecifics; they were no longer behaving proceptively towards potential mates. Failure to respond preferentially to the scent mark of a potential mate may have serious fitness consequences for voles that consume a low phytoestrogen content diet. Such voles may have fewer mating opportunities despite the fact that these voles still produce scent marks that are attractive to opposite-sex conspecifics. As the levels of phytoestrogens in plants vary with both environmental conditions (e.g., drought) and location, it may be that meadow voles would experience differing phytoestrogen exposure even when residing in the same area (Eldridge and Kwolek 1983). In that meadow voles could mate with multiple partners (Boonstra et al. 1993), the net result would be a potential decrease in fitness among voles that consume vegetation that is low in phytoestrogen content relative to those that consume vegetation that is higher in phytoestrogen content.

In the present study, we used voles that were raised from birth on a diet containing moderate levels of phytoestrogens. It was only during adulthood were some voles switched to a diet containing low levels of phytoestrogens. It is possible that we may have gotten different results if voles were fed a diet low in phytoestrogens during lactation or prior to puberty. These may be sensitive periods for exposure to phytoestrogens. Thus, it is possible that the timing of phytoestrogen treatment or removal (Pierson and Ferkin 2014) could impact its effects on sexual behavior. Such a conjecture is supported by the fact that female Phayre's leaf monkeys (*Trachypithecus phayrei crepusculus*), have longer reproductive cycle lengths, and greater likelihood of conception during times when a plant high in phytochemicals was abundant (Lu et al. 2011). Conversely, California quail showed more successful breeding during wet years when phytoestrogens are largely absent from their available vegetation (Leopold et al. 1976), and male Japanese quail exposed to the phytoestrogen genistein during their embryonic stage showed decreased reproductive behavior as adults (Viglietti-Panzica et al. 2007). In addition, exposure to phytoestrogens during the sensitive periods of development may induce long lasting effects on reproductive behavior and physiology, which has been previously observed in mice (Thigpen et al. 2003) and rats (Whitten et al. 1995; Odum et al. 2001; Lewis et al. 2003).

Acknowledgements We thank Malle Carrasco-Harris for reading an earlier version of this manuscript and providing constructive comments and suggestions. This research was supported by a Faculty Research Grant to M.H.F., a grant from the Vice Provost of Research, and funding from the Department of Biological Sciences at The University of Memphis.

References

Adams NR (1995) Detection of the effects of phytoestrogens on sheep and cattle. J Anim Sci 73:1509–1515

Albone ES (1984) Mammalian semiochemistry: the investigation of chemical signals between mammals. Wiley and Sons, New York, NY

Anthony MS, Clarkson TB, Hughes CL Jr, Morgan TM, Burke GL (1996) Soybean isoflavones improve cardiovascular risk factors without affecting the reproductive system of peripubertal rhesus monkeys. J Nutr 126:43–50

Asaba A, Tatsuya H, Mogi K, Kikusui T (2014) Sexual attractiveness of male chemicals and vocalizations in mice. Front Neurosci 8:1–13

Batzli GO (1985) Nutrition. In Tamarin RH (ed) Biology of the new world *Microtus*. Amer Soc Mammal, Sp Publ 8, Lawrence, p 779–811

Beach FA (1976) Sexual attractivity, proceptivity, and receptivity in female mammals. Horm Behav 7:105–138

Bergeron JM, Jodoin L (1987) Defining "high quality" food resources of herbivores: the case for meadow voles (*Microtus pennsylvanicus*). Oecologia 71:510–517

Bergeron JM, Brunet R, Jodoin L (1990) Is space management of female voles (*Microtus pennsylvanicus*) related to nutritive quality of plants? Oecologia 82:531–536

Boonstra R, Xia XH, Pavone L (1993) Mating system of the meadow vole, *Microtus pennsylvanicus*. Behav Ecol 4:83–89

Brown RE, Macdonald DW (eds) (1985) Social odours in mammals. Oxford University Press, Oxford

Brown AC, Stevenson LM, Leonard HM, Nieves-Puigdoller K, Clotfelter ED (2014) Phytoestrogens β-sitosterol and genistein have limited effects on reproductive endpoints in a female fish, *Betta splendens*. Biomed Res Int 2014:7 pages

Brzezinski A, Debi A (1999) Phytoestrogens: the "natural" selective estrogen receptor modulators? Eur J Obstet Gynecol Reprod Biol 85:47–51

Cameron JL (2003) Hormonal mediation of physiological and behavioral processes that influence fertility. In: Wachter KW, Bulatao RA (eds) Offspring: human fertility behavior in biodemographic perspective. National Academies Press, Washington, DC, pp 104–139

Cline JM, Paschold JC, Anthony MS, Obasanjo IO, Adams MR (1996) Effects of hormonal therapies and dietary soy phytoestrogens on vaginal cytology in surgically postmenopausal macaques. Fertil Steril 65:1031–1035

Corbitt C, Satre D, Adamson LA, Cobbs GA, Bentley GE (2007) Dietary phytoestrogens and photoperiodic response in a male songbird, the Dark-eyed Junco (Junco hyemalis). Gen Comp Endocrinol 154:16–21

Cos P, De Bruyne T, Apers S, Berghe DV, Pieters L, Vlietinck AJ (2003) Phytoestrogens: recent developments. Planta Med 69:589–599

Eldridge AC, Kwolek WF (1983) Soybean isoflavones: effect of environment and variety on composition. J Agric Food Chem 31:394–396

Ferkin MH (2011) Odor-related behavior and cognition in meadow voles, *Microtus pennsylvanicus* (Arvicolidae, Rodentia). Folia Zool 60:262–276

Ferkin MH, Gorman MR (1992) Gonadal hormones influence the odor preferences of male meadow voles, *Microtus pennsylvanicus*. Physiol Behav 51:1087–1091

Ferkin MH, Johnston RE (1993) Roles of gonadal hormones on controlling sex-specific odors in meadow voles (*Microtus pennsylvanicus*). Horm Behav 27:523–538

Ferkin MH, Johnston RE (1995) Meadow voles, *Microtus pennsylvanicus*, use multiple sources of scent for sex recognition. Anim Behav 49:37–44

Ferkin MH, Zucker I (1991) Photoperiod and ovarian hormones control seasonal odour preferences of meadow voles, *Microtus pennsylvanicus*. J Reprod Fertil 92:433–441

Ferkin MH, Gorman MR, Zucker I (1992) Gonadal hormones affect the odors broadcasted by male meadow voles (*Microtus pennsylvanicus*). J Reprod Fertil 95:729–736

Ferkin MH, Sorokin ES, Johnston RE, Lee CJ (1997) Attractiveness of scents varies with protein content of the diet in meadow voles. Anim Behav 53:133–141

Ferkin MH, Mech SG, Paz-y-Miño C (2001) Scent marking in meadow voles and prairie voles: a test of three hypotheses. Behaviour 138:1319–1336

Foth D, Cline JM (1998) Effects of mammalian and plant estrogens on mammary glands and uteri of macaques. Am J Clin Nutr 68:1413S–1417S

Getz LL (1985) Habitats. In Tamarin RH (ed) Biology of new world *Microtus*. Amer Soc Mammal, Sp Publ 8, Lawrence, p 286-309

Gore AC (2008) Developmental programming and endocrine disruptor effects on reproductive neuroendocrine systems. Front Neuroendocrinol 29:358–374

Henry LA, Witt DM (2002) Resveratrol: phytoestrogen effects on reproductive physiology and behavior in female rats. Horm Behav 41:220–228

Hobbs NJ, Ferkin MH (2012) Effects of food availability on proceptivity: a test of the reproduction at all costs and metabolic fuels hypotheses. Behav Proc 91:192–197

Howell WM, Black DA, Bortone SA (1980) Abnormal expression of secondary sex characters in a population of mosquitofish, Gambusia affinis holbrooki: evidence for environmentally-induced masculinization. Copeia 1980:676–681
Johnson DF, Phoenix CH (1976) Hormonal control of female sexual attractiveness, proceptivity, and receptivity in rhesus monkeys. J Comp Physiol Psychol 90:473–483
Johnston RE (1979) Olfactory preferences, scent marking, and "proceptivity" in female hamsters. Horm Behav 13:21–39
Keller BL (1985) Reproductive patterns. In: Tamarin RH (ed) Biology of new world *Microtus*. Amer Soc Mammal, Sp Publ 8, Lawrence, pp 725–778
Kiparissis Y, Balch GC, Metcalfe TL, Metcalfe CD (2003) Effects of the isoflavones genistein and equol on the gonadal development of Japanese medaka Oryzias latipes. Environ Health Perspect 111:1158–1163
Leonard ST, Ferkin MH (1999) Prolactin titers affect a vole's responses to scents of conspecifics. In: Johnston DM, Műller-Schwarze D, Sorenson PW (eds) Chemical signals in vertebrates, vol 8. Plenum Press, New York, NY, pp 437–444
Leopold AS, Erwin M, Oh J, Browning B (1976) Phytoestrogens: adverse effects on reproduction in California quail. Science 191:98–100
Lephart ED, West TW, Weber KS, Rhees RW, Setchell KDR, Adlercreutz H, Lund TD (2002) Neurobehavioral effects of dietary soy phytoestrogens. Neurotoxicol Teratol 24:5–16
Lephart ED, Setchell KD, Had RJ, Lund TD (2004) Behavioral effects of endocrine-disrupting substances: phytoestrogens. ILAR J 45:443–454
Lephart ED, Setchell KDR, Lund TD (2005) Phytoestrogens: hormonal action and brain plasticity. Brain Res Bull 65:193–198
Lewis RW, Brooks N, Milburn GM, Soames A, Stone S, Hall M, Ashby J (2003) The effects of the phytoestrogen genistein on the postnatal development of the rat. Toxicol Sci 71:74–83
Lu A, Beehner JC, Czekala NM, Koenig A, Larney E, Borries C (2011) Phytochemicals and reproductive function in wild female Phayre's leaf monkeys (*Trachypithecus phayrei crepusculus*). Horm Behav 59:28–36
Luine V, Attalla S, Mohan G, Costa A, Frankfurt M (2006) Dietary phytoestrogens enhance spatial memory and spine density in the hippocampus and prefrontal cortex of ovariectomized rats. Brain Res 1126:183–187
Madison DM (1980) Space use and social structure in meadow voles, *Microtus pennsylvanicus*. Behav Ecol Sociobiol 7:65–71
McGraw KJ, Correa SM, Adkins-Regan E (2006) Testosterone upregulates lipoprotein status to control sexual attractiveness in a colorful songbird. Behav Ecol Sociobiol 60:117–122
Milligan SR (1982) Induced ovulation in mammals. Oxford Rev Reprod 4:1–46
Mills JN, Childs JE (1998) Ecologic studies of rodent reservoirs: their relevance for human health. Emerg Infect Dis 4:529–537
Musey PI, Adlercreutz H, Gould KG, Collins DC, Fotsis T, Bannwart C, Mäkelä T, Wähälä K, Brunow G, Hase T (1995) Effect of diet on lignans and isoflavonoid phytoestrogens in chimpanzees. Life Sci 57:655–664
Newsome FE, Kitts WD (1980) The effects of feeding coumestrol on the reproductive organs of prepubertal lambs. Can J Anim Sci 60:53–58
Odum J, Tinwell H, Jones K, Van Miller JP, Joiner RL, Tobin G, Kawasaki H, Deghenghi R, Ashby J (2001) Effect of rodent diets on the sexual development of the rat. Toxicol Sci 61:115–127
Opalka DM, Kaminska B, Piskula MK, Puchajda-Skowronska H, Dusza L (2006) Effects of phytoestrogens on testosterone secretion by Leydig cells from Bilgoraj ganders (Anser anser). Br Poult Sci 47:237–245
Patisaul HB, Luskin JR, Wilson ME (2003) A soy supplement and tamoxifen inhibit sexual behavior in female rats. Horm Behav 45:270–277
Pierce AA, Ferkin MH, Williams TK (2005) Food-deprivation-induced changes in sexual behaviour of meadow voles, *Microtus pennsylvanicus*. Anim Behav 70:339–348
Pierce AA, Iwueke I, Ferkin MH (2007) Food deprivation and the role of estradiol in mediating sexual behaviors in meadow voles. Physiol Behav 90:353–361

Pierson LM, Ferkin MH (2014) The impact of phytoestrogens on sexual behavior and cognition in rodents. Mamm Biol 80(2):148–154. doi:10.1016/j.mambio.2014.11.006

Rich TJ, Hurst JL (1999) The competing countermarks hypothesis: reliable assessment of competitive ability by potential mates. Anim Behav 58:1027–1037

Roberts SC (2007) Scent marking. In: Wolff JO, Sherman PW (eds) Evolution of life histories of mammals, theory and pattern. Chicago University Press, Chicago, IL, pp 255–267

Santell RC, Chang YC, Nair MG, Helferich WG (1997) Dietary genistein exerts estrogenic effects upon the uterus, mammary gland and the hypothalamic/pituitary axis in rats. J Nutr 127: 263–269

Song G, Kochman L, Andolina E, Herko RC, Brewer KJ, Lewis V (2006) Beneficial effects of dietary intake of plant phytoestrogens on semen parameters and sperm DNA integrity in infertile men. Fertil Steril 86:S49

Stevenson LM, Brown AC, Montgomery TM, Clotfelter ED (2011) Reproductive consequences of exposure to waterborne phytoestrogens in male fighting fish *Betta splendens*. Arch Environ Contam Toxicol 60:501–510

Stopka P, MacDonald DW (1998) Signal interchange during mating in the wood mouse (*Apodemus sylvaticus*): the concept of active and passive signaling. Behaviour 135:231–249

Thigpen JE, Haseman JK, Saunders HE, Setchell KDR, Grant MG, Forsythe DB (2003) Dietary phytoestrogens accelerate the time of vaginal opening in immature CD-1 mice. Comp Med 53:607–615

Tou JCL, Chen J, Thompson LU (1998) Flaxseed and its lignan precursor, secoisolariciresinol diglycoside, affect pregnancy outcome and reproductive development in rats. J Nutr 128: 1861–1868

Tsao R, Papadopoulos Y, Yang R, Young JC, McRae K (2006) Isoflavone profiles of red clovers and their distribution in different parts harvested at different growing stages. J Agric Food Chem 54:5797–5805

Viglietti-Panzica C, Mura E, Panzica G (2007) Effects of early embryonic exposure to genistein on male copulatory behavior and vasotocin system of Japanese quail. Horm Behav 51:355–363

Wasserman MD, Milton K, Chapman CA (2013) The roles of phytoestrogens in primate ecology and evolution. Int J Primatol 34:861–878

Whitten PL, Naftolin F (1992) Effects of a phytoestrogen diet on estrogen-dependent reproductive processes in immature female rats. Steroids 57:56–61

Whitten PL, Lewis C, Russell E, Naftolin F (1995) Potential adverse effects of phytoestrogens. J Nutr 125:771S–776S

Wuttke W, Jarry H, Westphalen S, Christoffel V, Seidlová-Wuttke D (2002) Phytoestrogens for hormone replacement therapy? J Steroid Biochem 83:133–147

Wyatt TD (2014) Pheromones and animal behavior: chemical signals and signatures, 2nd edn. Cambridge University Press, Cambridge

Zanoli P, Zavatti M, Geminiani E, Corsi L, Baraldi M (2009) The phytoestrogen ferutinin affects female sexual behavior modulating ERα expression in the hypothalamus. Behav Brain Res 199:283–287

Zavatti M, Montanari C, Zanoli P (2006) Role of ferutinin in the impairment of female sexual function induced by *Ferula hermonis*. Physiol Behav 89:656–661

Zavatti M, Benelli A, Montanari C, Zanoli P (2009) The phytoestrogen ferutinin improves sexual behavior in ovariectomized rats. Phytomedicine 16:547–554

Chapter 16
Dominance Hierarchy in Indian Blackbuck (*Antelope cervicapra* L.): Sources, Behavior and Role of Pheromone Signals

Thangavel Rajagopal and Govindaraju Archunan

16.1 Introduction

Blackbuck (*Antilope cervicapra* Linn. 1758) is the only living representative of the genus *Antilope* Pallas 1766. The present geographical distribution of this species is limited to a few states of India. This unique animal has figured prominently in Indian ancient literature, art and also in the memoirs of British colonials and the hunters' note book (Schaller 1967; Mungall 1977). Blackbuck has been categorized as vulnerable by the IUCN endangered commission (ZSI 1994). Wild Blackbuck population is likely to continue to decline as a consequence of poaching, hunting, shrinkage of natural habitats and deforestation, and predation with probably fewer than 25,000 individuals in their native range (Kar 2001). This is critical in that the Blackbuck acts as an indicator for natural disturbances in the forest habitat (Rajagopal 2009). However, pollution and disturbance by man has affected the reproductive biology and odor communication of Indian Blackbucks.

Blackbuck shows a regular estrous cycle similar to domestic ewe and cow with the cycle length of 15–21 days. However, the receptive period is short, lasting only 24 h (Schmied 1973). Therefore, the possibility of chance for mating is unlikely. Moreover, under seminatural conditions, there is intense competition among males

T. Rajagopal
Centre for Pheromone Technology, Department of Animal Science,
Bharathidasan University, Tiruchirappalli, Tamil Nadu, 620 024, India

Department of Zoology, Thiagarajar College (Autonomous), Madurai,
Tamil Nadu, 625 009, India

G. Archunan (✉)
Centre for Pheromone Technology, Department of Animal Science,
Bharathidasan University, Tiruchirappalli, Tamil Nadu, 620 024, India
e-mail: archunan@bdu.ac.in

B.A. Schulte et al. (eds.), *Chemical Signals in Vertebrates 13*,
DOI 10.1007/978-3-319-22026-0_16

to gain access to estrus females and many males fail to mate, which contributes to further decline in the number of offspring. The behavior, growth, and reproduction of the captive Indian Blackbuck, also varies within captive populations. The reasons for such variation have yet to be identified (Ranjitsingh 1989; Kar 2001), but there is a need to increase their breeding in captivity.

Since the International symposium held in 1971 on ungulate behavior and management in Canada, our understanding of behavioral mechanisms of chemical communication in this group has progressed from description to experiments and modeling. Primer pheromones have been identified in a few domestic and non-domestic ungulates. Since urine, feces, and scent glands are the major sources of pheromonal communication there might contain information for communication (Archunan 2009), we have successfully identified pheromone compounds by gas chromatography-linked mass spectrometry from urine, saliva, feces and scent glands from farm animals like cow, buffaloes, and sheep, and few rodents (Archunan 2009, 2014; Archunan et al. 2014). However, we are not aware of any publications reporting the chemical identification of the pheromone compounds in the Indian Blackbuck. Thus, we initiated research into identifying such compounds in the Indian Blackbuck. In concert with these studies, we also investigated the social behavior of the Indian Blackbuck. We did so because literature about the social and reproductive behaviors of Indian Blackbuck is scarce. Most of this limited information comes from reports on the behavior of captive Blackbucks (Prasad and Ramana Roa 1984; Ranjitsingh 1989; Isvaran and Jhala 1999; Rajagopal 2009), Schaller's (1967) brief authoritative account on some aspects of their behavior and ecology of Blackbuck, and their population dynamics (Prasad 1983; Bhattacharya and Chattopadhya 1984). Therefore, we describe in this chapter our work in investigating the social behaviors and pheromones of the Indian Blackbuck. Our goal was to facilitate the development of procedures that would increase their breeding and conservation.

16.2 Social Dominance and Behavior of Indian Blackbuck

16.2.1 Social Organization of Blackbuck

Blackbucks are primarily grazers and herds which are characteristically loose and unstable associations that can range from less than ten individuals to several hundred. The main social units in blackbuck are (1) solitary (usually territorial) males, (2) all-male or bachelor groups-composed of two or more juvenile, subadult or adult males, (3) female groups-composed of females of all age-classes, fawns, and juvenile and subadult males and (4) mixed groups in which the whole range of age classes of both sexes may be represented (Mungall 1978). It is found that territorial males do not keep a harem, and that a female groups (three categories) would passes through the territories of one or more territorial males in the course of a day's wandering, associating temporarily with the resident male (Mungall 1978).

16.2.2 Dominance Hierarchy of Indian Blackbuck

Social hierarchy system in Indian Blackbuck is established by fighting or display behavior and results in a ranking of the animals in their territory (Menon 2000; Rajagopal and Archunan 2008). Generally, the hierarchies involve more than two animals and are composed of linear dominant–subordinate relationship. In this relationship, the alpha male dominates all group members. The beta male is submissive to the alpha male, but dominates the remaining group members, and so on. The lowest-rank animal in the hierarchy is subordinate to all group members (Rajagopal 2009).

We found that the repeated occurrence of leadership by the alpha male and the formation of hierarchy occurred when the number of female Blackbucks entered into estrus. This suggests that the formation and maintenance of dominance hierarchy system is associated with breeding (Rajagopal and Archunan 2008; Rajagopal et al. 2010).

16.2.3 Territoriality and Territory Size

The Indian Blackbuck is territorial and gregarious. It generally lives in a group of 30–100 individuals, and the regulation of its social life depends to a large extent on chemical communication as reflected by its various odoriferous skin glands and other sources of chemical signals (Menon 2000; Rajagopal et al. 2010). Each group of Blackbucks has a territory within which a linear dominance hierarchy through aggression interactions forms. For example, the dominant male makes grunting sound and chases the subordinate males to prevent them from entering the territory.

Blackbuck often use open canopy for grazing and resting areas for territories. The territory size range varies due to climatic conditions, topography, availability of food and water (Kar 2001; Rajagopal 2009). The size of territorial range is between a minimum limit of 1 ha (2.5 ac) and maximum limit of 20 ha (Mungall 1978). The bachelor male group or subordinate males have bigger home range having more than 5 km^2 (Kar 2001). Maximum range of home size was established by dominant males during the months of August, September, October, March, and April (Rajagopal 2009). Kar (2001) found that territorial size range decreased during the rainy season, but in winter and summer months the territory size increased to 2.5–4 km^2.

16.2.4 Lekking of Indian Blackbuck

Lekking is a rare behavioral phenomenon exhibited by less than 2 % of birds and 1 % of mammalian species (Bradbury 1981; Menon 2000; Isvaran 2005). It is an unusual mating system in which territorial male Blackbucks aggregate in an open area where they defend clustered territories, and are sought out by females for the sole purpose of mating. Apart from the males, these territories are typically devoid of other resources (Isvaran and Jhala 1999; Herlekar 2014).

16.2.5 Dominance Influences Scent Marking Behaviors

The role of scent marking behavior in mammalian communication has received considerable attention (Thiessen and Rice 1976; Gosling 1985; Ferkin 2015). Scent marking behaviors of many ungulates have already been described (Gosling 1985; Bowyer et al. 1994) and urine, feces, and scent glands are the major sources of the odors. Functions attributed to scent marking include the advertisement and defense of territories and resources, advertisement of social status, and regulation of social relationships, mate attraction and advertisement of reproductive condition, and spatial orientation (Smith et al. 2001; Arteaga et al. 2008). For the Indian Blackbuck scent marking includes urination, pelleting or defecation, and preorbital and interdigital gland marking aimed at specific conspicuous objects (Ranjitsingh 1989; Rajagopal 2009).

16.2.5.1 Preorbital Gland Scent Marking

In antelopes, both males and females possess preorbital organs, which consist of a glandular region in a pouch adjacent to the nasal (medial) corner of the eye (Muller-Using and Schloeth 1967; Rajagopal and Archunan 2011). There is no clear evidence of a role across species for the preorbital gland but most frequently reported role of preorbital secretions is in territorial marking (Gosling 1987). It is also reported that the preorbital glands are usually open during roaring in red deer stags (Butzler 1974) and expansion of the preorbital gland opening may be an expression of aggressive intent in brow-antlered deer (Blakeslee et al. 1979).

In the Indian blackbuck, the preorbital gland markings are associated with territorial and sexual behavior (Rajagopal and Archunan 2008). The male Blackbucks perform scent marking from their well-developed preorbital glands, while in females these glands are vestigial or functionless (Rajagopal and Archunan 2011). It is also reported that preorbital gland marking behavior occurs more often between 10.00–12.00 h and 16.00–18.00 h in the territorial males than non-territorial males of Indian Blackbuck (Rajagopal et al. 2011). The frequency of preorbital gland marking behavior by the dominant male Blackbuck is associated with the formation of hierarchy (Rajagopal 2009). Captive dominant male Blackbucks often prefer to deposit preorbital gland marks close to the food trough and peripheral regions of the territory. Male ungulates typically chose such places to deposit their scent marks and to delineate territorial borders (Gosling 1985; Alonso and Languth 1989). This suggests that the dominant Blackbuck male may also use the preorbital gland secretion as a signal for territory demarcation. In addition, preorbital gland marking by the dominant male may help to control the scent marking behavior of subordinates and reduce the likelihood of them from challenging the dominant male (Rajagopal et al. 2011).

16.2.5.2 Urinary Scent Marking

Major functions of urinary scent marking are defense of territory and resources, signaling social status, regulation of social relationships, mate attraction, and advertisement of reproductive condition (Brennan and Kendrick 2006; Thonhauser et al. 2013). The frequency of urine marking by the dominant male Blackbuck during hierarchy formation period is higher than before and after the hierarchy period as compared to bachelor males (Rajagopal et al. 2010). This observation is consistent with earlier investigations, which have described that the dominant male urine scent marking behavior may adversely affect the fighting ability, suppressing the social and sexual behavior of subordinate males in several species (Derix et al. 1993; Brant et al. 1998; Clutton-Brock et al. 2001).

16.2.5.3 Fecal and Interdigital Gland Scent Marking

Dfecation and scratching or interdigital gland marking behaviors are involved in odor communication in canidae and antelopes (Peters and Mech 1975; Rajagopal and Archunan 2008). High rate of defecation and interdigital gland marking behaviors were exhibited by the territorial male blackbucks compared to that of non-territorial males of adult and subadults. The dominant males use deposited interdigital gland secretions, feces, and urine close to the food trough and resting places (Rajagopal and Archunan 2008; Rajagopal et al. 2011). It is interesting to note that the dominant male deposited interdigital gland marks more frequently at the breeding lek or resting place than they deposited urine and feces (Rajagopal and Archunan 2008). Our findings are consistent with reports showing that the *Lagostomus maximus* scent mark near their resting place (Branch 1993), *Lutra lutra* scent marks near the feeding sites (Kruuk 1992), and the *Callithrix jacchus* deposits scent marks on food (Smith et al. 2001). Interdigital gland and feces marking may also may function to possession of a resource as well as indicate occupation of a territory.

16.2.6 Dominance Influences the Aggressive Behavior

Agonistic behaviors are reported in many types of social interactions, but are commonly used to establish either dominance–sub-ordinance hierarchies or exclusive territories (Allen et al. 1990; Freeman et al. 1992). The dominant male Blackbuck shows more dramatic increase in agonistic behavior during his hierarchy period than that of his before and after dominant hierarchy period. In this respect, high frequency of aggressive behavior may confer an advantage in competition for determination of hierarchy. During feeding, the dominant male does not permit the subordinate access to the food trough (Rajagopal and Archunan 2008; Rajagopal et al. 2011). These observations suggest that the agonistic behavior may help to establish dominant hierarchy and maintain the hierarchy system which always differs in respect to duration, intensity, and consequences.

16.2.7 Testosterone Influences Dominance

Testosterone is an important hormone in regulating many aspects of male reproductive physiology (Hews and Moore 1995; Tarasawa and Fernandez 2001). Testosterone secretion correlates with many behavioral traits related to competition for mates, such as reproductive motivation, territory, mate guarding, heightened aggression and display behavior (Arteaga et al. 2008). We found that testosterone level was the highest in the dominant male Blackbuck during his hierarchy period than before and after the formation of the hierarchy (Rajagopal 2009). Brockman et al. (2001) reported that the dominant male sifaka has higher levels of testosterone and scent marking behavior with male physiological sexual inhibition in subordinate males, suggesting a positive association between testosterone concentration and duration of leadership period in the dominant males (Klomberg et al. 2002).

16.3 Dominance Pheromones: Sources and Identification of Chemosignals in Blackbucks

This preselection of candidate substances has further resulted in successful characterization of several urinary pheromones in mammalian species. In the Indian Blackbuck twenty eight urinary volatiles and forty three preorbital gland secretion volatile compounds were identified in the dominant, bachelor and subadult males. The volatiles identified in the urine and preorbital gland secretion belong to the alkanes, alkenes, alcohols, aldehydes, ketones, phenols, pyridine, and carboxylic acid classes of compounds. Among the different constituents, alkanes (11 volatiles) and carboxylic acids (16 volatiles) are predominantly present in the urine and preorbital glandular secretion respectively (Rajagopal 2009; Rajagopal et al. 2010). These classes of compounds have already been reported in the urine and preorbital gland secretion of other mammals, including white-tailed deer (Jemiolo et al. 1995), cow (Rameshkumar et al. 2000), buffalo (Rajanarayanan and Archunan 2011), klipspringer (Burger et al. 1997), steenbok (Burger et al. 1999), and suni (Stander et al. 2002).

Among 28 urinary volatile compounds, 22, 26, and 13 compounds appeared in males before, during, and after the formation of dominance hierarchy period respectively. Furthermore, 12 compounds in bachelor and 10 compounds in subadult males were present in the urine sample. There were more detectable peaks, i.e., 26 compounds, during the formation of dominance hierarchy period in dominant male urine sample. Among them, three specific urinary volatile compounds, 3-hexanone, 6-methyl-5-hepten-2-one, and 1-phenyl-1,2-butanediol had molecular weight range between 84 and 166 and carbon atoms C_6–C_{10} (Rajagopal et al. 2010). The 3-hexanone has been reported in the white-tailed deer dominant male urine (Jemiolo et al. 1995). It is remarkable to note that the compounds 6-methyl-5-hepten-2-one and 1-phenyl-1, 2-butanediol have been reported as female-attracting pheromones in the novel moth [(Z)-4-hepten-2-one] and cigarette beetle insects [(2S,4S)-2,4-dimethyl-1,5-pentanediol] with slight variation of structure (Zhu et al. 1994; Ken and Kenji 2001). Schewende et al. (1986) reported that the urinary compound hepten-2-one acts as an aversive chemical signal in mouse. Hence, the appearance of these specific compounds in dominant male Blackbuck urine during hierarchy period may provide behaviorally important chemical cues suggesting their role in aggression to male and attraction to female.

The preorbital gland secretion showed 43 major volatile compounds, of which several were identified during the formation of dominance hierarchy period (i.e., 39 volatiles) compared to before (i.e., 34 volatiles) and after (i.e., 27 volatiles) the formation of dominance hierarchy period of dominant male, and bachelor (i.e., 25 volatiles) and subadult male (i.e., 13 volatiles). Among 43 volatiles, four volatiles (2-methyl propanoic acid, 2-methyl-4-heptanon, 2,7-dimethyl-1-octanol, and 1,15-pentadecanediol) were unique in the dominant male gland secretion during the formation of dominance hierarchy period (Rajagopal 2009). In the Indian Blackbuck, 2-methyl propanoic acid appeared dominant-specific compound during the formation of dominance hierarchy period. Hence, 2-methyl propanoic acid may be considered as a female attractant (Rajagopal 2009). The volatile 2-methyl-4-heptanone was also identified in preorbital gland secretion in reindeer (Andersson 1979). However, compound having additional methyl group, i.e., 6-methyl-2-heptanone was found in reindeer interdigital gland secretion (Andersson et al. 1979), sika deer metatarsal gland secretion (Wood 2003) and ear secretion of African elephant (Riddle et al. 2000). Other dominant-specific volatiles 2,7-dimethyl-1-octanol and 1,15-pentadecanediol were also reported as aggregation pheromone in the palm weevil (5-methyl-4-octanol) and defensive pheromone in Nymphalid butterfly (1,16-hexadecanediol diacetate) with slight variation in structure (Giblin-Davis et al. 1997; Ross et al. 2001). Therefore, the findings suggest that these four dominance-specific preorbital glandular volatiles may play a wide ranging role in the territorial marking behavior i.e., defensive signal in the same sex and attractive signal in opposite sex of Indian Blackbuck.

16.4 Pheromone Binding Proteins (Pbps) in Blackbuck

Pheromone binding protein (PBP) plays a significant role in mammalian chemical communication through binding with the specific volatile compounds (Archunan et al. 2014). The pheromone binding protein is belonged to the lipocalin super family present in all the pheromone sources including urine, saliva, vaginal mucus, sweat, scent gland etc. (Cavaggioni and Mucignat-Caretta 2000).

It is recently reported the presence of PBP, i.e., alpha 2u globulin in the preorbital gland of Blackbuck, a pheromone carrier protein (Rajagopal et al. 2015). The presence of α_{2u}-globulin in the preorbital gland posts of Blackbuck is consistent with the earlier reports pertaining to adult male rat and mouse, which have shown this protein to serve as a pheromone-carrier (Cavaggioni and Mucignat-Caretta 2000). The lipocalin proteins in the pheromone sources have been reported as pheromone-carrier in several mammalian species (Cavaggioni et al. 1990; Beynon and Hurst 2003). For example, aphrodisin, a lipocalin protein abundantly secreted from the vaginal fluid of the hamster, has five volatiles bound to it, which facilitate the mating behavior of male hamsters (Briand et al. 2004). Similarly, salivary lipocalins (SAL) secreted from the submaxillary glands of the adult male boar contain two endogenous ligands, 5α-androst-16-en-3-one and 5α-androst-16-en-3-ol as components of the boar sex pheromone system (Scaloni et al. 2001). Lipocalin EquC1 is the abundant protein in horse sweat, and contains a putative pheromone oleamide (D'Innocenzo et al. 2006). In the human, the components of axillary odor are associated with apolipoprotein-D and the potent odoriferous substance was identified as 3-methyl-2-heptenoic acid (Zeng et al. 1996). It is important to note that the bound form of pheromones such as farnesol 1 and 2 with the alpha 2u globulin (18.54 kDa) of preputial gland of Indian commensal rat (Ponmanickam et al. 2010) and further binding efficiency of farnesol in alpha 2u globulin was confirmed by molecular docking and fluorescence analysis (Ilayaraja et al. 2014). In another study, we reported four volatile compounds viz., 2-methyl propanoic acid, 2-methyl-4-heptanone, 2,7-dimethyl-1-octanol, and 1,15-pentadecanediol in the preorbital posts of the dominant male specifically during the hierarchy period (Rajagopal 2009). As a whole, these findings suggest that the α_{2u}-globulin in the preorbital gland may carry the volatile compounds for release to the environment. These volatile compounds may help maintain the dominance hierarchy by allowing the dominant males to signal their residency in an area as well as suppress the activity of subordinate males.

16.5 Future Perspectives

Our understanding of the pheromones of Blackbucks may be used as a tool to improve their conservation and management in captivity. We suggest that future research focus on determining whether volatiles from the urine and preorbital gland are sources of information about dominance and mate attraction. It would be interesting to also know if such volatiles can induce estrus in females. This would enhance their breeding in captivity. Identification of α_{2u}-globulin, a 17 kDa polypeptide, in preorbital glandular secretion provides circumstantial evidence of presence of pheromone-carrier protein in the gland. It is possible that this protein may carry the volatile ligand in male Blackbuck preorbital gland that induces scent marking during the establishment of the hierarchy and in attracting female conspecifics.

Acknowledgements We thank Prof. M.A. Akbarsha, Director and Chair, MGDC, Bharathidasan University for critical reading of the manuscript, and the Chief Wildlife Warden and the Director of the Arignar Anna Zoological Park (AAZP), Vandalur, Chennai, for granting permission to carry out this study. TR is thankful to the UGC for awarding Rajiv Gandhi National Research fellowship, New Delhi. We acknowledge the facilities provided to us by Bharathidasan University, DBT, DST, UGC-SAP and DST-PURSE.

References

Allen T, Ruthberg AT, Greenberg SA (1990) Dominance, aggression frequencies and mode of aggressive competition in feral pony mares. Anim Behav 40:322–331

Alonso C, Languth A (1989) Ecologia e comportamento de *Callithrix jacchus* (Primates: Callirhichidae) numa ilha de floresta attantica. Rev Nordestina Biol 16:105–137

Andersson G (1979) Volatile ketones from the preorbital gland of reindeer (*Rangifer t. tarandus* L.). J Chem Ecol 5:629–634

Andersson G, Brundin A, Andersson K (1979) Volatile compounds from the interdigital gland of reindeer (*Rangifer t. tarandus* L.). J Chem Ecol 5:321–333

Archunan G (2009) Vertebrate pheromones and their biological importance. J Exp Zool India 12:227–239

Archunan G (2014) Stud male-originating chemosignals: a luteotrophic agent. Indian J Exp Biol 52:5–16

Archunan G, Rajanarayanan S, Karthikeyan K (2014) Cattle pheromones. In: Mucignat-Caretta C (ed) Neurobiology of chemical communication. CRC Press, Boca Raton, FL, pp 461–488

Arteaga L, Bautista A, Martinez-Gomez M, Nicolas L, Hudson R (2008) Scent marking, dominance and serum testosterone levels in male domestic rabbits. Physiol Behav 94:510–515

Beynon RJ, Hurst JL (2003) Multiple roles of major urinary proteins in the mouse, *Mus domesticus*. Biochem Soc Trans 31:142–146

Bhattacharya T, Chattopadhya B (1984) A study on the structure of the social organization of Blackbuck, *Antilope cervicapra* Linne, in Ballavpur wildlife sanctuary, West Bengal, India. Lynx 22:5–10

Blakeslee CK, Rice CG, Ralls K (1979) Behaviour and reproduction of captive brown-antlered deer *Cervus eldi thamin* (Thomas, 1918). Saegetierk Mitteil 27:114–127

Bowyer RT, Van Ballenberghe V, Rock KR (1994) Scent marking by Alaskan moose: characteristics and spatial distribution of rubbed trees. Can J Zool 72:2186–2192

Bradbury J (1981) The evolution of lek. In: Alexander KD, Tinkle TK (eds) Natural selection and social behavior. Carron, New York, NY, pp 138–169

Branch LC (1993) Social organization and mating system of the plain viscacha (*Lagostomus maximus*). J Zool 229:473–491

Brant CL, Schwab TM, Vandenbergh JG, Schaefer RL, Solomon NG (1998) Behavioural suppression of female pine voles after replacement of the breeding male. Anim Behav 55: 615–627

Brennan PA, Kendrick KM (2006) Mammalian social odours: attraction and individual recognition. Phil Trans R Soc Lond Ser B Biol Sci 361:2061–2078

Briand L, Blon F, Trotier D, Pernollet JC (2004) Natural ligands of hamster aphrodisin. Chem Senses 29:425–430

Brockman DK, Whitten PL, Richard AF, Benander B (2001) Birth season testosterone levels in male Verreaux's sifaka, *Propithecus verreauxi*: insights into socio-demographic factors mediating seasonal testicular function. Behav Ecol Sociobiol 49:117–127

Burger BV, Yang TP, Le Roux M, Brandt WF, Cox AJ, Hart PF (1997) Mammalian exocrine secretions. XI. Constituents of the preorbital secretion of klipspringer, *Oreotragus oreotragus*. J Chem Ecol 23:2383–2400

Burger BV, Nell AE, Spies HSC, Le Roux M, Bigalke RC (1999) Mammalian exocrine secretions. XII. Constituents of preorbital gland secretions of bontebok, *Damaliscus dorcas dorcas* and blesbok, *D. d. phillipis*. J Chem Ecol 25:2085–2097

Butzler W (1974) Kampf- und Paarungsverhalten, soziale Rangordnung und Aktivitätsperiodik beim Rothirsch (*Cervus elaphus* L.). Springer, Berlin

Cavaggioni A, Mucignat-Caretta C (2000) Major urinary proteins, α_{2u}-globulins and Aphrodisin. Biochim Biophys Acta 1482:218–228

Cavaggioni A, Findlay JBC, Tirindelli R (1990) Ligand binding characteristics of homologous rat and mouse urinary proteins and pyrazine binding protein of the calf. Comp Biochem Physiol B Biochem Mol Biol 96:513–520

Clutton-Brock TH, Brotherton PNM, Russell AF, O'Riain MJ, Gaynor D, Kansky R, Griffin A, Manser M, Sharpe L, McIlrath GM, Small T, Moss A, Monfort S (2001) Cooperation, control, and concession in meerkat groups. Science 291:447–450

D'Innocenzo B, Salzano AM, Ambrosio DC, Gazzano A, Sorce C, Dani FR, Scaloni A, Pelosi P (2006) Secretory proteins as potential semiochemicals carriers in the horse. Biochemistry 45:13418–13429

Derix R, Van Hoof J, De Vries H, Wensing J (1993) Male and female mating competition in wolves: female suppression versus male intervention. Behaviour 127:141–171

Ferkin MH (2015) The response of rodents to scent marks: four broad hypotheses. Horm Behav 68:43–52

Freeman LC, Freeman SC, Romney AK (1992) The implications of social structure fordominance hierarchies in red deer, *Cervus elaphus* L. Anim Behav 44:239–245

Giblin-Davis RM, Gries R, Gries G, Pen-Rojas E, Pinzon I, Pana JE, Perez AL, Pierce HD, Oehlschlager AC (1997) Aggregation pheromone of palm weevil, *Dynamis borassi*. J Chem Ecol 23:2287–2297

Gosling LM (1985) The even-toed ungulates: order Artiodactyla. Sources, behavioural context, and function of chemical signals. In: Brown RE, Macdonald DW (eds) Social odours in mammals, vol 2. Clarendon, Oxford, pp 550–618

Gosling LM (1987) Scent marking in an antelope lek territory. Anim Behav 35:620–622

Herlekar I (2014) Architect of one's destruction. Curr Sci 106:917–918

Hews DK, Moore MC (1995) Influence of androgens on differentiation of secondary sex characteristics in tree lizards, *Urosaurus ornatus*. Gen Comp Endocrinol 97:86–102

Ilayaraja R, Rajkumar R, Rajesh D, Muralidharan AR, Padmanabhan P, Archunan G (2014) Evaluating the binding efficiency of pheromone binding protein with its natural ligand using molecular docking and fluorescence analysis. Sci Rep 4:5201

Isvaran K (2005) Female grouping best predicts lekking in blackbuck (*Antilope cervicapra*). Behav Ecol Sociobiol 57:283–294

Isvaran K, Jhala Y (1999) Variation in lekking costs in Blackbuck (*Antilope cervicapra* L.): relationship to lek-territory location and female mating patterns. Behaviour 137:547–563

Jemiolo B, Miller KV, Wiesler D, Jelinek I, Novotny M, Marchinton RL (1995) Putative chemical signals from white tailed deer (*Odocoileus-virginianus*). Urinary and vaginal mucus volatiles excreted by females during breeding season. J Chem Ecol 12:869–879

Kar S (2001) Balipadars Blackbuck. Wildlife wing forest department publication, Government of Orissa, Bhubaneswar, pp 1–41

Ken F, Kenji M (2001) New synthesis of serricornin, the female sex pheromone of the cigarette beetle. Biosci Biotechnol Biochem 65:1429–1433

Klomberg KF, Garland T, Swallow JG, Carter PA (2002) Dominance, plasma testosterone levels, and testis size in house mice artificially selected for high activity levels. Physiol Behav 77:27–38

Kruuk H (1992) Scent-marking by otters (*Lutra lutra*): signalling the use of resources. Behav Ecol 3:133–140

Menon RK (2000) Nature watch-the quintessential Antelope – life of the Blackbuck. Resonance 19:69–79

Muller-Using D, Schloeth R (1967) Das Verhalten der Hirsche Kukenthal. Handbook Zool 10:1–60

Mungall EC (1977) Social development of the young blackbuck antelope (*Antilope cervicapra*). AAZPA Region Conf Proc Amer Assoc Zool Parks Aquarium, Washington, DC, pp 100–109

Mungall EC (1978) The Indian Blackbuck Antelope: A Texas view. Kleberg studies in Natural Resources. The Texas agr. Expt. station Nevelle, P. Clarks, Director. Texas AM University System, College Station, TX, p 184

Peters RP, Mech LD (1975) Scent marking in wolves. Am Sci 63:628–637

Ponmanickam P, Palanivelu K, Govindaraj S, Baburajendran R, Habara Y, Archunan G (2010) Identification of testosterone-dependent volatile compounds and proteins in the preputial gland of rat *Rattus norvegicus*. Gen Comp Endocrinol 167:35–43

Prasad NLNS (1983) Horn growth in Blackbuck (*Antilope cervicapra* L.). J Bombay Nat Hist Soc 80:634–635

Prasad NLNS, Ramana Roa JV (1984) Evaluation of habitat structure of Blackbuck, *Antilope cervicapra* in Andhra Pradesh with special reference to mudmal and its relevance to behavior. Geobios 11:17–21

Rajagopal T (2009) A study on the reproductive behaviour and pheromones of an endangered Indian Blackbuck (Antilope cervicapra L.) to enhance captive breeding and conservation. Ph.D. thesis, Bharathidasan University, Tiruchirappalli, India

Rajagopal T, Archunan G (2008) Scent marking by Indian Blackbuck: characteristics and spatial distribution of urine, pellet, preorbital and interdigital gland marking in captivity. In: Reddy MV (ed) Biodiversity and wildlife conservation. Daya Publ, New Delhi, pp 71–80

Rajagopal T, Archunan G, Geraldine P, Balasundaram C (2010) Assessment of dominance hierarchy through urine scent marking and its chemical constituents in male blackbuck Antilope cervicapra, a critically endangered species. Behav Proc 5:58–67

Rajagopal T, Archunan G (2011) Histomorphology of preorbital gland in territorial and non-territorial male blackbuck *Antilope cervicapra*, a critically endangered species. Biologia 66:370–378

Rajagopal T, Manimozhi A, Archunan G (2011) Diurnal variation in preorbital gland scent marking behaviour of captive male Indian Blackbuck (*Antilope cervicapra* L.) and its territorial significance. Biol Rhythm Res 42:27–38

Rajagopal T, Rajkumar R, Ponmanickam P, Achiraman S, Padmanabhan P, Archunan G (2015) Identification of pheromone-carrying protein in the preorbital gland post in the endangered species of Indian male Blackbuck *Antilope cervicapra* L. Indian J Exp Biol (in press)

Rajanarayanan S, Archunan G (2011) Identification of urinary sex pheromones in female buffaloes and their influence on bull reproductive behaviour. Res Vet Sci 91:301–305

Rameshkumar K, Archunan G, Jeyaramanm R, Narasimhan S (2000) Chemical characterization of bovine urine with special reference to oestrus. Vet Res Comm 24:445–454

Ranjitsingh MK (1989) The Indian blackbuck. Nadraj Publ, Dehradun, pp 1–155

Riddle HS, Riddle SW, Rasmussen LEL, Goodwin TE (2000) First disclosure and preliminary investigation of a liquid released from the ears of African elephants. Zoo Biol 19:475–480

Ross GN, Fales HM, Lloyd HA, Jones T, Sokoloski EA, Marshall-Batty E, Blum MS (2001) Novel chemistry of abdominal defensive glands of Nymphalid butterfly, *Agrulis vanilla*. J Chem Ecol 27:1219–1228

Scaloni A, Paolini S, Brandazza A, Fantacci M, Bottiglieri C, Marchese S, Navarrini A, Fini C, Ferrara L (2001) Purification, cloning and characterization of odorant and pheromone binding proteins from pig nasal epithelium. Cell Mol Life Sci 58:823–834

Schaller GB (1967) The deer and the tiger. Chicago University Press, Chicago

Schewende FJ, Wiesler D, Jorgenson JW, Carmack M, Novotny MV (1986) Urinary volatile constituents of the house mouse, *Mus musculus* and their endocrine dependency. J Chem Ecol 12:277–296

Schmied A (1973) Beitrage zu einem Aktionssystem der Hirschziegenantiole (*Antilope cervicapra* Linne, 1758). Z Tierpsychol 32:153–198

Smith TE, Tomlinson AJ, Mlotkiewicz JA, Abbott DH (2001) Female marmoset monkeys (*Callithrix jacchus*) can be identified from the chemical composition of their scent marks. Chem Senses 26:449–458

Stander MA, Burger BV, Le Roux M (2002) Mammalian exocrine secretions. XVII: chemical characterization of preorbital secretion of male suni, *Neotragus moschatus*. J Chem Ecol 28:89–101

Tarasawa E, Fernandez DL (2001) Neurobiological mechanisms of the onset of puberty in primates. Endocrinol Rev 22:111–151

Thiessen DD, Rice M (1976) Mammalian scent marking and social behavior. Psychol Bull 83:505–539

Thonhauser KE, Raveh S, Hettyey A, Beissmann H, Penn DJ (2013) Scent marking increases male reproductive success in wild house mice. Anim Behav 86:1013–1021

Wood WF (2003) Volatile components in metatarsal glands of sika deer, *Cervus Nippon*. J Chem Ecol 29:2729–2733

Zeng C, Spielman AI, Vowels BR, Leyden JJ, Biemaan K, Preti G (1996) A human axillary odorant is carried by apolipoprotein D. Proc Natl Acad Sci U S A 93:6626–6630

Zhu J, Kozlov MV, Philipp P, Francke W, Lofstedt C (1994) Identification of a noval moth sex pheromone in *Eriocrania cicatricella* (Zett.) (Lepidoptera: Eriocraniidae) and its phylogenetic implications. J Chem Ecol 21:29–43

ZSI (1994) The Red Data book of Indian animals (Part-I vertebrate). Zoological Survey of India, Calcutta, pp 195–197

Chapter 17
Asian Elephant Reflections: Chirality Counts

L.E.L. 'Bets' Rasmussen, David R. Greenwood, Thomas E. Goodwin, and Bruce A. Schulte

17.1 Prelude

This chapter contains some of the last information on elephants obtained and analyzed by the late Dr. L.E.L. 'Bets' Rasmussen (Goodwin and Schulte 2008). As her collaborators, we were involved in various aspects of this work. Some of the information presented is incomplete in that some specifics on the procedures followed cannot be fully ascertained. We have noted these instances and we have written the results and discussion with these gaps in mind.

L.E.L.'Bets' Rasmussen
Department of Environmental & Biomolecular Systems, OGI School of Science & Engineering, Oregon Health & Science University, 20000 NW Walker Road, Beaverton, OR, 97006-8921, USA

D.R. Greenwood (✉)
The New Zealand Institute for Plant & Food Research Limited, Private Bag 92 169, Auckland Mail Centre, Auckland 1142, New Zealand

School of Biological Sciences, University of Auckland, Private Bag 92 019, Auckland Mail Centre, Auckland 1142, New Zealand
e-mail: david.greenwood@auckland.ac.nz

T.E. Goodwin
Department of Chemistry, Hendrix College, 1600 Washington Avenue, Conway, AR, 72032, USA

B.A. Schulte
Department of Biology, Western Kentucky University, 1906 College Heights Blvd., Bowling Green, KY, 42101, USA

B.A. Schulte et al. (eds.), *Chemical Signals in Vertebrates 13*,
DOI 10.1007/978-3-319-22026-0_17

17.2 Introduction

Chirality is intrinsic in nature, reflecting the exquisite underlying asymmetry of matter. The tetrahedral geometry of substituents on fully substituted carbon atoms allows the existence in biological systems of nonsuperimposable mirror (i.e., chiral) images (enantiomers) of otherwise structurally identical molecules. Insect pheromone systems reveal the biological significance of chirality, both its diversity of messages and precision of evoked responses (Silverstein 1981, 1985, 1988; Mori 1998, 2007, 2011). The importance of chirality is extensive and particularly well established in insect chemical communication. At least ten species of bark beetles of the genus *Dendroctonus*, a fly *Medetera bistriata*, a heteropteran bug *Lyctocoris elongates*, and several other beetle species release the pheromone frontalin (1,5-dimethyl-6,8-dioxabicyclo[3.2.1]octane), a bicyclic ketal that exists in two enantiomeric forms (Stewart et al. 1977). Insects employ chirality to ensure greater specificity of chemical messages. This specificity often involves the exclusive or predominant release of one enantiomer. In four species (*D. brevicomis*, *D. jeffreyi*, *D. ponderosae*, and *D. rufipennis*), males release the minus (−) form [1*S*,5*R*] (Kinzer et al. 1969; Stewart et al. 1977; Browne et al. 1979; Byers and Woods 1980; Byers 1983; Byers et al. 1984), whereas in another species (*D. frontalis*) males release the plus (+) form [1R,5S] (Rudinsky et al. 1974; Grosman et al. 1997). In other bark beetles, the specific enantiomer released depends on the sex. In two species (*D. frontalis* and *D. simplex*), females release the (−) form (Kinzer et al. 1969; Renwick and Vité 1969; Rudinsky et al. 1974; Stewart et al. 1977; Bridges 1982; Grosman 1996; Grosman et al. 1997), and in another species (*D. rufipennis*) females release the (+) form (Gries et al. 1988, 1992; Perez et al. 1996; Barkawi et al. 2003). The Jeffrey pine beetle, *D. jeffreyi*, best exemplifies the specificity and importance of chirality as females of this species release frontalin in a ratio of 33 % (+) to 67 % (−); this ratio is required to elicit a full response from males (Paine et al. 1999).

The importance of chirality in mammalian chemical signaling has been noted (Novotny and Soini 2008), specifically in mice (*Mus domesticus*; Novotny et al. 1995; Cavaggioni et al. 2003), porcupines (Li et al. 1997), and brushtail possums (*Trichosurus vulpecula*; Carman and Klika 1992). Male Asian elephants (*Elephas maximus*) secrete temporal gland fluid containing both enantiomers of frontalin in a changing ratio dependent on maturity and a male-specific condition termed musth (an annual period of heightened sexual activity and potential aggressiveness), expanding the concept of chirality to proboscidean chemical signals (Greenwood et al. 2005).

In male Asian elephants, maturation and musth are linked, progressively culminating in overall sexual, social, and physical maturity exemplified by discharge of characteristic chemical signals (Rasmussen 2003; Rasmussen et al. 2005). Concurrent serum androgen hormone elevations and behavioral changes accompany these secretions (Jainudeen et al. 1972; Rasmussen and Perrin 1999). Young, small males whose testes may be producing mature sperm but who lack social and sexual experience, exhibit sporadic, non-aggressive behaviors, and their serum androgen hormones levels fluctuate widely (Rasmussen et al. 2002). When these young males first experience musth (termed "moda" musth), their temporal gland secretion (TGS) volatile compounds reflect these variable androgen levels; they contain many pleasant floral-like compounds and no frontalin (Rasmussen et al. 2002). As these young males mature, they release increasing amounts of the acrid bicyclic ketal, frontalin, from the temporal gland exudate. Our investigations showed that total frontalin, initially detected in late teenage males, exhibited a 15-fold rise over a 25-year span (Greenwood et al. 2005).

We established previously that a synthetic racemic mixture of frontalin at 100 μM elicits differential behavioral responses from conspecifics, dependent on their sex, age and reproductive status (Rasmussen and Greenwood 2003). Females in the luteal phase and older males were indifferent, while pregnant females and young males were apprehensive and often exhibited avoidance; in contrast, females in the follicular phase were attracted. Although the observable, countable, and repetitive responses to this synthetic racemic frontalin were robust, we wished to pursue which enantiomer was released from male temporal glands and whether a time dependence was observable during continuous musth episodes.

Gas chromatographic separation on a capillary chiral cyclodextrin column of 100 TGS samples from six male Asian elephants established that two sterically permissible enantiomeric forms, (+) [1R,5S] and (−) [1S,5R], of frontalin were present and quantifiable (Greenwood et al. 2005). In younger males the proportion of (+) enantiomer was significantly higher than that of the (−) enantiomer. With older males, the proportion evened to almost equal (racemic) proportions, especially at mid-musth (Greenwood et al. 2005). We considered that these differences in frontalin ratio and quantity may well be important in signaling musth status with the added benefit of promoting self preservation by avoidance behaviors in receiving individuals.

On the basis of this chirality information about the musth pheromone, frontalin, we posed several questions to clarify its specificity and precision capabilities. (1) Is the frontalin concentration, or the ratio of its enantiomers, more critical to observed behavioral reactions by follicular phase females, the primary attractive conspecific responders? (2) What are the enantiomeric proportions of the frontalin and other insect pheromones observed in female Asian elephant TGS as well as male and female African elephant (*Loxodonta africana*) TGS? (3) Are both enantiomers of frontalin found in the blood of musth elephants and in what proportions? (4) Are there correlations between levels of frontalin, its presumed biosynthetic precursor, and testosterone?

17.3 Methods

17.3.1 Temporal Gland Secretion Samples

[*Coauthors note*: we do not know the identity of every elephant mentioned.]

We examined TGS and blood (serum) or urine samples from the following: (1) Ten Asian male elephants in musth or moda musth, (2) and the same males in non-musth, (3) nine Asian female elephants, (4) six female African elephants, and

(5) 13 male African elephants in musth. The majority of the samples were initially snap frozen in liquid nitrogen (−196 °C), on dry ice (−78 °C) or dry ice in ethanol (−30 to −50 °C). Samples were stored at −80 °C in 2 or 5 ml amber glass vials with Teflon septa. The ten male Asian elephants ranged from 8 to 43 years of age. For five of the males we collected over a span of ages (12–17 years; 16–21 years; 22–25 years; 26–42 years; 35–43 years). One individual was a wild male from Nagarhole National Park, Karnataka, South India and collection occurred while he was sedated. Testosterone levels were measured on serum or urinary samples by the respective facilities on most of the captive males at the time of TGS collection. Samples of TGS from African elephants were obtained from captive individuals at Riddle's Elephant and Wildlife Sanctuary and at Wildlife Safari Park, as well as from wild elephants in South Africa at the Addo Elephant National Park and Kruger National Park through the auspices of Dr. Antony Hall-Martin and Dr. Irven Buss.

17.3.2 Chemical Analyses

[*Coauthors note*: we do not know if the five males were the same animals sampled in a musth and non-musth state or if this represents ten different males, five in each condition, or a combination thereof.]

All samples of TGS, and ten samples of male Asian elephant blood (five from musth males and five from non-musth males) were subjected to headspace extraction and analyzed by two different gas chromatographic/mass spectrometric (GC-MS) methods (Protocols I and II). Using external standards and an achiral column Protocol I allowed the integration of the frontalin peak in the total ion chromatogram (TIC). Protocol II separated the enantiomers of frontalin on a chiral GC column and allowed the comparison of the areas of each enantiomer from the TIC with flame ionization detection (FID). Adsorption of headspace volatiles on solid phase microextraction (SPME) Carboxen/polydimethylsiloxane (CAR/PDMS) fibers was achieved by exposing the fiber after lowering it through a pre-drilled Teflon septum so that the tip was 1 cm above 200 μl of TGS in a standard 2 ml amber Agilent autosampler vial, immersed 1/3 in a 38 °C water bath. Micro stir bars were used in some samples. Headspace volatiles were allowed to adsorb for 1 h before the SPME fiber was transferred to and lowered into the injection port of the GC at 250 °C. The first of two sequential SPME samplings was analyzed by Protocol I and the second by Protocol II. Comparisons demonstrated no quantitative differences whether Protocol I or II was conducted first as long as they were conducted on the same day. Protocol III was designed to approximate the natural lifetime of the release of frontalin from TGS. GC-MS conditions were as discussed in Protocol I below.

For Protocols I and II, all TGS blood or serum aliquots were subjected to headspace sampling. Analyses were performed on the same day for these samples. For GC-MS references, racemic frontalin, *exo*- and *endo*-brevicomin, and enriched samples of (+) and (−) *exo*- and *endo*-brevicomin and other terpene standards were obtained from Pherotech.

17.3.2.1 Protocol I

The compounds were desorbed from the fiber in separate sequential analyses. These compounds were separated by gas chromatography on a DB-1 polymethyl silicone-coated capillary column (60 m × 0.25 mm ID, 1.0 μm film thickness; J&W Scientific) using an Agilent 6890 GC coupled to an Agilent mass spectrometer (MS) 5973 mass selective detector (MSD). The mass spectrum of frontalin was compared with its published spectrum (Gore et al. 1976) and the spectrum obtained on our system with our synthesized standards of racemic frontalin (Mori 1975; Turpin and Weigel 1992; Santiago and Soderquist 1992). Total frontalin was quantitated using both a GC/MS system recording total ion current (TIC) and a GC equipped with a flame ionization detector (FID).

17.3.2.2 Protocol II

To resolve and quantitate the two enantiomers of frontalin the compounds desorbed from the SPME fiber were separated by gas chromatography using a modified cyclodextrin as the stationary phase (γ-cyclodextrin trifluoroacetyl capillary column, Alltech Chiraldex GTA, 30 m × 0.25 mm ID, 0.125 μm film thickness). An Agilent 6890N GC was coupled to an Agilent MS 5973 MSD with the column eluant passing through a flow splitter so that quantitation and identification could be confirmed, employing flame ionization detection (FID) simultaneously with MS. Standards of synthesized enriched (+) [1R,5S] and enriched (−) [1S,5R] frontalin enantiomers (Mori 1975; Turpin and Weigel 1992; Santiago and Soderquist 1992; Greenwood et al. 2005) exhibited 21 and 30 % enantiomeric enrichment respectively with consistent ^{1}H NMR, ^{13}C NMR and MS spectra (Santiago and Soderquist 1992). Together with the synthesized racemic mixture (Pherotech), these were used to detect the retention time of frontalin. TIC values as well as the area values of extracted ion chromatograms for *m/z* 142 and 100 were measured to ensure accurate quantitation of frontalin as co-eluting peaks were detected occasionally.

17.3.2.3 Protocol III

This protocol involved a series of extended analyses. Selected TGS samples were subjected to a sequential collection treatment to demonstrate improved extraction by removing any possible interactions with carrier proteins. The first set of experiments (Protocol IIIA) mimicked the natural release of frontalin and the second (Protocol IIIB) was aimed at effecting its maximal release from a putative carrier protein. During Protocol IIIA, the natural lifetime of frontalin in TGS was assessed by monitoring the progressive release of frontalin over time utilizing periodic SPME extractions for 1 h at 37 °C, followed by quantitation using Protocols I and II (Table 17.3).

These experiments were carried out in amber vials to reduce degradation by light. Oxygen levels were that of the natural environment. One sample was kept at 25 °C and analyzed repeatedly over a 2-week period. Protocol IIIB utilized more extensive stirring and extended sorption times (1–24 h), as well as the addition of Protease K (200 μg/200 μl; Invitrogen #25530-015) to digest potential pheromone carrier proteins so as to enhance frontalin release (Table 17.4; Rasmussen 2001).

17.3.3 Bioassay Procedures

[*Coauthor note*: the detailed data of responses by each female elephant have been lost.]

New sets of bioassays were performed using three concentrations of total frontalin (0.1, 1.0 and 100 μM), each with three aliquots at different ratios of enantiomers (70:30, 50:50 and 30:70) prepared from our synthetic frontalin mixes. These represented ratios in the TGS of young teenage males at mid-musth, older teenage males, and males older than 25 years. Only females in the follicular stage of estrus, the highest responding group (Greenwood et al. 2005), were assessed in this study.

The bioassay methods have been described in detail previously (Rasmussen et al 1997; Rasmussen and Greenwood 2003). Briefly, samples were placed using a double-blind protocol in which the observer was not aware of the identity of the samples, and the elephant did not see the samples being placed. Nine follicular phase Asian elephants were observed for approximately one hour each from a public vantage point with good visibility but without distracting the animals. A focal location sampling supplemented by video recording was used to obtain all responses to a sample. Typical controls included water, water and buffer and/or water with vanilla extract (Schulte and Rasmussen 1999). The focus was on primary trunk tip behaviors (sniff, check, place and flehmen) along with accessory behaviors (trunk blow, flick, pinch, suck and wriggle) of the nine follicular phase females (see Bagley et al. 2006 for the high correlation between primary and accessory responses).

17.3.4 Data Manipulation and Analysis

Data are expressed mainly as means ± standard error. We calculated the Pearson product–moment correlation coefficient where appropriate. Most of the measurements were on individual samples and are depicted as single points or ranges.

17.4 3. Results

17.4.1 Responses to Frontalin by Females in Follicular Phase of Estrus

The responses of follicular phase females were affected by both the ratios of enantiomers and the concentration of frontalin (Table 17.1). The highest median hourly rate of bioresponses by these females occurred to the racemic ratio of frontalin at all three concentrations and the overall highest response rate occurred to the greatest concentration assayed (100 μM). Interestingly, the elephants did not seem to differentiate between the two non-racemic ratios.

17.4.2 Comparison of the Enantiomeric Proportions for Frontalin and Other Insect Pheromones in Female Asian TGS, and Male and Female African Elephant TGS

Frontalin and the brevicomins were not observed in TGS of teenage female elephants; however, secretions from older females contained some frontalin, primarily the (−) enantiomer, and all four of the brevicomin enantiomers (Table 17.2). In Asian female TGS essentially all (99.99 %) of frontalin was the (−) form. The brevicomin composition comprised 38.5 % (+)-*exo*-brevicomin, 37.6 % (−)-*exo-brevicomin*, 13.4 % (−)-*endo*-brevicomin, and 10.5 % (+)-*endo*-brevicomin. For African elephants, frontalin was predominantly in the minus form and the

Table 17.1 Bioresponse levels[a] by nine follicular phase female Asian elephants to three concentrations of total frontalin with three proportions of enantiomers

[Frontalin], μM	Enantiomeric ratio (+/−)	Bioresponse levels
0.1	70:30	Low
	50:50	Moderate
	30:70	Low
1.0	70:30	Low
	50:50	High
	30:70	Low
100	70:30	Moderate
	50:50	High
	30:70	Moderate

[a]Rasmussen's notes summarized these levels as total chemosensory behaviors per hour with low = 1 h^{-1}, moderate = 2–3 h^{-1}, and high = 4–5 h^{-1}

Table 17.2 Proportion of frontalin and brevicomin enantiomers in TGS of female Asian elephants and both sexes of African elephants

	Locale and		Age		Frontalin		Brevicomin			
N	species	Sex	(years)	*T*, ng/ml	+	−	+*exo*	−*exo*	−*endo*	+*endo*
6	Captive Asian	F	14–20	NA	nd	nd	nd	nd	nd	nd
3	Captive Asian	F	30-40	NA	0.001	0.999	0.385	0.376	0.134	0.105
3	Wild African	F	25–60	<1.0	trace	nd	nd	nd	nd	nd
						nd	nd	nd	nd	nd
						nd	nd	nd	nd	nd
3	Captive African	F	18–27	<1.0	nd	nd	nd	trace	trace	nd
					nd	trace	nd	1.0	trace	nd
					nd	trace	trace	nd	nd	nd
3	Wild African	M	25–30	21–87	0.249	0.751	0.001	0.378	0.621	nd
					nd	nd	nd	1.0	nd	nd
					nd	nd	nd	1.0	nd	nd
4	Wild African	M	30–40	71–98	nd	trace	0.002	0.353	0.664	0.001
					nd		nd	0.32	0.62	?
					nd		nd	0.68	0.24	0.08
					nd		nd	0.67	0.33	nd
1	Captive African	M	20	123	nd	nd	nd	1.0	nd	nd
2	Captive African	M	18	1–10	nd	trace	0.35	0.59	trace	0.05
					nd	1.0	0.05	0.70	0.20	0.05

For the three older female Asian elephants, only average proportion data were recoverable
N number of elephants and samples, *T* testosterone, *NA* not available, *nd* not detected, *trace* principal ions present (brevicomin, *m/z* 156, 114, 85; frontalin, *m/z* 142, 100, 72) but non-quantitative

brevicomin proportions changed with age and testosterone levels, with all four enantiomers detected in a wild male with high testosterone (Table 17.2). The finding of (−)-brevicomin as the major stereoisomer for African elephants aligns with what was published for brevicomin in female African elephant urine (Goodwin et al. 2006).

17.4.3 Enantiomers of Frontalin in the Blood of Asian Elephants in Musth

The analyses of headspace of whole, fresh, non-musth blood from male Asian elephants ($n=8$) did not detect any frontalin. In contrast, plentiful frontalin was revealed in the blood from older mid-musth male Asian elephants. The proportion of

Table 17.3 Longevity of frontalin enantiomers in musth Asian elephant blood at ambient temperature (25 °C)

Headspace analysis of musth blood over time (h)	*N*	Total frontalin % of total TIC	Frontalin enantiomer (%)	
			+	–
1	3	7±0.15	52.1	47.9
2–24	3	2.8±0.12	50.3	49.7
24–48	2	0.8±0.01	51.8	48.2

Table 17.4 Total changes in TGS frontalin at 37 °C from an aliquot of a TGS sample from an older male Asian elephant

Conditions and time	Total frontalin TIC area	Frontalin (ng/ml)
37 °C—1 h	600,009	30
37 °C+ protease—1 h	6,281,891	354
37 °C+ protease—24 h	16,592,766	2877
37C+ protease—2 weeks	nd	nd

nd not detectable

the two enantiomers was almost equal and remained similar when blood headspace was analyzed for two days (Table 17.3). *Exo*-brevicomin and *endo-brevicomin* were not detected.

At 37 °C, the release of frontalin from TGS was 15-fold higher from mature 30–40 year-old males than from young 13–14 year-old males, reaching levels of 30 ng/ml (Table 17.4). To begin to assess whether proteins might be binding frontalin in the TGS, we demonstrated that the addition of protease to this TGS aliquot resulted in a tenfold increased release of frontalin (Table 17.4) within a 1-h timeframe. With extended sampling times (24 h) at 37 °C after the addition of protease, the released frontalin increased another several-fold. After 2 weeks, however, no additional released frontalin was detectable (Table 17.4).

17.4.4 Correlations Between Levels of Frontalin, the Precursor, and Testosterone

Asian musth TGS samples with high levels of frontalin contained 6-methyl-6-hepten-2-one, which is a biochemical precursor of frontalin (Francke and Schröder 1999; Keeling et al. 2013). The measured level of this precursor and the testosterone concentration of the male Asian elephant were highly correlated ($r=0.96$; Table 17.5). The ratio of 6-methyl-6-hepten-2-one to frontalin ranged between 10.3 %:89.7 % and 25.6 %:74.4 %, regardless of the levels of total frontalin. In contrast, 6-methyl-5-hepten-2-one was inversely correlated to testosterone levels ($r=-0.82$).

Table 17.5 Relationship of the putative biochemical frontalin precursor (6-methyl-6-hepten-2-one) and an isomeric non-precursor (6-methyl-5-hepten-2-one) to musth TGS frontalin and serum testosterone using the GC-MS total ion chromatogram (TIC)

Frontalin, % TIC	6-Methyl-6-hepten-2-one, % TIC	6-Methyl-5-hepten-2-one, % TIC[a]	Ratio precursor/frontalin (%)	Serum testosterone, ng/ml
2.20	0.72	nd	24.7/75.3	157.1
1.82	0.51	nd	21.9/78.1	125.3
1.62	0.21	nd	11.5/88.5	75.6
1.31	0.15	0.01	10.3/89.7	61.3
1.02	0.13	3	11.3/88.7	55.4
0.32	0.11	5	25.6/74.4	15.4
0.04	0.01	7	20/80	3.2
none	None	0.9–15		

[a]For the purpose of running a correlation, nd was given as 0.001

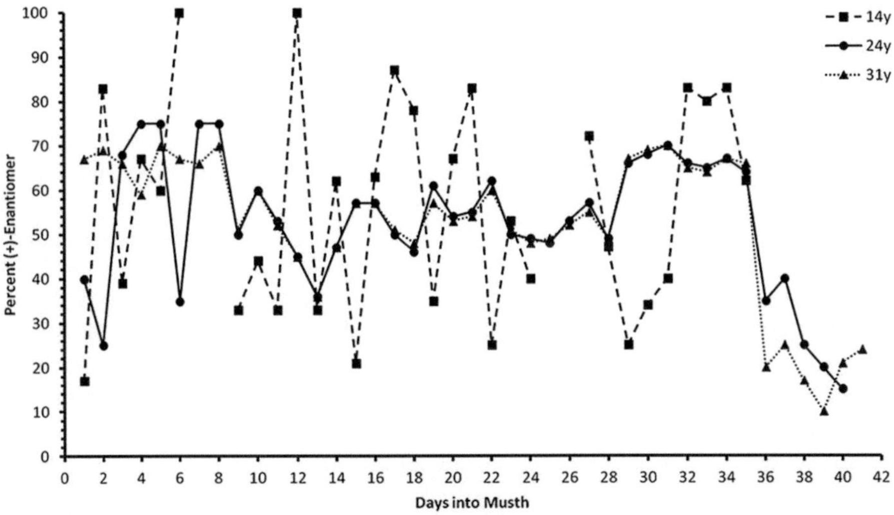

Fig. 17.1 Daily measurement of the proportion of the (+) frontalin enantiomer during the musth episodes of male Asian elephants of three different ages (14, 24 and 31 years). Note that on some days the 14 year old male produced no TGS and his musth episode was of shorter duration

17.4.5 Comparison of Frontalin in Musth Episodes of Three Male Asian Elephants

Frontalin enantiomers in TGS were monitored daily during three separate musth episodes from three individuals of three age categories. Samples were collected at approximately the same time each day and showed that both enantiomeric forms were generally present (Fig. 17.1). For the young 14 year old male the proportion of the (+) enantiomer ranged widely from 17 to 100 % and on several occasions no secretions were evident from this male resulting in a null score. The duration of his musth episode was shorter too. For the two older males (24 and 31 years) the ratios mostly remained close to 50 % (25–75 %), except at the end of musth when the (+) enantiomer was observed to drop to between 10 and 40 %. Overall, there is marked difference in the maintenance of a stable enantiomeric ratio between the younger male and the older males during the course of musth episodes.

17.5 Discussion

Frontalin serves as a chemical signal of musth in male Asian elephants (Rasmussen and Greenwood 2003). The enantiomeric ratio of frontalin changes as male elephants get older, and the change in the ratio affects the responses of conspecifics (Greenwood et al. 2005). We explored this sender-receiver relationship further by investigating how females in the follicular phase of estrus responded to three ratios of the enantiomers at three concentrations. The general results indicate that the racemic ratio at the highest concentration tested (100 μM) evoked the greatest behavioral responses. Unfortunately, the details of the responses by these nine follicular phase females were lost with the death of Bets Rasmussen. Thus, further testing of concentration levels with more types of receivers is warranted.

Since the identification of frontalin in 1969 (Kinzer et al. 1969), studies on bark beetles (Scolytidae) and especially the genus *Dendroctonus* have shown the importance of its role in blends of chemical signals as well as the signal value of precise enantiomeric ratios (Paine et al. 1999). In some species, apparently only females release frontalin and exclusively the (–) form (Francke et al. 1995). In other species, the females again produce frontalin, but in an 85:15 mixture of (–) and (+), whereas males of another species produce predominantly the (–) form (Stewart et al. 1977). With even more intricacy, in one species paired females and males released a single specific enantiomeric form but only after mating (Rudinsky et al. 1974). These records of exact ratios eliciting particular behaviors in insects using frontalin are especially relevant in view of our similar findings in elephants. Our discovery of distinct enantiomeric ratios with the brevicomins requires further investigation. Silverstein stated in referring to the *Ips* bark beetle species studies of Birch et al. (1980) that "Pheromonal specificity at the enantiomeric level is a principal isolation mechanism among sympatric species of *Ips*" (Silverstein 1981, 1988).

The genus *Loxodonta* never left the African continent, and during previous geological periods other progenitor species of *Elephas* coexisted with more ancient *Loxodonta* species. Although their habitat was presumably somewhat diverse, a degree of sympatry existed. Then *Elephas* migrated into Asia and vanished from Africa (the last recorded species being *E. iolensis* in the late Pleistocene period, about 10,000 years ago (Maglio 1973)). Thus, Asian and African elephants became geographically isolated, eliminating the need for totally separate and specific blends of pheromonal messages. The presence of frontalin in TGS from Asian female and

African male elephants suggests that it may have a general communicative function. Bioassays of 100 μL frontalin in 0.5 L water did not elicit bioresponses significantly different from a vanilla in water control or from the urine of a prepuberty female (Castelda 2008). Further testing of frontalin (in buffer) would help to clarify its potential signal value in African elephants. The specificity in the male Asian elephant is provided by precise enantiomeric ratios, especially during maturation and culminating in episodes of musth. Whether frontalin or another enantiomeric compound or compound blend plays a similar role in male African elephants requires further study.

We further investigated the occurrence of frontalin in male Asian elephant TGS (Greenwood et al. 2005) and extended the analysis to the presence of frontalin in blood, as well as the correlation of the biochemical frontalin precursor (6-methyl-6-hepten-2-one) (Francke et al. 1996; Francke and Schröder 1999; Francke and Schulz 1999; Keeling et al. 2013) and an isomeric non-precursor (6-methyl-5-hepten-2-one) to musth TGS frontalin and serum testosterone. A previous study identified frontalin in the headspace of blood from Asian male elephants in musth (Rasmussen and Perrin 1999); the enantiomeric ratio in the blood (one source for temporal gland secretions) has biosynthetic implications. The appreciable ratio of the precursor 6-methyl-6-hepten-2-one to frontalin, especially elevated in secretions with high frontalin concentrations, supports a common biosynthetic pathway between elephants and bark beetles (Barkawi et al. 2003; Goodwin et al. 2006). Rasmussen and Greenwood (2003, p. 444) offered the following conjecture: "Of interest therefore is the source of frontalin in elephants. Is it synthesized by males de novo or, more likely, is it transformed from plant-derived terpenic precursors by microbial action in the fermentative hind gut of the elephant and transported in the blood for release in the TGS? Evidence for the latter scenario comes from the finding of frontalin in the blood of musth males (L.E.L. Rasmussen, unpublished results)." However our finding of the isomeric compound 6-methyl-5-hepten-2-one as well in the blood of males but at levels elevated in the non-musth state (i.e., showing an inverse correlation with testosterone levels) suggests a more metabolically controlled scenario for frontalin biosynthesis. 6-Methyl-5-hepten-2-one is the product of a metabolic shunt that would effectively remove isoprenoid precursors from forming testosterone and frontalin. Through the likely action of an isomerase, excess precursors would be channeled through to this compound therefore diverting away from de novo frontalin synthesis and similarly testosterone that would be formed via the same precursor from which 6-methyl-6-hepten-2-one is synthesized. This scenario would provide an additional level of control over pheromone and hormone levels during their biosynthesis. Whatever the source, it is clear that Asian elephants have considerable olfactory acuity to distinguish between enantiomers as has been demonstrated recently in the study by Rizvanovic et al. (2013).

At least some of these unpublished results are presented herein. We have yet to demonstrate whether in elephants, like in insects (Francke et al. 1996; Perez et al. 1996; Barkawi et al. 2003; Seybold and Vanderwel 2003; Tittiger 2003), the presence of particular precursors affects frontalin production.

The clear differences observed in the daily ratios of frontalin enantiomers in TGS between young and older males during the duration of musth episodes is supportive of the idea that young inexperienced males with their ratios all over the place are to be ignored by conspecifics, who are likely sensing volatile frontalin levels (and ratios) to ascertain the state of musth of the emitting individual. This observation is mirrored in the corresponding responses of male and female conspecifics to TGS samples from males of different age classes. This in turn would likely impact significantly in the wild where broadcasting from, and similarly detecting the state of musth in a male at a distance, would help minimize agonistic encounters which are heightened during musth episodes of older males (Rasmussen and Greenwood 2003; Greenwood et al. 2005). Our work on the specificity of frontalin enantiomers moves the field of mammalian chemical communication forward into a more defined understanding of the links between pheromones and overt behavioral responses both in conspecifics and also a related species. Such work can have important implications for the conservation and management of elephants and mammals in general (Schulte et al. 2007). Moreover, our findings create a new perspective for pheromone receptor studies, since mirrored in such chirality is the inherent stereoselectivity and asymmetry of pheromone receptors—a focus of many investigations in mammalian olfaction. The olfactory-mediated world of bark beetles and of mammals may be somewhat superimposable. As Mori stated in 1998 (p. 578), "Understanding the relationship between structure and biological effect requires that the absolute configurations of naturally occurring chiral compounds must be determined." Our identification of the two released enantiomers from male Asian elephant temporal glands and the connection to behavioral responses by conspecifics suggests a reexamination of mammalian chemical signals is warranted and that the elucidation of the basis of receptor handling is a critical next step. The recent study by Rizvanovic et al (2013) is an encouraging step in the right direction.

Acknowledgments We thank Riddle's Elephant and Wildlife Sanctuary, Ringling Center for Elephant Conservation, Dickerson Park Zoo, Tulsa Zoo, Oregon Zoo, and Auckland Zoo for TGS and blood samples. This work was supported by Biospherics Research Corporation and International Science and Technology (ISAT) Linkages Fund.

References

Bagley KR, Goodwin TE, Rasmussen LEL, Schulte BA (2006) Male African elephants, *Loxodonta africana*, can distinguish oestrus status via urinary signals. Anim Behav 71:1439–1445

Barkawi LS, Francke W, Blomquist GJ, Seybold SJ (2003) Frontalin: de novo biosynthesis of an aggregation pheromone component by *Dendroctonus* spp. bark beetles (Coleoptera: Scolytidae). Insect Biochem Mol Biol 33:773–788

Birch MC, Light DM, Wood DL, Browne LE, Silverstein RM, Bergot BJ, Ohloff G, West JR, Young JC (1980) Pheromonal attraction and allomonal interruption of *Ips pini* in California by the two enantiomers of ipsdienol. J Chem Ecol 6:703–717

Bridges JR (1982) Effects of juvenile hormone on pheromone synthesis in *Dendroctonus frontalis* (Coleoptera: Scolytidae). Environ Entomol 11:417–420

Browne LE, Wood DL, Bedard WD, Silverstein RM, West JR (1979) Quantitative estimates of the western pine beetle attractive pheromone components, *exo*-brevicomin, frontalin, and myrcene in nature. J Chem Ecol 5:397–414

Byers JA (1983) Influence of sex, maturity, and host substances on pheromones in the guts of bark beetles, *Ips paraconfusus* and *Dendroctonus brevicomin*. J Insect Physiol 29:5–13

Byers JA, Woods DL (1980) Interspecific inhibition of the response of the bark beetles, *Dendroctonus brevicomis* and *Ips paraconfusus*, to their pheromones in the field. J Chem Ecol 6:149–164

Byers JA, Woods DL, Craig J, Hendry LB (1984) Attractive and inhibitory pheromones produced in the bark beetle *Dendroctonus brevicomis*, during host cololonization: regulation of inter- and intraspecific competition. J Chem Ecol 10:861–877

Carman RM, Klika KD (1992) Partially racemic compounds as brushtail possum urinary metabolites. Aust J Chem 45:651–657

Castelda SM (2008) Waterhole dynamics and chemical signals of African elephants (*Loxodonta africana*). Electronic theses and dissertation, paper 723, Georgia Southern University

Cavaggioni A, Mucignat-Caretta C, Zagotto G (2003) Absolute configuration of 2-sec-butyl-4,5-dihydrothiazole in male mouse urine. Chem Senses 28:791–797

Francke W, Schröder W (1999) Bicyclic acetals in systems of chemical communication. Curr Org Chem 3:407–443

Francke W, Schulz S (1999) Pheromones. In: Barton DHR, Nakanishi K (eds) Comprehensive natural products chemistry, vol 8. Elsevier, New York, NY, pp 197–261

Francke W, Bartels J, Meyer H, Schröder F, Kohnle U, Baader E, Vité JP (1995) Semiochemicals from bark beetles: new results, remarks and reflections. J Chem Ecol 21:1043–1063

Francke W, Schröder F, Philipp P, Meyer H, Sinnwell V, Gries G (1996) Identification and synthesis of new bicyclic acetals from the mountain pine beetle, *Dendroctonus ponderosae* Hopkins (Coleoptera: Scolytidae). Bioorg Med Chem 4:363–374

Goodwin TE, Schulte BA (2008) A tribute to L.E.L. "Bets" Rasmussen (1938-2006). In: Hurst JL, Beynon RJ, Roberts SC, Wyatt TD (eds) Chemical signals in vertebrates, vol 11. Springer, New York, NY, pp 1–9

Goodwin TE, Eggert MS, House SJ, Weddell ME, Schulte BA, Rasmussen LEL (2006) Insect pheromones and precursors in female African elephant urine. J Chem Ecol 32:1849–1853

Gore WE, Pearce GT, Silverstein RM (1976) Mass spectrometric studies of the dioxabicyclo[3.2.1] octanes multistriatin, frontalin and brevicomin. J Org Chem 41:607–610

Greenwood DR, Comesky D, Hunt MB, Rasmussen LEL (2005) Chemical communication: chirality in elephant pheromones. Nature 438:1097–1098

Gries G, Pierce HD, Lindgren BS, Borden JH (1988) New techniques for capturing and analyzing semiochemicals for scolytid beetles. J Econ Entomol 81:1715–1720

Gries G, Borden JH, Gries R, LaFontaine JP, Dixon EA, Wieser H, Whitehead AT (1992) 4-Methylene-6,6-dimethylbicyclo[3.1.1]hept-2-ene (verbenene): new aggregation pheromone of the scolytid beetle, *Dendroctonus rufipennis*. Naturwissenschaften 79:367–368

Grosman DM (1996) Chemical ecology of the southern pine beetle *Dendroctonus frontalis* Zimmerman (Coleoptera: Scolytidae). Online workshop: a technology transfer project of Virginia Tech, Texas Forest Service, University of Georgia & USDA Forest Service (online preparation by McClellen, QC), Blacksburg, VA: Virginia Tech, Dept. of Entomology (http://www.ento.vt.edu/~salom/Chemecology/chemecol.html)

Grosman DM, Salom SM, Ravlin FW, Young RW (1997) Geographic and gender differences in semiochemicals in emerging adult southern pine beetles (Coleoptera: Scolytidae). Ann Entomol Soc Am 90:438–446

Jainudeen MR, Katongole CB, Short RV (1972) Plasma testosterone levels in relation to musth and sexual activity in the male Asiatic elephant, *Elephas maximus*. J Reprod Fertil 29:99–103

Keeling CI, Chiu CC, Aw T, Li M, Henderson H, Tittiger C, Weng H-B, Blomquist GJ, Bohlmann J (2013) Frontalin pheromone biosynthesis in the mountain pine beetle, *Dendroctonus ponderosae*, and the role of isoprenyl diphosphate synthases. Proc Natl Acad Sci U S A 110: 18838–18843

Kinzer GW, Fentiman AF Jr, Page TF Jr, Foltz RL, Vité JP, Pitman GB (1969) Bark beetle attractants: identification, synthesis and field bioassay of a new compound isolated from *Dendroctonus*. Nature 221:477–478

Li G, Roze U, Locke DC (1997) Warning odor of the North American porcupine (*Erethizon dorsatum*). J Chem Ecol 23:2737–2754

Maglio VJ (1973) Origin and evolution of the Elephantidae. Trans Am Phil Soc Philadelphia 63:1–149

Mori K (1975) Synthesis of optically active forms of frontalin. Tetrahedron 31:1381–1384

Mori K (1998) Chirality and insect pheromones. Chirality 10:578–585

Mori K (2007) Significance of chirality in pheromone science. Bioorg Med Chem 15:7505–7523

Mori K (2011) Bioactive natural products and chirality. Chirality 23:449–462

Novotny MV, Soini HA (2008) Volatile mammalian chemosignals – structural and quantitative aspects. In: Hurst J, Beynon R, Wyatt T, Roberts C (eds) Chemical signals in vertebrates, vol 11. Springer, New York, NY, pp 13–24

Novotny MV, Xie TM, Harvey S, Wiesler D, Jemiolo B, Carmack M (1995) Stereoselectivity in mammalian chemical communication: male mouse pheromones. Experientia 51:738–743

Paine TD, Millar JG, Hanlon CC, Hwang JS (1999) Identification of semiochemicals associated with Jeffrey pine beetle, *Dendroctonus jeffreyi*. J Chem Ecol 25:433–453

Perez AL, Gries R, Gries G, Oehlschlager AC (1996) Transformation of presumptive precursors to frontalin and *exo*-brevicomin by bark beetles and the West Indian sugarcane weevil (Coleoptera). Bioorg Med Chem 4:445–450

Rasmussen LEL (2001) Source and cyclic release pattern of (Z)-7-dodecenyl acetate, the preovulatory pheromone of the female Asian elephant. Chem Senses 26:611–623

Rasmussen LEL (2003) Why musth elephants use pheromones. Biologist 50:195–196

Rasmussen LEL, Greenwood DR (2003) Frontalin: a chemical message of musth in Asian elephants (*Elephas maximus*). Chem Senses 28:433–446

Rasmussen LEL, Perrin TE (1999) Physiological correlates of musth: lipid metabolites and chemosignal composition of exudates. Physiol Behav 67:539–549

Rasmussen LEL, Lee TD, Zhang A, Roeloffs WL, Daves GD Jr (1997) Purification, identification, concentration and bioactivity of (*Z*)-7-dodecen-1-yl acetate: sex pheromone of the female Asian elephant, *Elephas maximus*. Chem Senses 22:417–437

Rasmussen LEL, Riddle HS, Krishnamurthy V (2002) Mellifluous matures to malodorous in musth. Nature 415:975–976

Rasmussen LEL, Krishnamurthy V, Sukumar R (2005) Behavioural and chemical confirmation of the preovulatory pheromone, (*Z*)-7-dodecenyl acetate, in wild Asian elephants: Its relationship to musth. Behaviour 142:351–396

Renwick JAA, Vité JP (1969) Bark beetle attractants: mechanisms of colonization by *Dendrotonus frontalis*. Nature 224:1222–1223

Rizvanovic A, Amundin M, Laska M (2013) Olfactory discrimination ability of Asian elephants (*Elephas maximus*) for structurally related odorants. Chem Senses 38:107–118

Rudinsky JA, Morgan ME, Libbey LM, Putman TB (1974) Antiaggregative rivalry pheromone of the mountain pine beetle, and a new arrestant of the southern pine beetle. Environ Entomol 3:90–98

Santiago B, Soderquist JA (1992) Nonracemic frontalin via copper and palladium-based sketetal construction and the asymmetric dihydroxylation. J Org Chem 57:5844–5847

Schulte BA, Rasmussen LEL (1999) Signal-receiver interplay in the communication of male condition by Asian elephants. Anim Behav 57:1265–1274

Schulte BA, Freeman E, Goodwin TE, Hollister-Smith J, Rasmussen LEL (2007) Honest signaling through chemicals by elephants with applications for conservation and care. Appl Anim Behav Sci 102:344–363

Seybold SJ, Vanderwel D (2003) Biosynthesis and endocrine regulation of pheromone production in Coleoptera. In: Blomquist GJ, Vogt RG (eds) Insect pheromone biochemistry and molecular biology: the biosynthesis and detection of pheromones and plant volatiles. Academic, London, pp 137–200

Silverstein RM (1981) Pheromones: background and potential for use in insect pest control. Science 213:1326–1332

Silverstein RM (1985) Chiral semiochemicals. In: Acree TE, Soderlund DM (eds) Semiochemicals: flavors and pheromones. Walter de Gruyter, New York, NY, pp p121–p140

Silverstein RM (1988) Chirality in insect communication. J Chem Ecol 14:1981–2004

Stewart TE, Plummer EL, McCandless LL, West JR, Silverstein RM (1977) Determination of enantiomer composition of several bicyclic ketal insect pheromone components. J Chem Ecol 3:27–43

Tittiger C (2003) Molecular biology of bark beetle pheromone production and endocrine regulation. In: Blomquist GJ, Vogt RG (eds) Insect pheromone biochemistry and molecular biology: the biosynthesis and detection of pheromones and plant volatiles. Academic, London, pp 201–230

Turpin JA, Weigel LO (1992) Enantioselective synthesis of the 6,8-dioxabicyclo-[3.2.1]octane skeleton by asymmetric dihydroxylation. Tetrahedron Lett 33:6563–6564

Part IV
Interspecific Signaling

Chapter 18
Detection of Fish and Newt Kairomones by Ovipositing Mosquitoes

Lauren L. Eveland, Jason R. Bohenek, Alon Silberbush, and William J. Resetarits Jr.

18.1 Introduction

Oviposition site choice is reproductive behavior in which oviparous organisms actively choose a habitat for their offspring based on its expected quality (Rausher 1983; Singer 1984). Many colonizing aquatic organisms with complex life cycles must choose between discrete aquatic habitats for their offspring, typically offering little parental care other than selecting a suitable oviposition site (Resetarits and Wilbur 1989). Since parental fitness depends on habitat quality, and the biotic and abiotic components of aquatic habitats can vary widely, selection should favor evolution of selective oviposition (Resetarits 1996). This is especially true for short-lived species such as insects since they have limited time to find a suitable oviposition site and may even only reproduce once (Blaustein 1999).

Selective oviposition has been documented in many aquatic insects (Chesson 1984; Petranka and Fakhoury 1991; Blaustein and Kotler 1993; Lowenberger and Rau 1994; Berendonk 1999; Resetarits 2001; Abjornsson et al. 2002; Binckley and Resetarits 2005) and amphibians (Resetarits and Wilbur 1989; Crump 1991; Kats and Sih 1992; Binckley and Resetarits 2002; Rieger et al. 2004; Vonesh and Buck 2007). Habitat choices are influenced by both abiotic (Bentley and Day 1989; Binckley and Resetarits 2007, 2008; Hocking and Semlitsch 2007) and biotic factors (Chesson 1984; Resetarits and Wilbur 1989; Petranka and Fakhoury 1991; Blaustein and Kotler 1993) that determine the potential quality of oviposition sites. The relative importance of these factors varies depending on the ecology and life-history of the species in question (Resetarits and Wilbur 1989; Berendonk 1999; Binckley and

L.L. Eveland (✉) • J.R. Bohenek (✉) • W.J. Resetarits Jr.
Department of Biology, The University of Mississippi, University, MS, 38677, USA
e-mail: leveland87@gmail.com; Jason.bohenek@gmail.com

A. Silberbush
Department of Life Sciences, Center for Biological Control (CBC),
Ben-Gurion University of the Negev, 653, Beer-Sheva 8410501, Israel

B.A. Schulte et al. (eds.), *Chemical Signals in Vertebrates 13*,
DOI 10.1007/978-3-319-22026-0_18

Resetarits 2007, 2008; Walton et al. 2009; Vonesh and Blaustein 2010). If individuals can reliably predict offspring performance, then selective oviposition will ultimately enhance parental fitness (Resetarits and Wilbur 1989; Blaustein et al. 2004; Hocking and Semlitsch 2007).

Mosquitoes, especially the genera *Culex* and *Culiseta*, offer a valuable model for studying oviposition site choice in response to predation because they meet criteria proposed by multiple authors (Rausher 1983; Singer 1984; Resetarits and Wilbur 1989; Resetarits 1996; Blaustein 1999): (1) egg-raft laying mosquitoes have few lifetime reproductive events, (2) they lay their eggs together in a clutch, and (3) their larvae are susceptible to predators. These three characteristics illustrate that a single, poor decision can lead to zero reproductive output. Also, predator characteristics may promote selective oviposition in prey, if predators are (4) heterogeneously distributed among patches, (5) remain in those patches during the prey larval period, and (6) reliably detectable (Resetarits and Wilbur 1989; Blaustein 1999; see also Blaustein et al. 2004). In addition, selectable patches must differ in a meaningful way in terms of predator distribution and abundance: if a predator is highly mobile, then selective oviposition may prove ineffective as a prey strategy. Given these criteria, one can predict which organisms should evolve selective oviposition in the context of effective, predictable predators.

It has long been assumed that semiochemicals, specifically kairomones, cued predator presence to ovipositing mosquitoes (Chesson 1984; Petranka and Fakhoury 1991; Angelon and Petranka 2002). A kairomone is an interspecific semiochemical released by an organism to the benefit of the receiver but not the emitter (Brown et al. 1970). Silberbush and Blaustein (2008) and Silberbush et al. (2010) established that ovipositing *Culiseta longiareolata* detect the presence of the predatory hemipteran *Notonecta maculata* via kairomones. A kairomone mechanism for fish deterrence has yet to be definitively established. Previous studies that suggested a kairomones mechanism had design issues, such as (1) using predators in cages (Petranka and Fakhoury 1991), which can still provide visual or tactile cues (Berendonk 1999), (2) using late instar larval counts instead of counting egg rafts, which is the only method to accurately assess oviposition choice (Petranka and Fakhoury 1991; Angelon and Petranka 2002), or (3) studying the interaction only with captive mosquitoes in artificial lab conditions (Van Dam and Walton 2008), which may not correlate with natural behavior (Bentley and Day 1989). We sought to improve upon these studies by definitively assaying natural mosquito oviposition in response to fish chemical cues.

Additionally, fish and notonectids have been used as model predators in 73 % of mosquito-predator oviposition research with 40 % of studies using only two species, the western mosquitofish (*Gambusia affinis*) and *Notonecta irrorata* (Vonesh and Blaustein 2010). However, other potential predators may affect mosquito oviposition. Central newts (*Notophthalmus viridescens louisianensis*) are aquatic salamanders that feed on mosquito larvae (DuRant and Hopkins 2008). They fill a similar role to small predatory fish in fishless habitats, and they function as keystone predators by preferentially feeding on superior competitors, thereby enhancing diversity (Morin 1981).

In this context, we wanted to answer two different questions: (1) Are kairomones alone responsible for fish avoidance during oviposition in *Culex* mosquitoes? (2) Do ovipositing *Culex* also avoid other predators during oviposition? In order to answer these questions, we conducted a series of three experiments using artificial pools to assay oviposition in natural mosquito populations.

18.2 Materials and Methods

18.2.1 Study Location

Our research was conducted at Tyson Research Center (795.8 ha) of Washington University located along the Meramec River in St. Louis County, MO. Tyson lies on the Ozark border and is comprised of oak and hickory secondary forest with sycamore, maple, and cottonwood in the bottomlands. It has patches of old fields and also a variety of permanent and temporary ponds and streams.

18.2.2 Fish Experiments

Two field experiments were conducted at Tyson during July–August 2013 to assay oviposition of natural mosquito populations. The experiments were constructed in two open fields with minimal tree canopy, 152 m apart. We constructed two separate arrays of black plastic tubs (66 × 45 × 16 cm), one array for each test species (16 pools place in eight pairs for each). All tubs were bleached, scrubbed, and power-washed between uses. Pools were separated by 1 m and each pair was 3 m from its nearest neighboring pair (Fig. 18.1). Pools were filled with tap water and left to age for 2 days. We then added 10 g of rabbit chow (Small World Rabbit Food—Mannapro, St. Louis, MO; 40 % protein) to facilitate pool detection.

Western mosquitofish (*G. affinis*) and green sunfish (*Lepomis cyanellus*) were selected for the experiment because they are native to Missouri and are both known predators of mosquito larvae (DuRant and Hopkins 2008; Silberbush and Resetarits *in prep*). Both species are opportunistic feeders that will also consume mosquito egg rafts (Eveland *personal observation*; Silberbush *personal observation*). Fish were collected from ponds at Tyson and kept in separate 1200 L holding tanks. Eight individuals of each fish species were haphazardly removed from their covered holding tanks and added to indoor 10 gal glass aquariums (51 × 28 × 30.5 cm) for 2 days. The fish were fed fish flakes (TetraMin® Tropical flakes—Tetra Holding inc., Germany) for the first 24-h then gut-cleared (no feeding) for an additional 48-h before being transferred to the pools. While in the pools, the fish had no access to food.

We used one predator species per array—western mosquitofish in one and green sunfish in another. In each array, we introduced a single fish to one pool in a pair that was randomly designated as the predator treatment. Each pool contained a cage that consisted of a black plastic plant pot with two screened sides (~1 mm mesh) and a screened lid. The fish were placed inside the cages of predator pools for 3 days and the pools covered to prevent any oviposition during the conditioning period. Before dusk on day 3, the fish were removed and the pools were opened to allow oviposition. All egg rafts were collected on the morning of day 4 and transferred to the laboratory. Egg rafts were individually hatched and larvae were reared to fourth instar and identified morphologically (Darsie and Ward 2005). The conditioning process was repeated four times for each site with different individual fish in the same pools.

18.2.3 Newt Experiment

From August 8th–14th we conducted a newt experiment at Tyson of much the same design as the fish experiments. We used eight pairs of the same type of plastic tubs with predator cages used in the fish experiments (Fig. 18.1). All tubs were cleaned in like manner as in the fish experiment. Each pool in a pair was separated by 0.91 m and pairs were dispersed over an area of 0.6 km^2. The minimum distance between two pairs was greater than 75 m. This design prevented interaction between pairs. All pools were positioned off dirt roads adjacent and parallel to the forest edge. The pools were filled with aged tap water and 5 g of rabbit chow (Small World Rabbit Food—Mannapro, St. Louis, MO; 40 % protein) was added to facilitate pool detection. Two randomly selected adult newts were introduced into the cage of one

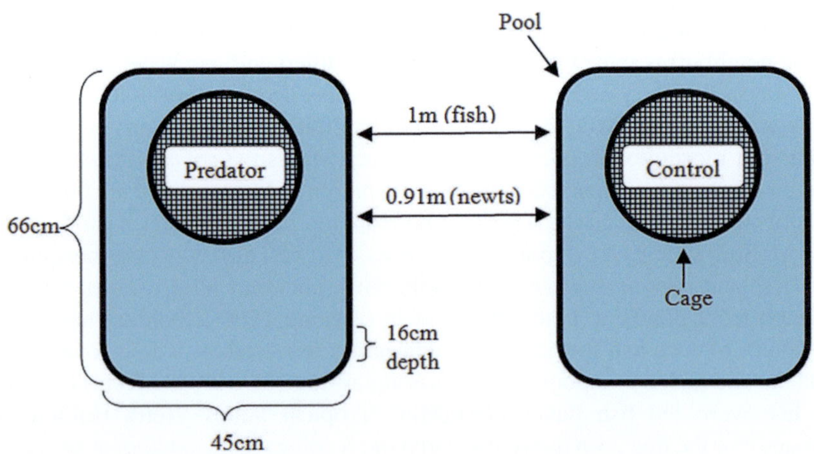

Fig. 18.1 The experimental design for all three experiments consisted of paired pools with cages. The fish experiments were conditioned with fish but only opened for oviposition when the fish were removed. The newt experiment contained caged newts. Newt pools were positioned 0.91 m (3 ft) apart and fish pools were separated by 1 m

randomly selected pool in each pair. The cage allowed the newts to interact with the water of the pool, but prevented prey consumption. Dead or consumed conspecific prey can produce cues that can potentially confound predator-released kairomones (Kats and Dill 1998). Curved plastic squares were placed in the cages to provide refuge for the newts. Unlike the previous experiment, newts were housed in the pools for the duration of the study, though it is unlikely they were visible or identifiable to species through the screen mesh. Pools were open for oviposition by natural mosquito populations. Since the pools were filled and had predators added on the evening on the same day, the first round of egg rafts were discarded without quantification. This was done in order to ensure mosquito oviposition took place only when the water was adequately conditioned with predator cue.

Egg raft collection and species identification mirrored that of the fish experiment, except egg rafts were collected daily since there was no conditioning period. The experiment was terminated after seven collection days, which was decided a priori because oviposition sharply declines after pools age for a week (Bohenek *unpublished data*; Bohenek and Silberbush *personal observation*). Newts were removed and weighed at the termination of the experiment in order to determine total predator biomass per pool.

18.2.4 Data Analysis

For each of the three experiments, we calculated a mean number of egg rafts per pool per day (daily average), which was square-root transformed [$\sqrt{(x+0.5)}$; (Yamamura 1999)]. We used paired, one-tailed Student's t-test to assess whether the daily average number of egg rafts laid in control pools was greater than predator pools. Lastly, a Pearson's product–moment correlation was performed for the newt experiment to determine if total newt biomass could predict the strength of the oviposition deterrence. Oviposition deterrence was calculated as the difference between the number of egg rafts in control pools and predator pools. All statistical analyses were performed in RStudio version 0.98.994.

18.3 Results

In total we collected 2006 egg rafts from our three experiments. A large sample from another experiment in the same study site was morphologically identified as *C. pipiens* complex (Barr 1957; Harbach 2012) and all (100 %) were subsequently identified using PCR as *C. pipiens* × *C. quinquifasciatus* hybrids (Silberbush and Resetarits *unpublished data*), hereafter "*C.* pxq." Our sample consisted of 75.12 % *Culex restuans* and 22.68 % *C.* pxq. Other species were ignored due to very low abundance (2.2 % of total).

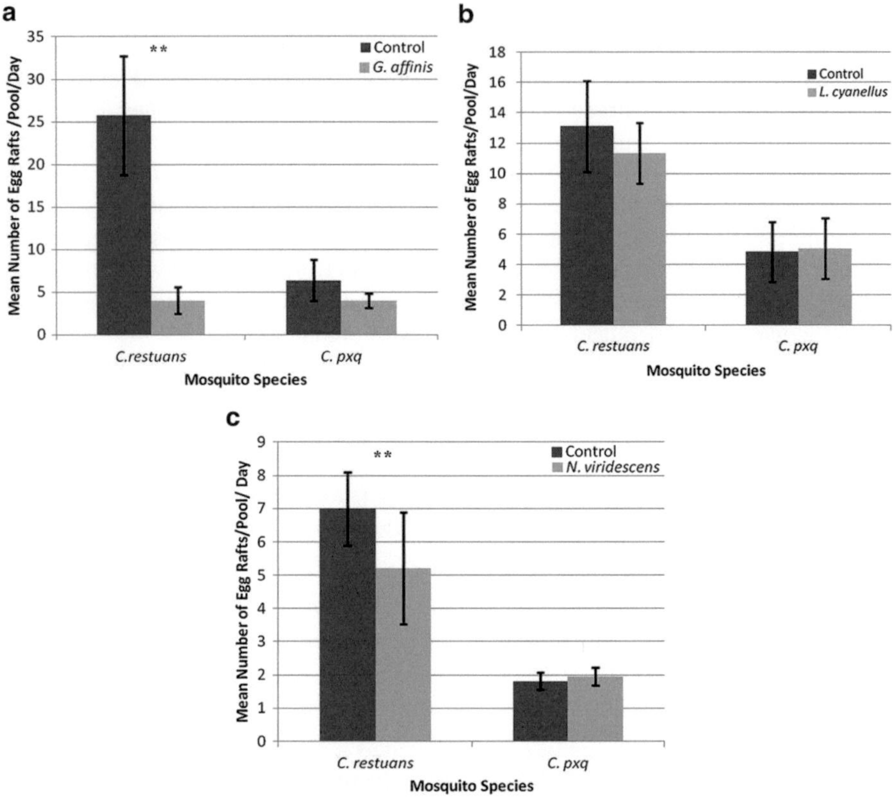

Fig. 18.2 Mean number (±SE) of egg rafts by mosquito species and predator treatment. The *dark gray bars* represent controls and the *light gray bars* represent predator pools with (**a**) western mosquitofish (*Gambusia affinis*), (**b**) green sunfish (*Lepomis cyanellus*) and (**c**) central newts (*Notophthalmus viridescens louisianensis*). $^{*}p<0.05$ and $^{**}p<0.01$

18.3.1 Fish Experiment

In the mosquitofish array, we collected 321 egg rafts; 238 (74.1 %) were *C. restuans* and 83 (25.8 %) were *C.* pxq. Mosquitofish had a significant effect on *C. restuans* oviposition ($t=4.00$, df=7, p-value=0.0026), with the mean number of egg rafts greater in the controls (Fig. 18.2a). Treatment had no effect on *C.* pxq oviposition ($t=0.71$, df=7, p-value =0.25). *C. restuans* laid 206 (87 %) egg rafts in the control pools and 32 (13 %) egg rafts in the mosquitofish pools. *Culex pipiens* laid 51 (61 %) egg rafts in the control pools and 32 (39 %) egg rafts in the mosquitofish pools.

There were two occurrences of fish mortality in the green sunfish array and the fish were immediately replaced. We collected 940 egg rafts; 692 (73.6 %) were *C. restuans* and 248 (26.4 %) were *C.* pxq. Green sunfish had no effect on the oviposition of either mosquito species—*C. restuans* ($t=1.15$, df=7, p-value=0.14) and *C.* pxq ($t=0.199$, df=7, p-value=0.42) (Fig. 18.2b). *C. restuans* laid 411 (59 %)

egg rafts in the control pools and 281 (41 %) egg rafts in the green sunfish pools. *C. pipiens*laid 128 (52 %) egg rafts in the control pools and 120 (48 %) egg rafts in the green sunfish pools.

18.3.2 Newt Experiment

We collected 913 egg rafts; 651 (71.3 %) were *C. restuans* and 149 (16.3 %) were *C.* pxq. One newt escaped from a pool, but the pool still contained the remaining newt. We detected a strong and significant species-specific response (Fig. 18.2c). Newts had a significant effect on *C. restuans* oviposition ($t=3.298$, df=7, p-value=0.0066), but not on *C.* pxq oviposition ($t=-0.84$, df=7, p-value=0.79). *C. restuans* laid 384 (59.0 %) egg rafts in control pools and 267 (41.0 %) egg rafts in predator pools. *Culex* pxq laid 74 (49.7 %) egg rafts in control pools and 75 (50.3 %) egg rafts in predator pools. Pearson's product–moment correlation revealed no significant relationship between total newt biomass per pool (range: 3.53–6.11 g) and oviposition deterrence for *C. restuans* ($r=-0.500$; $t=-1.42$, df=6, p-value=0.21) or *C.* pxq ($r=-0.296$; $t=-0.759$, df=6, p-value=0.48).

18.4 Discussion

The detection and avoidance of predators by female mosquitoes when selecting an oviposition site is of critical importance to offspring survival and is the only parental care they offer. Oviposition behavior can affect species distributions, species interactions, offspring survival, and community structure (Resetarits and Wilbur 1989). Therefore, determining the mechanisms that mosquitoes use to detect predators aids in understanding the complex arms race between mosquitoes and aquatic predators. We focused on whether or not *Culex* mosquitoes can detect fish predators and central newts through predator-released kairomones.

Our results demonstrate that *C. restuans* use kairomones to detect and avoid western mosquitofish and central newts, but they do not avoid green sunfish. All three predator species are efficient consumers of mosquito larvae (DuRant and Hopkins 2008; Silberbush and Resetarits *in prep*). Predator efficiency depends on numerous factors such as availability of alternative prey and prey refuges (Webb and Joss 1997; Willems et al. 2005; Juliano 2009). Mosquitofish may have a greater impact on mosquito larvae compared with green sunfish resulting in greater selection pressure for detection of a mosquitofish kairomone. Western mosquitofish are tolerant of harsh abiotic conditions and are often found in large numbers (Offill and Walton 1999; Willems et al. 2005; Dam and Walton 2007; Walton 2007). A single mosquitofish can consume thousands of mosquito larvae in a 24 h period (DuRant and Hopkins 2008); thus, high numbers will most likely completely eliminate mosquito cohorts. Even when a mosquito is desperate to locate an oviposition site, it is probably more advantageous to continue searching than to settle for a habitat with mosquitofish. Alternatively, a few green sunfish might not necessarily have the same effect and can even be an indicator of reduced numbers of insect predators (Knight et al. 2005). Predator size may be another important factor since mosquitofish remain relatively small at adulthood compared to green sunfish. Mosquitofish may pose a constant predation threat while green sunfish predation on mosquito larvae occurs mainly in the smaller fish size classes. Mosquitofish and green sunfish may also forage in different microhabitats with mosquitofish favoring the water surface and green sunfish foraging lower in the water column and in the vegetation. Lastly, the pirate perch (*Aphredoderus sayanus*) is a freshwater fish that is not avoided by ovipositing *Culex* spp., nor by treefrogs and colonizing beetles, due to a form of chemical camouflage (Resetarits and Binckley 2013, Silberbush and Resetarits *in prep*). It is possible that green sunfish also possess some form of camouflage.

Despite considerable overlap in ecological niche, *C.* pxq and *C. restuans* vary in oviposition responses to predators. *C. restuans*, but not *C.* pxq, were selective, which was unexpected given that *C.* pxq also meet all criteria for evolving selective oviposition (Resetarits and Wilbur 1989; Resetarits 1996; Blaustein 1999). The simplest explanation is that *Culex* pxq may not be able to detect the predator-released kairomones in question, or require greater concentrations for detection. *Culex quinquefasciatus* has shown lack of mosquitofish cue avoidance (Walton et al. 2009) and *Culex* pxq has shown limited sensitivity (Angelon and Petranka 2002), so our results are not surprising. When considered from a life history standpoint, *Culex* pxq larvae may be more predator tolerant than *C. restuans* larvae. Or perhaps, *C.* pxq larvae are poor competitors, and thus risk predation in favor of reducing competition with *C. restuans* larvae. This tradeoff has been shown in damselflies where *Enallagma* species found in fish lakes were better able to avoid predation than species found in fishless lakes (McPeek 1990). Another possibility is that the hybrids have different behavioral algorithms than either pure species. It is typical for hybrids to display some intermediate behavior, even if it is maladaptive. Peach-face lovebirds (*Agapornis roseicollis*) transport nest-material by tucking it into their tail feathers while Fischer's lovebirds (*Agapornis fischeri*) carry nest-material in their beaks (Buckley 1969). Hybrids lovebirds display considerable difficulty choosing between a behavior and will repeatedly tuck and untuck nest-material in their tail feathers (Dilger 1962). Similarly, hybrid tree frogs (*Hyla* spp.) produce a vocalization that is intermediate between the parent vocalizations and less attractive to females (Gerhardt 1974). Similar chimeric behavior could limit the effectiveness of *C.* pxq behavioral responses. Aggression between ovipositing mosquitoes should also be considered since pool surface area is limited and *C. pipiens* will actively defend their oviposition site (Krause et al. 1992). If this aggression operates interspecifically, and *C. restuans* is superior in this respect, then *C.* pxq may be forced out of predator-free habitats. This variation between congeners demonstrates the potential for hidden complexity in life-histories that can determine community structure.

The detection and avoidance of fish is a potentially adaptive capability that can lead to a decrease in offspring mortality and an increase in reproductive success. Artificial lab studies have previously demonstrated that *C. tarsalis* (Van Dam and Walton 2008) and phantom midges (Berendonk 1999) use kairomones to detect predatory fish. Silberbush et al. (2010) found *C. longiareolata* detect predatory backswimmers (*N. maculata*) via predator-released kairomones when selecting an oviposition site. However, our field experiment is the first evidence of natural mosquito populations responding to fish kairomones. The next step is to identify the fish-released kairomones. Silberbush et al. (2010) identified n-tricosane and n-heneicosane as the chemicals *C. longiareolata* use to detect *N. maculata*. The responses to these hydrocarbons are also species-specific since *Anopheles gambiae* avoid *N. maculata*-conditioned water, but not water containing n-tricosane and n-heneicosane (Warburg et al. 2011). Thus, mosquito species may vary in their responses to specific kairomones and/or mixtures of kairomones and behaviors of specific mosquito species should not imposed onto other species, even if they are closely related.

Our results also describe the first evidence of adult amphibians deterring mosquito oviposition via kairomones. Newts are efficient predators of larval mosquitoes (DuRant and Hopkins 2008). Combining both oviposition deterrence and predation, newts can function similarly to mosquitofish in fish-free habitats but without the consequences attendant to fish introductions into previously fishless habitats (Hecnar and McLoskey 1997; Kats and Ferrer 2003; Schilling 2008). Not only can newts suppress mosquito populations, but they can even increase diversity in aquatic habitats due to their keystone predator effect (Morin 1981), thus providing a dual benefit over more generally destructive fish predators (Hecnar and McLoskey 1997; Kats and Ferrer 2003; Schilling 2008).

Given that newts were caged in the pools, the possibility of mosquitoes using tactile and/or visual cues instead of chemical cues remains. *C. pipiens* readily flees a water surface when disturbed (Schober 1966; Meillon et al. 1967). Since we observed an even number of *C.* pxq egg rafts between newt and newt-free pools, it is unlikely that any tactile or visual cues were sufficient to drive away ovipositing mosquitoes. This is true unless *C.* pxq responds differently than *C. restuans* to tactile or visual cues. Additionally, mosquitoes oviposit at night in shady areas and the predators were housed in cages that had three screened surfaces and two large opaque sides. Thus, visual cues seem unlikely especially since the newts often hid under a plastic refuge in each cage; however, an additional study using only newt cues may be worthwhile. The strength of newt deterrence was lower than that of mosquitofish. This could be due to a weaker chemical signature but also to predator behavior. Newts are lethargic when compared to constantly swimming mosquitofish, which should translate to lower kairomone production due to lower metabolic rates. Newt body mass was a negligible factor and fish biomass was not recorded, but mosquitofish we worked with at Tyson typically weigh less than 1 g, which is substantially less than mean newt biomass per pool (4.79 g; two newts per pool), while the green sunfish we worked with typically weighed 7–12 g. *C. restuans* avoided a single 1 g mosquitofish, but not a single 7–12 g green sunfish, which emphasizes the relative importance of predator identity over predator biomass. These findings are in accordance with past studies with fish that showed predator identity and its presence/absence are much more important than predator size and number, once a threshold for detection is reached (Rieger et al. 2004).

Since mosquitoes are an important disease vector (Gubler 1998; Gratz 1999), mosquitofish have been widely introduced as an attempt to biologically control their populations (Courtenay and Meffe 1989). However, fish introductions can dramatically degrade amphibian and invertebrate biodiversity (Hecnar and McLoskey 1997; Kats and Ferrer 2003; Schilling 2008). This is especially troublesome because many amphibian species are declining worldwide (Stuart et al. 2004) and fish introductions have been implicated as a major cause of declines in some species (Bradford 1989; Hecnar and McLoskey 1997; Knapp and Matthews 2000; Denoel et al. 2005). The potential efficacy of native wildlife as biocontrol agents has largely been ignored in favor of predators (often exotic) with known or presumed effects (Courtenay and Meffe 1989). Thus, characterizing predator-released kairomones in order to create natural, chemical mosquito oviposition deterrents and implementing the use of native predators (e.g., amphibian and insect predators) could lead to the control of disease vectors whilst reducing the deleterious effects of fish introductions. Semiochemicals are widely used in push-pull pest strategies where pests are spatially manipulated (Cook et al. 2007). In our system, kairomones act as a deterrent, which can *push* mosquitoes from a habitat of interest (e.g., temporary ponds near human populations). Whichever habitats the mosquitoes are pushed toward can then be managed to control populations with cryptic predators or other control methods.

It is now evident that mosquito congeners vary in their abilities to detect, or their predilection to avoid, predators and that not all predators are treated equally. The question remains as to how and why only certain mosquitoes would evolve avoidance of predators, especially since there is such adaptive potential for all mosquitoes. The answer may lie in the evolution of specific sensory capabilities, but it is also possible that predators are manipulating their own cues. This situation could result in an ongoing evolutionary arms race between predator and prey where mosquitoes are evolving finer sensory capabilities while predators are evolving different chemical signatures. For example, pirate perch (*A. sayanus*) are chemically camouflaged to oviposting treefrogs and colonizing beetles (Resetarits and Binckley 2013). It could be the case that newts and mosquitofish are currently ahead in the arms races with *C.* pxq, but behind in the race with *C. restuans*. In that context, green sunfish could be ahead with both mosquito species. In order to further understand the specificity of this system, we are currently investigating how the variation in predator effects could relate to phylogenetic signal and phylogenetic distance, as well as habitat use patterns. In addition, we plan on comparing the chemical signature of repellant predators to the non-repellant pirate perch in order to characterize predator kairomones. With the identification and proper implementation of a control strategy based on fish kairomones, mosquito populations may be controlled in such a way that minimizes the impact on resident aquatic biodiversity.

References

Abjornsson K, Bronmark C, Hansson LA (2002) The relative importance of lethal and non-lethal effects of fish on insect colonisation of ponds. Freshw Biol 47:1489–1495

Angelon KA, Petranka JW (2002) Chemicals of predatory mosquitofish (*Gambusia affinis*) influence selection of oviposition site by *Culex mosquitoes*. J Chem Ecol 28:797–806

Barr AR (1957) The distribution of *Culex p. pipiens* and *C. p. quinquefasciatus* in North America. Am J Trop Med Hyg 6:153–165

Bentley MD, Day JF (1989) Chemical ecology and behavioral-aspects of mosquito oviposition. Annu Rev Entomol 34:401–421

Berendonk T (1999) Influence of fish kairomones on the ovipositing behavior of *Chaoborus imagines*. Limnol Oceanogr 44:454–458

Binckley CA, Resetarits WJ (2002) Reproductive decisions under threat of predation: squirrel treefrog (*Hyla squirella*) responses to banded sunfish (*Enneacanthus obesus*). Oecologia 130:157–161

Binckley CA, Resetarits WJ (2005) Habitat selection determines abundance, richness and species composition of beetles in aquatic communities. Biol Lett 1:370–374

Binckley CA, Resetarits WJ Jr (2007) Effects of forest canopy on habitat selection in treefrogs and aquatic insects: implications for communities and metacommunities. Oecologia 153:951–958

Binckley CA, Resetarits WJ Jr (2008) Oviposition behavior partitions aquatic landscapes along predation and nutrient gradients. Behav Ecol 19:552–557

Blaustein L (1999) Oviposition site selection in response to risk of predation: evidence from aquatic habitats and consequences for population dynamics and community structure. In: Wasser SP (ed) Evolutionary theory process: modern perspectives. Springer, Dordrecht, pp 441–456

Blaustein L, Kotler BP (1993) Oviposition habitat selection by the mosquito, *Culiseta longiareolata*: effects of conspecifics, food and green toad tadpoles. Ecol Entomol 18:104–108

Blaustein L, Kiflawi M, Eitam A, Mangel M, Cohen JE (2004) Oviposition habitat selection in response to risk of predation in temporary pools: mode of detection and consistency across experimental venue. Oecologia 138:300–305

Bradford DF (1989) Allotopic distribution of native frogs and introduced fishes in high Sierra Nevada lakes of California: implication of the negative effect of fish introductions. Copeia 1989:775–778

Brown WL, Eisner T, Whittaker RH (1970) Allomones and kairomones: transspecific chemical messengers. Bioscience 20:21–22

Buckley PA (1969) Disruption of species: typical behavior patterns in F1 hybrid Agapornis parrots. Z Tierpsychol 26:737–743

Chesson J (1984) Effect of Notonectids (Hemiptera, Notonectidae) on mosquitos (Diptera, Culicidae) – predation or selective oviposition? Environ Entomol 13:531–538

Cook SM, Khan ZR, Pickett JA (2007) The use of push-pull strategies in integrated pest management. Annu Rev Entomol 52:375–400

Courtenay WR Jr, Meffe GK (1989) Small fishes in strange places: a review of introduced poeciliids. In: Meffe GK, Snelson FF Jr (eds) Ecology and evolution of livebearing fishes: Poeciliidae. Prentice Hall, Englewood Cliffs, NJ, pp 319–331

Crump ML (1991) Choice of oviposition site and egg load assessment by a treefrog. Herpetologica 47:308–315

Darsie RF Jr, Ward RA (2005) Identification and geographical distribution of the mosquitoes of North America, North of Mexico. The University Press of Florida, Gainesville, FL

de Meillon B, Sebastian A, Khan ZH (1967) Time of arrival of gravid *Culex pipiens fatigans* at an oviposition site, the oviposition cycle and the relationship between time of feeding and time of oviposition. Bull World Health Org 36:39–46

Denoel M, Dzukic G, Kalezic ML (2005) Effects of widespread fish introductions on paedomorphic newts in Europe. Conserv Biol 19:162–170

Dilger WC (1962) The behavior of lovebirds. Sci Am 206:88–98
DuRant SE, Hopkins WA (2008) Amphibian predation on larval mosquitoes. Can J Zool 86:1159–1164
Gerhardt HC (1974) The vocalizations of some hybrid treefrogs: acoustic and behavioral analyses. Behaviour 49:130–151
Gratz NG (1999) Emerging and resurging vector-borne diseases. Annu Rev Entomol 44:51–75
Gubler DJ (1998) Resurgent vector-borne diseases as a global health problem. Emerg Infect Dis 4:442–450
Harbach RE (2012) *Culex pipiens*: species versus species complex – taxonomic history and perspective. J Am Mosq Control Assoc 28:10–23
Hecnar SJ, McLoskey RT (1997) The effects of predatory fish on amphibian species richness and distribution. Biol Conserv 79:123–131
Hocking DJ, Semlitsch RD (2007) Effects of timber harvest on breeding-site selection by gray treefrogs (*Hyla versicolor*). Biol Conserv 138:506–513
Juliano SA (2009) Species interactions among larval mosquitoes: context dependence across habitat gradients. Annu Rev Entomol 54:37–56
Kats LB, Dill LM (1998) The scent of death: chemosensory assessment of predation risk by prey animals. Ecoscience 5:361–394
Kats LB, Ferrer RP (2003) Alien predators and amphibian declines: review of two decades of science and the transition to conservation. Divers Distrib 9:99–110
Kats LB, Sih A (1992) Oviposition site selection and avoidance of fish by streamside salamanders (*Ambystoma barbouri*). Copeia 1992:468–473
Knapp RA, Matthews KR (2000) Non-native fish introductions and the decline of the mountain yellow-legged frog from within protected areas. Conserv Biol 14:428–438
Knight TM, McCoy MW, Chase JM, Mcoy KA, Holt RD (2005) Trophic cascades across ecosystems. Nature 437(7060):880–883
Krause J, Brown D, Corbet SA (1992) Spacing behaviour in resting *Culex pipiens* (Diptera, Culicidae): a computer modelling approach. Physiol Entomol 17:241–246
Lowenberger CA, Rau ME (1994) Selective oviposition by *Aedes aegypti* (Diptera: Culicidae) in response to a larval parasite, *Plagiorchis elegans* (Trematoda: Plagiorchiidae). Environ Entomol 23:1269–1276
McPeek MA (1990) Behavioral differences between *Enallagma* species (Odonata) influencing differential vulnerability to predators. Ecology 71:1714–1726
Morin PJ (1981) Predatory salamanders reverse the outcome of competition among three species of anuran tadpoles. Science 212:1284–1286
Offill YA, Walton WE (1999) Comparative efficacy of the threespine stickleback (*Gasterosteus aculeatus*) and the mosquitofish (*Gambusia affinis*) for mosquito control. J Am Mosq Control Assoc 15:380–390
Petranka JW, Fakhoury K (1991) Evidence of a chemically-mediated avoidance response of ovipositing insects to blue-gills and green frog tadpoles. Copeia 1991:234–239
Rausher MD (1983) Ecology of host-selection behavior in phytophagous insects. In: Denno RF, MCClure MS (eds) Variable plants herbivores in natural and managed systems. Academic, New York, pp 223–257
Resetarits WJ (1996) Oviposition site choice and life history evolution. Am Zool 36:205–215
Resetarits WJ (2001) Colonization under threat of predation: avoidance of fish by an aquatic beetle, *Tropisternus lateralis* (Coleoptera: Hydrophilidae). Oecologia 129:155–160
Resetarits WJ Jr, Binckley CA (2013) Is the pirate really a ghost? Evidence for generalized chemical camouflage in an aquatic predator, pirate perch *Aphredoderus sayanus*. Am Nat 181:690–699
Resetarits WJ Jr, Wilbur HM (1989) Choice of oviposition site by *Hyla chrysoscelis*: role of predators and competitors. Ecology 1989:220–228
Rieger JF, Binckley CA, Resetarits WJ Jr (2004) Larval performance and oviposition site preference along a predation gradient. Ecology 85:2094–2099

Schilling EG (2008) Effects of fish introductions on the geographic distribution and native invertebrate biodiversity of naturally fishless lakes in Maine. Dissertation, The University of Maine

Schober H (1966) Agitation of water surfaces by sprinkling to prevent mosquito breeding. Mosq News 26:144–149

Silberbush A, Blaustein B (2008) Oviposition Habitat Selection by a mosquito in response to predator: are predator-released kairomones air-borne cues?. J. of Vector Ecol 33:208-211

Silberbush A, Markman S, Lewinsohn E, Bar E, Cohen JE, Blaustein L (2010) Predator-released hydrocarbons repel oviposition by a mosquito. Ecol Lett 13:1129–1138

Singer MC (1984) Butterfly-hostplant relationships: host quality, adult choice and larval success. In: Vane-Wright RI, Ackery PR (eds) Symposium of the royal entomological society of London, vol 11. Academic, London, pp 81–88

Stuart SN, Chanson JS, Cox NA, Young BE, Rodrigues AS, Fischman DL, Waller RW (2004) Status and trends of amphibian declines and extinctions worldwide. Science 306:1783–1786

Van Dam AR, Walton WE (2007) Comparison of mosquito control provided by the arroyo chub (*Gila orcutti*) and the mosquitofish (*Gambusia affinis*). J Am Mosq Control Assoc 23:430–441

Van Dam AR, Walton WE (2008) The effect of predatory fish exudates on the ovipostional behaviour of three mosquito species: *Culex quinquefasciatus*, *Aedes aegypti* and *Culex tarsalis*. Med Vet Entomol 22:399–404

Vonesh JR, Blaustein L (2010) Predator-induced shifts in mosquito oviposition site selection: a meta-analysis and implications for vector control. Isr J Ecol Evol 56:123–139

Vonesh JR, Buck JC (2007) Pesticide alters oviposition site selection in gray treefrogs. Oecologia 154:219–226

Walton WE (2007) Larvivorous fish including *Gambusia*. J Am Mosq Control Assoc 23:184–220

Walton WE, Van Dam AR, Popko DA (2009) Ovipositional responses of two *Culex* (Diptera: Culicidae) species to larvivorous fish. J Med Entomol 46:1338–1343

Warburg A, Faiman R, Shtern A, Silberbush A, Markman S, Cohen JE, Blaustein L (2011) Oviposition habitat selection by *Anopheles gambiae* in response to chemical cues by *Notonecta maculata*. J Vector Ecol 36:421–425

Webb C, Joss J (1997) Does predation by the fish *Gambusia holbrooki* (Atheriniformes: Poeciliidae) contribute to declining frog populations? Aust Zool 30:316–324

Willems KJ, Webb CE, Russell RC (2005) A comparison of mosquito predation by the fish *Pseudomugil signifier* (Kner) and *Gambusia holbrooki* (Girard) in laboratory trials. J Vector Ecol 30:87–90

Yamamura K (1999) Transformation using (x + 0.5) to stabilize the variance of populations. Res Popul Ecol 41:229–234

Chapter 19
Evolutionary Aspects of the Use of Predator Odors in Antipredator Behaviors of Lumholtz's Tree-Kangaroos (*Dendrolagus lumholtzi*)

Sigrid R. Heise-Pavlov

19.1 Introduction

For prey species to survive predation the recognition and effective response to predatory threats is essential (Cruz et al. 2013). Most prey species live in multi-predator environments, which force them to recognize and respond to different predators. In many prey–predator systems, prey may be able to apply general anti-predator responses that are effective against all sympatric predator species (Cruz et al. 2013). These responses may include increased vigilance (Bednekoff and Blumstein 2009), decreased movement to avoid detection by the predator (Sih 1984), or increased movement to escape from predators (Cooper et al. 1990).

Multi-predator environments often consist of predators which use different hunting strategies (Cruz et al. 2013). Because these predators differ morphologically and behaviorally, Cox and Lima (2006) introduced the term 'archetype' to allow the distinction between them. Arboreal and terrestrial hunters are considered different 'archetype' predators. The application of general antipredator responses in a multi-archetype predator system can result in increased vulnerability of prey because the responses may only be effective against one of the predators (Cruz et al. 2013). Therefore the evolution of archetype-specific antipredator responses in a multi-predator system is of an advantage (Jędrzejewski et al. 1993). Bank voles, for example, select an 'arboreal escape' when threatened by specialized terrestrial predators while they apply crypsis for other terrestrial predators (Jędrzejewski et al. 1993; Banks and Dickman 2007).

Predator archetype-specific responses of prey can become lost when sources of their selection are removed or altered (Lahti et al. 2009). Tammar wallabies (*Macropus eugenii*), for example, lost traits of predator recognition and antipredator

S.R. Heise-Pavlov (✉)
Centre for Rainforest Studies at the School for Field Studies, Yungaburra, QLD, Australia
e-mail: sheise-pavlov@fieldstudies.org

B.A. Schulte et al. (eds.), *Chemical Signals in Vertebrates 13*,
DOI 10.1007/978-3-319-22026-0_19

behavior when they have been living on an island in complete isolation from predators for just 130 years, known as a consequence of "relaxed predator selection" (Blumstein and Daniel 2002; Blumstein et al. 2004). On the other hand, antipredator responses can persist when some of the selective sources remain, which is expressed in Blumstein's "multipredator hypothesis" (Blumstein 2006). However, the consequences of an altered or removed selective source on a particular trait are hard to predict since other sources of selection may still remain and influence the expression of a trait (Blumstein et al. 2004; Lahti et al. 2009).

Native Australian prey species have experienced a "relaxed predator selection" in fairly recent times during post-Pleistocene extinctions in which the Australian biota lost a number of large predators (Roberts et al. 2001; Long et al. 2002). This is likely to have altered their predator recognition and antipredator behavior. The alteration may have been manifested in a complete loss, modification or persistence of these traits. Loss and weakening of these traits have been identified as potential cause for the inability of extant prey species to adjust to re-introduced predators (Berger et al. 2001). Moreover, absent or weakened antipredator behaviors of Australian native prey species have been described as major cause for the extinction of many prey species when facing novel, introduced predators which had not been in their environment before (Woinarski et al. 2012). The vulnerability of extant prey species to novel predators is therefore likely to depend on the degree of similarity between the archetype of an extinct predator and the novel predator (Dunlop-Hayden and Rehage 2011), and the extent at which ancestral antipredator responses of prey have been modified during the relaxed predator selection phase following the Pleistocene extinction.

For some of the Australian prey species the inability to recognize novel predators has been reported. This inability is described as recognition-based naiveté level 1 (Blumstein et al. 2001; Cox and Lima 2006) and has been identified as major reason for the recent loss of many Australian prey species (Russell and Banks 2007). Other studies suggest that native Australian mammals apply insufficient strategies against novel predators and, consequently, succumb to the predation risk, a condition described as naiveté level 2 by Banks and Dickman (2007). Differences in predator 'archetypes' between the introduced eutherian predators and ancestral marsupial predators to which many Australian prey species had adapted may be responsible for this condition (Cox and Lima 2006).

This study will investigate whether an arboreal Australian marsupial is able to recognize and respond to terrestrial and arboreal predators and can transmit ancestral antipredator behaviors to novel, introduced predators.

The Lumholtz's tree-kangaroo (LTK) (*Dendrolagus lumholtzi*) is one of the largest arboreal folivore marsupials endemic to the rainforests of the Wet Tropics of northeastern Australia (Martin 2005). Its main extant predator is the python, specifically the amethystine python (*Morelia amethystina*), an ambush predator that has also been observed to prey on juvenile individuals of the Australian Bennett's tree-kangaroo (*Dendrolagus bennettianus*) (Martin 1995). When the dingo (*Canis lupus dingo*) reached Australia about 3500 years ago (Corbett 1995), it also became a predator of LTKs. However, the dingo does not seem to be a major predator of the species as a study from Vernes et al. (2001) reports that only 2.6 % of all investigated dingo scats contain remains of LTK. With the arrival of the Europeans, domestic and feral dogs (*Canis lupus familiaris*) have been added to the predators of LTKs.

Within the present highly urbanized and fragmented landscape of LTKs' distribution range, encounters of LTKs with modern canines (dogs, dingoes) are not a rare event. These encounters usually end in fatalities or LTK succumb to their internal injuries afterward (Newell 1999) suggesting that LTK is inefficient in dealing with novel terrestrial predators (Tree-kangaroo and Mammal Group 2000).

The main response of LTKs to threats consists of descending from the canopy to the forest floor (Procter-Gray and Ganslosser 1986). Animals also have been observed to jump from tree heights of up to 22 m to the ground (Martin 2005). This threat evasion method is adequate for arboreal predators, the pythons (Procter-Gray and Ganslosser 1986), but appears ill adapted to terrestrial predators such as canines.

The evolution of LTKs was not free from terrestrial predators. They evolved with predatory *Thylacines* such as *Thylacoleo carnifex*, a large, leopard-sized carnivore that possessed the ability to climb, and *Thylacinus cynocephalus* which went extinct about 3500 years ago (Flannery et al. 1996; Jones and Stoddart 1998; Wroe et al. 2007).

Given the species' evolutionary history with different 'archetypes' of predators (terrestrial and arboreal), it is hypothesized that tree-kangaroos developed archetype-specific antipredator responses. It is expected that LTKs would descend rapidly to the ground and flee when threatened in the canopy by arboreal predators, but would climb up and remain still when threatened from the ground by terrestrial predators to reduce detectability. If this is the case, LTKs should be well adapted to cope with novel introduced, terrestrial predators, but incidences of dog/dingo related fatalities of LTKs show otherwise.

This study aims to investigate whether (1) LTK lacks the ability to recognize novel terrestrial predators in contrast to arboreal predators (naiveté level 1), or (2) LTK recognizes both predator types and shows the same response to them (naiveté level 2), or (3) LTK recognizes both predator types and shows archetype-specific responses.

The ability to recognize the presence of a predator should manifest itself in increased vigilance (Blumstein and Daniel 2002; Pays et al. 2009) and decreased comfort behaviors (Apfelbach et al. 2005). Due to the relatively short time since the extinction of its ancestral native terrestrial predators, it is expected that LTKs are still able to recognize predators of this archetype. Different antipredator responses of this arboreal species should be manifested in differences in the frequency and duration of its movements. It is hypothesized that LTKs will limit frequency and duration of movements in the presence of a terrestrial predator since it is restricted in its movements in the canopy due to its relatively large body size. Decreased movements will also reduce its detectability by a terrestrial predator. In contrast, tree-kangaroos will increase their movements in the presence of an arboreal predator in an attempt to flee from this predator to the ground. It is postulated that LTKs had developed archetype-specific antipredator strategies against arboreal and terrestrial predators and that it has retained them during times of relaxed predation selection (Blumstein 2006).

If LTK has evolved archetype-specific antipredator responses, the reported fatalities of LTK when encountering novel canines suggest that LTK is either not able to transmit ancestral archetype-specific antipredator responses to novel predators, or, that these responses have been weakened due to the extinction of its ancestral terrestrial predators during the Pleistocene.

19.2 Methods

19.2.1 Animal Facilities and Study Subjects

Five captive LTKs were used as study subjects. Four animals were kept at a wildlife sanctuary, 'The Wildlife Habitat' in Port Douglas, Queensland, Australia. One animal was kept at a wildlife rehabilitation center, the Lumholtz's Lodge near Atherton on the Atherton Tablelands, Queensland, as it could not be released due to its blindness.

At the wildlife sanctuary the subjects were kept in separate enclosures. The enclosures measured 7×4×3.5 m, had concrete floors and were surrounded by metal mesh fencing. Wooden shelving and logs were provided as climbing structures. Enclosures were cooled by sprinklers and fans several times during the day. Subjects received freshly cut browse every second day and were provided with a small variety of fruits and vegetables as supplementary food twice a day (for further details see Heise-Pavlov et al. 2014).

At the wildlife rehabilitator center the subject was kept in an enclosure that measured 10×8×3.05 m and contained a 3.1×2.9 m metal shed in one corner. The enclosure floor was covered by the legume pinto peanuts (*Arachis pintoi*) which the subject occasionally consumed. The enclosure contained interconnected logs and a wooden shelf in the shed. The subject received freshly cut browse every second day and a variety of fruit and vegetables once a day (M. Cianelli pers. communication).

The five study subjects consisted of two males and three females. Two of them came into care as joeys, which had not yet left the pouch of the mother; one female came into care when she was approximately 3 years old. Two of the subjects were born at the wildlife sanctuary.

Three of the studied subjects had no prior experience with any predators while the experience with predators was unknown for the female that came into care at an age of 3 years. However, subject number 5 had been attacked by a dog some years before the study.

At the time of the study subjects were between 3 and 8 years of age, being classified as adults and subadults (Table 19.1).

Table 19.1 Sex, age and weight of the study subjects at the time of the study

Subject	1	2	3	4	5
Sex	Female	Male	Male	Female	Female
Location	Port Douglas Wildlife Habitat	Port Douglas Wildlife Habitat	Port Douglas Wildlife Habitat	Port Douglas Wildlife Habitat	Wildlife Rehabilitation Center
Age at study (date of birth)	Approx. 96 months	Thirty three months (20/7/2009)	Approx. 36 months	Sixty five months (2/11/2006)	Approx. 69 months
Maturity level at time of data collection	Adult	Subadult	Subadult	Adult	Adult
Weight	3.38 kg (as of 25/11/2011)	7.74 kg (as of 9/4/2012)	7.63 kg (as of 9/4/2012)	6.82 kg (as of 9/4/2012)	6.94 kg (as of 25/11/2011)

19.2.2 Observations

Observations of the subjects took place in November 2011 and April 2012. In November 2011 subjects 1, 2, 4 and 5 were available for observations. By April 2012 subject 1 had deceased while subject 3 had been added to the tree-kangaroo stock of the wildlife sanctuary.

Observations were done by videotaping the subjects for 10 min. Due to the large dimension of the enclosures one observer remained in the enclosure during observation times to ensure that the video camera stayed focused on the subject. The observer maintained a maximum distance to the subject to minimize disturbance of the subject due to the presence of the observer.

19.2.3 Experimental Design

The study consisted of sets of control and corresponding treatment trials. Each trial lasted 10 min, and trials were separated by a 5 min break.

During a 10 min control trial a rectangular plastic container with a base size of 13 cm × 8 cm, an opening size of 17.5 cm × 12 cm, and a height of 6 cm containing a paper towel soaked with approximately 3.5 mL of water was inserted into the enclosure.

During a 10 min treatment trial a same-sized container with fecal material from a python, a dingo or a domestic dog was inserted into the enclosure. Approximately 100 g of fecal material was placed on a paper towel to which 3.5 mL water was added to increase the intensity of the odor (Müller-Schwarze 2006).

Table 19.2 Providers of fecal material and the diet of animals from which feces were collected

Odor origin	Provider of fecal material	Diet of animals from which feces were collected
Domestic dog	Resident near Malanda, North Queensland who owns several hunting dogs which are used regularly for feral pig hunting	Animals receive fresh meat and bones and a minimum of dry dog food
Dingo	Dingo Discovery Sanctuary and Research Center, Toolern Vale, Victoria	Animals receive fresh meat and bones
Amethystine python	Accredited reptile breeder in Biboohra, North Queensland	Animals receive rodents and other fresh meat

Fecal material was obtained from captive dingoes (*C. lupus dingo*), domestic dogs (*C. lupus familiaris*) and amethystine pythons (*M. amethystina*). It was ensured that the animals which provided the fecal material received a species-appropriate diet (Table 19.2). Fecal material was kept refrigerated after collection and was between 3 and 24 h old when presented to the subjects.

Lavender oil was used as neutral odor. In these treatment trials 3–5 drops of 100 % essential lavender oil were dropped onto a paper towel placed in a plastic container of the above mentioned dimensions. As in previous treatment trials 3.5 mL of water was added to increase the intensity of the odor.

In both control and treatment trials the container was waved several times in the enclosure prior to the recording of the subject's behavior to facilitate the dispersal of odor components. The container was placed on the ground during the control and its corresponding treatment trial when lavender or fecal material from domestic dogs or dingoes was presented. This was done to simulate the position of a terrestrial predator. When fecal material from a python was presented, containers during control and corresponding treatment trials were placed at a height of about 2 m on a suitable log in each enclosure. This was to simulate better the position of arboreal predators which, specifically pythons, use a range of branches in a forest.

Sets of control and corresponding treatment trial were conducted between 9 a.m. and 12:30 p.m. and again between 1:30 p.m. and 4:30 p.m. on 4 days. Between sets of different odor sources was a 10 min break. Each set of control and corresponding treatment trials was run four times with each of the subjects at different times of the observation days.

19.2.4 Data Analyses

Recorded videos of the subjects were analyzed using JWatcher version 1.0 (Blumstein et al. 2006). Frequencies and durations of vigilant behavior and various behaviors of 'comfort' and 'movements' of the subjects were retrieved from the recorded videos. Durations of behaviors were obtained as proportion of the time the subject was in sight. Table 19.3 contains the description of the behaviors for which frequencies and durations were recorded.

Table 19.3 Ethogram of behaviors making up the behavioral categories 'comfort' and 'movements' and description of the behavior "vigilance' (expanded and modified from an ethogram provided by Procter-Gray and Ganslosser 1986)

Behavioral category	Behavior	Description
Comfort	Feeding	Chewing on a leaf or twig, or using hands to pull a food object close to mouth
	Grooming	Behaviors such as scratching, biting skin, or licking fur
	Resting	Sitting upright or with body leaning down while eyes are half closed, dozing off or sleeping
	Vigilance	Sitting still with eyes wide open; ear movement often occurring; individual appears more aware of surroundings
Movements	Climbing down	Moving vertically down trunk either headfirst or posterior-first
	Climbing up	Arms sliding up a vertical surface followed by hind feet hopping up simultaneously
	Horizontal movement	Hopping or walking horizontally on provided structures
	Jumping	Either "springing" from branch to branch; ground to branch, or branch to ground

For each subject frequencies and durations of all behaviors contributing to the category 'comfort' and all behaviors contributing to the category 'movements' were summed respectively for each control and each treatment trial separately.

Values of frequencies and durations of the recorded behavior 'vigilance' and the summed behaviors of the categories 'comfort' and 'movements' for each of the four control and four corresponding treatment trials per set and for each subject were used in statistical analyses. Trials with uncontrollable disturbances, such as sudden noises, were eliminated from the analyses.

Wilcoxon's paired tests were performed to compare the mean frequencies and mean durations of the behavior 'vigilance' and the behavioral categories 'comfort' and 'movements' respectively between control and corresponding treatment trials per set using the four replicates of each subject. After testing the distributions of the data for normality, one-way ANOVAs with repeated measurements were performed to compare the mean frequencies and durations of the behavior 'vigilance' and the behavioral categories 'comfort' and 'movements' between the treatment trials of the different odor sets. Post-hoc Tukey-tests were used. All statistical analyses were run in PAST (Hammer et al. 2001).

19.3 Results

All subjects showed a significantly higher mean duration of comfort behaviors in control trials compared to the corresponding treatment trials when odors from a python or from a dog were presented (Wilcoxon's paired tests: python: $W = 69.0$,

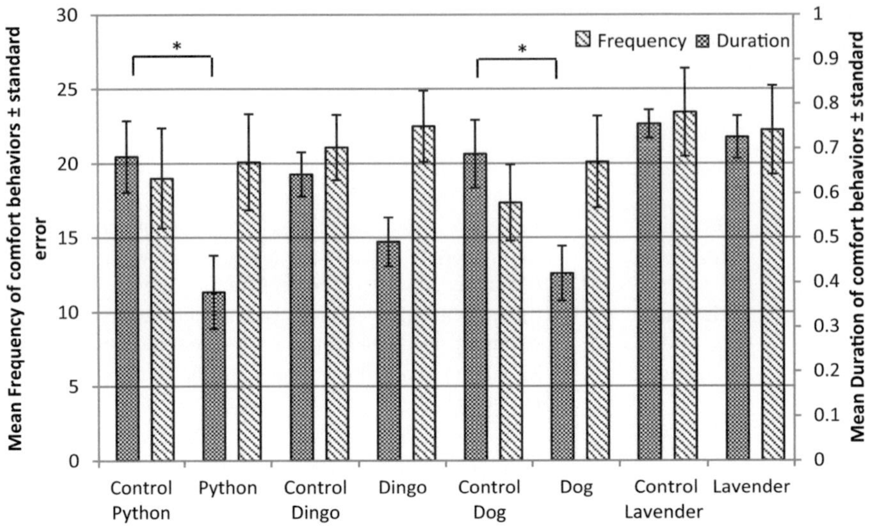

Fig. 19.1 Mean frequencies and durations of comfort behaviors of all subjects in control and corresponding treatment trials (±standard error). Significant differences at $p<0.05$ are indicated with *bars* and *asterisks*

$p=0.019$, $n=12$; dog: $W=98.0$, $p=0.004$, $n=14$). The mean duration of comfort behaviors was higher, but not significantly different, in control trials compared to treatment trials when odors from a dingo were presented ($W=83.0$, $p=0.055$, $n=14$). There were no significant differences in the mean duration of comfort behaviors between control and corresponding lavender treatments trials ($W=72.0$, $p=0.836$, $n=16$) (Fig. 19.1).

The mean frequency of comfort behaviors did not differ significantly between control and corresponding treatment trials regardless of the type of odor presented (Wilcoxon's paired tests: python: $W=43.5$, $p=0.724$, $n=12$; dingo: $W=54.0$, $p=0.552$, $n=14$; dog: $W=57.0$, $p=0.421$, $n=14$; lavender: $W=80.5$, $p=0.518$, $n=16$) (Fig. 19.1).

All subjects remained vigilant for a longer mean duration in treatment trials compared to the corresponding control trials when odors from a dog were presented (Wilcoxon's paired tests: $W=88.0$, $p=0.026$, $n=14$). Subjects also showed a higher, although not significantly different, mean duration of vigilance in treatment trials compared to the corresponding control trials when odors from a python or a dingo were presented (Wilcoxon's paired tests: python: $W=49.0$, $p=0.155$, $n=12$; dingo: $W=74.0$, $p=0.177$, $n=14$). No significant differences were obtained between control and corresponding lavender treatments trials with respect to the mean duration of vigilance shown by the subjects ($W=75.0$, $p=0.717$, $n=16$).

The mean frequency of vigilance was significantly higher in treatment trials compared to the corresponding control trials when odors from a dingo or from a dog were presented (Wilcoxon's paired tests: dingo: $W=86.0$, $p=0.035$, $n=14$; dog: $W=85.0$, $p=0.041$, $n=14$). Subjects also showed a higher, although not

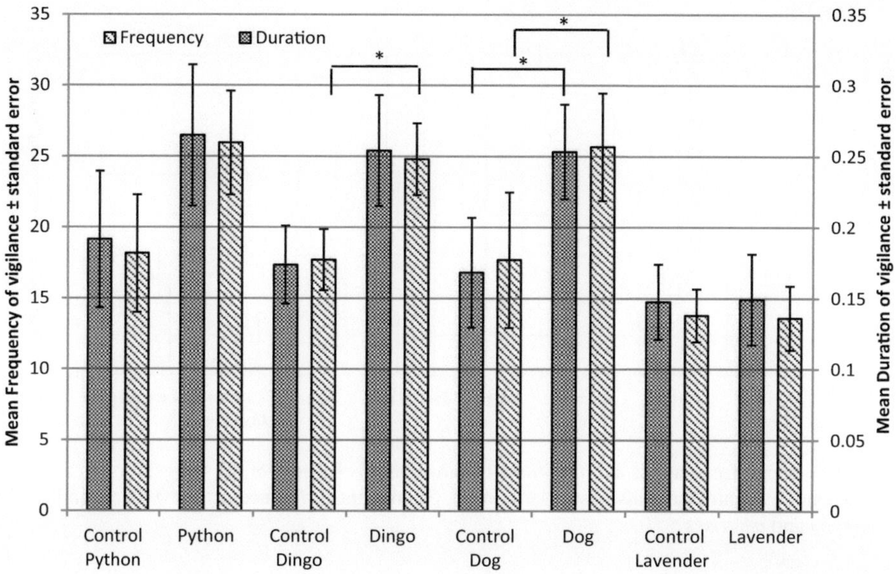

Fig. 19.2 Mean frequencies and durations of vigilance of all subjects in control and corresponding treatment trials (±standard error). Significant differences at $p < 0.05$ are indicated with *bars* and *asterisks*

significantly different, mean frequency of vigilance in treatment trials compared to the corresponding control trials when odors from a python were presented (Wilcoxon's paired tests: python: $W = 52.0$, $p = 0.091$, n = 12). There were no significant differences in the mean frequencies of vigilance of subjects between control and corresponding lavender treatments trials (Wilcoxon's paired tests: $W = 68.0$, $p = 0.649$, $n = 16$) (Fig. 19.2).

All subjects spent significantly more time on movement behaviors in treatment trials compared to the corresponding control trials when odors from a python were presented (Wilcoxon's paired tests: $W = 56.0$, $p = 0.041$, $n = 13$). Subjects showed a slightly higher, but not significantly different, mean duration of movement behaviors in treatment trials compared to the corresponding control trials when odors from a dingo or lavender were presented (Wilcoxon's paired tests: dingo: $W = 62.0$, $p = 0.551$, $n = 14$; lavender: $W = 78.0$, $p = 0.109$, $n = 16$). When odors from a dog were presented, subjects showed a higher, but not significantly different, mean duration of movement behaviors in control trials compared to the corresponding treatment trials (Wilcoxon's paired tests: $W = 64.0$, $p = 0.470$, $n = 14$).

All subjects showed a significantly higher mean frequency of movement behaviors in treatment trials compared to the corresponding control trials when odors from a python were presented (Wilcoxon's paired tests: $W = 42.0$, $p = 0.021$, $n = 13$). Subjects showed a slightly higher, but not significantly different, mean frequency of movement behaviors in treatment trials compared to the corresponding control trials when odors from a dingo or lavender were presented (Wilcoxon's paired

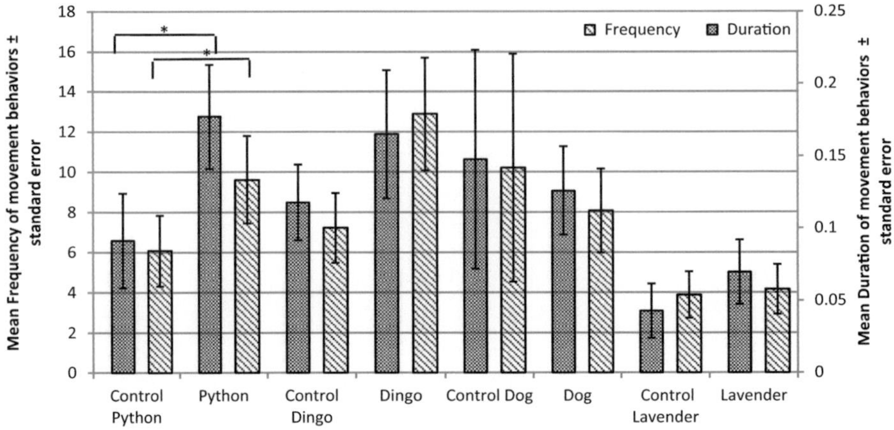

Fig. 19.3 Mean frequencies and durations of movement behaviors of all subjects in control and corresponding treatment trials (±standard error). Significant differences at $p<0.05$ are indicated with *bars* and *asterisks*

tests: dingo: $W=79.0$, $p=0.096$, $n=14$; lavender: $W=59.0$, $p=0.682$, $n=16$). When odors from a dog were presented, subjects showed a higher, but not significantly different, mean frequency of movement behaviors in control trials compared to the corresponding treatment trials (Wilcoxon's paired tests: $W=58.0$, $p=0.382$, $n=14$) (Fig. 19.3).

When comparing treatment trials only, subjects spent significantly more time on comfort behaviors when lavender was presented compared to treatment trials in which odors from a python, a dingo or a dog were presented (ANOVA: $F_{3,55}=5.02$, $p=0.004$; Tukey test: lavender vs python: $p=0.030$; lavender vs dingo: $p=0.026$; lavender vs dog: $p=0.017$) (Fig. 19.4). Frequencies, subjects spent on comfort behaviors, did not differ significantly between treatment trials (ANOVA: $F_{3,53}=0.28$, $p=0.840$) (Fig. 19.4).

When comparing treatment trials only, subjects remained for a longer mean duration vigilant in treatment trials in which predator odors were presented than in treatment trials in which lavender was presented. However, the differences were not significant (ANOVA: $F_{3,52}=2.21$, $p=0.097$); (Fig. 19.5). Subjects were significantly more frequently vigilant in treatment trials in which the odors from a python or a dog were presented compared to treatment trials in which lavender was presented (ANOVA: $F_{3,52}=4.14$, $p=0.011$; Tukey test: lavender vs python: $p=0.031$; lavender vs dog: $p=0.036$) (Fig. 19.5).

Subjects did not spent significantly different times on movement behaviors when treatment trials were compared (ANOVA: $F_{3,55}=1.45$; $p=0.238$) (Fig. 19.6). However, subjects exhibited significantly more often movement behaviors in treatment trials in which the odor from a dingo was presented compared to treatment trials in which lavender was presented (ANOVA: $F_{3,55}=4.76$, $p=0.005$; Tukey test: lavender vs dingo: $p=0.002$) (Fig. 19.6).

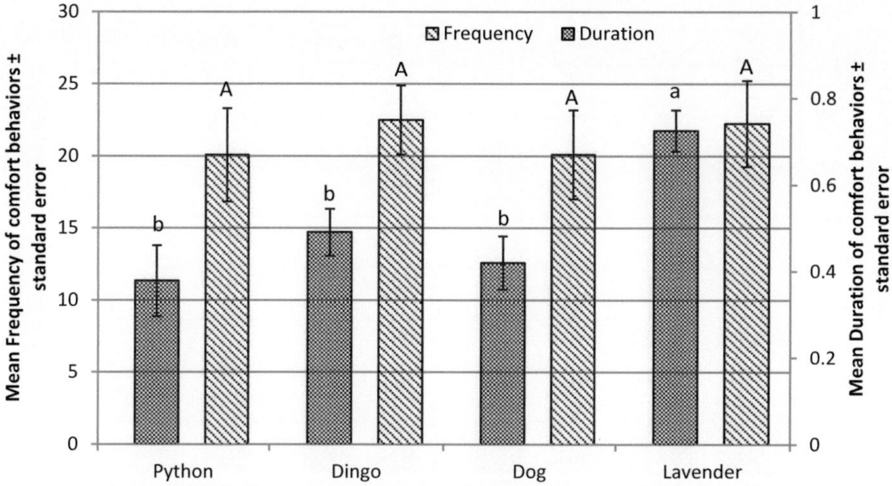

Fig. 19.4 Mean frequencies and durations of comfort behaviors of all subjects in treatment trials (±standard error). Significant differences at $p < 0.05$ are indicated with *letters*. *Capital letters* refer to frequencies, *small letters* refer to durations. Different letters indicate significant differences between the means

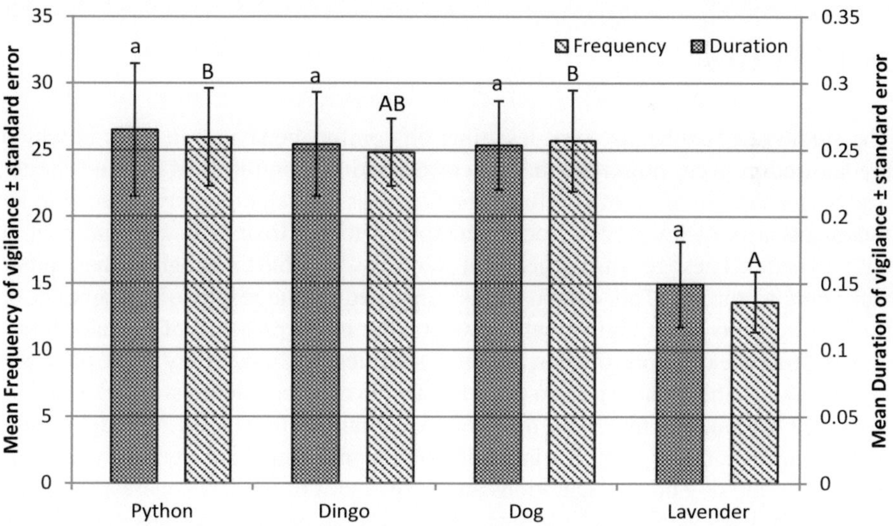

Fig. 19.5 Mean frequencies and durations of vigilance of all subjects in treatment trials (±standard error). Significant differences at $p < 0.05$ are indicated with *letters*. *Capital letters* refer to frequencies, *small letters* refer to durations. Different letters indicate significant differences between the means

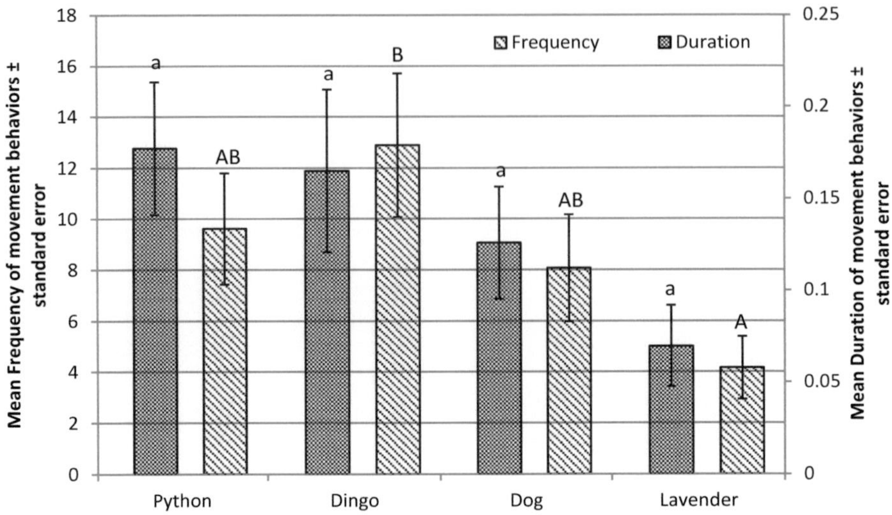

Fig. 19.6 Mean frequencies and durations of movement behaviors of all subjects in treatment trials (±standard error). Significant differences at $p<0.05$ are indicated with *letters*. *Capital letters* refer to frequencies, *small letters* refer to durations. Different letters indicate significant differences between the means

19.4 Discussion

The studied LTK subjects spent less time on comfort behaviors in trials in which predator odors were presented than in corresponding control trials in which water was presented. Duration of comfort behaviors was also significantly reduced in all treatments trials with predator odors compared to the treatment with the neutral odor lavender. These results suggest that subjects were able to recognize these predator odors as threats. This conclusion is supported by the results of the amount of time the subjects spent vigilant which was higher in the presence of predator odors than in presence of a control odor, although significantly higher only in the presence of a dog odor. This trend was also demonstrated in the frequencies at which subjects displayed vigilance. They were more often vigilant in trials with a predator odor than in corresponding control trials (although significantly different only in trials with dingo and dog odors). Durations and frequencies of vigilance were also lower, albeit not always significantly, in treatments with the neutral odor lavender than in treatments with predator odors.

Increased vigilance aids an organism in detecting, localizing, and avoiding potential threats (Apfelbach et al. 2005; Blumstein et al. 2010). Observations of LTKs in the wild showed that they spend a large amount of time alerted (Procter-Gray and Ganslosser 1986) which may have contributed to the clear response in vigilance of the studied subjects to the presented predator odors.

Changes in comfort behaviors and vigilance were demonstrated differently in duration and frequency of these behaviors. Increase in comfort behavior was more manifested in the duration than in the frequency of this behavior which reflects changes in comfort behaviors more appropriately. Likewise, increased vigilance was more manifested in the frequency than in the duration of this behavior, which reflects more appropriately changes in the degree of alertness of a subject.

In the presence of the neutral odor lavender, subjects did not show any changes in durations and frequencies of comfort behaviors and vigilance in comparison to the control odor water. Extended times and frequencies of comfort behaviors and decreased vigilance in the presence of lavender indicate that subjects do not feel threatened by this neutral and novel odor. This response suggests that LTKs do not seem to have a generalized avoidance response to novel odors or stimuli, which is known as neophobia (Brown et al. 2013). Neophobia, however, has been reported for other species (Węsierska et al. 2003; Brown et al. 2013).

The apparent lack of response to a novel odor by the studied subjects can have several reasons. First, the subjects live in a predator-free environment which has been shown to reduce the plasticity of prey species' neophobia in the context of predator avoidance (Brown et al. 2013). Second, LTK may not rely on olfactory cues to assess its environment. However, this seems to be unlikely for an arboreal marsupial that lives in a dense and foliated canopy that may restrict vision and hearing. Third, LTK may detect novel odors, but does not respond to them in an adverse manner. This would be in accordance with observations of LTK leaving forest fragments and utilizing non-forested matrices and small isolated forested patches (Laurance 1990; Tree-kangaroo and Mammal Group 2000).

Since the majority of subjects had no prior experience with different predators, but recognize predator odors as threats, this recognition must be an inherited or hard-wired trait (Blumstein et al. 2004). The only subject that experienced a dog attack prior to the experiment did not show a more profound response in her comfort behaviors and vigilance to dog odors compared to the other subjects.

The recognition of odors from predators as threats is attributed to the presence of high amounts of sulfurous metabolites, which are typically linked with excretions from meat digesting predators (Nolte et al. 1994). All predator donors received meat regularly before the experiment commenced. Therefore, the similarity in the composition of their feces is likely to be responsible for the general recognition of the odor cues by the subjects. The reliability on chemical cues in predator recognition by marsupials has often been debated since some studies conclude that the use of olfactory cues from predators is not hard-wired but relies on experience (Blumstein et al. 2002). It also has been argued that olfactory cues emitted from feces are rather indirect, depend on the age of the cue and may not pose an immediate threat to the prey (Dickman and Doncaster 1984) while visual and acoustic cues signal the proximity of a predator (Bouskila and Blumstein 1992; Nernesian et al. 2012). However, other studies show that marsupials modify their foraging behaviors in response to olfactory cues from predators (Gesser 1996; Morgan and Woolhouse 1997). These contrasting results may arise from differences in the environments in which the studied species are living. Tree-kangaroos, spending most of their time high up in dense rainforest canopies, may not be able to use acoustic and visual cues and therefore rely more on chemical cues instead. Furthermore, the humid environment in rainforest habitats promotes the distribution of volatiles which facilitates their detection (Müller-Schwarze 2006).

Because our subjects clearly recognize all three predatory odors as threats, a failure in recognizing novel predators, and therefore naiveté level 1, must be excluded as possible reason for the observed fatalities of LTKs when encountering novel predators such as modern canines (dog/dingo) (Cruz et al. 2013). The co-evolution of LTKs with terrestrial predators of similar 'archetypes' to the novel predators may enable LTK to recognize novel predators as threats (Kovacs et al. 2012). Studies of other marsupials have, however, repeatedly reported the failure of native Australian prey to recognize novel predators (and displaying naiveté level 1) (Russell and Banks 2007; Mella et al. 2010). For instance, southern brown bandicoots (*Isoodon obesulus*) and common brushtail possums (*Trichosurus vulpecula*) do not alter their foraging behavior when odor cues from novel predators are presented in traps (Mella et al. 2010). Similarly, Russell and Banks (2007) showed that marsupial antechinus (*Antechinus stuartii*) does not avoid traps scented with the odors from a native or an introduced, novel predator. These results were often interpreted by a general lack of the ability of native Australian marsupials to recognize odors originating from novel predators (Mella et al. 2010) and may therefore demonstrate a kind of inferiority of marsupials within the mammalian class (Coulson 1996). The result of the present study, however, shows that this conclusion cannot be transferred to LTKs, and that the recognition of odors from novel predators by marsupials may not be fully recognizable through trapping experiments (Russell and Banks 2007).

Several movement behaviors were combined and analyzed to assess whether LTK subjects possess archetype-specific antipredator responses. It was predicted that LTK would respond to a threat from an arboreal predator by increasing its movements in an attempt to descend from the canopy and flee on the ground, but would decrease its movements in the presence of a threat from a terrestrial predator as it remains in the canopy avoiding detection. If this is the case, incidences of fatalities of LTKs due to dog attacks could be attributed to the potential weakening of LTKs' response to terrestrial predators due to a time of relaxed predation selection since the extinction of their ancestral terrestrial predators, or to the inability of LTKs to transmit their ancestral response to a terrestrial predator archetype to novel terrestrial predators.

The results show that subjects significantly increased the frequency as well as the duration of their movement behaviors when presented with the odors of an arboreal predator (python) compared to a control odor (water). Subjects did increase, but not significantly, frequency and duration of movement behaviors in the presence of the odor from dingoes. However, when odors from a dog were presented subjects showed a decrease, but not significant, in frequency and duration of movements in comparisons to control odor (water). These results do not support the prediction that LTKs had evolved archetype-specific antipredator responses, but rather indicate that LTKs may respond to all predator archetypes in a similar way. This conclusion is supported by the lack of differences in durations and frequencies of movement behaviors between treatments with odors from different predator archetypes, and generally higher durations and frequencies of movement behaviors in the presence of predator odors compared to treatments with the neutral odor lavender.

Since the majority of the studied subjects had no prior experience with the predators from which the odor sources came, and showed a general increase in movements in response to all predator odors, it can be concluded that they possess a general antipredator response and that this response is inherited and hard wired (Blumstein et al. 2010).

Increased movements as a general antipredator 'escape' strategy towards all predator archetypes in LTK may originate from LTKs ancestors. Tree-kangaroos evolved from rock-wallabies (*Petrogale spp.*) which adapted to dry, open rocky habitats (Flannery et al. 1996; Martin 2005). Early detection of terrestrial predators is facilitated by group living in rock-wallabies, while predator avoidance is based on escape at high speed (Blumstein et al. 2001). Long legs enable rock-wallabies to achieve fast movements while their long tail propelled them across rocky outcrops and boulders (Sharman et al. 1995). When tree-kangaroos evolved from rock-wallabies by moving back into rainforest habitat they re-acquired the original arboreality of marsupials as a secondary trait (Flannery et al. 1996), but probably retained the escape strategy of rock-wallabies as a general antipredator response. The transmission and subsequent application of this general antipredator response to modern terrestrial predators, such as dingoes and dogs, makes it an inappropriate strategy against these predators and therefore demonstrates a case of naiveté level 2 (Banks and Dickman 2007) [which is described as prey applying wrong, and therefore inappropriate antipredator behaviors to novel predators (Cox and Lima 2006)]. In contrast, the persistence of antipredator responses from ancestors has been favorable in adaptation to modern terrestrial predators in the case of the brushtail possums (*T. vulpecula*) which typically respond to terrestrial predators by fleeing up a tree, reflecting their arboreal ancestry (Pickett et al. 2005).

The general antipredator response of LTK to flee from predators may have been, however, appropriate in the past. The adaptation of LTKs to a life in the canopy involved morphological changes (Barnett and Napier 1953) related to tibio-fibular articulations resulting in a higher ability to grasp around tree trunks. While these morphological changes facilitated the ability to climb they were accompanied by a reduction of the primitive freedom of movement of the foot (Jones 1929) resulting in disadvantaged mobility on the ground. This reduced mobility on the ground may lead ultimately to succumbing to modern predators which are superior on the ground, and may have more advanced hunting strategies than previous terrestrial predators of LTKs. Additionally, fragmentation of LTKs' habitat due to human-caused impacts increases the likelihood of LTKs to move across un-forested areas where they can become victims of attacks from modern canines (Laurance 1990).

Specific antipredator responses to terrestrial predator archetypes may not have evolved in LTKs due to the short time since LTKs have adapted to the arboreal life and since they have been exposed to novel terrestrial predators. While tree-kangaroos' ancestors appear to have evolved from terrestrial kangaroos some time during the late Miocene (Flannery 1989), first records of tree-kangaroos come from fossil findings in New South Wales, the Nullarbor Plain in Western Australia, and near Rockhampton in Queensland which were dated from the late Pliocene and Pleistocene (Flannery and Szalay 1982; Hocknull et al. 2007; Prideaux and Warburton 2008). During this time large terrestrial predators started to disappear from the Australian biota (Roberts et al. 2001; Long et al. 2002) leaving prey in a relaxed predator environment with weak selection for antipredator behavior against terrestrial predator (Lahti et al. 2009). The more recent introduction of novel terrestrial predators in form of the dingo some 5000–3500 years ago (Corbett 1995; Savolainen et al. 2004), and the domestic dog some 200 years ago did not pose a strong enough selection pressure on tree-kangaroos to evolve specific antipredator behaviors against these predators during the short time since their introduction. In common ringtail possums (*Pseudocheirus peregrinus*), on the other hand, selection pressure from a novel predator, the red fox (*Vulpus vulpus*) has been sufficient for this semi-arboreal marsupial to develop an efficient antipredator behavior over a few generations since foxes became established (Anson and Dickman 2013). This stresses the possibility that marsupials are able to develop specific antipredator strategies towards novel predators which has also been reported for placental prey in America which show swift adaptations to newly re-introduced predators by learned experience (Berger et al. 2001).

A potential evolution of specific responses of LTKs to terrestrial predators by climbing higher into the canopy and remain hidden cannot be excluded after their more recent adaptation to the arboreal life and the emergence of selective predation pressure due to the arrival of novel terrestrial predators. However, due to the short time since arrival of these predators, a specific response against them cannot be expected to be manifested in genes, but rather may depend on experiences and learning processes instead. This would correspond with the lack of display of archetype specific responses in the studied, captive LTK subjects, which had no prior experience with terrestrial predators. The exceptional long mother-joey relationship in LTK suggests that various traits are learned by juvenile LTKs from their mothers (Procter-Gray 1985). This may be a valuable prerequisite for the potential adaptation of LTK to novel terrestrial predators.

Our results cannot completely exclude the possibility that LTK may have had different antipredator responses to different predator 'archetypes'. The non-significant increase in movement behaviors when the odor from dingoes was presented and the inconclusive results with respect to changes in movement behaviors of the subjects when odors from dogs were presented would support those speculative conclusions. If LTKs possessed specific antipredator behaviors against terrestrial predators, it can be speculated that these behaviors may have got lost due to predator relaxation post Pleistocene extinction. LTKs' ancestral terrestrial predators such as *T. carnifex*, a large, leopard-sized carnivore that possessed the ability to climb, and *T. cynocephalus*, went extinct about 3500 years ago (Flannery et al. 1996; Jones and Stoddart 1998; Wroe et al. 2007). Additionally, the continuous presence of an arboreal predator during the relaxed predation selection against terrestrial predators could have re-enforced antipredator responses which LTKs had inherited from their ancestors, the rock-wallabies (Blumstein 2006).

Further studies are required to test this possibility. This will be done by more detailed analyses of separate movement behaviors, such as climbing down and climbing up in response to presented predator odors. Furthermore, the response of the studied subjects to an extant terrestrial predator, the Tasmanian Devil (*Sarcophilus harrisii*) with which they co-evolved, but which has been isolated from LTK for about 3500 years (Brown 2006), may help to confirm conclusions drawn from the current study.

Currently, the LTK is classified as "near threatened" under the Queensland's Nature Conservation Act of 1992 (Queensland Government 2011). Whether fatalities of tree-kangaroos due to domestic dogs and dingoes contribute to the species' abundance and demography remains unknown. Unless we can properly assess the reasons and extent of these incidences, novel predators have to be considered a potential impact factor on the species (Tree-kangaroo and Mammal Group 2000; Kanowski et al. 2003).

Acknowledgments This study would not have been possible without the caretakers of the wildlife sanctuary 'The Wildlife Habitat' in Port Douglas and wildlife rehabilitator Margit Cianelli who granted access to the animals. Many students of the Center for Rainforest Studies at The School for Field Studies assisted in data collection while the Center provided logistic support. A special thanks to Jay Goldberg who was willing to present parts of this study at CSiV 2014. The study was done under the Ethics Permit CA 2011/02/487 and the scientific permit WISP089610011 granted to Sigrid Heise-Pavlov.

References

Anson J, Dickman C (2013) Behavioral responses of native prey to disparate predators: naiveté and predator recognition. Oecologia 171:367–377

Apfelbach R, Blanchard C, Blanchard R, Hayes R, McGregor I (2005) The effects of predator odors in mammalian prey species: a review of field and laboratory studies. Neurosci Biobehav Rev 29:1123–1144

Banks P, Dickman C (2007) Alien predation and the effects of multiple levels of prey naiveté. Trends Ecol Evol 22:229–230

Barnett C, Napier J (1953) The form and mobility of the fibula in metatherian mammals. J Anat 87:207–213

Bednekoff P, Blumstein D (2009) Peripheral visibility influences marmot vigilance. Behav Ecol 20:1111–1117

Berger J, Swenson JE, Persson I-L (2001) Recolonizing carnivores and naïve prey: conservation lessons from Pleistocene extinctions. Science 291:1036–1039

Blumstein D, Daniel J, Evans C (2001) Yellow-footed rock-wallaby group size effects reflect a trade-off. Ethology 107:655–664

Blumstein D, Daniel J (2002) Isolation from mammalian predators differentially affects two congeners. Behav Ecol 13:657–663

Blumstein D, Mari M, Daniel J, Ardron J, Griffin A, Evans C (2002) Olfactory predator recognition: wallabies may have to learn to be wary. Anim Conserv 5:87–93

Blumstein D, Daniel J, Springett B (2004) A test of the multi-predator hypothesis: rapid loss of antipredator behavior after 130 years of isolation. Ethology 110:919–934

Blumstein D (2006) The multipredator hypothesis and the evolutionary persistence of antipredator behavior. Ethology 112:209–217

Blumstein D, Daniel J, Evans C (2006) JWatcher. UCLA. Accessed via: www.jwatcher.ucla.edu
Blumstein D, Lea A, Olson L, Martin J (2010) Heritability of anti-predatory traits: vigilance and locomotor performance in marmots. J Evol Biol 23:879–887
Bouskila A, Blumstein D (1992) Rules of thumb for predation hazard assessments: predictions from a dynamic model. Am Nat 139:161–176
Brown O (2006) Tasmanian devil (*Sarcophilus harrisii*) extinction on the Australian mainland in the mid-Holocene: multicausality and ENSO intensification. Alcheringa 31:49–57
Brown G, Ferrari M, Elvidge C, Ramnarine I, Chivers D (2013) Phenotypically plastic neophobia: a response to variable predation risk. Proc R Soc B Biol Sci 280:1756
Cooper W, Vitt L, Hedges R, Huey R (1990) Locomotor impairment and defense in gravid lizards (*Eumeces laticeps*): behavioral shift in activity may offset costs of reproduction in an active forager. Behav Ecol Sociobiol 27:153–157
Corbett L (ed) (1995) The dingo in Australia and Asia. University of New South Wales Press, Sydney
Coulson G (1996) Anti-predator behaviour in marsupials. In: Coulson G, Croft D (eds) Comparison of marsupials and placental behaviour. Filander Verlag, Fürth, pp 158–186
Cox J, Lima S (2006) Naiveté and an aquatic–terrestrial dichotomy in the effects of introduced predators. Trends Ecol Evol 21:674–680
Cruz J, Sutherland D, Anderson D, Glen A, Tores P, Leung L (2013) Antipredator responses of koomal (*Trichosurus vulpecula hypoleucus*) against introduced and native predators. Behav Ecol Sociobiol 67:1329–1338
Dickman C, Doncaster C (1984) Responses of small mammals to red fox (*Vulpes vulpes*) odour. J Zool 204:521–531
Dunlop-Hayden KL, Rehage JS (2011) Antipredator behavior and cue recognition by multiple Everglades prey to a novel cichlid predator. Behaviour 148:795–823
Flannery TF, Szalay FS (1982) *Bohra paulae*: a new giant fossil tree kangaroo (Marsupialia: Macropodidae) from New South Wales, Australia. Aust Mamm 5:83–94
Flannery TF (1989) Phylogeny of the Macropodoidea: a study on convergence. In: Grigg GC, Jarman PJ, Hume ID (eds) Kangaroos, wallabies and rat-kangaroos. Surrey Beatty & Sons, Chipping Norton, NSW, pp 1–46
Flannery T, Martin R, Szalay A (eds) (1996) Tree-kangaroos: a curious natural history. Reed Books Australia Port Melbourne, Victoria
Gesser S (1996) Anti-predator behaviour of the common brushtail possum (*Trichosurus vulpecula*) at Burrendon Dam. N.S.W. Hons Thesis, University of Sydney, Sydney
Hammer O, Harper D, Ryan P (2001) PAST: paleontological statistics software package for education and data analysis. Palaeontol Electron 4(1):9 pp. http://palaeo-electronica.org/2001_1/past/issue1_01.htm
Heise-Pavlov S, Anderson C, Moshier A (2014) Studying food preferences in captive cryptic folivores can assist on conservation planning: the case of the Lumholtz's tree-kangaroo (*Dendrolagus lumholtzi*). Aust Mamm 36:200–211
Hocknull S, Zhao J, Feng Y, Webb G (2007) Responses of quaternary rainforest vertebrates to climate change in Australia. Earth Planet Sci Lett 264:317–331
Jędrzejewski W, Rychlik L, Jędrzejewska B (1993) Responses of bank voles to odours of seven species of predators: experimental data and their relevance to natural predator-vole relationships. Oikos 68:251–257
Jones F (1929) Man's place among the mammals. Edward Arnold and Co., London
Jones M, Stoddart M (1998) Reconstruction of the predatory behaviour of the extinct marsupial thylacine (*Thylacinus cynocephalus*). J Zool 246:239–246
Kanowski J, Winter J, Simmons T, Tucker N (2003) Conservation of Lumholtz's tree-kangaroo on the Atherton Tablelands. Ecol Manag Restor 4:220–221
Kovacs E, Crowther M, Webb J, Dickman C (2012) Population and behavioural responses of native prey to alien predation. Oecologia 168:947–957

Lahti DC, Johnson NA, Ajie BC, Otto SP, Hendry AP, Blumstein DT, Coss RG, Donohue K, Foster SA (2009) Relaxed selection in the wild. Trends Ecol Evol 24:487–496

Laurance W (1990) Comparative responses of five arboreal marsupials to tropical forest fragmentation. J Mammal 71:641–653

Long J, Archer M, Flannery T, Hand S (2002) Prehistoric mammals of Australia and New Guinea: one hundred million years of evolution. University of New South Wales Press, Adelaide

Martin R (1995) Field observations of predation on Bennett's tree-kangaroo (*Dendrolagus bennettianus*) by an Amethystine python (*Morelia amethystina*). Herpetol Rev 26:74–76

Martin R (ed) (2005) Tree-Kangaroos of Australia and New Guinea. The University of Chicago Press, Chicago

Morgan D, Woolhouse A (1997) Predator odors as repellents to brushtail possums and rabbits. In: Mason J (ed) Repellents in wildlife management. National Wildlife Research Center, Ft. Collins, pp 241–252

Müller-Schwarze D (ed) (2006) Chemical ecology of vertebrates. Cambridge University Press, Cambridge

Mella V, Cooper C, Davies S (2010) Predator odour does not influence trappability of southern brown bandicoots (*Isoodon obesulus*) and common brushtail possums (*Trichosurus vulpecula*). Aust J Zool 58:267–272

Nernesian C, Banks P, McArthur C (2012) Behavioural responses to indirect and direct predator cues by a mammalian herbivore, the common brushtail possum. Behav Ecol Sociobiol 66:47–55

Newell G (1999) Australia's tree-kangaroos: current issues in their conservation. Wildl Res 87:1–12

Nolte D, Mason J, Epple G, Aronov E, Campbell D (1994) Why are predator urines aversive to prey? J Chem Ecol 20:1505–1516

Pays O, Dubot A-L, Jarman PJ, Loisel P, Goldizen AW (2009) Vigilance and its complex synchrony in the red-necked pademelon, *Thylogale thetis*. Behav Ecol 20:22–29

Pickett K, Hik D, Newsome A, Pech R (2005) The influence of predation risk on foraging behaviour of brushtail possums in Australian woodlands. Wildl Res 32:121–130

Prideaux GJ, Warburton NM (2008) A new Pleistocene tree-kangaroo (Diprotodontia: Macropodidae) from the Nullarbor Plain of South-Central Australia. J Vertebr Paleontol 28:463–478

Procter-Gray E (1985) The behavior and ecology of Lumholtz's tree kangaroo, *Dendrolagus lumholtzi* (Marsupialia: Macropodidae). Ph.D. Thesis, Harvard University, Cambridge, USA

Procter-Gray E, Ganslosser U (1986) The individual behaviors of Lumholtz's tree-kangaroo: repertoire and taxonomic implications. J Mammal 67:343–352

Queensland Government (2011) Lumholtz's tree-kangaroo; Department of Environment and Resource Management. http://www.derm.qld.gov.au/wildlife-ecosystems/wildlife/threatened_plants_and_animals.html

Roberts RG, Flannery TF, Ayliffe LK, Yoshida H, Olley JM, Prideaux GJ, Laslett GM, Baynes A, Smith MA, Jones R, Smith BL (2001) New ages for the last Australian megafauna: continent-wide extinction about 46,000 years ago. Science 292:1888–1892

Russell B, Banks P (2007) Do Australian small mammals respond to native and introduced predator odours? Aust J Ecol 32:277–286

Savolainen P, Leitner T, Wilton AN, Matisoo-Smith E, Lundeberg J (2004) A detailed picture of the origin of the Australian dingo, obtained from the study of mitochondrial DNA. Proc Natl Acad Sci U S A 101:12387–12390

Sharman G, Maynes G, Eldridge M (1995) Yellow-footed rock-wallaby. In: Strahan R (ed) The mammals of Australia. Reed Books, Sydney, pp 391–393

Sih A (1984) The behavioral response race between predator and prey. Am Nat 123:143–150

Tree-kangaroo and Mammal Group (2000) Tree-kangaroos on the Atherton Tablelands: rainforest fragments as wildlife habitats. Information for shire councils, land managers and the local community. Available at http://www.tree-kangaroo.net/documents/tkinfo

Vernes K, Dennis A, Winter J (2001) Mammalian diet and broad hunting strategy of the dingo (*Canis familiaris dingo*) in the wet tropical rain forests of Northeastern Australia. Biotropica 33:339–345

Węsierska M, Walasek G, Kilijanek J, Djavadian R, Turlejski K (2003) Behavior of the gray short-tailed opossum (*Monodelphis domestica*) in the open field and in response to a new object, in comparison with the rat. Behav Brain Res 143:31–40

Woinarski J, Burbidge A, Harrison P (eds) (2012) The action plan for Australian mammals 2012. CSIRO Publishing, Collingwood, VIC

Wroe S, Clausen P, McHenry C, Moreno K, Cunningham E (2007) Computer simulation of feeding behaviour in the thylacine and dingo as a novel test for convergence and niche overlap. Proc Biol Sci 274:2819–2828

Chapter 20
Small Emissions with Major Consequences: Specialized Malodorous Defenses in Birds

Gregory Röder, Vittorio Baglione, Diana Bolopo, José M. Marcos, Alfréd Trnka, and Daniela Canestrari

20.1 Introduction

Nowadays, birds are among the most conspicuous living organisms on Earth. Their large size, often colorful plumage, breeding behavior, migration, as well as recognizable song and potential for domestication have made them probably the best studied group of vertebrates. Ornithologists, biologists, bird watchers, or simple amateurs have exhaustively studied their anatomy, physiology, distribution, ecology, behavior, and classification.

Nevertheless, it would be presumptuous to consider birds as a fully known territory. Much more remains to be studied among the nearly 10,000 avian species. Now, a fresh look at their intra- and interspecific interactions might be profitable, as old beliefs need to be revised. Largely underestimated, the avian sense of smell, as well

G. Röder (✉)
Laboratory of Fundamental and Applied Research in Chemical Ecology (FARCE), Institute of Biology, University of Neuchâtel, Neuchâtel, Switzerland
e-mail: gregory.roeder@unine.ch

V. Baglione
Department of Agro-forestry, University of Valladolid, Valladolid, Spain

Sustainable Forest Management Research Institute, Palencia, Spain

D. Bolopo • J.M. Marcos
Department of Agro-forestry, University of Valladolid, Valladolid, Spain

A. Trnka
Department of Biology, University of Trnava, Trnava, Slovakia

D. Canestrari
Department of Biology of Organisms and Systems (BOS), University of Oviedo, Oviedo, Spain

Research Unit of Biodiversity (UMIB, CSIC), University of Oviedo, Oviedo, Spain

B.A. Schulte et al. (eds.), *Chemical Signals in Vertebrates 13*,
DOI 10.1007/978-3-319-22026-0_20

as taste to some extent, belong to this category. For decades, this bias has negatively impacted potential observations and research on the importance of airborne signals in birds. The moment to catch up has arrived.

20.1.1 The Sense of Smell in Birds

If a complete review on avian olfactory abilities would be inappropriate, an overview is indispensable to place our work in context.

Until recently, most experts considered that birds made little, if any, use of olfaction. John James Audubon, the famous nineteenth century naturalist, was not the only scientist to make the argument, but he was particularly influential in arguing against the idea of a developed sense of smell in birds. His work on the turkey vulture (*Cathartes aura* L.), a scavenger from the New World feeding on carrion, suggested the absence of odorous attraction by rotten carcasses hidden in an open field. This became the central pillar of a sort of avian anosmia theory, which was cited regularly during the next 140 years (Audubon 1820). Among other points, it has been argued that the extreme diffusion of volatile compounds in the air makes the use of smell unreliable to localize food, prey, predators, or mating partners for any bird species. Even the fact that myriad other organisms (e.g. thousands of insect species) use volatile cues on a daily basis to cope with both complex life cycles and diverse environments did not modify such preconceived ideas. Furthermore, the exceptionally efficient sight and hearing capacities of birds were naturally mentioned as the major means to orientate and locate. All these elements explain why this "insensitive" theory dominated for so long, leading to consideration of the olfactory aptitudes of birds as anecdotal, and of little relevance.

Within the scientific community and beyond, there was real astonishment when diverse studies revealed a functional and well developed sense of smell in many unrelated avian species. In the 1960s, Kenneth Stager and Bernice Wenzel first broke this taboo. Stager tested again the olfactory responsiveness of the turkey vulture. He showed that odors emitted by fresh dead animals did in fact attract the vulture (Stager 1964). Further experiments on the same model proved its ability to find fresh cadavers without visual cues, relying mainly on the typical chemical components emitted naturally by dead animals in the early stages of decomposition (i.e. butanoic acid, ethanethiol, and trimethylamine). Later, monitoring of individual heart frequency and further bioassays revealed the limits of detection for those chemicals (Applegate 1990; Smith and Paselk 1986). Thus, Audubon's conclusions were incorrect, mainly because carcasses in an advanced state of decomposition are not the favored diet of the turkey vulture. Surprisingly, this discovery by Stager went almost unnoticed by scientific and public opinion. A few years later, Bernice Wenzel published multiple results on olfaction in kiwis (*Apteryx sp.*), pigeons (*Columba sp.*), and procellariiformes such as the Northern fulmar (*Fulmarus glacialis* L.) (Wenzel 1968, 1986; Wenzel and Rausch 1977). Her pioneering observations proved that unrelated bird species rely on olfactory information to navigate, to forage, or to interact socially. Since this challenge to the dogma, field observations, bioassays, anatomical and physiological studies, as well as modern genetic tools have been involved to restore the truth about the sense of smell in birds. Studies on procellariiform seabirds unite all these complementary aspects and propose a complete view of the role and importance of odors in this group. Albatrosses and petrels have in common interesting features such as colonial life, monogamy and long-term pairing, philopatry, limited reproductive capacities, and astonishing foraging skills. Pelagic seabirds localize directly and precisely fish banks kilometers away from their colonies along the shore. Olfactory navigation from nest to food resources has been observed in several albatross, petrel, and shearwater species (Grubb 1972, 1974; Nevitt et al. 2008). Bioassays confirmed the role of dimethyl sulfide (DMS), a foraging cue processed by procellariiform seabirds. Phytoplankton, if consumed by zooplankton or other small grazers, release dimethylsulfionopropionate (DMSP), which can be transformed into volatile DMS by aquatic microorganisms. Associative learning between DMS scent and areas offering abundant prey is frequent in these pelagic birds. Furthermore, fledglings of these species learn the importance of DMS before they have to forage themselves (Nevitt et al. 1995; Nevitt and Bonadonna 2005). During the same period, olfactory aptitudes in many non-procellariiform birds were observed, such as in the owl parrot or kakapo (*F. glacialis*, Gray 1845) which is endemic to New Zealand (Hagelin 2004).

On the morphological and functional side, complete olfactory systems including nasal cavities, tubular nasal passages, and olfactory bulbs have been studied in many species (Bang 1960; Wenzel 1971), and most extensively in procellariiformes (Bang and Cobb 1968; Bang and Wenzel 1985; Wenzel and Meisami 1987). Avian olfactory bulbs found close to the brain are responsible for the processing and integration of odorous volatiles. Their size and shape varies among species, with a positive correlation between a prominent bulb and a well-developed sense of smell. With a bulb proportionally larger than most of the other birds studied so far, the flightless kiwi is a perfect example of this morphological adaptation (Bang and Cobb 1968). In this species, chemoreception is crucial, with nostrils placed at the apex of a long beak used to “scent” earthworms and small arthropods in soil litter. On a physiological level, measurements of heart rate and electrical impulses in olfactory nerves have been used to identify some of the molecules that birds of diverse groups can smell (Clark et al. 1993). Finally, modern genetic tools have identified hundreds of functional olfactory receptor genes in birds (OR genes), with some species showing equivalent numbers of OR genes to other vertebrate species. Additionally, many of these avian OR genes are exclusive to this taxon, and they demonstrate a high degree of functionality, unlike in humans where 60 % of OR genes are not expressed. These results are indisputable, especially in kiwi and kakapo species with an estimated total number of OR genes close to 600 (Steiger et al. 2008). More surprisingly, even the common canary (*Serinus canaria* L.), described as a typical visual species, exhibits more than 150 OR genes. Finally, specific differences in genetic patterns in olfactory receptor genes have been observed between nocturnal and diurnal species, stressing again the existence of a fully functional and specialized avian sense of smell (Steiger et al. 2009).

Nowadays, after 40 years of multidisciplinary scientific studies, all this evidence should convince the last skeptics that an olfactory sensibility in birds can no longer be denied. But as in other vertebrates, the avian sense of smell shows significant interspecific disparities.

20.1.2 Roles and Importance of Odors in Birds

With these new convictions in mind, the ornithological community has begun to carry out investigations into the roles and the importance of odors in bird behavior. Themes such as (1) navigation, (2) foraging, and (3) reproduction became the most familiar and well documented.

1. The influence of airborne signals for avian navigation gained real credibility with comprehensive studies on homing in pigeons, which add olfaction to other senses in order to orientate (Papi 1976; Wallraff 1982). In addition, procellariiformes brought new evidence that the localization of their colonies and nests may rely on odorous cues. In harsh and night conditions, Leach's storm petrels (*Oceanodroma leucorhoa*, Vieillot 1818) showed such ability (Grubb 1973), whereas other species were able to reach their nest burrows in very poor visibility (Bonadonna and Bretagnolle 2002). Furthermore, the global role of odors in navigation has been considered in many other species (Papi 1990; Jorge et al. 2009).
2. On the foraging side, albatrosses and other pelagic seabirds have been observed and tested meticulously. As described previously, captivating studies pointed out an essential olfactory assistance used to reach restricted nutritive resources in vast oceans (Hutchison and Wenzel 1980; Verheyden and Jouventin 1994; Nevitt et al. 1995).
3. As in other organisms, probable links between avian-produced odors and reproduction have been suspected in birds. Early research has proved that Leach's petrels use volatile indicators to join potential partners in precise places of reproduction (Grubb 1979), while diverse Antarctic seabirds rely on olfactory signatures to identify appropriate sexual partners (Bonadonna and Nevitt 2004; Jouventin et al. 2007). In 2003, Freeman-Gallant et al. studied the effect of similar or dissimilar major histocompatibility complexes (MHCs) on female mating fidelity in savannah sparrows (*Passerculus sandwichensis*, Gmelin 1789). MHCs present very diverse genes involved in immunity, and complementary genes brought by each parent should result in an optimal immunocompetency in their descendants. The classic hypothesis suggests that individual odors in vertebrates contain information related to this genetic assortment. Birds are not an exception (Zelano and Edwards 2002). If no precise odorants were identified in the aforementioned study, females in MHC-similar pairs were revealed to mate more regularly with extra-pair males, whereas breeding pairs showing MHC-dissimilarities exhibited higher fidelity, as expected (Freeman-Gallant et al. 2003). Likewise, specific odors released by different body parts and glands may

serve as a perfect signaling tool to assess the status of a mating partner. In the preen oil of dark-eyed juncos (*Junco hyemalis* L.), the quantities of volatiles and semivolatiles have been correlated with hormonal variations, such as for testosterone. The resulting chemical signal may attest to the receptiveness of a potential mating partner during the breeding season (Whittaker et al. 2011). Going further, overall individual fitness and potential reproductive performance might be evaluated directly through perfumed messages emitted intentionally or unconsciously by both partners (Whittaker et al. 2013). Obviously, assessing this function is not an easy task because both the emission and perception of odors may vary considerably, among species, seasons, ages, migration areas, and populations (Clark and Smeraski 1990; Whittaker et al. 2010). Variability in uropygial gland secretions of budgerigars (*Melopsittacus undulates*, Gould 1840) (Zhang et al. 2010), gray catbirds (*Dumetella carolinensis* L.) (Shaw et al. 2011), and starlings (*Sturnus unicolor*, Temminck 1820) (Amo et al. 2012) have been investigated in this context. Linked with earlier observations, the existence of true pheromones in birds has also been debated (Balthazart and Schoffeniels 1979; Caro and Balthazart 2010), adding credit to the overall role of odors in bird reproduction (Lambrechts and Hossaert-McKey 2006).

In addition to the above areas, further hints support a major role of avian-produced volatile organic compounds (VOCs) in antagonist interactions as predation, parasitism, or microbial infections. To avoid or to repel those enemies is of the main importance for all avian species. Moreover, several birds are predators themselves, and regularly on other birds. In these circumstances, both sense of smell and production of VOCs act directly in various defensive or offensive avian interactions.

20.1.3 Chemical Defenses in Birds

Birds are among the most detectable inhabitants of most ecosystems, so they can represent a perfect target for predators. While the ability to fly is certainly the best strategy to flee from predators, some species are flightless, and the risk of predation can itself come from the sky. Moreover, both eggs and hatchlings show great vulnerability, with the greatest risks when the parents are foraging far from the nest. These risks might have led birds to evolve repulsive defenses, aiming to increase survival at any step of their life cycles. In this context, diverse avian defensive behaviors and strategies have been described in the literature, including some examples relying on chemical compounds. In the 1940s, Cott observed dead laughing doves (*Streptopelia senegalensis* L.) being consumed by scavenger hornets (*Vespa orientalis* L.), whilst the fresh carcass of pied kingfisher (*Ceryle rudis* L.) stayed intact because of likely unpalatable components contained in its tissues (Cott 1947). The same naturalist conducted experiments on 200 species from 57 families in order to assess the palatability of the eggs and meat for humans, hedgehogs, rats, ferrets, and cats.

Without any identification of putative repellent compounds, the results suggested a correlation between the visibility of a bird species and its unpalatability (Cott 1954; Müller-Schwarze 2012). Nevertheless, re-analysis of Cott's data questioned whether the criteria used to assess conspicuousness were appropriate. Indeed, some bias may have been introduced because mainly females with usually less bright plumages than related males were considered in the past (Götmark 1994). Almost 40 years after the publications of Cott, observations of the New Guinean pitohuis (fam. Oriolidae and Pachycephalidae), and especially the hooded pitohui (*Pitohui dichrous*, Bonaparte 1850), revealed the poisonous birds to the world (Dumbacher et al. 1992). Shortly after this initial discovery, the blue-capped ifrit (*Ifrita kowaldi*, De Vis 1890) was described as another toxic passerine (Dumbacher et al. 2000). Both studies highlighted the presence of batrachotoxin alkaloids (e.g. homobatrachotoxin) as defensive chemicals in these species. In all likelihood, these compounds are sequestered in the skin and feathers of noxious birds. The repellent chemicals are obtained by feeding on soft-wing flower beetles of the *Choresine* genus (Coleoptera: Melyridae) (Dumbacher et al. 2004). Aposematism is present in these bird genera in the form of warning coloration of the plumage. In the same way, but on other continent, the African spur-winged goose (*Plectropterus gambensis* L.) exhibits toxic properties when it eats blister beetles (Coleoptera: Meloidae). This time, cantharidin is the sequestered compound, which is a familiar toxin found in several insect species (Bartram and Boland 2001). All these observations prove that chemical defenses in birds might be more widespread than expected at first glance. Undoubtedly, toxins play a role in chemical protection of various bird species, whereas avian volatile-produced compounds identified recently open the door to a further question: are the odoriferous VOCs emitted by birds involved in defensive purposes?

20.1.4 Avian Odorous Defenses

The production and release of odors in birds has not suffered the same negative prejudice as their sense of smell. For centuries, it has been established that many species emit perfumes, with written scientific reports, popular books, or even sacred texts mentioning the malodorous hoopoe (*Upupa epops* L.). Scientific studies have rapidly shown that odors released by birds can contribute to their protection in many ways.

The scent of the predator itself may indicate its presence to avian prey. This has been documented both in Eurasian blue tits (*Cyanistes caeruleus* L.) and house finches (*Haemorhous mexicanus*, Müller 1776) exposed to mustelid odors and mammalian predator feces, respectively (Amo et al. 2008; Roth et al. 2008). Nevertheless, this aptitude might be scarce: untroubled Eastern bluebirds (*Sialia sialis* L.) build their nests close to the natural odors of their main predators (Godard et al. 2007), while both native New Zealand rifleman (*Acanthisitta chloris*, Sparrman

1787) and South Island robin (*Petroica australis*, Sparrman 1788) are unable to detect and to react appropriately toward odors from their introduced mammalian predators (Stanbury and Briskie 2015).

Otherwise, avian chemical emissions have been shown to keep away various ectoparasites. The crested auklet (*Aethia cristatella*, Pallas 1769) produces a strong tangerine-like odor based on two aldehydes with remarkable repulsive properties against ticks (Douglas et al. 2004; Douglas 2006). In seabirds, some of the VOCs produced by the colony deter mosquitoes (Douglas et al. 2005). Antiparasitic and antimicrobial properties might occur in nests perfumed with aromatic plants, as in the Eurasian blue tit (Petit et al. 2002). Many species exhibit obvious malodorous defenses against intruders. The fulmars and some other procellariiform birds spit stinking stomach oil on avian or mammalian invaders (Swennen 1974; Clarke and Prince 1976). Unsurprisingly, hatchlings are especially prompt to release such an unpleasant-smelling liquid in order to defend themselves. Still about the nests, eider ducks (*Somateria mollissima* L.) and shovelers (*Anas sp.*) have been observed using feces to protect their own (Swennen 1968), while insectivorous Eurasian roller nestlings (*Coracias garrulus* L.) emit an orange oral secretion as defense against potential predators (Parejo et al. 2013). Through olfaction, the same vomit can even inform the parents about recent attacks at their nests and modulate their parental cares (Parejo et al. 2012). Currently, the classic example of repulsive secretion in birds is found in hoopoes. Both adults and young rely on pungent odors and smelly substances to dissuade potential predators. In this species, extreme malodorous VOCs and other chemicals are produced in the uropygial gland, some through bacterial transformation of ubiquitous organic precursors (Ligon 2001). Multiple functions have been assessed or hypothesized for such secretions: repellent effect against predators of both adults and hatchlings (nests are covered with uropygial secretion by parents); direct attraction of small invertebrate prey (e.g. insects) for fledglings in the nest; and antimicrobial activity protecting young hoopoes against numerous infections (Martin-Platero et al. 2006; Martín-Vivaldi et al. 2010).

As suggested earlier, studies on odorous chemical defenses in birds are complex to carry out, as the studied volatiles may fluctuate considerably. Uropygial and other secretions show great variability, depending on several environmental and genetic parameters (Ruiz-Rodríguez et al. 2014). Comprehensive investigations are required to obtain better knowledge of the importance and use of odoriferous protections in birds. Axes of research may include juvenile-adult specificities, types of nests, seasonality, or migration behaviors. For instance, the possible development of aposematic colors linked to avian malodorous defensive abilities has only been recently investigated in four European Coraciiformes and Upupiformes species, suggesting that their colorful plumages do not reduce attacks by potential predators (Ruiz-Rodríguez et al. 2013). Finally, in-depth approaches are required to disentangle the respective roles of the same avian volatile. This remains a real challenge, because the same group of odors might be used in defense, intraspecific information, and reproduction. Such multiple functions have already been observed in the crested auklets (Hagelin and Jones 2007).

20.2 Foul-Smelling Defense of a Brood Parasitic Bird and Its Consequences

20.2.1 Avian Interspecific Brood Parasitism

Among all the various ecological interactions observable between birds, brood parasitism has drawn the attention of scientists for decades. Interspecific avian brood parasites lay eggs and leave their progeny in the nests of other species. Many but not all cuckoos (fam. Cuculidae) are commonly associated with typical brood parasites in birds. Usually, hosts raise the foreigner hatchling and undergo massive reduction of their own reproductive success. This failure can be due to either evicting parasites, which clear the nest of native eggs and offspring, or from non-evicting parasites, whose chicks can monopolize parental care and drive native offspring to starve to death (Spottiswoode et al. 2012). In evolutionary biology, the arms race between brood parasites and their hosts has been studied thoroughly. Host parents are known to harass parasitic adults in the proximity of the nest, or to remove parasitic eggs (Kilner and Langmore 2011). Generally, this leads to a costly waste of time and effort. In addition, harassment attempts present some risks, as adult parasites are often bigger and stronger than parasitized species. On the other side, many alien eggs imitate the legitimate ones in terms of size, shape, and color pattern, limiting efficient ejection by hosts, whereas under particular conditions some parasitic species can discourage their ejection with punitive actions against the entire host nest (the mafia hypothesis or mafia-like behavior) (Zahavi 1979). Despite this arms race, it has been hypothesized that the absence of defense might be seen between a brood parasite and its host, if: (1) interaction is too recent for there to have been the necessary time for defenses to evolve; (2) costs of defenses exceed their benefits; (3) presence of the parasite might be beneficial to its host, as suggested half a century ago (Smith 1968). This last counterintuitive possibility, tested successfully on giant cowbirds (*Molothrus oryzivorus*, Gmelin 1788) in the late 1960s, was never replicated (Davies 2000). Recently, our new observations in a non-evicting cuckoo-host system with the release of avian-produced volatile defense have revealed remarkable outcomes supporting the controversial Smith's hypothesis (Canestrari et al. 2014). More than an isolated case, these observations strengthen the crucial role and the importance of VOCs emitted by birds at various ecological levels.

20.2.2 Malodorous Secretion Produced by a Juvenile Brood Parasite Modifies Host–Parasite Interaction: A Concrete Case

For both avian connoisseurs and the population in general, cuckoos are infamous for their brood parasitic members. This label is partly wrongfully appropriated, the majority of the cuckoo species raising their own chicks like other birds. Among the parasitic

Pict. 20.1 An example of a non-evicting brood parasite that does not monopolize all crucial resources. In this nest, one great spotted cuckoo chick and three carrion crow fledglings are raised together successfully. This contributes to reduce the negative outcomes for the host. Courtesy of D. Canestrari and V. Baglione

cuckoos, more than 50 species are obligate brood parasites, laying their eggs in nests built by unrelated bird species. Few are non-obligate brood parasites, but occasionally place their eggs in the nests of other members of the same species (Payne 2005).

In southern Europe, the great spotted cuckoo (*Clamator glandarius* L.) is a common brood parasite of corvids, mainly magpies (*Pica pica* L.). This non-evicting cuckoo species frequently outcompetes magpie nestlings for food, and is therefore responsible for major reproductive failure in this host, which has evolved defensive strategies to discourage adult cuckoos from laying their eggs (Arías de Reyna 1998). As far as possible, parent magpies also try to remove cuckoo eggs (Soler et al. 1999a, b). In northern Spain, great spotted cuckoo chicks are also found in the nests of carrion crows (*Corvus corone corone* L.), which, unlike magpies, are larger than the parasite. In these nests, legitimate and parasitic hatchlings are frequently raised together (Pict. 20.1), and crows surprisingly do not show any defensive arsenal against their brood parasite (Canestrari et al. 2009). As in the controversial hypothesis of Smith, a beneficial brood parasite might explain the lack of host-parasite defenses observed in this system, opening the door to further investigations.

Those researches were carried by Daniela Canestrari and her colleagues during the last 16 years. During this period, intensive observations and monitoring on crow reproductive success in hundreds of nests ($n=741$) and territories ($n=109$) showed that parasitized and non-parasitized host nests produce, on average, an equivalent number of young crows annually (1.584 ± 0.149 vs. 1.379 ± 0.068, respectively; $z=0.390$, $p=0.694$, $n=550$) (Canestrari et al. 2014). This unusual outcome derives from the combination of two counterbalancing effects: surprisingly, parasitized nests are more likely to successfully complete the reproductive cycle (i.e. are more likely to fledge at least one crow) compared with non-parasitized ones, whereas in successful nests fewer crow fledglings are produced if a cuckoo chick is present, as a consequence of competition for food among host and parasitic chicks (Canestrari et al. 2014). Experimental translocations of cuckoo hatchlings from parasitized to non-parasitized nests showed that nests where a cuckoo was added experienced an

Pict. 20.2 Upon harassment, a juvenile great spotted cuckoo releases its defensive foul-smelling secretion. The odors from this emission can repel most nest predators. Courtesy of D. Canestrari and V. Baglione

increased probability of fledging at least one crow chick compared to naturally non-parasitized nests, while nests where the cuckoo was removed failed more often than naturally parasitized nests. In a parallel control treatment, crow chicks were translocated among naturally non-parasitized nests, in order to exclude a possible confounding effect of brood size change (Canestrari et al. 2014). These results demonstrate that having a cuckoo chick in the nest is responsible for the advantageous effect on nest success, discarding the possibility that cuckoo adults defending the nest where they have laid eggs (Soler et al. 1999a, b) is the mechanism behind the patterns observed in the long-term dataset. The most plausible explanation for the positive effect of parasitism on crow nest success is increased nest protection against potential predators, through a copious, foul-smelling secretion released by juvenile cuckoo chicks when frightened by an intruder in the vicinity of the nest (Pict. 20.2). Interestingly, only nestlings produce such secretion in this species, adults being unable to emit this mixture even when threatened. The supposed repulsiveness of this pungent emission and its beneficial effect were tested in several ways: (1) gas chromatography and mass spectrometry (GC-MS) were used to identify and quantify the volatile blend from the cuckoo secretion; (2) an artificial transparent solution mimicking the odor of juvenile cuckoo excretion was synthetized in order to assess the exact role of volatiles in repulsiveness; (3) deterring tests of both natural secretion and artificial solution were carried out on representative nest predator species (Canestrari et al. 2014; Röder et al. 2014).

1. First of all, VOCs released by the juvenile parasite secretion were trapped with the solid phase microextraction (SPME) technique and analyzed with GC-MS (Röder et al. 2014). The results revealed a caustic mixture of malodorous organic volatiles, which must deter most of the usual mammalian and avian nest predators (Fig. 20.1). Several short-chain fatty acids (e.g. propanoic, butyric, valeric acid, etc.), organosulfur compounds (dimethyl disulfide, dimethyl trisulfide, etc.), phenolic compounds (e.g. *p*-cresol), and indoles (e.g. indole and scatole) were

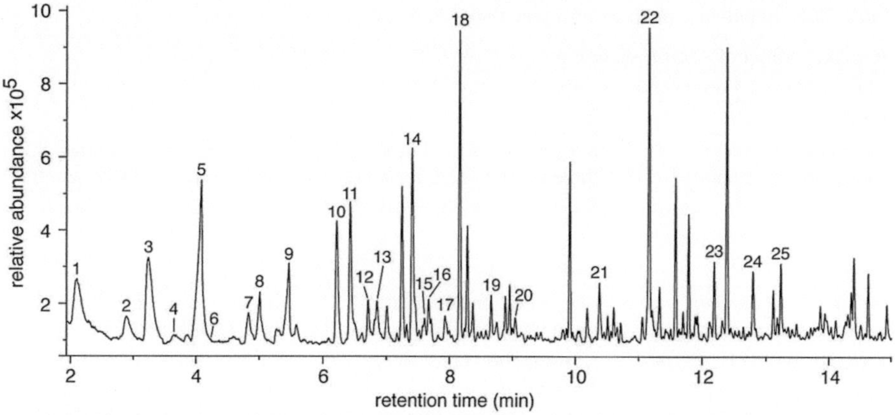

Fig. 20.1 Representative chromatographic profile of the volatiles found in defensive excretion of the great spotted cuckoo hatchling. The volatiles were analyzed by various methods (Canestrari et al. 2014), including the SPME method shown here. The major compounds are: 1, acetic acid; 2, propanoic acid; 3, dimethyl disulfide; 4, isobutyric acid; 5, butyric acid; 6, pivalic acid; 7, isovaleric acid; 8, 2-methylbutanoic acid; 9, valeric acid; 10, α-pinene; 11, dimethyl trisulfide; 12, phenol; 13, caproic acid; 14, 3-carene; 15, 2-ethylhexanol; 16, limonene; 17, acetophenone; 18, *p*-cresol; 19, nonanal; 20, camphor; 21, dimethyl tetrasulfide, 22, indole; 23, skatole; 24, longicyclene; 25, longifolene

well represented in this emission and create rancid and rotten perfumes, known to repel predators such as coyotes, dogs, or wolverines (Lehner et al. 1976; Landa and Tømmerås 1997).

2. In the overall repellent effect of the secretion, an evaluation of the efficiency of volatiles alone was not possible, as the natural secretion is dark tinted and sticky. To disentangle these traits, a liquid synthetic blend that mimicked the odor of juvenile cuckoo secretion was obtained by mixing together the 22 pure standards of the dominating volatile compounds found in the natural emission. In this solution, the ratios between the 22 compounds were based on the analytical results for four different defensive secretions of young great spotted cuckoos (Röder et al. 2014).
3. Finally, repellent properties of the juvenile cuckoo secretion were tested in wild, semi-captive and captive mammals and birds used as models of potential predators of crow nests. As expected, both natural secretion and artificial solution caused a clear deterrence in *quasi* feral cats that usually hunt in the neighborhood, carrion crows which show intraspecific predation, and hand-raised raptor birds (*Falco sp.*) (Table 20.1) (Canestrari et al. 2014; Röder et al. 2014). In artificial solution trials, one exception has to be mentioned regarding the carrion crows: they consumed the same amount of treated and control chicken baits. In this species, visual cues might play a role in repulsiveness, whilst a non-repellent effect by the odor of the secretion alone would be logical in terms of feeding habits (carrion feeders), as well as the care at the nest that adoptive crow parents must offer even after a cuckoo hatchling has released the secretion.

Table 20.1 Repellant effects against potential nest predators

Repellent effect of *natural* secretion		
Mammals (*quasi* feral cats)	Corvids (captive crows)	Raptors (captive raptors)
$n=17$	$n=7$	$n=7$
Ten pieces of chicken meat, treated or control	Six pieces, one at a time, three treated and three control	Six pieces, one at a time, three treated and three control
Strongly avoided; $p=0.01$	Strongly avoided; $p=0.008$	Strongly avoided; $p=0.001$
Repellent effect of *artificial* blend		
$n=6$	$n=7$	$n=7$
Ten pieces of chicken meat, treated or control	Six pieces, one at a time, three treated and three control	Six pieces, one at a time, three treated and three control
Strongly avoided; $p=0.014$	Consumption equivalent; $p=0.755$	Strongly avoided; $p=0.008$

Both natural defensive secretion of juvenile great spotted cuckoo and synthetic solution mimicking its pungent odors were applied on chicken baits. Fisher's Exact tests were used for statistical interpretation. Complete experiments are described in Canestrari et al. (2014) and Röder et al. (2014)

These recent findings offer a significant breakthrough, being one of the first studies to identify and test exhaustively avian-produced volatiles and their repellent effects on potential predators. Furthermore, they confirm that juvenile great spotted cuckoo use such preventive emissions for protection, and by extension the entire crow nest against intruders. The annual outcome of crow/cuckoo interaction is context-dependent. As the presence of cuckoos reduces nest failure due to predation, raising a cuckoo chick conveys a net benefit in terms of annual reproductive success for crow foster parents only in years with high predator pressure, while in years with low predation, parasitized crows produce, on average, fewer young and the net effect of parasitism is negative. Additional records on laying date, dependency period, annual adult survival, effort to raise a cuckoo chick, the condition of crow fledglings growing alongside cuckoos, or subsequent reproductive success of parasitized parents indicate that the costs of raising a parasitic chick in terms of adult provisioning effort do not erode the net benefit on crow reproductive success in years of high predation, because raising a cuckoo requires significantly less effort than a crow chick, due to the smaller size and shorter dependence period of the parasite (Canestrari et al. 2014). Annually, all these parameters lead to a fluctuation from parasitic to mutualistic interaction in this system. Because the abundance and composition of the predator community varies annually and geographically within the same host, and also differs among host species, the effect of the presence of a great spotted cuckoo chick on nest success is likely to vary with space and time within the same host, and among different host species.

It may be argued that alternative mechanisms, such as the "farming" and "mafia" hypotheses, which suggest a role of adult cuckoos as potential predators of non-parasitized nests ("farming") or of parasitized nests where cuckoo eggs or chicks have been tossed out by host parents ("mafia") in order to force hosts to re-lay, may explain the increased success of crow nests with cuckoos (Zahavi 1979; Soler et al. 1995; Arcese et al. 1996). However, both hypotheses can be discarded (Canestrari et al. 2014). A "farming strategy" is not consistent with the patterns found in the long-term dataset, where failure rate of non-parasitized nests was higher at the hatching stage but not at the egg stage, when crows are more likely to re-lay upon failure and thus adult cuckoos are expected to destroy host clutches. Furthermore, annual failure rate of non-parasitized nests did not increase with annual parasitism pressure (which is expected under the "farming hypothesis" scenario). Theoretically, a "mafia tactic" by adult cuckoos may be responsible only for the increased failure rate of experimental nests where cuckoo chicks were removed (as crows do not toss out alien eggs or chicks from the nest), but it cannot explain why nests where cuckoos were added experienced an increased success.

Based on our long-term field observations, carrion crow parents are not monopolized if they raise a foreigner cuckoo chick in their nests, suppressing the likely development of costly but useless defenses. The simple chemical signals produced by this juvenile brood parasite species illustrates the role and the importance that avian odors and sense of smell can have in precise ecological interspecific interactions. The system offers a significant advance in brood parasitism, as well as in research considering the importance of volatile cues in vertebrates.

20.3 New Perspectives in Malodorous Avian Chemical Defenses

Pioneering studies carried out on great spotted cuckoos and their carrion crow hosts have brought as many questions as answers. This opens new perspectives of research in avian chemical ecology. In many circumstances, both interspecific and intraspecific interactions are of ecological importance. In the paradoxical interactions between a brood parasitic bird and its host, what might be the various consequences when VOCs are emitted by one species cohabiting with another? Most of the brood parasite–host interactions have been well studied and documented, but never from the point of view of ecological effects ruled by odorants released from parasites or hosts. Based on the great spotted cuckoo—carrion crow system, many questions are currently being evaluated: (1) In the great spotted cuckoo, is the specialized secretion very different from characteristic excrements? (2) Which anatomical structure(s) is/are responsible for the production of the mixture? (3) Is the secretion based on metabolic byproducts or biosynthesized by symbiotic bacterial strains? (4) Can an equivalent secretion be found in other brood parasitic bird species, and what is its effect? (5) Is the production and the efficiency of such malodorous excretion correlated with a specific pattern of brood parasitism (e.g. evicting or non-evicting avian parasites; only in juvenile or at different ages; sedentary or migrant habits; host–parasite size ratio, etc.)? Further experiments will explore these paths. In the meantime, our recent investigations provide some hint of answers:

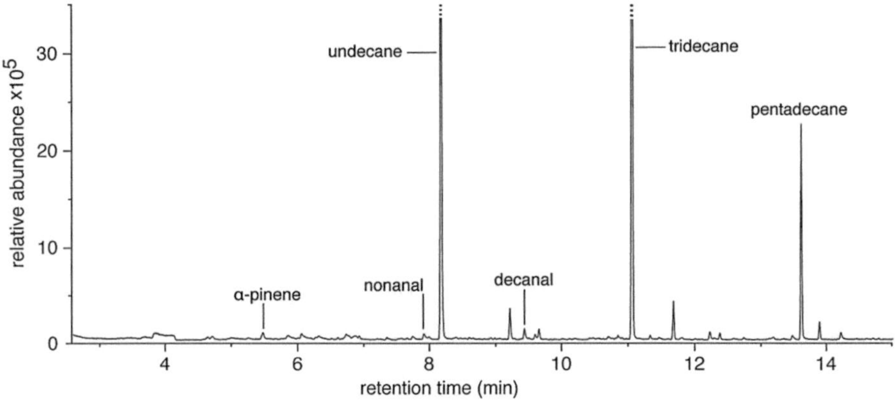

Fig. 20.2 This chromatographic profile of the VOCs emitted by excrements of great spotted cuckoo chicks was obtained by the SPME method. Comparison with Fig. 20.1 underlines the drastic differences between the foul-smelling volatiles of the defensive secretion and the less odoriferous compounds found in the feces

1. After initial batches of VOCs analysis carried out on the secretion produced by juvenile great spotted cuckoo, it has been necessary to test whether the defensive emission was a special excretion or a sort of natural excrements. New repulsive secretions of the great spotted cuckoo chicks were analyzed at the University of Neuchâtel (Switzerland). In addition, samples of excrements from each protagonist involved in this brood parasite–host system were added. All the different vials of fresh secretions and feces were sent from Spain in a frozen container at –20 °C. Immediately after arrival, volatile components were trapped with a solid phase microextraction (SPME) technique and analyzed with GC-MS, exactly as described in Röder et al. 2014. The odors found in these secretions were then compared with those found in excrements of juvenile great spotted cuckoos, carrion crow hatchlings, and parental carrion crows (Fig. 20.1). The comparison of the odorant profiles showed evident differences both in composition and quantity, as illustrated in Fig. 20.2. This chromatogram presents the volatiles trapped from the excrements of juvenile great spotted cuckoo. For a human nose, the repulsive secretion produced by cuckoo chicks is undoubtedly the worst smelling. Various vomit and rancid perfumes are issued by carboxylic acids whilst remarkable amounts of *p*-cresol or skatole are responsible for the standard odors of mammalian dirty hair and feces, respectively. On the other hand, excrements produced by birds showed diverse odorous profiles, but with few volatile compounds and limited malodorous properties. After exhaustive data processing with MSD ChemStation Enhanced Data Analysis (Agilent), XCMS online (Scripps Center for Metabolomics), and MassLynx Mass Spectrometry Software (Waters), a principal component analysis (PCA) based on the constitutive ions of VOCs linked with GC-MS retention time attest the dissimilarities between

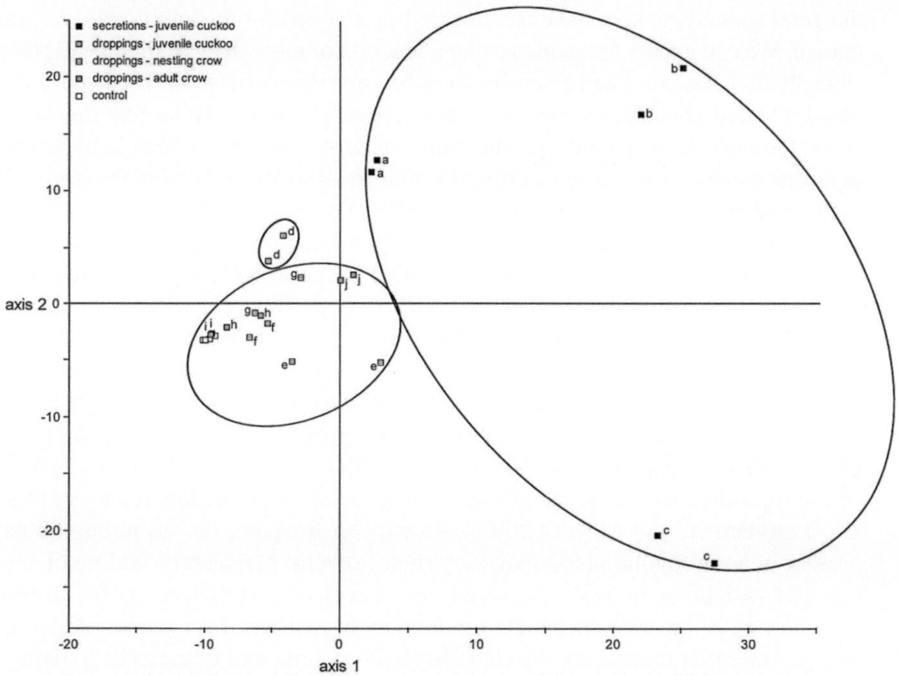

Fig. 20.3 Unsupervised principal component analysis (PCA) based on aligned features (ions) obtained with GC-MS analysis carried out on three great spotted cuckoo chick defensive secretions, seven excrements (1 juvenile cuckoos—3 young crows—3 adult crows), and five control samples. Each sample has been split in two parts and analyzed individually (same letter in the graphic). The more a sample is close to the controls (*white squares*), the less it smells. Variability exists in defensive secretion of the juvenile great spotted cuckoo but cannot be mistaken for any other classic droppings

specialized repulsive emissions of the great spotted cuckoo chicks, and the feces produced by both cuckoos and carrion crows (Fig. 20.3). Identical lettering in this PCA shows that two different portions of each sample were analyzed separately to assess the homogeneity of such production. Finally, five corresponding control analyses were run aiming to exclude ghost peaks and to determine which of the chemicals were not of avian origin (Fig. 20.3). Without further analysis, our results prove that the defensive secretion emitted by the great spotted cuckoo chicks is special, and distinct from the usual excrements. This trait supports the defensive function of this secretion produced only in great spotted cuckoo nestlings.

2. Until now, avian-produced malodorous volatiles have been mainly associated with uropygial glands, as in hoopoes (Cramp 1998), or with oils expelled from digestive organs, as in procellariiformes (Swennen 1974; Clarke and Prince 1976) and nestlings of Eurasian rollers (Parejo et al. 2013). Interestingly, these two common origins are not the source of the defensive secretion of cuckoo nestlings. In

the great spotted cuckoo, the exact place of production for this mixture is not yet known. We can expect that with a sudden ejection of more than 10 ml, the responsible gland or structure will be easily identified and described soon. The avian anal gland, cloacal gland, or cecum are the best candidates to assume this function. Birds typically have paired ceca hosting numerous bacteria, which help in the digestion of various nutriments. Presently, this organ seems to show all the required traits to produce this foul-smelling secretion.

3. The constitutive compounds of the secretion produced by the juvenile great spotted cuckoo have a biochemical origin which remains unclear. The cecum and glands contain a rich bacterial fauna, which is commonly associated with the production of malodorous volatiles. Hence, a bacterial biotransformation of metabolic precursors is a logical explanation, as described in the uropygial gland of hoopoes and red-billed woodhoopoe (*Phoeniculus purpureus*, Miller 1784). In hoopoe, the production of odoriferous excretion requires the presence of *Enterococcus faecalis* (Martín-Vivaldi et al. 2010), whereas the uropygial gland of the red-billed woodhoopoe houses *Enterococcus phoeniculicola* among others (Law-Brown and Meyers 2003). As we can imagine, the bacterial strains present in an uropygial gland may vary under several parameters, adding diversity and variability in avian-produced volatiles (Lucas and Heeb 2005). In the secretion of *C. glandarius*, high production of scatole (i.e. 3-methylindole), indole, and other unpleasant odorants supposes a comparable bacterial transformation, especially because avian excrements do not emit such compounds. Within bacterial communities, the enterobacteria (e.g. *Enterobacter cloacae*) are known to generate such molecules (Jensen et al. 1995). More interestingly, our previous repellence tests revealed an optimal protection against mammals. Here, we can hypothesize that evolution has developed an ideal odoriferous protection for great spotted cuckoo chicks. Indeed, predators remain strongly repelled by their own excrements, mainly to avoid disease and intoxication. Therefore, a juvenile cuckoo smelling like their own dung surely deters most mammals. It would not be a surprise if classic mammalian fecal bacteria would be present in the avian structure producing the defensive secretion, corroborating the expected bacterial origin of this protection.
4. Interspecific parasitic cuckoo chicks cannot rely on their own parents for protection. Usually, smaller and weaker host parents raise them. For a juvenile cuckoo, the development of efficient defenses might be a plus. Although the young great spotted cuckoos exhibit such malodorous defense, nothing is known in other parasitic bird species. Is the secretion of the great spotted cuckoo unique, or is there any equivalence in other species? Currently, this question is being evaluated by comparison with the common cuckoo (*Cuculus canorus* L.), whose nestlings also produce a sticky secretion upon harassment (Pict. 20.3). Preliminary volatile analyses have been carried out on the secretion produced by young common cuckoos. As previously described, GC-MS was used to obtain chromatographic descriptions of both the likely new defensive secretion and classic excrements of juvenile common cuckoo chicks. The final publication is in preparation, but some informative results can be presented here. The unlabeled chro-

Pict. 20.3 A common cuckoo fledgling in defensive posture, before the emission of its foul-smelling secretion. Courtesy of A. Trnka

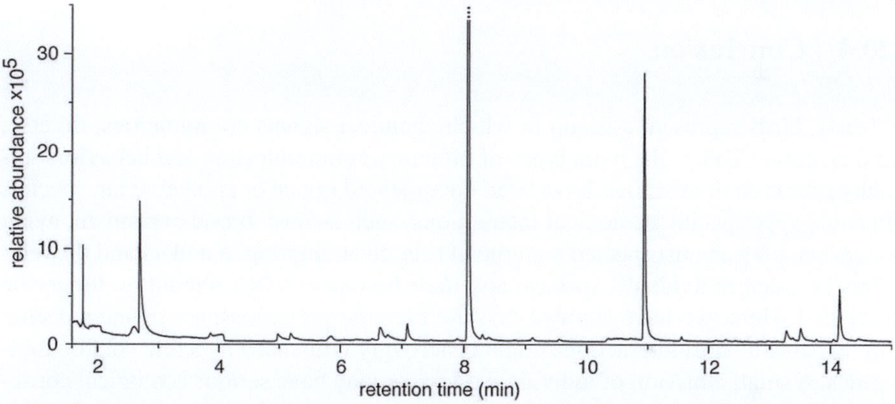

Fig. 20.4 Another defensive secretion in a brood parasite: Preliminary chromatographic results of the volatiles released by juvenile common cuckoo upon threat. Visually, the secretion produced by common cuckoo hatchling looks like the one of the great spotted cuckoo chick, but chemically the constitution is very different. This odorant profile was obtained by the SPME method

matographic profile shows the odorants found in the secretion released by young common cuckoos (Fig. 20.4). It suggests that odors are far from identical with those emitted by *C. glandarius*. Compared with this species (see Fig. 20.1), the main malodorous volatiles are absent or very weak in the common cuckoo hatchlings. This highlights the great variability of such secretions, and suggests variation in the effectiveness of defense. The common cuckoo is an evicting brood parasite, which reduces strongly the reproductive success of its hosts. Compared with the great spotted cuckoo, both the strategy and needs are different. Consequently, we can expect that being alone in the nest with full parental care might have led to the evolution of different traits, including less fetid secretion ensuring that host parents will stay after emission. Obviously, the evolution of these defensive secretions might be under many selective parameters, and may have diversified as a function of the main brood parasitic strategies.

5. Finally, an extensive study of odorous defensive secretions in parasitic cuckoos may reveal a correlation of various types of secretions with particular ecological patterns. In our studied organisms, only juvenile great spotted and common cuckoos are able to produce and release a specialized secretion against intruders. Is this true for all other parasitic cuckoo species, in which adults would be unable to use any malodorous defense? As described previously, the different strategies between evicting and non-evicting cuckoos may have consequences for the production of odoriferous defenses. Moreover, many brood parasitic adults are migrant species, which may have an impact on the production, use, and efficiency of likely smelly protections. With many other future questions, all these current interrogations are waiting for answers. These new perspectives will surely lead to new captivating findings in the field of avian-produced volatile defenses.

20.4 Conclusion

Clearly, birds represent a group in which chemical signals are numerous, diverse, and essential. Today, the main types of olfactory communication and behaviors initially observed in mammals have been documented in one or another avian species. In some very specific ecological interactions, such as host–brood parasitism, avian odors can play an unsuspected but crucial role. In attempting to understand the relations between individuals, species, and their biotopes, VOCs should no longer be excluded. Here, we have showed that the extreme precariousness of interspecific brood parasite–host interactions might be strongly influenced by a few volatile molecules. A small emission of individual odorants may have serious ecological consequences for an entire system. For researchers and ornithologists, this constitutes a great challenge, as well as some fantastic opportunities to work with emblematic systems and species, but with a new approach. Chemical ecology in birds is surely at the dawn of major discoveries.

Acknowledgments We are very grateful for the help and unconditional support of Aline Verdon for figures, and Dr. Russell E. Naisbit (Albion Science Editing) for revision of the manuscript. We sincerely thank two anonymous reviewers for their helpful comments and suggestions.

References

Amo L, Galvan I, Tomas G, Sanz JJ (2008) Predator odour recognition and avoidance in a songbird. Funct Ecol 22:289–293

Amo L, Avilés JM, Parejo D, Peña A, Rodríguez J, Tomás G (2012) Sex recognition by odour and variation in the uropygial gland secretion in starlings. J Anim Ecol 81:605–613

Applegate RD (1990) Can vultures smell? Turkey vulture caught in cage trap. N Am Bird Bander 15:141–142

Arcese P, Smith JN, Hatch MI (1996) Nest predation by cowbirds and its consequences for passerine demography. Proc Natl Acad Sci U S A 93:4608–4611
Arias de Reyna L (1998) Coevolution of great spotted cuckoo and its hosts. In: Rothstein SI, Robinson SK (eds) Parasitic birds and their hosts, studies in coevolution, Oxford University Press, Oxford, p 129–142
Audubon JJ (1820) In: Corning H (ed) Journal of John James Audubon: made during his trip to New Orleans in 1820-1821, Cambridge, Business Historical Society, pp 260
Balthazart J, Schoffeniels E (1979) Pheromones are involved in the control of sexual behaviour of birds. Naturwissenschaften 66:55–56
Bang BG (1960) Anatomical evidence for olfactory function in some species of birds. Nature 188:547–549
Bang BG, Cobb S (1968) The size of the olfactory bulb in 108 species of birds. Auk 85:55–61
Bang BG, Wenzel BM (1985) Nasal cavity and olfactory system. In: King AS, McLelland J (eds) Form and function in birds. NY Academic Press, New York, pp 195–225
Bartram S, Boland W (2001) Chemistry and ecology of toxic birds. Chem Bio Chem 2:809–811
Bonadonna F, Bretagnolle V (2002) Smelling home: a good solution for burrow-finding in nocturnal petrels? J Exp Biol 205:2519–2523
Bonadonna F, Nevitt GA (2004) Partnerspecific odor recognition in an Antarctic seabird. Science 306:835
Canestrari D, Marcos JM, Baglione V (2009) Cooperative breeding in carrion crows reduces the rate of brood parasitism by great spotted cuckoos. Anim Behav 77:1337–1344
Canestrari D, Bolopo D, Turlings TCJ, Röder G, Marcos JM, Baglione V (2014) From parasitism to mutualism: unexpected interactions between a cuckoo and its host. Science 343:1350–1352
Caro SP, Balthazart J (2010) Pheromones in birds: myth or reality? J Comp Physiol A 196: 751–766
Clark L, Smeraski CA (1990) Seasonal shifts in odor acuity by starlings. J Exp Zool 177:673–680
Clark L, Avilova KV, Beans NJ (1993) Odor thresholds in passerines. Comp Biochem Physiol A 104:305–312
Clarke A, Prince PA (1976) The origin of stomach oil in marine birds: Analyses of the stomach oil of six species of subantarctic procellariiform birds. J Exp Mar Biol Ecol 23:15–30
Cott HB (1947) The edibility of birds: illustrated by five years' experiments and observations (1941–1946) on the food preferences of the hornet, cat and man: and considered with special reference to the theories of adaptive coloration. Proc Zool Soc Lond 116:371–524
Cott HB (1954) The palatability of the eggs of birds: mainly based upon observations of an egg panel. Proc Zool Soc Lond 124:335–464
Cramp S (1998) The complete birds of the Western Palearctic on CD-ROM. Software Optimedia/ Oxford University Press, Oxford
Davies NB (ed) (2000) Cuckoos, cowbirds and other cheats. T & AD Poyser, London
Douglas HD (2006) Measurement of chemical emissions in crested auklets (*Aethia cristatella*). J Chem Ecol 32:2559–2567
Douglas HD, Co JE, Jones TH, Conner WE (2004) Interspecific differences in *Aethia spp.* auklet odorants and evidence for chemical defense against ectoparasites. J Chem Ecol 30: 1921–1935
Douglas HD, Co JE, Jones TH, Conner WE, Day JF (2005) Chemical odorant of colonial seabird repels mosquitoes. J Med Entomol 42:647–651
Dumbacher JP, Beehler BM, Spande TF, Garraffo HM, Daly JW (1992) Homobatrachotoxin in the genus *Pitohui*: chemical defense in birds? Science 258:799–801
Dumbacher JP, Spande TF, Daly JW (2000) Batrachotoxin alkaloids from passerine birds: a second toxic bird genus (*Ifrita kowaldi*) from New Guinea. Proc Natl Acad Sci U S A 97:12970–12975
Dumbacher JP, Wako A, Derrickson SR, Samuelson A, Spande TF, Daly JW (2004) Melyrid beetles (*Choresine*): a putative source for the batrachotoxin alkaloids found in poison-dart frogs and toxic passerine birds. Proc Natl Acad Sci U S A 101:15857–15860

Freeman-Gallant CR, Meguerdichian M, Wheelright NT, Sollecito SV (2003) Social pairing and female mating fidelity predicted by restriction fragment length polymorphism similarity at the major histocompatibility complex in a songbird. Mol Ecol 12:3077–3083

Godard RD, Bowers BB, Morgan Wilson C (2007) Eastern bluebirds *Sialia sialis* do not avoid nest boxes with chemical cues from two common nest predators. J Avian Biol 38:128–131

Götmark F (1994) Are bright birds distasteful? A reanalysis of H B Cott's data on the edibility of birds. J Avian Biol 25:184–197

Grubb TC (1972) Smell and foraging in shearwaters and petrels. Nature 237:404–405

Grubb TC (1973) Colony location by Leach's petrel. Auk 90:78–82

Grubb TC (1974) Olfactory navigation to the nesting burrow in Leach's petrel (*Oceanodroma leucorhoa*). Anim Behav 22:192–202

Grubb TC (1979) Olfactory guidance of Leach's storm petrel to the breeding island. Wilson Bull 91:141–143

Hagelin JC (2004) Observations on the olfactory ability of the Kakapo *Strigops habroptilus*, the critically endangered parrot of New Zealand. Ibis 146:161–164

Hagelin JC, Jones IL (2007) Bird odors and other chemical substances: a defense mechanism or overlooked mode of intraspecific communication? Auk 124:741–761

Hutchison LV, Wenzel BM (1980) Olfactory guidance in foraging by procellariiforms. Condor 82:314–319

Jensen MT, Cox RP, Jensen BB (1995) 3-Methylindole (skatole) and indole production by mixed populations of pig fecal bacteria. Appl Environ Microbiol 61:3180–3184

Jorge PE, Marques PAM, Phillips JB (2009) Activational effects of odours on avian navigation. Proc R Soc B 276:45–49

Jouventin P, Mouret V, Bonadonna F (2007) Wilson's storm petrels *Oceanites oceanicus* recognise the olfactory signature of their mate. Ethology 113:1228–1232

Kilner RM, Langmore NE (2011) Cuckoos versus hosts in insects and birds: adaptations, counter-adaptations and outcomes. Biol Rev Camb Philos Soc 86:836–852

Lambrechts MM, Hossaert-McKey M (2006) Olfaction, volatile compounds and reproduction in birds. Acta Zool Sini 52:284–287

Landa A, Tømmerås BA (1997) A test of aversive agents on wolverines. J Wildl Manag 61:510–516

Law-Brown J, Meyers PR (2003) *Enterococcus phoeniculicola* sp. nov., a novel member of the enterococci isolated from the uropygial gland of the Red-billed Woodhoopoe, *Phoeniculus purpureus*. Int J Syst Evol Micr 53:683–685

Lehner PN, Krumm R, Cringan AT (1976) Tests for olfactory repellents for coyotes and dogs. J Wildl Manag 40:145–150

Ligon JD (2001) Family Phoeniculidae (woodhoopoes). In: del Hovo J, Elliot A, Sargatal J (eds) Handbook of the birds of the world. Lynx Edicions, Barcelona, pp 412–434

Lucas FS, Heeb P (2005) Environmental factors shape cloacal bacterial assemblages in great tit *Parus major* and blue tit *P. caeruleus* nestlings. J Avian Biol 36:510–516

Martín-Platero AM, Valdivia E, Ruíz-Rodríguez M, Soler JJ, Martín-Vivaldi M, Maqueda M, Martínez-Bueno M (2006) Characterization of antimicrobial substances produced by Enterococcus faecalis MRR 10-3, isolated from the uropygial gland of the Hoopoe (Upupa epops). Appl Environ Microbiol 72:4245–4249

Martín-Vivaldi M, Peña A, Peralta-Sánchez JM, Sánchez L, Ananou S, Ruiz-Rodríguez M, Soler JJ (2010) Antimicrobial chemicals in hoopoe preen secretions are produced by symbiotic bacteria. Proc Biol Sci 277:123–130

Müller-Schwarze D (ed) (2012) Chemical ecology of vertebrates. Cambridge University Press, New York, pp 259–261

Nevitt GA, Bonadonna F (2005) Sensitivity to dimethyl sulphide suggests a mechanism for olfactory navigation by seabirds. Biol Lett 1:303–305

Nevitt GA, Veit RR, Kareiva P (1995) Dimethyl sulphide as a foraging cue for Antarctic procellariiform seabirds. Nature 376:681–682

Nevitt GA, Losekoot M, Weimerskirch H (2008) Evidence for olfactory search in wandering albatross (*Diomedea exulans*). Proc Natl Acad Sci U S A 105:4576–4581
Papi F (1976) The olfactory navigation system of the homing pigeon. Verh Dtsch Zool Ges 69:184–205
Papi F (1990) Olfactory navigation in birds. Experientia 46:352–363
Parejo D, Amo L, Rodríguez J, Avilés JM (2012) Rollers smell the fear of nestlings. Biol Lett 8:502–504
Parejo D, Avilés JM, Peña A, Sánchez L, Ruano F, Zamora-Muñoz C, Martín-Vivaldi M (2013) Armed rollers: does nestling's vomit function as a defence against predators? PLoS One 8, e68862
Payne RB (ed) (2005) The cuckoos. NY Oxford University Press, New York
Petit C, Hossaert-McKey M, Perret P, Blondel J, Lambrechts MM (2002) Blue Tits use selected plants and olfaction to maintain an aromatic environment for nestlings. Ecol Lett 5:585–589
Röder G, Canestrari D, Bolopo D, Marcos JM, Villard N, Baglione V, Turlings TCJ (2014) Chicks of the great spotted cuckoo may turn brood parasitism into mutualism by producing a foul-smelling secretion that repeals predators. J Chem Ecol 40:320–324
Roth TC, Cox JG, Lima SL (2008) Can foraging birds assess predation risk by scent? Anim Behav 76:2021–2027
Ruiz-Rodríguez M, Avilés JM, Cuervo JJ, Parejo D, Ruano F, Zamora-Muñoz C, Sergio F, López-Jiménez TA, Martín-Vivaldi M (2013) Does avian conspicuous colouration increase or reduce predation risk? Oecologia 173:83–93
Ruiz-Rodríguez M, Soler JJ, Martín-Vivaldi M, Martín-Platero AM, Méndez M, Peralta-Sánchez JM, Ananou S, Valdivia E, Martínez-Bueno M (2014) Environmental factors shape the community of symbionts in the hoopoe uropygial gland more than genetic factors. Appl Environ Microbiol 80:6714–6723
Shaw CL, Rutter JE, Austin AL, Garvin MC, Whelan RJ (2011) Volatile and semivolatile compounds in gray catbird uropygial secretions vary with age and between breeding and wintering grounds. J Chem Ecol 37:329–339
Smith NG (1968) The advantage of being parasitized. Nature 219:690–694
Smith AS, Paselk AR (1986) Olfactory sensitivity of the turkey vulture (*Cathartes aura*), to three carrion-associated odorants. Auk 103:586–592
Soler M, Soler JJ, Martínez G, Møller A (1995) Magpie host manipulation by great spotted cuckoos: evidence for an avian mafia? Evolution 49:770–775
Soler JJ, Martínez JG, Soler M, Møller AP (1999a) Genetic and geographic variation in rejection behavior of cuckoo eggs by European magpie populations: an experimental test of rejecter-gene flow. Evolution 53:947–956
Soler M, Martínez JG, Soler JJ (1999b) Nest defence by magpies (*Pica pica*) and the brood parasitic great spotted cuckoos (*Clamator glandarius*) in parasitized and unparasitized nests. J Ornithol 140:199–205
Spottiswoode C, Kilner R, Davies NB (2012) Brood parasitism. In: Royle NJ, Smiseth PT, Kölliker M (eds) The evolution of parental care. Oxford University Press, Oxford, pp 226–356
Stager KE (1964) The role of olfaction in food location by the turkey vulture, *Cathartes aura*. Los Angeles Co Mus Contrib Sci 81:1–63
Stanbury M, Briskie JV (2015) I smell a rat: can New Zealand birds recognize the odor of an invasive mammalian predator? Curr Zool 61:34–41
Steiger SS, Fidler AE, Valcu M, Kempenaers B (2008) Avian olfactory receptor gene repertoires: evidence for a well-developed sense of smell in birds? Proc R Soc B 275:2309–2317
Steiger SS, Fidler AE, Kempenaers B (2009) Evidence for increased olfactory receptor gene repertoire size in two nocturnal bird species with well-developed olfactory ability. BMC Evol Biol 9:117
Swennen C (1968) Nest protection of eiderducks and shovelers by means of faeces. Ardea 56:248–258
Swennen C (1974) Observation on the effect of ejection of stomach oil by the fulmar (*Fulmarus glacialis*) on other birds. Ardea 62:111–117

Verheyden C, Jouventin P (1994) Olfactory behavior of foraging procellariiforms. Auk 111:285–291

Wallraff HG (1982) The homing mechanism of pigeons. Nature 300:293–294

Wenzel BM (1968) Olfactory prowess of the kiwi. Nature 220:1133–1134

Wenzel BM (1971) Olfaction in birds. In: Beidler LM (ed) Handbook of sensory physiology. Springer, New York, pp 432–448

Wenzel BM (1986) The ecological and evolutionary challenges of procellariiform olfaction. In: Duvall D, Müller-Schwarze D, Silverstein RM (eds) Chemical signals in vertebrates, vol 4. Plenum Press, New York, pp 357–368

Wenzel BM, Meisami E (1987) Number, size and density of mitral cells in the olfactory bulbs of the Northern Fulmar and Rock Dove. In: Roper SD, Atema J (eds) Olfaction and taste, vol 9. New York Academy of Sciences, New York, pp 700–701

Wenzel BM, Rausch LJ (1977) Does the olfactory system modulate affective behavior in the pigeon? Ann NY Acad Sci 290:314–330

Whittaker DJ, Soini HA, Atwell JW, Hollars C, Novotny MV, Ketterson ED (2010) Songbird chemosignals: Preen oil volatile compounds vary among individuals, sexes, and populations. Behav Ecol 21:608–614

Whittaker DJ, Soini HA, Gerlach NM, Posto AL, Novotny MV, Ketterson ED (2011) Role of testosterone in stimulating seasonal changes in a potential avian chemosignal. J Chem Ecol 37:1349–1357

Whittaker DJ, Gerlach NM, Soini HA, Novotny MV, Ketterson ED (2013) Bird odour predicts reproductive success. Anim Behav 86:697–703

Zahavi A (1979) Parasitism and nest predation in parasitic cuckoos. Am Nat 113:157–159

Zelano B, Edwards SV (2002) An MHC component to kin recognition and mate choice in birds: predictions, progress, and prospects. Am Nat 160:S225–S237

Zhang JX, Wei W, Zhang JH, Yang WH (2010) Uropygial gland-secreted alkanols contribute to olfactory sex signals in budgerigars. Chem Senses 35:375–382

Part V
Frog Chemosignals

Chapter 21
A Review of Chemical Defense in Poison Frogs (Dendrobatidae): Ecology, Pharmacokinetics, and Autoresistance

Juan C. Santos, Rebecca D. Tarvin, and Lauren A. O'Connell

21.1 Introduction

Chemical defense has evolved multiple times in nearly every major group of life, from snakes and insects to bacteria and plants (Mebs 2002). However, among land vertebrates, chemical defenses are restricted to a few monophyletic groups (i.e., clades). Most of these are amphibians and snakes, but a few rare origins (e.g., *Pitohui* birds) have stimulated research on acquired chemical defenses (Dumbacher et al. 1992). Selective pressures that lead to defense are usually associated with an organism's limited ability to escape predation or conspicuous behaviors and phenotypes that increase detectability by predators (e.g., diurnality or mating calls) (Speed and Ruxton 2005). Defended organisms frequently evolve warning signals to advertise their defense, a phenomenon known as aposematism (Mappes et al. 2005). Warning signals such as conspicuous coloration unambiguously inform predators that there will be a substantial cost if they proceed with attack or consumption of the defended prey (Mappes et al. 2005). However, aposematism is likely more complex than the simple pairing of signal and defense, encompassing a series of traits (i.e., the aposematic syndrome) that alter morphology, physiology, and behavior (Mappes and

J.C. Santos (✉)
Department of Zoology, Biodiversity Research Centre, University of British Columbia, #4200-6270 University Blvd, Vancouver, BC, Canada, V6T 1Z4
e-mail: infraguttatus@gmail.com

R.D. Tarvin
University of Texas at Austin, 2415 Speedway Stop C0990, Austin, TX 78712, USA
e-mail: rdtarvin@gmail.com

L.A. O'Connell
Center for Systems Biology, Harvard University, Northwest Labs 359.30, 52 Oxford Street, Cambridge, MA 02138, USA
e-mail: aloconnell@fas.harvard.edu

B.A. Schulte et al. (eds.), *Chemical Signals in Vertebrates 13*,
DOI 10.1007/978-3-319-22026-0_21

Alatalo 1997; Hagman and Forsman 2003; Darst et al. 2005; Despland and Simpson 2005; Ozel and Stynoski 2011; Santos and Cannatella 2011; Zhen et al. 2012).

Several lineages within five families of anurans have chemical defense and are known as poison frogs (i.e., Bufonidae, Dendrobatidae, Eleutherodactylidae, Mantellidae, and Myobatrachidae); see references for accounts in non-dendrobatid groups (e.g., Daly et al. 2005; Rodriguez et al. 2011). However, the most studied of these frogs are those within Dendrobatidae; our use of the term "poison frog" in this review will refer only to this group. For some time, defensive compounds in dendrobatids were thought to be products of biosynthetic pathways that evolved once in this family (Myers et al. 1991). However, ecological and phylogenetic studies revealed that poison frogs sequester alkaloids from their diet and that this capacity has evolved at least four times across Dendrobatidae (Fig. 21.1). This evolutionary complexity makes dendrobatids an unparalleled model clade for the study of acquired defenses among vertebrates. Although much research effort has focused on natural product discovery and characterization of chemicals found in the skin of poisonous anurans, other aspects of poison frog ecology remain poorly studied. In this review, we present an integrative summary of dendrobatid chemical ecology with an evolutionary perspective, highlighting some of these poorly studied aspects and exploring potential avenues of future research.

21.2 Evolutionary Ecology of Dendrobatid Poison Frogs

Dendrobatidae is a monophyletic clade containing more than 300 species. All members of this lineage are Neotropical endemics and common leaf litter inhabitants across tropical and mountainous forests in Central and South America (Santos et al. 2009). One species (*Dendrobates auratus*), however, occurs outside of this range in O'ahu and Maui (Hawaii, USA) since its introduction as a pest control in 1932 (McKeown 1996; Kraus and Duvall 2004). Most dendrobatid species have lowland distributions at altitudes of less than 2000 m and they tend to be diurnal, terrestrial, and locally abundant near streams. Many are popular pet trade animals, and many color morphs exist in the pet trade that do not exist in the wild (Lötters et al. 2007).

Approximately two-thirds of dendrobatids are considered to be nontoxic and cryptically colored. The remaining ~100 species are regarded as aposematic because they have both visual warning signals and defensive compounds (Summers and Clough 2001; Santos et al. 2003; Vences et al. 2003). Of the total 12 dendrobatid genera, six contain aposematic species: *Ameerega* (31 species), *Colostethus* (1 sp.), *Epipedobates* (6 spp.), *Dendrobates* (50 spp.), *Hyloxalus* (2 spp.), and *Phyllobates* (5 spp.). The genus *Dendrobates* is the most studied and the focus of frequent taxonomic reviews (Myers 1987; Bauer 1994; Grant et al. 2006; Brown et al. 2011). Seven lineages within *Dendrobates* have been proposed as new genera (Fig. 21.1), which we consider as subgenera until further taxonomic work is provided. Other chemically defended but non-aposematic dendrobatids include members of *Aromobates*, *Epipedobates*, and *Colostethus* (Myers et al. 1991; Daly et al. 1994b;

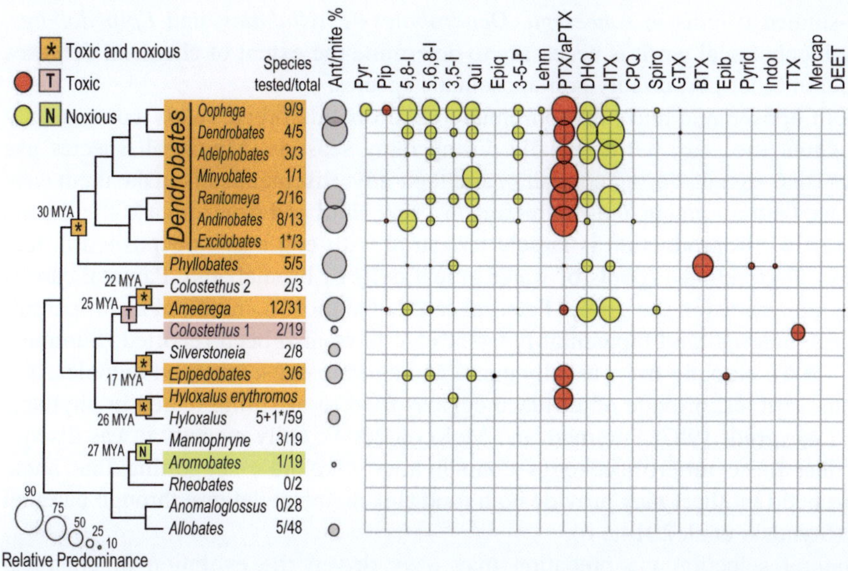

Fig. 21.1 Relative predominance (quantity and diversity) of different compound classes found in skin extracts of poison frogs of the family Dendrobatidae. Toxic and unpalatable genera are indicated by approximate age in millions of years (MYA) since diversification from their last common ancestor. *Colostethus* is split in two clades (i.e., *Colostethus* 1 and *Colostethus* 2) because it is paraphyletic with respect to *Ameerega* (i.e., *Colostethus* 2 is the sister taxa to *Ameerega*). The number of species in each genus/subgenus that has been tested for alkaloid presence is indicated over the total number of species in that genus. Diet is indicated by the percentage of the most representative prey items that are known sources of alkaloids: ant and mites. The relative predominance of each alkaloid class was determined by the sum of all the different compounds in that class multiplied by its quantity (i.e., trace, minor, major), then divided by the total number of species in the genus and finally adjusted to 100 %. Some uncertainties exist for chemical defense characterization. For example, *Dendrobates* (*Excidobates*) *captivus* and *Hyloxalus azureiventris* have tested positively for presence or ability to sequester alkaloids but the identity of alkaloids sequestered in wild populations of these species is unknown (indicated by *). The abbreviation of alkaloid classes is as follows: *Pyr* Pyrrolidines, *Pip* Piperidines, *5,8-I* 5,8-disubstituted indolizidines, *5,6,8-I* 5,6,8-trisubstituted indolizidines, *3,5-I* 3,5-disubstituted indolizidines, *Qui* Quinolizidines, *Epiq* Epiquinamide, *3,5-P* 3,5-disubstituted pyrrolizidine, *Lehm Lehmizidines*, *PTX/aPTX* Pumiliotoxins/Allopumiliotoxins, *DHQ* Decahydroquinolines, *HTX* Histrionicotoxins, *CPQ* Cyclopentaquinolizidines, *Spiro* Spiropyrrolizidines, *GTX* Gephyrotoxins, *BTX* Batrachotoxins, *Epib* Epibatidines, *Pyrid* Pyridinic alkaloids, *Indol* Indolic alkaloids, *TTX* Tetrodotoxin, *Mercap* Mercaptan-odor, *DEET* N,N-diethyltoluamide

Cipriani and Rivera 2009). The remaining dendrobatids in the genera *Allobates*, *Anomaloglossus*, *Hyloxalus*, *Mannophryne*, *Rheobates*, and *Silverstoneia* are largely considered to lack chemical defenses (Grant et al. 2006). However, at least one species of *Hyloxalus* (Santos et al. 2014) has defensive alkaloids (i.e., *H. erythromos*) and another (*H. azureiventris*) might be able to sequester them (Saporito et al. 2009; Santos and Cannatella 2011). In both instances, each species is closely related to non-defended *Hyloxalus* species, suggesting that there might be two independent origins of chemical defense in this clade, in addition to the three other well-studied origins in *Ameerega*, *Dendrobates*+*Phyllobates*, and *Epipedobates*. More biochemical work is necessary to determine the extent of chemical defenses present in other purportedly non-aposematic species.

Both aposematic and non-aposematic dendrobatids breed nearly continuously throughout the year, but especially during rainy seasons. Almost all species use audiovisual signals during courtship, and their advertisement calls make them easy to locate (Zimmermann and Zimmermann 1988; Hödl and Amezquita 2001; Santos et al. 2014). Reproduction is largely terrestrial with egg masses deposited in leaf litter or phytotelmata (pools of water in leaf axils or bromeliads). Parental care is nearly ubiquitous in the group. Paternal care is the most common mode of parenting, but maternal and biparental types of care have also been reported (Summers et al. 2006). Nursing frogs usually guard and hydrate eggs, transport tadpoles, and in some species, provide unfertilized eggs as food to offspring reared in phytotelmata (Weygoldt 1987; Summers and McKeon 2004). Only recently it was discovered that these unfertilized eggs contain trace alkaloids, suggesting that some dendrobatid mothers may provide both food and chemical defense through parental care (Stynoski et al. 2014a, b).

Natural selection via predation may have driven the evolution of chemical defense in poison frogs. There are few reports of predation on aposematic dendrobatids, suggesting that their chemical defense is highly effective (Poulin et al. 2001; Santos and Cannatella 2011). Nevertheless, experimental evidence using clay model frogs suggests that avian predators are relatively important (Summers and Clough 2001; Saporito et al. 2007b; Noonan and Comeault 2009; Rojas et al. 2014) and visual perception models support that birds are likely to recognize brightly colored poison frog species as conspicuous (Maan and Cummings 2012). However, natural history anecdotes suggest a diversity of predators including birds (e.g., *Baryphthengus martii*; see Alvardo et al. 2013) and snakes (e.g., *Rhadinaea decorate*; see Lenger et al. 2014). In fact, most accounts refer to snake predation (69 % or 25/36 events), followed by spiders (17 % or 6/36), then birds (6 % or 2/35), and a few others (9 % including ants, fish, and crabs). Consequently, poison frogs might be under selection by multiple predators with diverse sensory biases and varied tolerances to dendrobatid defenses (Santos and Cannatella 2011). Alternatively, it is possible that aposematic species do not have specialized predators and that most predation events come from inexperienced individuals sampling aposematic frogs for the first time. Such learned avoidance is a critical prediction of the evolution and maintenance of aposematism (Speed and Ruxton 2005).

Dendrobatid defensive compounds appear to deter diverse organisms (Tables 21.1, 21.2 and 21.3). However, the toxic and repellent effects of these substances on predators and their possible accumulation at higher trophic levels are relatively unstudied. Dendrobatid species with significant quantities of toxic alkaloids, such as members of *Phyllobates*, *Dendrobates* sensu lato, *Ameerega*, and *Epipedobates* should have a large impact on their ecological communities. Organisms that consume these poison frogs may become toxic to their own predators, cascading the effects of alkaloid accumulation up the trophic chain. Likewise, poison frogs may impact lower trophic levels by altering the diversity and abundance of their arthropod prey because most

Table 21.1 Diversity of dendrobatid poison frog compounds, environmental source, and their defensive/antipredator properties

	Compounds[a]		Environmental source[a]		Number of species by compound predominance[b]			Defensive/antipredator properties[c]			
	N	*N* Unique[d]	Demonstrated	Proposed	Major	Minor	Trace	Tox	Unp	ATB	AF
Pyrrolidines	10	7	Ants|mites	–	4	0	1	–	R	A	A
Piperidines	26	23	Ants	–	1	5	3	T	R	A	A
5,8-Disubstituted indolizidines	63	27	Ants|mites	–	10	8	13	t	B*,R	A	–
5,6,8-Trisubstituted indolizidines	47	24	Mites	Ants	4	12	13	–	B*	–	–
3,5-Disubstituted indolizidines	15	7	Ants|mites	–	6	2	10	t	B*	–	–
1,4-Disubstituted quinolizidines	11	3	Mites	–	2	13	11	–	B*	–	–
4,6-Disubstituted quinolizidines	2	1	Ants|mites	–				–	B*	–	–
Epiquinamide	1	1	–	–	0	0	1	No	No	–	–
3,5-Disubstituted pyrrolizidines	13	4	Ants	–	1	6	9	–	B*	–	–
Lehmizidines	9	9	–	Ants	0	0	7	–	B*	–	–
Pumiliotoxins	31	7	Ants|mites	–	25	7	7	T	R	A	A
Allopumiliotoxins	18	12	Ants|mites	–				T	R	–	–
Deoxypumiliotoxins	5	2	Mites	Ants				T	–	–	–
Desmethylpumiliotoxins	1	1	–	Ants|mites				T	–	–	–
Homopumiliotoxins	4	1	Mites	Ants				T	–	–	–
Desmethylhomopumiliotoxins	1	0	–	Ants|mites				T	–	–	–
Deoxyhomopumiliotoxins	3	2	–	Ants|mites				T	–	–	–
Decahydroquinolines	40	28	Ants	–	14	12	7	t	B	A	–
Histrionicotoxins	16	16	Ants	–	20	10	2	t	B,R	A	–
Cyclopentaquinolizidines	9	9	–	–	1	0	1	–	B*	–	–
Spiropyrrolizidines	5	1	Mites|millipedes	–	3	2	1	–	R	–	–
Gephyrotoxins	2	2	–	Ants	0	0	1	t	–	–	–
Batrachotoxins	6	6	–	Beetles	3	0	2	T	B,R	–	–

(continued)

Table 21.1 (continued)

	Compounds[a]		Environmental source[a]		Number of species by compound predominance[b]			Defensive/antipredator properties[c]			
Epibatidines	4	4	–	Plant to arthropod	0	0	2	T	B	–	–
Pyridinic alkaloids[e]	3	2	–	Plant to arthropod	0	0	4	T	B,R	–	–
Indolic alkaloids[f]	2	2	–	Plant to arthropod	0	0	4	t	R	–	–
Tetrodotoxin	1	0	–	Bacteria|endogenous	1	0	0	T	–	–	–
Mercaptan-odor	1	–	–	Endogenous	0	0	1	–	R	–	–
N,N-diethyltoluamide	1	–	–	Moths	0	0	1	t	R	–	–
Other tricyclics	31	21	Mites|beetles	–	1	2	3	–	–	–	–
Unclassified alkaloids	144	115	–	–	–	–	–	–	–	–	–
Total	525	337									

[a]Alkaloid/compound diversity, quantity, and environmental sources for dendrobatid poison frogs derived from literature (Jones and Jacobson 1968; Daly et al. 1987, 1994a, b, 1999, 2005, 2009; Myers et al. 1991; Ujvary 1999; Saporito et al. 2003, 2007a, 2012; Dumbacher et al. 2004; Fitch et al. 2009, 2010; Mebs et al. 2010, 2014; Chau et al. 2011; Jones et al. 2012)

[b]The designation of major, minor, and trace indicates the amount of each type of substance found in frog skin (Daly et al. 1987). Major represents >50 μg, minor >5 μg, and trace <5 μg per 100 mg frog skin (Daly et al. 2009). We used this information to assign dendrobatid species to each predominance category

[c]Antipredator properties are mostly based on tests using mouse/rat models. We used these experiments to estimate toxicity (Tox): more toxic "T" or less toxic "t". Defensive properties are indicated by unpalatable (Unp): demonstrated as bitter "B", presumed bitter "B*", or repellent "R". Antibiotic (ATB) or antifungal (AF) properties are indicated by "A". All unknown states are indicated by "–"

[d]Number of unique alkaloids found in dendrobatid poison frogs is based on (Saporito et al. 2012)

[e]Nicotine, noranabasamine, pyridylnicotine

[f]Chimonanthine and calycanthine

Table 21.2 Effects of dendrobatid compounds on mammals and actions on muscular and neuronal receptors and channels

		Mouse model			Receptor/ion channels				
	Toxicity	LD50 ug/ mouse	Compound Tested	Effect	Compound Tested	nACh	Na	K	References
Piperidines	High	9.375	253 J (solenopsin)	Depressed cardiorespiratory function, seizures, and death	253 J 241D	Block[a]	–	–	Howell et al. (2005)
Pyrrolidines	–	–	–	–	197B	Block[a]	–	–	Daly et al. (1999)
3,5-Disubstituted indolizidines	Low	>200	239CD	Locomotor difficulties, piloerection, prostration with recovery > 4 h	223AB 239AB 239CD	Block[a]	–	–	Daly and Spande (1986), Daly et al. (1999)
Other indolizidines	–	–	–	–	205A, 207A 209B 235B′, 235B″	Block[a]	–	–	Daly et al. (1999), Tsuneki et al. (2004)
Spiropyrrolizidine	–	–	–	–	222, 236 238	Block[a]	–	–	Daly et al. (1999)
Coccinelline-like	–	–	–	–	205B[b]	Block[a]	–	–	Daly et al. (1999)
Epiquinamide	None	–	196	Inactive	196	None	None	None	Fitch et al. (2009)

(continued)

Table 21.2 (continued)

		Mouse model			Receptor/ion channels				
	Toxicity	LD50 ug/ mouse	Compound Tested	Effect	Compound Tested	nACh	Na	K	References
Pumiliotoxins	Medium to high	200	251D[c]	Pain; hyperactivity; convulsions; cardiac depressant; death < 6 min	251D	–	Block	Block Alter deactivation	Daly and Spande (1986), Daly et al. (1999)
		20–50	307A (PTX A) 323A (PTX B)	Locomotor difficulties; paralysis; convulsions; death < 20 min; Potentiates/prolongs muscle contraction; repetitive neuronal firing	307A 323A	–	Prevent inactivation	–	Daly et al. (1990), Vandendriessche et al. (2008)
Allopumiliotoxins	High	50[d]	323B 339A 339B	Potentiate and prolong muscle contraction; cardiotonic	323B 339A 339B	–	Prevent inactivation	–	Daly and Spande (1986), Daly et al. (1999)
		40	267A	Pain; hyperactivity; locomotor difficulties; death in < 6 min[e]		–	–	–	
Decahy-droquinolines	Low	>400	219A	Locomotor difficulties at >125 mg/kg; convulsions and death in 10 min	195A 219A	Block[a]	Block	Block	Daly and Spande (1986), Daly et al. (1999)

Histrionicotoxins	Low	>1000	283A 285A	Locomotor difficulties; prostration	283A 285C 285B 287A 287B 291A	Block[a]	Block	Block	Daly and Spande (1986), Daly et al. (1999)
Gephyrotoxins	Low	>500	287C 289B	Weak muscarinic antagonist (e.g., cognitive impairment)	287C 289B	Block[a,f]	Block[g]	Block (at high levels)	Daly and Spande (1986), Daly et al. (1999)
Batrachotoxins	Very high	0.1	BTX h-BTX BTX A	Arrhythmia; cardiac arrest; anesthesia; death; depolarization of muscle and neuronal membranes	BTX h-BTX BTX A	–	Prevent inactivation, alter Activation	Prevent inactivation[g]	Daly and Spande (1986)
Pyridine alkaloids	High to very high	66.8	162 (nicotine)	Cardiac depression; paralysis/ convulsion; death	162 222/224	Agonist Agonist	–	–	Lazutka et al. (1969), Daly et al. (1999), Fitch et al. (2010)
		0.4	208/210 (epibatidine)	Analgesia; locomotor depression; reduced body temperature; changes in blood pressure; bradycardia; death	208/210	Agonist	–	–	
Indole alkaloids	Low	880	346C (calycanthine)	Cardiac depression, paralysis, and convulsion	346C[h]	–	–	–	Chebib et al. (2003)

(continued)

Table 21.2 (continued)

		Mouse model			Receptor/ion channels				
	Toxicity	LD50 ug/ mouse	Compound Tested	Effect	Compound Tested	nACh	Na	K	References
Tetrodotoxin	Very high	0.1–0.214	TTX	Respiratory failure; death	TTX	–	Block	–	Daly and Spande (1986), Bane et al. (2014)
N,N-diethyltoluamide	Low	>2000	DEET	Irritant; neurological damage	DEET[i]	–	–	–	Sudakin and Trevathan (2003)

[a]Non-competitive blocker
[b]Results are from a test of a synthetic enantiomer
[c]PTX 251D is more toxic for insects and less toxic for mammals than PTX A or PTX B
[d]Tests were done with pig atria, so lethal dose is extrapolated from other studies of comparably toxic compounds (i.e., PTX A)—see Daly and Spande (1986)
[e]Daly et al. (2003) state that aPTX 267A is toxic to mice while Daly and Spande (1986) state that it is not
[f]May also block glutamatergic NMDA receptors
[g]Effect speculated by cited authors
[h]Blocks calcium channels, GABA receptors
[i]Inhibits cholinesterases

Table 21.3 Effects of dendrobatid compounds on insects, bacteria, and fungi

	Arthropods				Bacteria/fungi			
	Compound Tested	mM	Organisms	Effect	Compound Tested	μg	Effect	References
Pyrrolidines	–	–	Ants Termites	Repellent	Synthetic	10–30	Antibiotic antifungal	Jones et al. (1989), Daly et al. (1999), Macfoy et al. (2005)
Piperidines	253 J (solenopsin)	11–33	*Solenopsis invicta*	Reduced ambulation; deterrent	253 J Synthetic	10–50	Antibiotic antifungal	Daly et al. (2005), Macfoy et al. (2005), Weldon et al. (2013)
5,8-Disubstituted indolizidines	205A 235B	33–100	*S. invicta*	Reduced ambulation	235B′	100	Antibiotic	Macfoy et al. (2005), Weldon et al. (2013)
3,5-Disubstituted indolizidines	223AB	100	*S. invicta*	None	239AB	50	Inactive	Macfoy et al. (2005), Weldon et al. (2013)
Pumiliotoxins	267C 323A 251D 307A	0.001–33	*S. invicta* *Heliothis virescens* *Aedes aegypti*	Convulsions; reduced ambulation	251D 307A Synthetic	50 50 30–50	None or Antibiotic antifungal	Bargar et al. (1995), Macfoy et al. (2005), Weldon et al. (2006, 2013)
	237A	100	*S. invicta*	None	–	–	–	Weldon et al. (2013)
Allopumiliotoxins	267A	0.01–0.04	*S. invicta*	Reduced ambulation	–	–	–	Weldon et al. (2013)
Decahydroquinolines	cis-223 F	100	*S. invicta*	None	trans-243A	100	Antibiotic	Macfoy et al. (2005), Weldon et al. (2013)
Histrionicotoxins	291A 259A 285A	0.33–100	*S. invicta*	Convulsions; reduced ambulation	Synthetic	100	Antibiotic	Macfoy et al. (2005), Weldon et al. (2013)
Spiropyrrolizidines	236	100	*S. invicta*	None	236	200	None	Macfoy et al. (2005), Weldon et al. (2013)

(continued)

Table 21.3 (continued)

	Arthropods				Bacteria/fungi			
	Compound Tested	mM	Organisms	Effect	Compound Tested	μg	Effect	References
	151B (Polyzonamine)	0.1–10	*Formica exsectoides Periplaneta americana Phormia regina*	Repellent; irritant	–	–	–	Smolanoff et al. (1975), Ujvary (1999)
Gephyrotoxins	287C	100	*S. invicta*	None	–	–	–	Weldon et al. (2013)
Batrachotoxins	BTX	33–100	*S. invicta*	Convulsions; reduced ambulation	BTX	20	None	Macfoy et al. (2005), Weldon et al. (2013)
	BTX-A	100	*S. invicta*	None	BTX-A	50	None	Macfoy et al. (2005), Weldon et al. (2013)
	hBTX	–	Chewing lice (Phthiraptera)	Reduced life span	–	–	–	Dumbacher (1999)
Pyridinic alkaloids	162 (Nicotine)	0.33–1	*S. invicta*	Reduced ambulation; distasteful	Synthetic	50–200	None	Macfoy et al. (2005), Weldon et al. (2013)
Tetrodotoxin	TTX	–	*Musca domestica*	High affinity for Na+ channels; likely insecticide	–	–	–	Pauron et al. (1985), How et al. (2003), Gordon et al. (2007)
N,N-diethyltoluamide	191C (DEET)	0.48–10	*A. aegypti*	Repellent	–	–	–	Jones and Jacobson (1968), Bohbot and Dickens (2010)

defensive substances sequestered by these frogs come from specialized diets (Daly et al. 1994a; Caldwell 1996). More ecological studies will reveal the impact of toxic dendrobatids on the ecological dynamics of their communities.

21.2.1 Diet and Defense in Dendrobatidae

There is a tight association between diet and defense within Dendrobatidae (Daly 1998). In the wild, dendrobatids feed continuously during the day and actively defend territories that are likely associated with food and chemical defense resources (Caldwell 1996; Pröhl 2005). Diet specialization on alkaloid-bearing arthropods such as ants and mites is phylogenetically correlated with origins of chemical defense (Darst et al. 2005); both have evolved in parallel in at least three clades of aposematic dendrobatids (Fig. 21.1; *Epipedobates*, *Ameerega*, and *Dendrobates*). These lineages include diet specialists that have morphological and biomechanical adaptations that allow them to consume large quantities of diminutive prey (i.e., microphagy), including changes in tongue shape and use (elongated, narrow, fast and shooting), reduction of teeth, and compaction of cranial shape (Toft and Duellman 1979; Toft 1980, 1981, 1995; Emerson 1985; Lieberman 1986; Donnelly 1991; Simon and Toft 1991; Vences et al. 1998).

Most dendrobatid alkaloids appear to have a dietary origin. Identical alkaloids have been found in leaf litter arthropods (e.g., ants, mites, and millipedes) and dendrobatids; moreover, wild-caught frogs kept in captivity show a marked reduction in alkaloid diversity and quantity (Daly et al. 1992, 2000, 2002; Saporito et al. 2003, 2004; Jones et al. 2012). However, a few chemicals found in dendrobatids may have another source (Tables 21.1, 21.2 and 21.3). For example, one cryptic dendrobatid *Colostethus panamansis* has tetrodotoxin (TTX) (Daly et al. 1994b). The mechanisms of acquisition, transport, and storage of TTX by this frog are unknown, but in other TTX-defended systems this substance is usually produced by endosymbiont bacteria or sequestered following the consumption of other TTX-defended prey (Daly et al. 1997; Chau et al. 2011; Wood et al. 2012; Bane et al. 2014). Another interesting case is *Aromobates nocturnus*, a nocturnal species that releases a pungent (mercaptan) defensive odor when handled (Myers et al. 1991). Unfortunately, individuals of this species have not been found for at least 20 years (Barrio-Amoros et al. 2011). However, *Aromobates* relatives are common at lower elevations (~2000 m) in geographic regions close to the known distribution of *A. nocturnus* in the Merida Andes in Venezuela. Interestingly, some of these sympatric *Aromobates* (e.g., *A. saltuensis*) release a similar mercaptan odor when manipulated (Barrio-Amoros and Santos 2012; JCS *pers. obs.*).

The ultimate origin of dendrobatid alkaloids is a topic of continuous research, although many are found in arthropods, including ants (formicine and myrmicine), coccinellid beetles, siphonotid millipedes, and oribatid mites (Table 21.1; see also Saporito et al. 2009). Some alkaloids might be produced by plants, which are consumed by arthropods and then taken up by poison frogs (Daly et al. 1999; Saporito et al. 2012). For example, the chimonanthine and calycanthine alkaloids found in

Phyllobates terribilis (Daly et al. 1999) might originate from syntopic *Psychotria* (Rubiaceae) plants (Verotta et al. 1998). Several genera of ants are attracted to and consume *Psychotria* including *Solenopsis* (*Diplorhoptrum*) sp., which is an ant genus found in stomachs of several poison frog species (Born et al. 2010; Arce-Dominguez and Rengifo-Mosquera 2013; Bieber et al. 2013). Alternatively, evidence also supports that ants (and to a lesser extent, mites) are in fact producing alkaloids, rather than acting as intermediaries (Jones et al. 2012; Saporito et al. 2011). Recent observations (LAO unpublished data) of the diet of *D.* (*Oophaga*) *sylvaticus* show a prevalence of several species of fungus-growing ants, which are known to be in close association with diverse bacteria and fungi symbionts (Mueller et al. 1998; Currie et al. 1999). Hence, some dendrobatid alkaloids may actually originate in plants and microbes, but so far they have only been traced to their arthropod diet.

21.2.2 Toxicity and Unpalatability

Poison frog alkaloids can be toxic (i.e., causing damage to the consumer), unpalatable (i.e., distasteful or repellent to predators), or both. Among the first to notice the toxicity of dendrobatids were the Native Americans of the Chocoan Emberá and Noanamá tribes in Western Colombia (Myers et al. 1978). Several anthropologists, ethnologists, and zoologists have described in detail how these indigenous people use skin extracts of *Phyllobates* frogs to poison blowgun darts to hunt large game (Cochrane 1825; Posada-Arango 1883; Wassen 1935; Marki and Witkop 1963). These observations attracted the attention of biochemists, who were interested in understanding the basis of poison frog toxicity. Early work focused on the physiological effects of dendrobatid alkaloids in model organisms such as mice, rats, and frogs (Daly and Myers 1967; Daly and Spande 1986; Daly et al. 1999). Thereafter, the rapid development of gas chromatography–mass spectrometry stimulated intense work on alkaloid structure, classification, synthesis, and mechanism of action (Daly et al. 1978, 1987, 1999; Saporito et al. 2012). A summary of the toxic and unpalatable effects of these dendrobatid compounds is provided in Tables 21.1, 21.2 and 21.3.

An early method used to detect chemical defense in anurans was tasting (e.g., licking) their skin secretions (Neuwirth et al. 1979). Species rich in alkaloids (e.g., *Dendrobates* sensu lato) were usually described as bitter and causing burning and numbing sensations (Myers and Daly 1976; Neuwirth et al. 1979). These observations led researchers to propose that some dendrobatid chemical defenses may not actually be lethal (Daly et al. 2005). Certain classes of dendrobatid compounds with low toxicity instead appear to be distasteful to a broad diversity of predators, even at small quantities (Daly et al. 2005). Such repellents may allow potential predators to rapidly develop an aversion to consuming poison frogs because they will recall their unpleasant taste (Darst and Cummings 2006; Bassoli et al. 2007). Among these unpalatable compounds are histrionicotoxins (HTX) and decahydroquinolines (DHQ), which have relatively low toxicity in mammals (Daly and Spande 1986), and several types of izidines (e.g., indolizidines, pyrrolizidines, and quinolizidines), which are considered bitter or unpalatable (Table 21.1). However, aside from their unpleasant taste, some HTXs, DHQs, and izidines are surprisingly toxic to arthropods that might parasitize poison frogs (Table 21.3). Interestingly, these substances also appear to have antibiotic and antifungal properties although this should be further investigated (Macfoy et al. 2005).

The evolutionary significance of toxicity versus unpalatability in dendrobatids is a topic of continuous research. To frame this hypothesis phylogenetically, we determined relative measurements of both properties based on how predominant each compound class is among the genera of poison frogs (Fig. 21.1). However, we emphasize that many of the species that potentially have defensive compounds have not been analyzed and some surveys show significant variation among species and even individuals (Daly et al. 1978, 1992; Saporito et al. 2007a). Given these caveats, we summarize the following observations. Most species within *Dendrobates* sensu lato are both toxic and unpalatable as many of their members have PTX/aPTXs (toxic) and DHQ, HTX, and izidines (unpalatable) as predominant alkaloid classes (i.e., >50 μg/100 mg of frog skin). Most notable in this group is the subgenus *Oophaga*, >90 % species of which have PTX/aPTX as its major alkaloid type. *Phyllobates* and *Epipedobates*, on the other hand, are toxic with predominance of BTX, PTX 251D and epibatidine alkaloids respectively, but they are relatively poor in unpalatable compounds such as DHQs, HTXs, and izidines. *Ameerega* is mostly unpalatable and less toxic, with a predominance of DHQ and HTX alkaloids. Insect repellent compounds have also been isolated from *Ameerega* species, including DEET (N,N-diethyltoluamide) and polyzonamine, which might function against a broad spectrum of ectoparasites.

21.3 Acquiring Chemical Defenses: A Pharmacokinetics Perspective

Although very little is known about the physiological mechanisms of toxin sequestration and chemical modification in poison frogs, recent technological advances have opened the possible exploration of these unique biological mechanisms like never before. The question of how poisonous amphibians have evolved physiological mechanisms to sequester dietary chemical defenses is essentially a study in pharmacokinetics. Here we discuss current and future directions in advancing our understanding of the mechanistic basis of toxin acquisition and storage in amphibians. Approaching this topic of toxin physiology will require pursuing questions within an integrative framework and combining methods in genomics, proteomics, and pharmacology.

A dendrobatid's ability to sequester alkaloids is genetic, as captive-raised dendrobatids fed with alkaloid-dusted fruit flies are able to uptake, modify, and accumulate most alkaloids (Daly et al. 1994a, 2003; Saporito et al. 2009). Dermal granular glands are the main alkaloid storage organs and are responsible for their release, but some reports also suggest that small traces of alkaloids are present in other tissues including liver, muscle, and oocytes (Neuwirth et al. 1979; Delfino et al. 2010; Saporito et al. 2010b, 2012; Prates et al. 2012; Stynoski et al. 2014a). The number and size of granular skin glands increases allometrically with body size, allowing larger and older frogs to accumulate more alkaloids than juveniles (Saporito et al. 2010b). However, the physiological mechanisms of alkaloid sequestration are topics of ongoing research.

As poison frogs acquire chemical defenses through their diet (Saporito et al. 2012), the first candidate tissue to uptake small molecule lipophilic alkaloids is the intestine. The gut has long been known to play a major role in the oral bioavailability of compounds in the context of orally prescribed drug absorption in humans (Zhang and Benet 2001). Lipophilic compounds, like alkaloid toxins found on many poisonous amphibians, undergo passive absorption in the gut, but that does not imply that lipophilic compounds pass through the gut into the blood unhindered. In many organisms, the first line of defense against ingested toxic substances is a series of membranes between the gut epithelium and the blood. Within these barriers are highly expressed protein families that interact with dietary compounds, including the ABC (ATP-binding cassette) transporters that promiscuously bind compounds for transport either back into the lumen for excretion or into the blood for circulation, and the cytochrome p450 family of enzymes that metabolize compounds (Chan et al. 2004). Some ABC efflux proteins are known to transport alkaloid compounds that are used in cancer treatments (Chan et al. 2004), and it is likely they transport other lipophilic alkaloids as well, making them good candidates for sequestration of alkaloid toxins in amphibians. It is currently unknown if dietary alkaloids that successfully surpass these filtering mechanisms are sequestered in the skin (rather than being excreted) or if poisonous amphibians have altered expression patterns of these protein families in the gut to prevent loss and favor uptake of dietary alkaloids.

Much of the circulating blood is filtered through the liver, an environment rich in enzymes that metabolize and neutralize potentially toxic substances. The cytochrome p450 superfamily of enzymes is highly abundant in the intestines and especially in the liver, and is best known for metabolism of drugs through oxidation (Danielson 2002). Although many of the frog alkaloid toxins are sequestered unchanged through the diet, there is one example of a compound that is chemically modified by frogs. Some species of South American frogs can stereoselectively hydroxylize PTX (+)-251D to aPTX (+)-267A, which is roughly fivefold more toxic than its precursor (Daly et al. 2003). Although only a few species in the Dendrobatidae clade were tested, only frogs in the *Dendrobates* genus were able to metabolize pumiliotoxin into allopumiliotoxin, whereas this chemical modification was not observed in species in the genus *Phyllobates* or in the more distant *Epipedobates*. As the conversion from PTX (+)-251D to aPTX (+)-267A involves a 7-hydroxylation, obvious candidate enzymes for this metabolism are members of the cytochrome p450 family. The question of how dendrobatids can accomplish this conversion, but *Phyllobates* or *Epipedobates* cannot, remains to be determined. With next-generation sequencing technologies becoming common for nontraditional model species, a molecular evolution study comparing the cytochrome p450 family would be a step forward towards understanding these differences in toxin metabolism.

As alkaloid toxins are sequestered in various tissues, they must eventually be bound and transported in the blood, passing through various tissues to eventually accumulate in the skin; however the identity of such alkaloid transporters in poison frogs is unknown. There is little known about proteins that bind neurotoxins in general, with the exception of saxitoxin, a shellfish alkaloid neurotoxin that targets sodium channels in a manner similar to tetrodotoxin (Terlau et al. 1991). In plasma, saxitoxin is bound by saxiphilin, a unique member of the transferrin family of Fe^{3+}-binding proteins that has evolved to bind alkaloid toxins (Morabito and Moczydlowski 1994). Saxiphilin was initially discovered in the plasma of the North American bullfrog (*Rana catesbeiana*) (Mahar et al. 1991), but it was then determined that this protein is highly conserved throughout animals as a mechanism to bioaccumulate shellfish toxins encountered in the environment away from the animal's nervous system (Llewellyn et al. 1997). It is possible that a similar mechanism has evolved in poison frogs, with neofunctionalization (e.g., a paralog gene that takes on a totally new function) of transport carrier proteins to bind and transport alkaloids in plasma en route to the skin granular glands where toxins are stored until secretion (Neuwirth et al. 1979). Once more genetic information becomes available for poison frogs, molecular evolution methods could be employed to determine candidate genes that may play a similar role as saxiphilin in the binding and transport of alkaloid toxins compared to nontoxic amphibians.

Well-studied examples of toxin sequestration and storage mechanisms and their fitness consequences are insect herbivores that specialize on alkaloid-producing plants. In many of these cases, the host-plant derived alkaloids are pyrrolizidines coopted for defense by some moths, beetles, and other insects (see Opitz and Muller 2009 for detailed review). Some plants synthesize these pyrrolizidines in a non-toxic N-oxide form, which is unstable and easily reduced to a toxic form in the insect gut by cytochrome p450s (Hartmann et al. 1997, 1999; Narberhaus et al. 2005). Moths adapted to specialize on pyrrolizidine-producing plants have developed a monooxygenase that reoxidizes the toxic alkaloid into the N-oxide form where it is stored in tissues inaccessible to cytochrome p450s (Lindigkeit et al. 1997; Naumann et al. 2002). Interestingly, controlled feeding experiments with labeled alkaloids have shown beetles that specialize on pyrrolizidine-producing plants have a specific membrane carrier for the pyrrolizidine alkaloid senecionine (Narberhaus et al. 2004), although the exact protein has yet to be identified. In some arctiid insects, many of these alkaloids sequestered from the diet are stored and repurposed for critical roles in reproductive behavior (Conner et al. 1981, 2000). Females store alkaloids in eggs (Dussourd et al. 1988), which has recently been shown to occur in poison frogs (Stynoski et al. 2014a). Similarly, male arctiid insects incorporate alkaloids into spermatophores (Dussourd et al. 1988) and/or convert them into pheromones (Hartmann and Witte 1995; Hartmann et al. 2003). Females are highly attracted to these alkaloid-derived pheromones and prefer males with higher alkaloid content (Bogner and Boppre 1989), potentially selecting for more efficient sequestration mechanisms (Eisner and Meinwald 1995; Opitz and Muller 2009). These studies in insects can inform new directions for researchers interested in poison frog sequestration mechanisms, such as specific membrane transporters for alkaloids or how sequestration of alkaloids may play a role in poison frog reproduction and mate choice.

An example of the importance of considering pharmacokinetics in the study of frog chemical ecology is the strawberry dart frog [*Dendrobates* (*Oophaga*) *pumilio*]. In this species, females have a greater quantity and diversity of alkaloids present in the skin than males (Saporito et al. 2010a). Although such differences tend to be explained by ecologists as purely environmental (i.e., males and females may have different foraging preferences, although diet was not examined in the above study), there is evidence in humans that there are many gender effects in pharmacokinetics of orally administered compounds, especially involving drug binding and metabolism (Harris et al. 1995). It is possible then that there are also sex differences in absorption and/or metabolism of small lipophilic alkaloids in anurans and the relative contributions of diet and physiology need to be further dissected.

Although great and insightful work in chemical ecology of poisonous amphibians has mostly focused on descriptive ecological contributions (Saporito et al. 2012), the emergence of new high-throughput technologies has opened the doors to understanding the chemical ecology of poisonous amphibians in a mechanistic way. For example, RNA sequencing could be applied to dietary studies in poisonous frogs to determine how gene expression changes across various tissues with a toxic or non-toxic diet. Whole transcriptome sequencing for many species is now possible and creates many opportunities to examine sequence variation in protein families that may be involved in toxin sequestration and metabolism. Moreover, the ease of sequencing a transcriptome now makes high-throughout proteomics possible, enabling the identification of proteins that bind alkaloids, which would have been extremely difficult five to ten years ago. We predict that harnessing these technologies will lead to rapid and informative advances in the field of amphibian chemical ecology, moving the field past correlative ecology and into a deeper mechanistic understanding of how amphibians have evolved these unique traits. It is even more remarkable that toxin sequestration in anurans has independently evolved several times (Santos et al. 2003), and in similar ways (i.e., to sequester the same families of small molecule compounds from the diet). Once more is known about the physiological mechanisms of toxin sequestration in a model poison frog family, one can begin to ask if the convergent evolution of toxin sequestration among amphibians involves a convergence in underlying mechanism or entirely different molecular pathways.

The physiological adaptation of sequestering a large variety of small molecule alkaloids in poison frogs is distinctive among complex biological systems and represents a unique opportunity to investigate binding properties of protein systems. Poison frogs have presumably evolved a set of proteins that bind small molecule alkaloids with some selectivity and yet have promiscuous enough binding properties to bind a range of alkaloids. The evolution and kinetics of this yet-to-be-discovered protein family promises not only rich insights into the evolution of toxicity in amphibians but also general insights into protein systems that bind toxins targeting the nervous system. Such studies focusing on basic mechanisms could lead to more translational research of therapeutics for human pathologies and chemical defense strategies.

21.4 Autoresistance as a Component of Chemical Defense

Chemical defense is a complex phenotype that involves a suite of ecological, morphological, physiological, and genetic changes over time (Härlin and Härlin 2003; Santos and Cannatella 2011). One often-overlooked component is autoresistance—how do chemically defended organisms resist self-intoxication, and how does resistance evolve? Long-term exposure to low levels of toxins may select for some degree of toxin resistance in organisms without any chemical defense (Hua et al. 2013). In species with diet-derived defense, both chemical sequestration and diet specialization increase toxin exposure, which in turn would select for greater resistance (Dobler et al. 2011). Hence, autoresistance in aposematic organisms may evolve as a combination of pre-existing resistance and resistance that evolves in parallel with chemical defense. In this section, we describe the evolution of toxin resistance in a number of taxa and explore the potential complexity of this phenotype in dendrobatid poison frogs. These amphibians are an ideal study system for the evolution of autoresistance because within this clade chemical defense has evolved independently at least four times at various evolutionary timescales (Fig. 21.1).

21.4.1 Tetrodotoxin as an Example of the Evolution of Toxin Resistance

One of the best-studied naturally occurring toxins is tetrodotoxin (TTX). This neurotoxic alkaloid is found in many organisms including pufferfish, frogs, newts, molluscs, crabs and the blue-ringed octopus, among other animals (Noguchi et al. 1984, 1986; Tsuruda et al. 2002; Soong and Venkatesh 2006; Hwang et al. 2007). In anurans, TTX is found in *Brachycephalus*, *Polypedates*, *Atelopus*, and the dendrobatid *Colostethus panamansis* (Daly et al. 1994b; Tanu et al. 2001; Pires et al. 2005). The toxicity of TTX is well characterized: it binds to and blocks voltage-gated sodium channels (VGSCs), cell membrane proteins encoded by the Nav1 gene family that mediate neural communication and muscle contraction (Wang and Wang 1998; Cestele and Catterall 2000). The Nav1 genes are functionally important, evident by the maintenance of gene duplications and their subfunctionalization in a number of lineages (six genes in amphibians, eight in teleosts, and ten in mammals) and the extremely high conservation of amino acid (AA) sequences in these genes across animals (Zakon 2012). By disrupting VGSCs, TTX causes paralysis or respiratory failure; at low doses it slows muscle reaction and diminishes sensory input (Soong and Venkatesh 2006). However, AA substitutions in VGSCs at the TTX binding site prevent TTX from binding; hence, TTX resistance can be traced to specific AA substitutions in Nav1 genes. Alternatively, TTX resistance could potentially be conferred by compartmentalization or metabolic inactivation of TTX (Saporito et al. 2012). However, there is overwhelming support that high levels of TTX resistance is conferred by nonsynonymous (i.e., AA changing) substitutions in Nav1 genes.

VGSCs have tissue-specific expression and TTX can cross cell membranes, so whole-body resistance to TTX likely requires AA substitutions in all Nav1 paralogs. In pufferfish, TTX resistance has independently evolved in four species (*Takifugu rubripes*, *Tetraodon nigroviridis*, *Canthigaster solandri*, *Arothron nigropunctatus*) via convergent AA substitutions in all eight Nav1 paralogs (Jost et al. 2008). However, in *Thamnophis sirtalis*, a snake that consumes TTX-defended newts (*Taricha*), Nav1 genes expressed in the peripheral nervous system have TTX-resistant AA substitutions, but those expressed in the central nervous system do not, suggesting that selection for whole-body TTX resistance is less in garter snakes than in pufferfish (McGlothlin et al. 2014). TTX-defended newts (*Cynops pyrrhogaster*, *Taricha granulosa*) and other species of snakes (*T. couchii*, *T. sierrae*, *T. atratus*, *Liophis epinephelus*, *Amphiesma pryeri*, *Rhabdophis tigrinus*) that consume TTX-defended amphibians (*Taricha*, *Cynops*, *Polypedates*, *Atelopus*) have convergent AA substitutions in the skeletal muscle VGSC Nav1.4 (Kaneko et al. 1997; Hanifin et al. 1999; Feldman et al. 2009, 2012). However, other Nav1 paralogs in these species have not yet been sequenced, so patterns of whole-body resistance cannot be assessed across taxa.

Other defensive alkaloids and steroids affect ion channels and transporters in a similar way as TTX and resistance can also be traced via nonsynonymous AA substitutions in target proteins. Many organisms are exposed to low levels of these compounds, so pre-existing resistance may sometimes play a role in the evolution of chemical defense. For example, in milkweed butterflies (Nymphalidae, Danaini) and leaf beetles (Chrysomelidae, *Chrysochus*), some resistance to cardenolides via one AA substitution in the sodium-potassium pump evolved prior to diet specialization on and sequestration from cardenolide-rich plants (Dobler 2001; Dobler et al. 2011; Aardema et al. 2012). In diet specialized and toxic lineages, three additional substitutions conferring greater levels of resistance are found (Petschenka et al. 2013a). Similarly, in salamanders, one TTX-resistance-conferring AA substitution in the Nav1.4 pore evolved prior to the origin of chemical defense in this clade, followed by the accumulation of four other AA substitutions that provide higher levels of resistance in the toxic lineages (Hanifin and Gilly 2015). Pre-existing resistance may have also contributed to the multiple recurrent origins of TTX-defended newt consumption by garter snakes and of bufadienolide-defended toads by natricine snakes (Savitzky et al. 2012; McGlothlin et al. 2014). Resistance to plant secondary compounds in insects other than monarchs and beetles has arisen many times, but whether or not some of this resistance evolved prior to chemical defense remains largely uninvestigated (Dobler et al. 2011).

21.4.2 Evolution of Alkaloid Resistance in Dendrobatidae

Chemically defended poison frogs sequester alkaloids from arthropod prey, deposit them in granular dermal glands, and release them as defense (Neuwirth et al. 1979; Daly et al. 1994c; Saporito et al. 2009). These substances are generally unpalatable; some are toxic and highly effective deterrents against diverse predators and parasites (Tables 21.1, 21.2 and 21.3). Like TTX, dendrobatid alkaloids affect the function of ion channels and transporters causing neuromuscular disruption in sensitive organisms (Karalliedde 1995; Daly et al. 2003; Vandendriessche et al. 2008; Petschenka et al. 2013a). Although resistance could be conferred by metabolic inactivation (see Petschenka et al. 2013b), most alkaloids are taken up and released without any sort of modification (Daly et al. 2003), suggesting the absence of such a detoxification process. Hence, poison frog autoresistance should be conferred by key AA substitutions in target proteins like in other chemically defended systems (Wang et al. 2006; Jost et al. 2008; Dobler et al. 2011; Feldman et al. 2012; Zakon 2012; Petschenka et al. 2013a). We propose that this is the major mechanism of autoresistance in dendrobatids.

Dendrobatids have an outstanding diversity of alkaloids (Table 21.1; ~520 types), with substantial variation across species, populations, and individuals (Clark et al. 2005; Daly et al. 2005; Darst and Cummings 2006; Saporito et al. 2007a). The sheer variation of alkaloid profiles among species is perhaps the most predictable aspect of poison frog defense. This makes dendrobatid clades difficult to categorize by their alkaloid profiles (Fig. 21.1). The only two types of alkaloids that appear to have a synapomorphic (unique to a single clade) distribution are batrachotoxin (BTX) in *Phyllobates* and epibatidines in *Epipedobates* (*Saporito* et al. 2009). However, the large diversity of alkaloid types within a single species of dendrobatids presents a unique problem for the evolution of autoresistance. This particularity of dendrobatid chemical ecology clearly contrasts with other chemical defense systems that predominately involve one toxin that targets only one type of ion channel.

We propose that because most, if not all, dendrobatids consume some amount of alkaloid-rich prey (Table 21.1: ants and mites) there must exist some common alkaloid resistance that evolved prior to and facilitated the four origins of chemical defense in the group (Fig. 21.1). The significant correlation between diet specialization on alkaloid-bearing arthropods and the evolution of chemical defense in dendrobatids highlights the evolutionary ties between alkaloid exposure, sequestration, and presumed resistance (Darst et al. 2005). Increasing exposure to alkaloids via diet specialization likely resulted in selection for increased resistance, closely timed with the evolution of chemical sequestration and defense (Saporito et al. 2012; Savitzky et al. 2012). From these patterns, two major questions arise: (1) how have dendrobatids responded to selection for diverse alkaloid resistance; and (2) what role has alkaloid resistance played in the evolution of dendrobatids?

21.4.3 How Have Dendrobatids Responded to Selection for Diverse Alkaloid Resistance?

Despite numerous studies of TTX resistance in multiple organisms, there is almost no information regarding autoresistance in poison frogs, perhaps because of its inherent complexity. Dendrobatid alkaloids target at least five families of ion channel genes (>30 genes) and bind to different regions of the channels, diversely affecting their function (Table 21.2). Moreover, many ion channel proteins are composed of multiple subunits, so one ion channel complex may be encoded by more than four distinct genes or isoforms (Hille 2001). Most dendrobatid alkaloids are lipophilic, meaning that they cross cell membranes and may permeate the blood–brain barrier. Hence, all expressed channels targeted by poison frog alkaloids should be subject to selection for resistance, even if their expression is limited to specific tissues. However, there may be variation in the level of selection on genes with tissue-specific expression correlated with alkaloid concentrations of each tissue, as is seen in garter snakes (McGlothlin et al. 2014). Moreover, the role of ion channel isoforms in autoresistance across alkaloid-defended taxa is virtually unstudied. Such attributes make in vitro studies of ion channel/receptor proteins challenging and imply that the genetic basis of autoresistance in Dendrobatidae is rather complex.

Most studies of dendrobatid alkaloids were performed in the 1970–1980s using mice or frog nerve and muscle preps (Daly and Spande 1986; Daly et al. 1999). These assays elucidate physiological effects of alkaloids but do not always reveal specific target channels and binding sites; for these, in vitro expression assays are required (e.g., Vandendriessche et al. 2008). Experiments of alkaloid resistance using dendrobatid tissues are limited to *Dendrobates* (*Oophaga*) *histrionicus*, *Phyllobates terribilis*, and *P. aurotaenia* (Albuquerque et al. 1973, 1974; Daly et al. 1980). These assays demonstrated that *Phyllobates* frogs are physiologically resistant to batrachotoxin (BTX) while *Rana pipiens* (not defended) and *D. histrionicus* (alkaloid defended but not by BTX) were not. For HTX toxicity, *D. histrionicus* were more resistant than *R. pipiens*, although at high concentrations, *D. histrionicus* were not completely resistant; *Phyllobates* were not tested. These authors also showed through a breeding experiment that the basis for alkaloid resistance in *P. terribilis* is genetic. We now know that BTX interacts with the inner pore of voltage-gated sodium channels and that BTX resistance is conferred by amino acid substitutions in these regions (Wang and Wang 1998, 2003; Wang et al. 2006). However, no published studies have investigated the genetics of alkaloid resistance in dendrobatids, although data suggest that various species of *Phyllobates*, *Dendrobates*, *Epipedobates*, and *Ameerega* have multiple AA substitutions in the inner pore of Nav1.4, of which at least two appear to be unique to *Phyllobates* and may provide BTX resistance (Frezza et al. 2010; Marquez and Amezquita 2014; Tarvin et al. 2014). No other information of sensitivity to BTX, PTX, or other compounds is available. Much more is to be learned from studying these channels across dendrobatids.

21.4.4 What Role Has Alkaloid Resistance Played in the Evolution of Dendrobatids?

Evolutionary patterns of poison frog alkaloid resistance likely influenced the evolutionary trajectories of alkaloid defense in dendrobatids. If some resistance is ancestral to Dendrobatidae (Darst et al. 2005; Savitzky et al. 2012), this exaptation (pre-adaptation) may have facilitated early stages of alkaloid sequestration and dietary specialization by diminishing the negative effects of increased alkaloid exposure. Therefore, phylogenetic patterns of autoresistance in ion channel gene families across the poison frog phylogeny should correlate with alkaloid profiles and toxicity of each lineage, reflecting their co-evolutionary history.

Alkaloid profiles are dependent on available arthropods, by the ability to sequester alkaloids, and variation in sequestration rates of the different compounds across species (Daly et al. 1994c; Clark et al. 2005; Saporito et al. 2007a, 2009). Similar alkaloid profiles in other defended anuran groups (i.e., Mantellidae: *Mantella*, Myobatrachidae: *Pseudophryne*, Bufonidae: *Melanophryniscus*, Eleutherodactylidae: *Eleutherodactylus*) suggest a common metabolic pattern of chemical sequestration (Daly et al. 1987; Smith et al. 2002; Clark et al. 2005; Rodriguez et al. 2011). Sequestration of the same classes of alkaloids is likely also evidence of either their broad dietary availability and/or a selective advantage in sequestering these particular compounds. For example, pumiliotoxins (PTX) are found in all five groups (Daly et al. 1999). One of these toxins, PTX 251D, modulates both voltage-gated sodium and potassium channels and is known to deter mosquitoes and ants (Table 21.2). Perhaps this compound is particularly effective in defense; however, resistance to PTX 251D would require genetic changes in more genes than is required for resistance to BTX or TTX. Alternatively, resistance to PTX 251D may require AA substitutions in less-conserved regions of ion channels that are subject to lower levels of purifying selection.

If alkaloid resistance affects alkaloid profiles, selection for resistance to compounds like PTX 251D with diverse physiological effects may promote diversification in alkaloid defense. The most toxic alkaloid found in dendrobatids, BTX, occurs in *Phyllobates* at the level of oldest origin of chemical defense (Dendrobatinae: *Phyllobates*+*Dendrobates*; Fig. 21.1). *Dendrobates* species presented with BTX-dusted fruit flies reject them as prey and are physiologically sensitive to BTX, suggesting that *Dendrobates* may either be incapable of sequestering BTX or not resistant enough to do so (Daly et al. 1980, 1999). It may be possible that the accumulation of AA substitutions providing resistance to various alkaloids in the *Dendrobates* sensu lato clade facilitated the evolution of resistance to and sequestration of BTX in *Phyllobates*. However, more rigorous experiments of alkaloid resistance, availability, and sequestration are needed to clarify what determines the alkaloid profile of a species and why certain alkaloids are more common than others. Further analyses of alkaloid diversity may reveal unexpected patterns in alkaloid resistance and vice versa.

21.5 Future Directions

Many questions remain regarding the chemical ecology of poison frogs. How are alkaloids transported from the digestive system to the skin? What physiological changes during metamorphosis facilitate the sequestration of alkaloids from diet? What are the relative roles of resistance and dietary specialization in the evolution of alkaloid sequestration and defense? Do defended diet generalists have different patterns of resistance than defended diet specialists, owing to their exposure to a broader array of alkaloids? How does dietary conservatism of predators (Thomas et al. 2003; Marples et al. 2005; Darst and Cummings 2006; Lee et al. 2010) affect these patterns? If AA substitutions are not found in all alkaloid-affected proteins, how is the expression of both sensitive and resistant forms modulated and what effect does this have on sequestration, defense, and resistance? Moreover, how does the evolution of autoresistance and sequestration in poison frogs compare to that of frogs that synthesize their own toxins (e.g., *Pseudophryne*; see Smith et al. 2002)? New next-generation sequencing methods will facilitate the genetic aspects of these studies (Li et al. 2013).

The large volume of pharmacological research on poison frog alkaloids highlights its significance. Some of these compounds have extremely promising applications in development of novel anesthetics (e.g., epibatidine) and basic research on neuromuscular functions (e.g., pumiliotoxins, batrachotoxins, and izidines) (Daly 2005). The work of John Daly and colleagues (Savitzky and Saporito 2012) has resulted in a prolific field of artificial synthesis of dendrobatid alkaloids and design of therapeutic agents based on their structures (Toyooka and Nemoto 2002). However, detailed in vitro assays of their function are scarce except for a few compounds (BTX, PTX 251D, epibatidine) and progress in studying these compounds in general has slowed under stricter legal regulations in response to concerns regarding bioprospecting for commercial drugs (Angerer 2011). Nevertheless, it is surprising that the synthesis of these compounds has not ameliorated some of these problems. New emphasis should be given to collaborative efforts between native and international researchers to further the exploration and causal bases of dendrobatid chemical defenses. Our labs are working towards uncovering the basis of resistance and sequestration across Dendrobatidae; in conjunction with a more complete survey of alkaloid profiles across dendrobatids and more in-depth physiological studies of these alkaloids, we should soon be able to answer many questions regarding dendrobatid evolution.

Acknowledgements We thank Anurag Agrawal for organizing the symposium "Evolutionary Ecology of Chemically Mediated Interactions" during the ISCE-CSiV 2014. We also thank Bruce A. Schulte for facilitating this special edition. JCS acknowledges the support of the NSERC-CREATE Training Program in Biodiversity Research at the University of British Columbia. LAO is supported by the Bauer Fellowship from Harvard University, the William F. Milton Fund from Harvard Medical School and the L'Oreal For Women in Science Fellowship. RDT is supported by the National Science Foundation and the University of Texas at Austin EEB program. We thank the two reviewers for their insights and suggestions that improved the quality of our review.

References

Aardema ML, Zhen Y, Andolfatto P (2012) The evolution of cardenolide-resistant forms of Na plus, K plus-ATPase in Danainae butterflies. Mol Ecol 21:340–349

Albuquerque EX, Warnick JE, Sansone FM, Daly JW (1973) The pharmacology of batrachotoxin. V. A comparative study of membrane properties and the effect of batrachotoxin on sartorius muscles of the frogs *Phyllobates aurotaenia* and *Rana pipiens*. J Pharmacol Exp Therapeut 184:315–329

Albuquerque EX, Kuba K, Lapa AJ, Daly JW, Witkop B (1974) Acetylcholine receptor and ionic conductance modulator of innervated and denervated muscle membranes. In: Milhorat AT (ed) Effect of histrionicotoxins. Exploratory concepts in muscular dystrophy II. Excerpta Medica, Amsterdam, pp 585–600

Alvardo JB, Alvarez A, Saporito RA (2013) *Oophaga pumilio* predation by *Baryphthengus martii*. Herpetol Rev 44:298

Angerer K (2011) Frog tales – on poison dart frogs, epibatidine, and the sharing of biodiversity. Innov Eur J Soc Sci Res 24:353–369

Arce-Dominguez F, Rengifo-Mosquera JT (2013) Dieta de *Phyllobates aurotaenia* y *Oophaga histrionica* (Anura: Dendrobatidae) en el municipio del Alto Baudó, Chocó, Colombia. Acta Zool Mex 29:255–268

Bane V, Lehane M, Dikshit M, O'Riordan A, Furey A (2014) Tetrodotoxin: chemistry, toxicity, source, distribution and detection. Toxins 6:693–755

Bargar TM, Lett RM, Johnson PL, Hunter JE, Chang CP, Pernich DJ, Sabol MR, Dick MR (1995) Toxicity of pumiliotoxin 251d and synthetic analogs to the cotton pest *Heliothis virescens*. J Agric Food Chem 43:1044–1051

Barrio-Amoros CL, Santos JC (2012) A phylogeny for Aromobates (Anura: Dendrobatidae) with description of three new species from the Andes of Venezuela, taxonomic comments on *Aromobates saltuensis*, *A. inflexus*, and notes on the conservation status of the genus. Zootaxa 3422:1–31

Barrio-Amoros CL, Rivero R, Santos JC (2011) A new striking dendrobatid frog (Dendrobatidae: Aromobatinae, *Aromobates*) from the Venezuelan Andes. Zootaxa 3063:39–52

Bassoli A, Borgonovo G, Busnelli G (2007) Alkaloids and the bitter taste. In: Fattorusso E, Taglialatela-Scafati O (eds) Modern alkaloids: structure, isolation, synthesis and biology. Wiley, Weinheim

Bauer L (1994) New names in the family Dendrobatidae (Anura, Amphibia). Ripa Netherlands Fall:1–6

Bieber AGD, Silva PSD, Oliveira PS (2013) Attractiveness of fallen fleshy fruits to ants depends on previous handling by frugivores. Ecoscience 20:85–89

Bogner F, Boppre M (1989) Single cell recordings reveal hydroxydanaidal as the volatile compound attracting insects to pyrrolizidine alkaloids. Entomol Exp Appl 50:171–184

Bohbot JD, Dickens JC (2010) Insect repellents: modulators of mosquito odorant receptor activity. PLoS One 5(8), e12138. doi:10.1371/journal.pone.0012138

Born M, Bongers F, Poelman EH, Sterck FJ (2010) Dry-season retreat and dietary shift of the dart-poison frog *Dendrobates tinctorius* (Anura: Dendrobatidae). Phyllomedusa 9:37–52

Brown JL, Twomey E, Amézquita A, Barbosa de Souza M, Caldwell JP, Lötters S, von May R, Melo-Sampaio PR, Mejía-Vargas D, Perez-Peña P, Pepper M, Poelman EH, Sanchez-Rodriguez M, Summers K (2011) A taxonomic revision of the Neotropical poison frog genus *Ranitomeya* (Amphibia: Dendrobatidae). Zootaxa 3083:1–120

Caldwell JP (1996) The evolution of myrmecophagy and its correlates in poison frogs (family Dendrobatidae). J Zool 240:75–101

Cestele S, Catterall WA (2000) Molecular mechanisms of neurotoxin action on voltage-gated sodium channels. Biochimie 82:883–892

Chan LMS, Lowes S, Hirst BH (2004) The ABCs of drug transport in intestine and liver: efflux proteins limiting drug absorption and bioavailability. Euro J Pharm Sci 21:25–51

Chau R, Kalaitzis JA, Neilan BA (2011) On the origins and biosynthesis of tetrodotoxin. Aquat Toxicol 104:61–72

Chebib M, Duke RK, Duke CC, Connor M, Mewett KN, Johnston GAR (2003) Convulsant actions of calycanthine. Toxicol Appl Pharm 190:58–64

Cipriani I, Rivera M (2009) Detección de alcaloides en la piel de cuatro especies de anfibios ecuatorianos (Anura: Dentrobatidae). Revista Ecuatoriana de Medicina y Ciencias Biológicas 30 (1–2):42–49

Clark VC, Raxworthy CJ, Rakotomalala V, Sierwald P, Fisher BL (2005) Convergent evolution of chemical defense in poison frogs and arthropod prey between Madagascar and the Neotropics. PNAS 102:11617–11622

Cochrane CS (1825) Journal of a residence and travels in Colombia, during the years 1823 and 1824. Printed for Henry Colburn, London

Conner WE, Eisner T, Vandermeer RK, Guerrero A, Meinwald J (1981) Pre-copulatory sexual interaction in an arctiid moth (*Utetheisa-ornatrix*) – role of a pheromone derived from dietary alkaloids. Behav Ecol Sociobiol 9:227–235

Conner WE, Boada R, Schroeder FC, Gonzalez A, Meinwald J, Eisner T (2000) Chemical defense: bestowal of a nuptial alkaloidal garment by a male moth on its mate. PNAS 97:14406–14411

Currie CR, Scott JA, Summerbell RC, Malloch D (1999) Fungus-growing ants use antibiotic-producing bacteria to control garden parasites. Nature 398:701–704

Daly JW (1998) Thirty years of discovering arthropod alkaloids in amphibian skin. J Nat Prod 61:162–172

Daly JW (2005) Nicotinic agonists, antagonists, and modulators from natural sources. Cell Mol Neurobiol 25:513–552

Daly JW, Myers CW (1967) Toxicity of Panamanian poison frogs (*Dendrobates*): some biological and chemical aspects. Science 156:970–973

Daly JW, Spande TF (1986) Amphibian alkaloids: chemistry, pharmacology, and biology. In: Pelletier SW (ed) Alkaloids: chemical and biological perspectives. Wiley, New York, pp 1–274

Daly JW, Brown GB, Mensah-Dwumah M, Myers CW (1978) Classification of skin alkaloids from Neotropical poison-dart frogs (Dendrobatidae). Toxicon 16:163–188

Daly JW, Myers CW, Warnick JE, Albuquerque EX (1980) Levels of batrachotoxin and lack of sensitivity to its action in poison-dart frogs (*Phyllobates*). Science 208:1383–1385

Daly JW, Myers CW, Whittaker N (1987) Further classification of skin alkaloids from neotropical poison frogs (Dendrobatidae), with a general survey of toxic/noxious substances in the Amphibia. Toxicon 25:1023–1095

Daly JW, Gusovsky F, McNeal ET, Secunda S, Bell M, Creveling CR, Nishizawa Y, Overman LE, Sharp MJ, Rossignol DP (1990) Pumiliotoxin alkaloids – a new class of sodium-channel agents. Biochem Pharm 40:315–326

Daly JW, Secunda SI, Garraffo HM, Spande TF, Wisnieski A, Nishihira C, Cover JF Jr (1992) Variability in alkaloid profiles in neotropical poison frogs (Dendrobatidae): genetic versus environmental determinants. Toxicon 30:887–898

Daly JW, Garraffo HM, Jaramillo C, Rand AS (1994a) Dietary source for skin alkaloids of poison frogs (Dendrobatidae)? J Chem Ecol 20:943–955

Daly JW, Gusovsky F, Myers CW, Yotsu-Yamashita M, Yasumoto T (1994b) First occurrence of tetrodotoxin in a dendrobatid frog (*Colostethus inguinalis*), with further reports for the bufonid genus *Atelopus*. Toxicon 32:279–285

Daly JW, Secunda SI, Garraffo HM, Spande TF, Wisnieski A, Cover JF Jr (1994c) An uptake system for dietary alkaloids in poison frogs (Dendrobatidae). Toxicon 32:657–663

Daly JW, Padgett WL, Saunders RL, Cover JF Jr (1997) Absence of tetrodotoxins in a captive-raised riparian frog, *Atelopus varius*. Toxicon 35:705–709

Daly JW, Garraffo HM, Spande TF (1999) Alkaloids from amphibian skins. In: Pelletier SW (ed) Alkaloids: chemical and biological perspectives. Pergamon Press, New York, pp 1–161

Daly JW, Garraffo HM, Jain P, Spande TF, Snelling RR, Jaramillo C, Rand AS (2000) Arthropod-frog connection: decahydroquinoline and pyrrolizidine alkaloids common to microsympatric myrmicine ants and dendrobatid frogs. J Chem Ecol 26:73–85

Daly JW, Kaneko T, Wilham J, Garraffo HM, Spande TF, Espinosa A, Donnelly MA (2002) Bioactive alkaloids of frog skin: combinatorial bioprospecting reveals that pumiliotoxins have an arthropod source. PNAS 99:13996–14001

Daly JW, Garraffo HM, Spande TF, Clark VC, Ma JY, Ziffer H, Cover JF (2003) Evidence for an enantioselective pumiliotoxin 7-hydroxylase in dendrobatid poison frogs of the genus Dendrobates. PNAS 100:11092–11097

Daly JW, Spande TF, Garraffo HM (2005) Alkaloids from amphibian skin: a tabulation of over eight-hundred compounds. J Nat Prod 68:1556–1575
Daly JW, Ware N, Saporito RA, Spande TF, Garraffo HM (2009) N-Methyldecahydroquinolines: an unexpected class of alkaloids from Amazonian poison frogs (Dendrobatidae). J Nat Prod 72:1110–1114
Danielson PB (2002) The cytochrome P450 superfamily: biochemistry, evolution and drug metabolism in humans. Curr Drug Metab 3:561–597
Darst CR, Cummings ME (2006) Predator learning favours mimicry of a less-toxic model in poison frogs. Nature 440:208–211
Darst CR, Menendez-Guerrero PA, Coloma LA, Cannatella DC (2005) Evolution of dietary specialization and chemical defense in poison frogs (Dendrobatidae): a comparative analysis. Am Nat 165:56–69
Delfino G, Giachi F, Nosi D, Malentacchi C (2010) Serous cutaneous glands in *Phyllobates bicolor* (Anura: Dendrobatidae): an ontogenetic, ultrastructural study on secretory product biosynthesis and maturation. Copeia 2010:27–37
Despland E, Simpson SJ (2005) Surviving the change to warning colouration: density-dependent polyphenism suggests a route for the evolution of aposematism. Chemoecology 15:69–75
Dobler S (2001) Evolutionary aspects of defense by recycled plant compounds in herbivorous insects. Basic Appl Ecol 2:15–26
Dobler S, Petschenka G, Pankoke H (2011) Coping with toxic plant compounds – the insect's perspective on iridoid glycosides and cardenolides. Phytochemistry 72:1593–1604
Donnelly MA (1991) Feeding patterns of the strawberry poison frog, *Dendrobates pumilio* (Anura, Dendrobatidae). Copeia 1991:723–730
Dumbacher JP (1999) Evolution of toxicity in pitohuis: I. Effects of homobatrachotoxin on chewing lice (order Phthiraptera). Auk 116:957–963
Dumbacher JP, Beehler BM, Spande TF, Garraffo HM, Daly JW (1992) Homobatrachotoxin in the genus *Pitohui*: chemical defense in birds? Science 258:799–801
Dumbacher JP, Wako A, Derrickson SR, Samuelson A, Spande TF, Daly JW (2004) Melyrid beetles (Choresine): a putative source for the batrachotoxin alkaloids found in poison-dart frogs and toxic passerine birds. PNAS 101:15857–15860
Dussourd DE, Ubik K, Harvis C, Resch J, Meinwald J, Eisner T (1988) Defense-mechanisms of arthropods. 86. biparental defensive endowment of eggs with acquired plant alkaloid in the moth *Utetheisa-ornatrix*. PNAS 85:5992–5996
Eisner T, Meinwald J (1995) The chemistry of sexual selection. PNAS 92:50–55
Emerson SB (1985) Skull shape in frogs: correlations with diet. Herpetologica 41:177–188
Feldman CR, Brodie ED, Pfrender ME (2009) The evolutionary origins of beneficial alleles during the repeated adaptation of garter snakes to deadly prey. PNAS 106:13415–13420
Feldman CR, Brodie ED, Pfrender ME (2012) Constraint shapes convergence in tetrodotoxin-resistant sodium channels of snakes. PNAS 109:4556–4561
Fitch RW, Sturgeon GD, Patel SR, Spande TF, Garraffo HM, Daly JW, Blaauw RH (2009) Epiquinamide: a poison that wasn't from a frog that was. J Nat Prod 72:243–247
Fitch RW, Spande TF, Garraffo HM, Yeh HJC, Daly JW (2010) Phantasmidine: an epibatidine congener from the Ecuadorian poison frog *Epipedobates anthonyi*. J Nat Prod 73:331–337
Frezza L, Castano S, Asencio H, Sandtner W, Fierro L, Bezanilla F, Correa AM (2010) Is the skeletal muscle sodium channel of the batrachotoxin (BTX)-producing *Phyllobates aurotaenia* poison dart frog resistant to BTX? Biophys J 98:113A
Gordon D, Karbat I, Ilan N, Cohen L, Kahn R, Gilles N, Dong K, Stuhmer W, Tytgat J, Gurevitz M (2007) The differential preference of scorpion alpha-toxins for insect or mammalian sodium channels: Implications for improved insect control. Toxicon 49:452–472
Grant T, Frost DR, Caldwell JP, Gagliardo R, Haddad CFB, Kok PJR, Means BD, Noonan BP, Schargel W, Wheeler WC (2006) Phylogenetic systematics of dart-poison frogs and their relatives (Anura: Athesphatanura: Dendrobatidae). Bull Am Mus Nat Hist 299:1–262
Hagman M, Forsman A (2003) Correlated evolution of conspicuous coloration and body size in poison frogs (Dendrobatidae). Evolution 57:2904–2910

Hanifin CT, Gilly WF (2015) Evolutionary history of a complex adaptation: tetrodotoxin resistance in salamanders. Evolution 69:232–244
Hanifin CT, Yotsu-Yamashita M, Yasumoto T, Brodie ED (1999) Toxicity of dangerous prey: variation of tetrodotoxin levels within and among populations of the newt *Taricha granulosa*. J Chem Ecol 25:2161–2175
Härlin C, Härlin M (2003) Towards a historization of aposematism. Evol Ecol 17:197–212
Harris RZ, Benet LZ, Schwartz JB (1995) Gender effects in pharmacokinetics and pharmacodynamics. Drugs 50:222–239
Hartmann T, Witte L (1995) Chemistry, biology, and chemoecology of the pyrrolizidine alkaloids, pp. 155–233, in S. W. Pelletier (ed.). Alkaloids: Chemical and Biological Perspectives, Vol. 9. Pergamon Press, Oxford.
Hartmann T, Witte L, Ehmke A, Theuring C, Rowell-Rahier M, Pasteels JM (1997) Selective sequestration and metabolism of plant derived pyrrolizidine alkaloids by chrysomelid leaf beetles. Phytochemistry 45:489–497
Hartmann T, Theuring C, Schmidt J, Rahier M, Pasteels JM (1999) Biochemical strategy of sequestration of pyrrolizidine alkaloids by adults and larvae of chrysomelid leaf beetles. J Insect Physiol 45:1085–1095
Hartmann T, Theuring C, Bernays EA (2003) Are insect-synthesized retronecine esters (creatonotines) the precursors of the male courtship pheromone in the arctiid moth *Estigmene acrea*? J Chem Ecol 29:2603–2608
Hille B (2001) Ion channels of excitable membranes, 3rd edn. Sinauer Associates, Inc., Sunderland
Hödl W, Amezquita A (2001) Visual signaling in anuran amphibians. In: Ryan MJ (ed) Anuran communication. Smithsonian Institution Press, Washington, pp 121–141
How CK, Chern CH, Huang YC, Wang LM, Lee CH (2003) Tetrodotoxin poisoning. Am J Emergen Med 21:51–54
Howell G, Butler J, deShazo RD, Farley JM, Liu HL, Nanayakkara NPD, Yates A, Yi GB, Rockhold RW (2005) Cardiodepressant and neurologic actions of *Solenopsis invicta* (imported fire ant) venom alkaloids. Ann Allergy Asthma Immunol 94:380–386
Hua J, Morehouse NI, Relyea R (2013) Pesticide tolerance in amphibians: induced tolerance in susceptible populations, constitutive tolerance in tolerant populations. Evol Appl 6:1028–1040
Hwang PA, Tsai YH, Lin HP, Hwang DF (2007) Tetrodotoxin-binding proteins isolated from five species of toxic gastropods. Food Chem 103:1153–1158
Jones WA, Jacobson M (1968) Isolation of N, N-diethyl-m-toluamide (deet) from female pink bollworm moths. Science 159:99
Jones TH, Blum MS, Escoubas P, Ali TMM (1989) Novel pyrrolidines in the venom of the ant *Monomorium indicum*. J Nat Prod 52:779–784
Jones TH, Adams RMM, Spande TF, Garraffo HM, Kaneko T, Schultz TR (2012) Histrionicotoxin alkaloids finally detected in an ant. J Nat Prod 75:1930–1936
Jost MC, Hillis DM, Lu Y, Kyle JW, Fozzard HA, Zakon HH (2008) Toxin-resistant sodium channels: parallel adaptive evolution across a complete gene family. Mol Biol Evol 25:1016–1024
Kaneko Y, Matsumoto G, Hanyu Y (1997) TTX resistivity of Na+ channel in newt retinal neuron. Biochem Biophys Res Comm 240:651–656
Karalliedde L (1995) Animal toxins. Br J Anaesth 74:319–327
Kraus F, Duvall F (2004) New records of alien reptiles and amphibians in Hawaii. In: Evenhuis NL, Eldredge LG (eds) Records of the Hawaii biological survey for 2003 Part 2: notes. Bishop Museum Press, Honolulu, pp 62–64
Lazutka FA, Vasiliauskene AP, Gefen SG (1969) On the toxicological assessment of the insecticide nicotine sulfate. Gig Sanit 34:30–33
Lee TJ, Marples NM, Speed MP (2010) Can dietary conservatism explain the primary evolution of aposematism? Anim Behav 79:63–74
Lenger DR, Berkey JK, Dugas MB (2014) Predation on the toxic *Oophaga pumilio* (Anura: Dendrobatidae) by *Rhadinaea decorata* (Squamata: Collubridae). Herpetol Notes 7:83–84
Li CH, Hofreiter M, Straube N, Corrigan S, Naylor GJP (2013) Capturing protein-coding genes across highly divergent species. Biotechniques 54:321–326

Lieberman SS (1986) Ecology of the leaf litter herpetofauna of a Neotropical rain forest: La Selva, Costa Rica. Acta Zool Mex 15:1–72

Lindigkeit R, Biller A, Buch M, Schiebel HM, Boppre M, Hartmann T (1997) The two faces of pyrrolizidine alkaloids: the role of the tertiary amine and its N-oxide in chemical defense of insects with acquired plant alkaloids. Eur J Biochem 245:626–636

Llewellyn LE, Bell PM, Moczydlowski EG (1997) Phylogenetic survey of soluble saxitoxin-binding activity in pursuit of the function and molecular evolution of saxiphilin, a relative of transferrin. Proc Roy Soc B Biol Sci 264:891–902

Lötters S, Jungfer KH, Henkel FW, Schmidt W (2007) Poison frogs. Biology, species and captive husbandry. Edition Chimaira, Frankfurt am Main

Maan ME, Cummings ME (2012) Poison frog colors are honest signals of toxicity, particularly for bird predators. Am Nat 179:E1–E14

Macfoy C, Danosus D, Sandit R, Jones TH, Garraffo HM, Spande TE, Daly JW (2005) Alkaloids of anuran skin: antimicrobial function? Z Naturforsch C - J Biosci 60:932–937

Mahar J, Lukacs GL, Li Y, Hall S, Moczydlowski E (1991) Pharmacological and biochemical-properties of saxiphilin, a soluble saxitoxin-binding protein from the bullfrog (*Rana catesbeiana*). Toxicon 29:53–71

Mappes J, Alatalo RV (1997) Effects of novelty and gregariousness in survival of aposematic prey. Behav Ecol 8:174–177

Mappes J, Marples N, Endler JA (2005) The complex business of survival by aposematism. Trends Ecol Evol 20:598–603

Marki F, Witkop B (1963) The venom of the Colombian arrow poison frog *Phyllobates bicolor*. Experientia 19:329–376

Marples NM, Kelly DJ, Thomas RJ (2005) Perspective: the evolution of warning coloration is not paradoxical. Evolution 59:933–940

Marquez R, Amezquita A (2014) Auto-toxicity and the evolution of the muscular voltage-gated sodium channel in Phyllobates poison frogs (Oral presentation). Evolution, Raleigh

McGlothlin JW, Chuckalovcak JP, Janes DE, Edwards SV, Feldman CR, Brodie EDJ, Pfrender ME, Brodie EDI (2014) Parallel evolution of tetrodotoxin resistance in three voltage-gated sodium channel genes in the garter snake *Thamnophis sirtalis*. Mol Biol Evol 31(11):2836–2846. doi:10.1093/molbev/msu237

McKeown S (1996) A field guide to reptiles and amphibians in the Hawaiian Islands. Diamond Head Publishing Inc., Los Osos

Mebs D (2002) Venomous and poisonous animals: a handbook for biologists, toxicologists and toxinologists, physicians and pharmacists. Medpharm Scientific Publications, Stuttgart

Mebs D, Jansen M, Kohler G, Pogoda W, Kauert G (2010) Myrmecophagy and alkaloid sequestration in amphibians: a study on *Ameerega picta* (Dendrobatidae) and *Elachistocleis* sp. (Microhylidae) frogs. Salamandra 46:11–15

Mebs D, Alvarez JV, Pogoda W, Toennes SW, Kohler G (2014) Poor alkaloid sequestration by arrow poison frogs of the genus *Phyllobates* from Costa Rica. Toxicon 80:73–77

Morabito MA, Moczydlowski E (1994) Molecular-cloning of bullfrog saxiphilin – a unique relative of the transferrin family that binds saxitoxin. PNAS 91:2478–2482

Mueller UG, Rehner SA, Schultz TR (1998) The evolution of agriculture in ants. Science 281:2034–2038

Myers CW (1987) New generic names for some neotropical poison frogs (Dendrobatidae). Pap Avul Zool (São Paulo) 36:301–306

Myers CW, Daly JW (1976) Preliminary evaluation of skin toxins and vocalizations in taxonomic and evolutionary studies of poison-dart frogs (Dendrobatidae). Bull Am Mus Nat Hist 157:173–262

Myers CW, Daly JW, Malkin B (1978) A dangerously toxic new frog (*Phyllobates*) used by Embera Indians of western Colombia, with discussion of blowgun fabrication and dart poisoning. Bull Am Mus Nat Hist 161:307–366

Myers CW, Paolillo A, Daly JW (1991) Discovery of a defensively malodorous and nocturnal frog in the family Dendrobatidae: phylogenetic significance of a new genus and species from the Venezuelan Andes. Am Mus Nov 3002:1–33

Narberhaus I, Papke U, Theuring C, Beuerle T, Hartmann T, Dobler S (2004) Direct evidence for membrane transport of host-plant-derived pyrrolizidine alkaloid N-oxides in two leaf beetle genera. J Chem Ecol 30:2003–2022

Narberhaus I, Zintgraf V, Dobler S (2005) Pyrrolizidine alkaloids on three trophic levels – evidence for toxic and deterrent effects on phytophages and predators. Chemoecology 15:121–125

Naumann C, Hartmann T, Ober D (2002) Evolutionary recruitment of a flavin-dependent monooxygenase for the detoxification of host plant-acquired pyrrolizidine alkaloid-defended arctiid alkaloids in the moth *Tyria jacobaeae*. PNAS 99:6085–6090

Neuwirth M, Daly JW, Myers CW, Tice LW (1979) Morphology of the granular secretory glands in skin of poison-dart frogs (Dendrobatidae). Tissue Cell 11:755–771

Noguchi T, Maruyama J, Narita H, Hashimoto K (1984) Occurrence of tetrodotoxin in the gastropod mollusk *Tutufa lissostoma* (frog shell). Toxicon 22:219

Noguchi T, Jeon JK, Arakawa O, Sugita H, Deguchi Y, Shida Y, Hashimoto K (1986) Occurrence of tetrodotoxin and anhydrotetrodotoxin in *Vibrio* sp. isolated from the intestines of a xanthid crab, *Atergatis floridus*. J Biochem 99:311–314

Noonan BP, Comeault AA (2009) The role of predator selection on polymorphic aposematic poison frogs. Biol Lett 5:51–54

Opitz SEW, Muller C (2009) Plant chemistry and insect sequestration. Chemoecology 19:117–154

Ozel LD, Stynoski JL (2011) Differences in escape behavior between a cryptic and an aposematic litter frog. J Herpetol 45:395–398

Pauron D, Barhanin J, Lazdunski M (1985) The voltage-dependent nNa+ channel of insect nervous-system identified by receptor-sites for tetrodotoxin, and scorpion and sea-anemone toxins. Biochem Biophys Res Comm 131:1226–1233

Petschenka G, Fandrich S, Sander N, Wagschal V, Boppre M, Dobler S (2013a) Stepwise evolution of resistance to toxic cardenolides via genetic substitutions in the Na(+)/K+ -ATPase of milkweed butterflies (Lepidoptera: Danaini). Evolution 67:2753–2761

Petschenka G, Pick C, Wagschal V, Dobler S (2013b) Functional evidence for physiological mechanisms to circumvent neurotoxicity of cardenolides in an adapted and a non-adapted hawkmoth species. Proc R Soc B Biol Sci 280:20123089. doi:10.1098/rspb.2012.3089

Pires OR, Sebben A, Schwartz EF, Morales RAV, Bloch C, Schwartz CA (2005) Further report of the occurrence of tetrodotoxin and new analogues in the Anuran family Brachycephalidae. Toxicon 45:73–79

Posada-Arango A (1883) Veneno de la rana. Usado por los salvajes del Chocó. Revista Medica de Bogota 7:451–456

Poulin B, Lefebvre G, Ibáñez R, Jaramillo C, Hernández C, Rand AS (2001) Avian predation upon lizards and frogs in a neotropical forest understorey. J Trop Ecol 17:21–40

Prates I, Antoniazzi MM, Sciani JM, Pimenta DC, Toledo LF, Haddad CFB, Jared C (2012) Skin glands, poison and mimicry in dendrobatid and leptodactylid amphibians. J Morph 273:279–290

Pröhl H (2005) Territorial behavior in dendrobatid frogs. J Herpetol 39:354–365

Rodriguez A, Poth D, Schulz S, Vences M (2011) Discovery of skin alkaloids in a miniaturized eleutherodactylid frog from Cuba. Biol Lett 7:414–418

Rojas B, Rautiala P, Mappes J (2014) Differential detectability of polymorphic warning signals under varying light environments. Behav Proc 109(LPt B):164–172

Santos JC, Cannatella DC (2011) Phenotypic integration emerges from aposematism and scale in poison frogs. PNAS 108:6175–6180

Santos JC, Coloma LA, Cannatella DC (2003) Multiple, recurring origins of aposematism and diet specialization in poison frogs. PNAS 100:12792–12797

Santos JC, Coloma LA, Summers K, Caldwell JP, Ree R, Cannatella DC (2009) Amazonian amphibian diversity is primarily derived from late Miocene Andean lineages. PLoS Biol 7, e56. doi:10.1371/journal.pbio.1000056

Santos JC, Baquero M, Barrio-Amoros C, Coloma LA, Erdtmann LK, Lima AP, Cannatella DC (2014) Aposematism increases acoustic diversification and speciation in poison frogs. Proc R Soc B Biol Sci 281(1796):20141761. doi:10.1098/rspb.2014.1761

Saporito RA, Donnelly MA, Hoffman RL, Garraffo HM, Daly JW (2003) A siphonotid millipede (*Rhinotus*) as the source of spiropyrrolizidine oximes of dendrobatid frogs. J Chem Ecol 29:2781–2786

Saporito RA, Garraffo HM, Donnelly MA, Edwards AL, Longino JT, Daly JW (2004) Formicine ants: an arthropod source for the pumiliotoxin alkaloids of dendrobatid poison frogs. PNAS 101:8045–8050

Saporito RA, Donnelly MA, Jain P, Garraffo HM, Spande TF, Daly JW (2007a) Spatial and temporal patterns of alkaloid variation in the poison frog *Oophaga pumilio* in Costa Rica and Panama over 30 years. Toxicon 50:757–778

Saporito RA, Zuercher R, Roberts M, Gerow KG, Donnelly MA (2007b) Experimental evidence for aposematism in the dendrobatid poison frog *Oophaga pumilio*. Copeia 2007:1006–1011

Saporito RA, Spande TF, Garraffo HM, Donnelly MA (2009) Arthropod alkaloids in poison frogs: a review of the 'dietary hypothesis'. Heterocycles 79:277–297

Saporito RA, Donnelly MA, Madden AA, Garraffo HM, Spande TF (2010a) Sex-related differences in alkaloid chemical defenses of the dendrobatid frog *Oophaga pumilio* from Cayo Nancy, Bocas del Toro, Panama. J Nat Prod 73:317–321

Saporito RA, Isola M, Maccachero VC, Condon K, Donnelly MA (2010b) Ontogenetic scaling of poison glands in a dendrobatid poison frog. J Zool 282:238–245

Saporito RA, Norton RA, Andriamaharavo NR, Garraffo HM, Spande TF (2011) Alkaloids in the mite *Scheloribates laevigatus*: further alkaloids common to oribatid mites and poison frogs. J Chem Ecol 37:213–218

Saporito RA, Donnelly MA, Spande TF, Garraffo HM (2012) A review of chemical ecology in poison frogs. Chemoecology 22:159–168

Savitzky AH, Saporito RA (2012) Sequestration of defensive toxins by tetrapod vertebrates: contributions in memory of John W. Daly OBITUARY. Chemoecology 22:139–140

Savitzky AH, Mori A, Hutchinson DA, Saporito RA, Burghardt GM, Lillywhite HB, Meinwald J (2012) Sequestered defensive toxins in tetrapod vertebrates: principles, patterns, and prospects for future studies. Chemoecology 22:141–158

Simon MP, Toft CA (1991) Diet specialization in small vertebrates: mite-eating in frogs. Oikos 61:263–278

Smith BP, Tyler MJ, Kaneko T, Garraffo HM, Spande TF, Daly JW (2002) Evidence for biosynthesis of pseudophrynamine alkaloids by an Australian myobatrachid frog (*Pseudophryne*) and for sequestration of dietary pumiliotoxins. J Nat Prod 65:439–447

Smolanoff J, Kluge AF, Meinwald J, McPhail A, Miller RW, Hicks K, Eisner T (1975) Polyzonimine – novel terpenoid insect repellent produced by a millipede. Science 188:734–736

Soong TW, Venkatesh B (2006) Adaptive evolution of tetrodotoxin resistance in animals. Trends Genet 22:621–626

Speed MP, Ruxton GD (2005) Aposematism: what should our starting point be? Proc R Soc B Biol Sci 272:431–438

Stynoski JL, Torres-Mendoza Y, Sasa-Marin M, Saporito RA (2014a) Evidence of maternal provisioning of alkaloid-based chemical defenses in the strawberry poison frog *Oophaga pumilio*. Ecology 95:587–593

Stynoski JL, Shelton G, Stynoski P (2014b) Maternally derived chemical defences are an effective deterrent against some predators of poison frog tadpoles (*Oophaga pumilio*). Biol Lett 10:4

Sudakin DL, Trevathan WR (2003) DEET: a review and update of safety and risk in the general population. J Toxicol Clin Toxicol 41:831–839

Summers K, Clough M (2001) The evolution of coloration and toxicity in the poison frog family (Dendrobatidae). PNAS 98:6227–6232

Summers K, McKeon CS (2004) The evolutionary ecology of phytotelmata use in neotropical poison frogs. Misc Publ Univ Mich Mus Zool 193:55–73

Summers K, McKeon CS, Heying H (2006) The evolution of parental care and egg size: a comparative analysis in frogs. Proc R Soc B Biol Sci 273:687–692

Tanu MB, Mahmud Y, Tsuruda K, Arakawa O, Noguchi T (2001) Occurrence of tetrodotoxin in the skin of a rhacophoridid frog *Polypedates* sp. from Bangladesh. Toxicon 39:937–941

Tarvin R, Santos JC, O'Connell LA, Cannatella DC, Zakon HH (2014) Genetic basis of alkaloid resistance in harlequin toads and poison frogs (Oral presentation). Evolution, Raleigh

Terlau H, Heinemann SH, Stuhmer W, Pusch M, Conti F, Imoto K, Numa S (1991) Mapping the site of block by tetrodotoxin and saxitoxin of sodium channel-Ii. FEBS Lett 293:93–96

Thomas RJ, Marples NM, Cuthill IC, Takahashi M, Gibson EA (2003) Dietary conservatism may facilitate the initial evolution of aposematism. Oikos 101:458–466

Toft CA (1980) Feeding ecology of thirteen syntopic species of anurans in a seasonal tropical environment. Oecologia 45:131–141

Toft CA (1981) Feeding ecology of Panamanian litter anurans: patterns of diet and foraging mode. J Herpetol 15:139–144

Toft CA (1995) Evolution of diet specialization in poison-dart frogs (Dendrobatidae). Herpetologica 51:202–216

Toft CA, Duellman WE (1979) Anurans of the lower Río Llullapichis, Amazonian Perú: a preliminary analysis of community structure. Herpetologica 35:71–77

Toyooka N, Nemoto H (2002) Synthetic studies on dart-poison frog alkaloids. Recent research developments in organic chemistry 6 (part 2). Research Signpost, Trivandrum, pp 611–624

Tsuneki H, You YR, Toyooka N, Kagawa S, Kobayashi S, Sasaoka T, Nemoto H, Kimura I, Dani JA (2004) Alkaloids indolizidine 235B′, quinolizidine 1-epi-2071, and the tricyclic 205B are potent and selective noncompetitive inhibitors of nicotinic acetylcholine receptors. Mol Pharm 66:1061–1069

Tsuruda K, Arakawa O, Kawatsu K, Hamano Y, Takatani T, Noguchi T (2002) Secretory glands of tetrodotoxin in the skin of the Japanese newt *Cynops pyrrhogaster*. Toxicon 40:131–136

Ujvary I (1999) Nicotine and other insecticidal alkaloids. In: Yamamoto I, Casida JE (eds) Nicotinoid insecticides and the nicotinic acetylcholine receptor. Springer, Tokyo, p 300

Vandendriessche T, Abdel-Mottaleb Y, Maertens C, Cuypers E, Sudau A, Nubbemeyer U, Mebs D, Tytgat J (2008) Modulation of voltage-gated Na+ and K+ channels by pumiliotoxin 251D: a "joint venture" alkaloid from arthropods and amphibians. Toxicon 51:334–344

Vences M, Glaw F, Böhme W (1998) Evolutionary correlates of microphagy in alkaloid-containing frogs (Amphibia: Anura). Zool Anz 236:217–230

Vences M, Kosuch J, Boistel R, Haddad CFB, La Marca E, Lötters S, Veith M (2003) Convergent evolution of aposematic coloration in neotropical poison frogs: a molecular phylogenetic perspective. Organ Div Evol 3:215–226

Verotta L, Pilati T, Tato M, Elisabetsky E, Amador TA, Nunes DS (1998) Pyrrolidinoindoline alkaloids from *Psychotria colorata*. J Nat Prod 61:392–396

Wang SY, Wang GK (1998) Point mutations in segment I-S6 render voltage-gated Na+ channels resistant to batrachotoxin. PNAS 95:2653–2658

Wang SY, Wang GK (2003) Voltage-gated sodium channels as primary targets of diverse lipid-soluble neurotoxins. Cell Signal 15:151–159

Wang SY, Mitchell J, Tikhonov DB, Zhorov BS, Wang GK (2006) How batrachotoxin modifies the sodium channel permeation pathway: computer modeling and site-directed mutagenesis. Mol Pharm 69:788–795

Wassen SH (1935) Notes on southern groups of Chocó Indians in Colombia. Ethnografiska Museet. Göteborg, Etnologiska Studier 1:35–182

Weldon PJ, Kramer M, Gordon S, Spande TF, Daly JW (2006) A common pumiliotoxin from poison frogs exhibits enantioselective toxicity against mosquitoes. PNAS 103:17818–17821

Weldon PJ, Cardoza YJ, Vander-Meer RK, Hoffmann WC, Daly JW, Spande TF (2013) Contact toxicities of anuran skin alkaloids against the fire ant (*Solenopsis invicta*). Naturwissenschaften 100:185–192

Weygoldt P (1987) Evolution of parental care in dart poison frogs (Amphibia: Anura: Dendrobatidae). Zeits Zool Syst Evol 25:51–67
Wood SA, Taylor DI, McNabb P, Walker J, Adamson J, Cary SC (2012) Tetrodotoxin concentrations in *Pleurobranchaea maculata*: temporal, spatial and individual variability from New Zealand populations. Mar Drugs 10:163–176
Zakon HH (2012) Adaptive evolution of voltage-gated sodium channels: The first 800 million years. PNAS 109:10619–10625
Zhang Y, Benet LZ (2001) The gut as a barrier to drug absorption: combined role of cytochrome P450 3A and P-glycoprotein. Clin Pharmacokinet 40:159–168
Zhen Y, Aardema ML, Medina EM, Schumer M, Andolfatto P (2012) Parallel molecular evolution in an herbivore community. Science 337:1634–1637
Zimmermann H, Zimmermann E (1988) Etho-taxonomie und zoogeographische artengruppenbildung bei pfeilgiftfröschen. Salamandra 24:125–160

Chapter 22
Semiochemicals in Anurans: Testing Different Categories with One Poison Frog Species

Lisa M. Schulte

22.1 Communication by Semiochemicals in Anurans

Adult anurans, i.e., frogs and toads, are particularly known for their acoustic communication (Gerhardt and Huber 2002; Narins et al. 2006). However, over 100 years ago, it was shown that they also have a sense of smell and are able to detect chemical compounds by olfaction (artificial smells in adults and smell of food in tadpoles; Risser 1914). Thirty-five years later, a further study regarding anuran semiochemicals was published (Eibl-Eibesfeldt 1949). Since then, over 200 studies regarding the use of semiochemicals in anurans have been published.

The term semiochemicals was introduced by Law and Regnier (1971) and describes a chemical substance that serves as a messenger for chemical communication, within and among species. Semiochemicals used in intraspecific communication are termed pheromones (Karlson and Lüscher 1959), and can be divided into releaser pheromones that trigger behavioral changes, and primer pheromones that trigger physiological changes in the receiving animal (compare Fig. 22.1).

Considering that pheromones are typically produced consistently across individuals of one species, semiochemicals that differ among individuals (or colonies) of the same species and allow the distinction between them are defined as signature mixtures (Wyatt 2010a). Interspecifically used semiochemicals are summarized as allelochemicals, which can be divided into kairomones (advantageous for the receiving animal), allomones (advantageous for the emitting animal), and synomones

L.M. Schulte (✉)
Department of Biology, East Carolina University, 1001 E Tenth Street, Greenville, NC 27858, USA

Department of Biogeography, Trier University, Universitätsring 15, 54286 Trier, Germany
e-mail: Lisa_Schulte@gmx.de

B.A. Schulte et al. (eds.), *Chemical Signals in Vertebrates 13*,
DOI 10.1007/978-3-319-22026-0_22

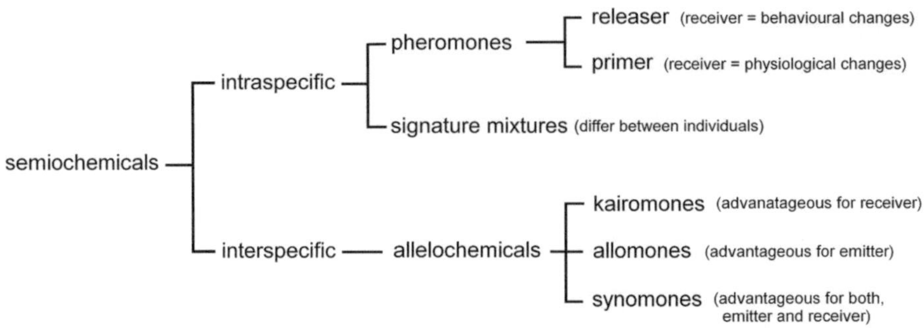

Fig. 22.1 Different categories of semiochemicals divided into substances used in intraspecific and interspecific communication. Modified after Nordlund et al. (1981)

(advantageous for both emitter and receiver). To the best of my knowledge, all but the latter two types of semiochemicals have been described for chemical communication in anuran amphibians (Nordlund et al. 1981; Wyatt et al. 2010b).

22.1.1 Releaser Pheromones in Anurans

A number of studies regarding chemical communication in anurans are focused on tadpole alarm substances (e.g., Kulzer 1954; Wilson and Lefcort 1993; Ferrari et al. 2007). Although in most studies the chemical alarm substances are described as unintentionally released alarm cues (Schreckstoffe), Fraker et al. (2009) showed that the release of these chemicals is actually an active secretory process, during which two different peptides are combined. In this case when already dead or killed tadpoles were crushed, no behavioral effect could be elicited in conspecific individuals, i.e., the alarm substance was not released. This means that the correct term would be alarm signals as cues provide information unintentionally whereas signals are intentional (Steiger et al. 2011). Most of these signals elicit behavioral changes between conspecific and not heterospecific tadpoles (e.g., Hews and Blaustein 1985; Petranka 1989; Hagman and Shine 2008). However, some species are able to also recognize heterospecific alarm signals (see Sect. 1.4). Other types of releaser pheromones described in anurans occur between adults and stimulate for example aggression among individuals of the same sex, (King et al. 2005; Poth et al. 2012) or attraction and agitation among individuals of the opposite sex (e.g., Rabb and Rabb 1963; Wabnitz et al. 1999; Asey et al. 2005).

22.1.2 Primer Pheromones in Anurans

Just as behavioral responses can be triggered by tadpole alarm pheromones, so can longer-term physiological alterations. However, the definitions of releaser and primer pheromones are not mutually exclusive, since physiological changes also trigger behavior. In anurans physiological effects are typically of morphological nature (e.g., Mandrillon and Saglio 2008; Hagman et al. 2009), but they can also influence time until hatching or metamorphosis (e.g., Kiesecker et al. 2002; Touchon et al. 2006; Mandrillon and Saglio 2007). Based on an extensive literature review, primer pheromones influencing the adult stage have not been described thus far in anurans.

22.1.3 Signature Mixtures in Anurans

Several studies have shown that anurans in different life stages were able to differentiate between kin and non-kin, or self and nonself (e.g., Blaustein et al. 1984; Waldman 1985; Lee and Waldman 2002; Eluvathingal et al. 2009). This indicates the existence of signature mixtures in anurans, which may be either endogenous or acquired from the environment (e.g., Pfennig 1990; compare Wyatt 2010b).

22.1.4 Kairomones in Anurans

For tadpoles there is abundant evidence that they are able to recognize and avoid kairomones of predators. In many cases, alarm cues of previously digested tadpoles are involved (e.g., Lefcort et al. 1999; Chivers and Mirza 2001), but there are also examples where the predators themselves are recognized (e.g., Kiesecker and Blaustein 1997; Polo-Cavia et al. 2010). The same is true for froglets and adults (e.g., Flowers and Graves 1997; Murray et al. 2004). Furthermore, some tadpoles are able to take advantage of alarm signals of heterospecifics tadpoles (e.g., Hrbáček 1950; Pfeiffer 1966; Adams and Claeson 1998). Considering that this does not benefit the releasing species, alarm signals in this context might be classified as a kairomones.

22.1.5 Allomones and Synomones in Anurans

As far as is known, allomones and synomones, the allelochemicals that are advantageous to either only the emitting or to both emitting and receiving species, have not been tested and described in any anuran species, neither for larval nor adult frogs.

22.2 The Poison Frog *Ranitomeya variabilis*

Poison frogs of the family Dendrobatidae are mainly diurnal frogs that occur in the Neotropics. Many of them are brightly colored and secrete skin alkaloids (Lötters et al. 2007; Saporito et al. 2012). Besides their coloration, poison frogs are especially known for their remarkably diverse parental care behaviors (e.g., Weygoldt 1987; Summers and McKeon 2004, 2006). The species *Ranitomeya variabilis* occurs in the rainforests east of the Andes in Peru, Ecuador, and Colombia and uses small phytotelmata (water bodies in plants such as leaf axils; Varga 1928) for both clutch and tadpole depositions (Brown et al. 2008b, 2011). Reproduction is initiated by the male, who usually starts calling in response to the presence of a female. Once the female is in close proximity, the male leads it to a phytotelm, where the eggs are attached to the inner wall, just at the water's surface (Brown et al. 2008b). After development, males retrieve tadpoles by either removing them from the eggs or the phytotelm into which they hatched. Because tadpoles of this species are highly cannibalistic and feed on eggs as well as on conspecific tadpoles when deposited together, tadpoles are usually transported individually and depositions with conspecific tadpoles are avoided (Summers 1999). Unlike some congeneric species, *R. variabilis* does not regularly feed its tadpoles with unfertilized eggs, but facultative feeding with fertilized eggs, as well as seasonal feeding with younger tadpoles can be observed for this species (Schulte and Lötters 2013; Schulte 2014). The main predators of the tadpoles of this species (besides conspecific cannibalistic tadpoles) are damselfly larvae of the phytotelmata breeding species *Microstigma rotundatum* (Brown et al. 2008a; Schulte et al. 2013).

22.3 Communication by Semiochemicals in *Ranitomeya variabilis*

Due to its very specific set of behaviors, *R. variabilis* is an excellent species to study chemical communication. On the one hand, the interaction of different life stages (e.g., cannibalism, tadpole transportation) creates an optimal precondition for possible interactions via pheromones. On the other hand, the multitude of species occurring in the same habitat with *R. variabilis*, favors testing for possible interactions via allelochemicals. The following passages give an overview of the semiochemicals already tested in this species, as well as those that still need to be studied (see also Fig. 22.2).

22.3.1 Releaser Pheromones in Ranitomeya variabilis

The behavioral reaction of adult *R. variabilis* towards potential releaser pheromones was studied under two different circumstances, during parental care behavior and in relation to conspecific adults:

During parental care behavior, the reaction of adults towards chemical compounds of conspecific tadpoles was tested. To answer the question how are tadpole-transporting males able to distinguish between free and already occupied

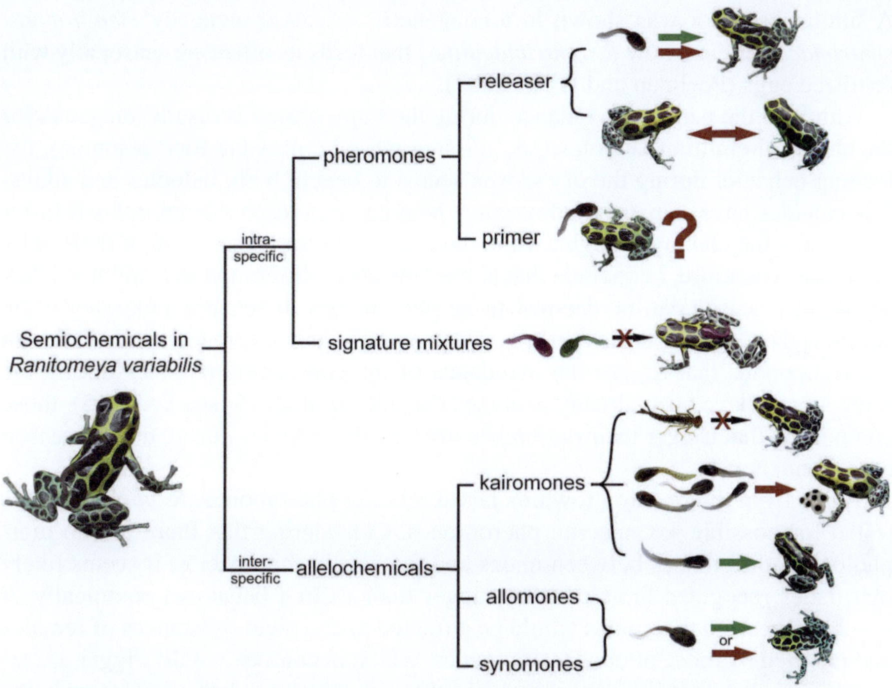

Fig. 22.2 Overview of semiochemicals used by *Ranitomeya variabilis*. Emitters are positioned on the *left*, receivers on the *right* side; *green arrows* represent positive reactions of the receiver towards the chemical substances of the emitter (e.g., attraction, offspring deposition); *red arrows* signify negative reactions towards the chemical substances (e.g., repulsion, avoidance for offspring deposition); *black*, crossed out *arrows* signify that no reaction towards possible chemical substances could be found; *question marks* mean that there are no data regarding this category of semiochemicals

phytotelmata (for tadpoles are often hidden on the ground, thus invisible to the frogs), in-situ pool-choice experiments were conducted. Artificial phytotelmata were hung up in pairs in the natural habitat of the frogs. One phytotelm of each pair was filled with rainwater, the other one with water used by conspecific tadpoles. It turned out that *R. variabilis* only requires the chemical compounds of conspecific tadpoles to locate and avoid them for tadpole and clutch deposition (Schulte et al. 2011), and visual tadpole cues are irrelevant in this decision (Schulte et al. 2013). However, this avoidance behavior could only be observed during the rainy season. In the dry season, the frogs completely changed their reaction towards the chemical tadpole compounds. Parental *R. variabilis* did not avoid but preferred conspecific cues during the dry season for tadpole depositions (Schulte and Lötters 2013). This may be interpreted as a way to provide the older tadpoles in their territories with food (i.e., with younger tadpoles), to accelerate their development when facing desiccation risk.

A similar behavior was shown in a congeneric species (previously *Dendrobates ventrimaculatus* and now *R. ventrimaculata*) that feeds its offspring seasonally with fertilized eggs (Poelman and Dicke 2007).

Although the parental avoidance during the rainy season is disadvantageous for the chemical-emitting tadpoles (i.e., missing out on a possible food resource), the feeding behavior during the dry season seems to benefit both, tadpoles and adults. The tadpoles have an extra food resource helping them reach metamorphosis faster and escape the coming drought, while the adults support the survival of their older offspring. Therefore, I conclude that at least the chemical compounds involved during the dry season can be deemed to be pheromones. It remains unknown if the compounds triggering these two opposite parental behaviors are identical. Although the compounds that trigger the avoidance of the conspecific tadpoles during the rainy season have been already analyzed (Schulte et al. 2015; see Sect. 3.5), those compounds that trigger their preference towards the tadpoles during the dry season are still unknown.

Aside from the reaction towards larval releaser pheromones, *R. variabilis* was tested for possible sex specific pheromones. Considering that there are no morphological differences between males and females in this species, it seems likely that males recognize females (that trigger their calling behavior) chemically. It was hypothesized that males would be attracted to chemical substances of females and repelled by those of conspecific males, which in contrast would trigger attraction in females. Instead, all individuals showed avoidance behavior towards the conspecific chemicals (Schulte and Rößler 2013). This might be an artifact due to confinement stress of the releaser animals that may cause the emission of disturbance cues, which trigger avoidance behavior in their conspecifics. However, it is unknown if these disturbance cues are merely excretory byproducts (e.g., urine), or if they are comparable with the alarm cues shown in tadpoles of other anuran species, which were just recently shown to be actively emitted signals (Fraker et al. 2009).

22.3.2 Primer Pheromones in Ranitomeya variabilis

Primer pheromones have not yet been described for *R. variabilis*. Tadpoles growing up together in their hatching phytotelm (when not transported by the male, as observed on several occasions by Brown et al. 2008b) may be influenced by chemical cues of their cannibalistic siblings, and thereby they might grow at a different rate and reach metamorphosis at a different time than usual (e.g., compare *Rhinella marina*: tadpoles exposed to conspecific cues during embryonic development and hatching phase had strongly reduced survival and growth rates and reached metamorphosis slightly later; Crossland and Shine 2012).

22.3.3 *Signature Mixtures in* Ranitomeya variabilis

To test if *R. variabilis* is able to distinguish its offspring chemically to feed only its own tadpoles (and not those of other males) during the dry season, frogs were confronted again with in situ pool-choice experiments during tadpole transportation. Pools of each cup-pair were either filled with plain rainwater or with chemical cues of single tadpoles that were found in the males' territories. Genetic kinship analyses were conducted between those tadpoles emitting the chemical cues and those deposited in the same cup-pair. Then, the distribution of the related and unrelated tadpoles (i.e., together with or in the cup next to the emitting tadpole) was compared with each other. However, the results showed that frogs did not choose to deposit their offspring with or without another tadpole due to relatedness, i.e., kin recognition by chemical cues and thereby the use of signature mixtures could not be confirmed in *R. variabilis* (Schulte and Veith 2014).

22.3.4 *Kairomones in* Ranitomeya variabilis

Besides the detection of conspecific tadpoles during parental care behavior, the reaction of *R. variabilis* was tested toward possible chemical cues of several heterospecific tadpoles (*Rhinella poeppigii*, *Osteocephalus mimeticus*, *Amereega trivittata*, *Hyloxalus nexipus*, and *H. azureiventris*) as well as towards predatory damselfly larvae (*M. rotundatum*). The latter, being the most dangerous heterospecific predator of *R. variabilis* tadpoles, was only avoided by visual but not by chemical cues (Schulte et al. 2013). There are two possible explanations: (1) the frogs are just unable to receive or recognize the cues of the damselflies or (2) the damselfly larvae are able to mask or break down their byproducts in a way that they are chemically cryptic in the phytotelmata. Therefore, predator and prey (i.e., their parents) that use the same phytotelmata for breeding might be involved in an evolutionary predator–prey arms race (Brodie and Brodie 1999).

The heterospecific tadpoles in contrast were all recognized chemically by parental *R. variabilis*. Although for tadpole depositions only chemical cues of dendrobatid tadpoles occurring in phytotelmata were avoided (i.e., *H. azureiventris* and *A. trivittata*; see Sect. 3.5), cues of all species were avoided for clutch deposition, no matter whether or not they shared the same breeding resource (Schulte et al. 2011; Schulte and Lötters 2014). Accordingly, it seems that *R. variabilis* recognizes all tadpoles as potential egg predators, and thus uses their chemical excretions as kairomones to protect their clutches.

Additionally, the chemical substances of tadpoles of the genus *Osteocephalus* were recognized by parental *R. variabilis*. However, they were not avoided but preferred for tadpole depositions. Considering that one species of this genus also breeds in phytotelmata, these tadpoles might be recognized as a potential prey for the predatory poison frog larvae—another way to use kairomones to the advantage of their own offspring (Schulte and Lötters 2014).

22.3.5 *Allomones or Synomones in* Ranitomeya variabilis

One of the poison frogs whose tadpoles were avoided chemically by adult *R. variabilis* for both tadpole and clutch depositions was the phytotelmata breeding, non-predatory species *Hyloxalus azureiventris*. To see if the chemical substances released by the tadpoles of this species are identical to those released by the *R. variabilis* tadpoles (i.e., possibly due to their evolutionary relationship), the active chemical substances of both species were analyzed. It was found that each species released two compounds responsible for the avoidance behavior in *R. variabilis*, of which one was the same ($C_9H_{18}N_2O$) and one differed between the two species (*R. variabilis*: C_8H_7NO, *H. azureiventris*: $C_{13}H_{25}NO_2$; Schulte et al. 2015). The species-specific chemical compounds of the *H. azureiventris* tadpoles may be defined as chemical signals since they are advantageous for their non-predatory releasers, preventing the addition of predatory tadpoles into their habitat. Depending on the effect the avoidance has on the predatory *R. variabilis* tadpoles, these inter-specific signals might be defined either as allomones or as synomones. On the one hand, allomones would be only advantageous to the *H. azureiventris*, not the *R. variabilis* tadpoles, which miss out on a potential food resource. On the other hand, synomones would be advantageous for the *R. variabilis* tadpoles, too. This might be the case because the *H. azureiventris* tadpoles outgrow *R. variabilis* tadpoles quickly and therefore might represent a potential competitor, rather than a food resource to the predatory tadpoles (Schulte et al. 2015).

22.4 Conclusion

The overview represented here of semiochemicals used (or not used) by *R. variabilis* (see Fig. 22.2) clearly illustrates how fruitful it can be to carefully analyze different categories of chemical communication to obtain a better overall understanding of the chemical abilities of a species. Although over 80 anuran species are known to react towards semiochemicals, most species have only been tested towards one or two different types of semiochemicals (mostly predator kairomones and alarm pheromones in tadpoles). Only a few species have been examined for the biological relevance of more than two different semiochemical categories. One of them is the American toad *Anaxyrus* (*Bufo*) *americanus*. This species is known to react towards alarm pheromones, both behaviorally (i.e., releaser pheromones; e.g., Mirza et al. 2006) and physiologically (i.e., primer pheromones; Touchon et al. 2006), as well as towards sex pheromones (Forester and Thompson 1998), predator kairomones (e.g., Gallie et al. 2001) and signature mixtures (differentiation between kin and non-kin; Waldman 1985). A similar wide range of the use of semiochemicals is described for the common frog *Rana temporaria*. Like *A. americanus*, this species is known to react toward alarm pheromones both behaviorally (e.g., Marquis et al. 2004) as well as physiologically (e.g., Laurila et al. 2001). It was further shown to react towards kairomones (e.g., Saglio and Mandrillon 2006), but the ability for kin recognition could not be confirmed (Kiseleva 1996). In addition, there is evidence that sexual pheromones might also exist in this species (Willaert et al. 2013).

These examples underline once more how the conduction of different experiments covering several or all categories of semiochemicals, might expand our often unilateral focus and offer us a broader view and a better understanding of chemical communication in animal species. Chemical communication is the evolutionary oldest way of intraspecific and interspecific interactions (Wilson 1970; Wyatt 2010b); the extent of the usage of this communication form over a multitude of closely related species might give us answers regarding the evolution of different communication modes between species (i.e., communication by pheromones versus acoustic communication) and within species (i.e., over different developmental stages).

Acknowledgements I am thankful to Jenna Danielle Kelly for proofreading the manuscript. This research was funded with grants of the "Studienstiftung des deutschen Volkes" and the "Deutsche Forschungsgemeinschaft" (to S. Lötters, M. Veith and W. Brack; LO 1681/1-1).

References

Adams MJ, Claeson S (1998) Field response of tadpoles to conspecific and heterospecific alarm. Ethology 104:955–961

Asey MJ, Harowicz PG, Su L (2005) Chemically mediated mate recognition in the tailed frog (*Ascaphus truei*). In: Mason RT, LeMaster MP, Müller-Schwarze D (eds) Chemical signals in vertebrates. Springer, New York, pp 24–31

Blaustein AR, O'Hara RK, Olson DH (1984) Kin preference behaviour is present after metamorphosis in *Rana cascadae* frogs. Anim Behav 32:445–450

Brodie ED, Brodie ED Jr (1999) Predator-prey arms races. Bioscience 49:557–568

Brown JL, Morales V, Summers K (2008a) Divergence in parental care, habitat selection and larval life history between two species of Peruvian poison frogs: an experimental analysis. J Evol Biol 21:1534–1543

Brown JL, Twomey E, Morales V, Summers K (2008b) Phytotelm size in relation to parental care and mating strategies in two species of Peruvian poison frogs. Behaviour 145:1139–1165

Brown JL, Twomey E, Amézquita A, Barbosa de Souza M, Caldwell JP, Lötters S, von May R, Melo-Sampaio PR, Mejía-Vargas D, Perez-Peña P, Pepper M, Poelman EH, Sanchez-Rodriguez M, Summers K (2011) A taxonomic revision of the neotropical poison frog genus *Ranitomeya* (Amphibia: Dendrobatidae). Zootaxa 3083:1–120

Chivers DP, Mirza RS (2001) Importance of predator diet cues in responses of larval wood frogs to fish and invertebrate predators. J Chem Ecol 27:45–51

Crossland MR, Shine R (2012) Embryonic exposure to conspecific chemicals suppresses cane toad growth and survival. Biol Lett 8:226–229

Eibl-Eibesfeldt I (1949) Über das Vorkommen von Schreckstoffen bei Erdkrötenquappen. Experientia 5:236

Eluvathingal LM, Shanbhag BA, Saidapur SK (2009) Association preference and mechanism of kin recognition in tadpoles of the toad *Bufo melanostictus*. J Biosci 34:435–444

Ferrari MCO, Messier F, Chivers DP (2007) Degradation of chemical alarm cues under natural conditions: risk assessment by larval woodfrogs. Chemoecology 17:263–266

Flowers MA, Graves BM (1997) Juvenile toads avoid chemical cues from snake predators. Anim Behav 53:641–646

Forester DC, Thompson KJ (1998) Gauntlet behaviour as a male sexual tactic in the American toad (Amphibia: Bufonidae). Behaviour 135:99–119

Fraker ME, Hu F, Cuddapah V, McCollum SA, Relyea RA, Hempel J, Denver RJ (2009) Characterization of an alarm pheromone secreted by amphibian tadpoles that induces behavioral inhibition and suppression of the neuroendocrine stress axis. Horm Behav 55:520–529

Gallie JA, Mumme RL, Wissinger SA (2001) Experience has no effect on the development of chemosensory recognition of predators by tadpoles of the American toad, *Bufo americanus*. Herpetologica 57:376–383

Gerhardt HC, Huber F (2002) Acoustic communication in insects and anurans. University of Chicago Press, Chicago

Hagman M, Shine R (2008) Tadpoles of invasive cane toads (*Bufo marinus*) do not respond behaviourally to chemical cues from tadpoles of four species of Australian frogs. Aust J Zool 56:211–213

Hagman M, Hayes RA, Capon RJ, Shine R (2009) Alarm cues experienced by cane toad tadpoles affect post-metamorphic morphology and chemical defences. Funct Ecol 23:126–132

Hews DK, Blaustein AR (1985) An investigation of the alarm response in *Bufo boreas* and *Rana cascadae* tadpoles. Behav Neural Biol 43:47–57

Hrbáček J (1950) On the flight reaction of tadpoles of the common toad caused by chemical substances. Cell Mol Life Sci 6:100–102

Karlson P, Lüscher M (1959) Pheromones: a new term for a class of biologically active substances. Nature 183:55–56

Kiesecker JM, Blaustein AR (1997) Population differences in responses of red-legged frogs (*Rana aurora*) to introduced bullfrogs. Ecology 78:1752–1760

Kiesecker JM, Chivers DP, Anderson M, Blaustein AR (2002) Effect of predator diet on life history shifts of red-legged frogs, *Rana aurora*. J Chem Ecol 28:1007–1015

King JD, Rollins-Smith LA, Nielsen PF, John A, Conlon JM (2005) Characterization of a peptide from skin secretions of male specimens of the frog, *Leptodactylus fallax* that stimulates aggression in male frogs. Peptides 26:597–601

Kiseleva EI (1996) Chemical interaction of frog tadpoles (*Rana temporaria* L., Anura, Amphibia) with conspecific and heterospecific anuran tadpoles. Zh Obshch Biol 57:740–746

Kulzer E (1954) Untersuchungen über die Schreckreaktion der Erdkrötenkaulquappen (*Bufo bufo*). Z Vergl Physiol 36:443

Laurila A, Crochet P-A, Merilä J (2001) Predation-induced effects on hatchling morphology in the common frog (*Rana temporaria*). Can J Zool 79:926–930

Law JH, Regnier FE (1971) Pheromones. Annu Rev Biochem 40:533–548

Lee JSF, Waldman B (2002) Communication by fecal chemosignals in an archaic frog, *Leiopelma hamiltoni*. Copeia 2002:679–686

Lefcort H, Thomson SM, Cowles EE, Harowicz HL, Livaudais BM, Roberts WE, Ettinger WF (1999) Ramifications of predator avoidance: predator and heavy-metal-mediated competition between tadpoles and snails. Ecol Appl 9:1477–1489

Lötters S, Jungfer K-H, Henkel F-W, Schmidt W (2007) Poison frogs: biology, species and captive husbandry. Chimaira, Frankfurt

Mandrillon AL, Saglio P (2007) Effects of embryonic exposure to conspecific chemical cues on hatching and larval traits in the common frog (*Rana temporaria*). Chemoecology 17:169–175

Mandrillon AL, Saglio P (2008) Developmental windows and origins of the chemical cues mediating hatching responses to injured conspecific eggs in the common frog (*Rana temporaria*). Can J Zool 86:447–455

Marquis O, Saglio P, Neveu A (2004) Effects of predators and conspecific chemical cues on the swimming activity of *Rana temporaria* and *Bufo bufo* tadpoles. Arch Hydrobiol 160:153–170

Mirza RS, Ferrari MCO, Kiesecker JM, Chivers DP (2006) Responses of American toad tadpoles to predation cues: behavioural response thresholds, threat-sensitivity and acquired predation recognition. Behaviour 143:877–889

Murray DL, Roth JD, Wirsing AJ (2004) Predation risk avoidance by terrestrial amphibians: the role of prey experience and vulnerability to native and exotic predators. Ethology 110:635–647
Narins PM, Feng AS, Fay RR, Popper AN (2006) Hearing and sound communication in amphibians. Springer, New York
Nordlund DA, Jones RL, Lewis WJ (1981) Semiochemicals: their role in pest control. Wiley, New York
Petranka JW (1989) Response of toad tadpoles to conflicting chemical stimuli: predator avoidance versus optimal foraging. Herpetologica 45:283–292
Pfeiffer W (1966) Die Verbreitung der Schreckreaktion bei Kaulquappen und die Herkunft des Schreckstoffes. J Comp Physiol A 52:79–98
Pfennig DW (1990) "Kin recognition" among spadefoot toad tadpoles: a side-effect of habitat selection? Evolution 44:785–798
Poelman EH, Dicke M (2007) Offering offspring as food to cannibals: oviposition strategies of Amazonian poison frogs (*Dendrobates ventrimaculatus*). Evol Ecol 21:215–227
Polo-Cavia N, Gonzalo A, Lopez P, Martin J (2010) Predator recognition of native but not invasive turtle predators by naive anuran tadpoles. Anim Behav 80:461–466
Poth D, Wollenberg K, Vences M, Schulz S (2012) Volatile amphibian pheromones: macrolides from Mantellid frogs from Madagascar. Angew Chem 51:2187–2190
Rabb GB, Rabb MS (1963) Additional observations on breeding behavior of the Surinam toad, *Pipa pipa*. Copeia 1963:636–642
Risser J (1914) Olfactory reactions in amphibians. J Exp Zool 16:617–652
Saglio P, Mandrillon AL (2006) Embryonic experience to predation risk affects tadpoles of the common frog (*Rana temporaria*). Arch Hydrobiol 166:505–523
Saporito RA, Donnelly MA, Spande TF, Garraffo HM (2012) A review of chemical ecology in poison frogs. Chemoecology 22:159–168
Schulte LM (2014) Feeding or avoiding? Facultative egg feeding in a Peruvian poison frog (*Ranitomeya variabilis*). Ethol Ecol Evol 26:58–68
Schulte LM, Lötters S (2013) The power of the seasons: rainfall triggers parental care in poison frogs. Evol Ecol 27:711–723
Schulte LM, Lötters S (2014) A danger foreseen is a danger avoided: how chemical cues of different tadpoles influence parental decisions of a neotropical poison frog. Anim Cogn 17:267–275
Schulte LM, Rößler DC (2013) Do poison frogs recognize chemical cues of the other sex or do they react to cues of stressed conspecifics? Behav Process 100:32–35
Schulte LM, Veith M (2014) Is the response of a neotropical poison frog (*Ranitomeya variabilis*) to larval chemical cues influenced by relatedness? Herpetol J 24:189–192
Schulte LM, Yeager J, Schulte R, Veith M, Werner P, Beck LA, Lötters S (2011) The smell of success: choice of larval rearing sites by means of chemical cues in a Peruvian poison frog. Anim Behav 81:1147–1154
Schulte LM, Schulte R, Lötters S (2013) Avoiding predation: the importance of chemical and visual cues in poison frog reproductive behaviour. In: East ML, Dehnhard M (eds) Chemical signals in vertebrates 12. Springer, New York, pp 309–321
Schulte LM, Krauss M, Lötters S, Schulze T, Brack W (2015) Decoding and discrimination of chemical cues and signals: avoidance of predation and competition during parental care behavior in sympatric poison frogs. PLoS One 10:E0129929
Steiger S, Schmitt T, Schaefer HM (2011) The origin and dynamic evolution of chemical information transfer. Proc R Soc B Biol Sci 278:970–979
Summers K (1999) The effects of cannibalism on Amazonian poison frog egg and tadpole deposition and survivorship in *Heliconia* axil pools. Oecologia 119:557–564
Summers K, McKeon CS (2004) The evolutionary ecology of phytotelmata use in neotropical poison frogs. Misc Publ Mus Zool Univ Mich 193:55–73

Summers K, McKeon S (2006) The evolution of parental care and egg size: a comparative analysis in frogs. Proc R Soc B Biol Sci 273:687

Touchon JC, Gomez-Mestre I, Warkentin KM (2006) Hatching plasticity in two temperate anurans: responses to a pathogen and predation cues. Can J Zool 84:556–563

Varga L (1928) Ein interessanter Biotop der Biocönose von Wasserorganismen. Biol Zentralbl 48:143–162

Wabnitz PA, Bowie JH, Tyler MJ, Wallace JC, Smith BP (1999) Animal behaviour: aquatic sex pheromone from a male tree frog. Nature 401:444–445

Waldman B (1985) Olfactory basis of kin recognition in toad tadpoles. J Comp Physiol A 156:565–577

Weygoldt P (1987) Evolution of parental care in dart poison frogs (Amphibia: Anura: Dendrobatidae). J Zool Syst Evol Res 25:51–67

Willaert B, Bossuyt F, Janssenswillen S, Adriaens D, Baggerman G, Matthijs S, Pauwels E, Proost P, Raepsaet A, Schoofs L (2013) Frog nuptial pads secrete mating season-specific proteins related to salamander pheromones. J Exp Zool 216:4139–4143

Wilson EO (1970) Chemical communication within animal species. In: Sondheimer E, Simeone JB (eds) Chemical ecology. Academic, New York, pp 133–155

Wilson DJ, Lefcort H (1993) The effect of predator diet on the alarm response of red-legged frog, *Rana aurora*, tadpoles. Anim Behav 46:1017–1019

Wyatt TD (2010a) Pheromones and signature mixtures: defining species-wide signals and variable cues for identity in both invertebrates and vertebrates. J Comp Physiol A 196:685–700

Wyatt TD (2010b) Pheromones and animal behaviour: communication by smell and taste. Cambridge University Press, Cambridge

Chapter 23
Chemical Communication in Archaic New Zealand Frogs

Bruce Waldman

23.1 Introduction

Studies of the social behavior of anuran amphibians traditionally have centered on the role of bioacoustic signaling in mate choice. Only in recent years have researchers begun to realize that frogs and toads communicate through multimodal channels including not only bioacoustic but also visual, seismic, and chemical signals (Starnberger et al. 2014a, b). Amphibians use chemical cues for detecting prey and predators, homing and navigation, territorial defense, alarm signaling, mate choice, and social recognition (reviewed in Waldman and Bishop 2004). By-products of physiological processes incidentally may inform conspecifics or heterospecifics of individuals' sex, diet, size, health, movements, reproductive state, or dominance status. When revealing this information benefits the sender, natural selection may favor the specialization of these cues into signals that foster social communication (Steiger et al. 2011).

While salamanders and caecilians are known to communicate through the chemosensory modality, the importance of chemical signaling in anurans only now is becoming appreciated (Belanger and Corkum 2009; Woodley 2010, 2014). Sex pheromones that may be important in male–male interactions and female mate choice have been identified in aquatic frog species (Wabnitz et al. 1999; Pearl et al. 2000). Some frogs secrete contact courtship pheromones, chemically similar to those produced by salamanders, from their nuptial pads (Willaert et al. 2013), while others release volatile pheromones to which conspecifics respond (Poth et al. 2012). Chemosignals modulate calling behavior in some terrestrial frogs (Byrne and Keogh 2007), and may be broadcast in concert with bioacoustic signals from vocal sac

B. Waldman (✉)
Laboratory of Behavioral and Population Ecology, School of Biological Sciences, Seoul National University, 1 Gwanak-ro, Gwanak-gu, Seoul 151-747, South Korea
e-mail: waldman@snu.ac.kr

B.A. Schulte et al. (eds.), *Chemical Signals in Vertebrates 13*,
DOI 10.1007/978-3-319-22026-0_23

glands (Starnberger et al. 2014b). Even for species that communicate primarily through bioacoustic signaling, chemical communication probably influences social interactions.

Anurans that lack the ability to communicate through bioacoustic signaling, such as the archaic New Zealand frogs (family Leiopelmatidae) and the North American tailed frogs (family Ascaphidae), appear to rely on chemical cues as a primary means of social communication (Lee and Waldman 2002; Waldman and Bishop 2004; Asay et al. 2005). These taxa comprise the most basal lineage of extant anurans (Pyron and Wiens 2011), possessing characters found in Mesozoic fossils (Roček 2000). Although leiopelmatid frogs occasionally emit calls that may startle predators (Bell 1978), these vocalizations lack the structural properties characteristic of anuran advertisement or contact calls and thus are unlikely to serve as signals to conspecifics (Waldman, unpublished data). Furthermore, these frogs lack external eardrums that are characteristic of modern frogs (Lewis and Lombard 1988). Their hearing thus lacks acuity, but anatomical studies suggest that their vomeronasal and olfactory systems are functional (Stephenson 1951, 1955). Possibly the earliest anurans never evolved mechanisms to communicate by bioacoustic signaling (Bogert 1960). Thus, study of their abilities to use chemosignals may offer a window into the early evolution of anuran social systems.

Of the three extant species of leiopelmatid frogs, Hamilton's frog, *Leiopelma hamiltoni*, has the most limited distribution, living only on Maud and Stephens Islands in the Marlborough Sounds, New Zealand. The species is one of the rarest frogs in the world and access to its habitat is strictly regulated. Although frogs on Maud Island are considered by some to be a separate species (named *L. pakeka*) from those on Stephens Island (Bell et al. 1998), molecular analyses suggest that they are best considered different populations of the same species (Holyoake et al. 2001). The frogs are fully terrestrial, occupying small home ranges on the forest floor. They demonstrate site tenacity to rocks, logs, and litter, under which they find shelter during the day. At night, depending on environmental conditions, they emerge and travel over short distances, forage, and periodically interact with conspecifics including potential mates (Webster 2004). The frogs are extremely long-lived, and individuals have been repeatedly censused over 40 years or more in the same area, even under the very same rocks (Bell and Pledger 2010; B.D. Bell, personal communication).

23.2 Materials and Methods

23.2.1 Animal Collection and Odor Sampling

To investigate whether *L. hamiltoni* communicates by chemical cues, my research group conducted a series of field experiments over several years on the social recognition abilities of these frogs. We collected frogs that we found on or under rocks during evening hours. Sometimes several frogs co-occupied particular areas and could be repeatedly found under the same rocks. In such cases, we collected all of the frogs. We mapped their locations and measured distances among them. Distances ranged from 0 to 90 m. We held the frogs for 72 h in plastic containers (33 × 20 × 8 cm) lined with clean, moist paper towels which we stored in a dark, quiet room in the field station on Maud Island. During this time, body secretions, urine, and feces from the subjects were collected on the paper substrates in each container. In this manner, subjects were allowed to "mark" the paper substrates. Different subjects were used in each experimental series, and no subject was tested more than once in any experiment.

23.2.2 Self/Nonself Recognition

First, we conducted tests to determine whether frogs could discriminate between chemical cues that they themselves had deposited on the substrate and those deposited by conspecific individuals. Three series of tests were run, using frogs collected in the same home range (under the same rock), nearby home ranges (less than 5 m apart), or distant home ranges (more than 5 m apart). Twenty frogs were tested in each series.

For all experiments, subjects were placed into a testing apparatus consisting of a plastic container (33 × 20 × 8 cm) divided by a line drawn in its center. On one side, we placed a paper towel that previously had been marked by the test subject itself; on the other side, we placed a paper towel that had been marked by another individual. We removed fecal matter from the towels prior to running the trials, but odors from feces, urine, as well as exocrine gland secretions may have persisted on the towels. Movements on either side of the testing apparatus were recorded for 60 min. Halfway through each trial, we switched sides by rotating the testing apparatus 180° to control for potential biases of subjects to move in particular directions.

23.2.3 Self Attraction vs. Conspecific Avoidance

Frogs might be attracted to odors of any conspecifics (e.g., Graves et al. 1993), but discrimination, if it occurs, may arise from attraction to one's own odor or avoidance of those of conspecifics. To distinguish between these possibilities, we placed frogs into containers and gave them a choice either between a paper towel with which they previously had had contact and an unmarked paper towel, or between an unmarked paper towel and one marked by a conspecific. Conspecifics were collected from a distant home range (more than 5 m away). Movements of subjects on either side of the apparatus were recorded for 60 min, following the same testing protocol as above. Twenty frogs were tested in each series.

23.2.4 Source of Chemical Cues

To determine the source of the chemical cues used in recognition, we next repeated these experiments but using paper towels marked only by specific cue sources. We tested subjects on paper substrate imbued with isolated samples of urine, feces, or skin secretions. Samples were collected from frogs from distant home ranges (more than 5 m apart). Urine was collected from frogs by gently inserting a blunt-ended cannula into each frog's cloaca until urine was released. The urine then was transferred onto the conspecific's side of the apparatus. Feces were collected from the containers in which subjects were housed and stored in sealed vials until needed. Feces were rubbed into the conspecific's substrate so that no tactile information was available to the experimental subject. Skin secretions were collected by swabbing the dorsal surface of a frog with a cotton bud. The contents of the cotton bud then were rubbed into the conspecific's substrate. Twenty-two frogs were tested in each series, following the same testing protocol as above.

23.2.5 Testing Conditions and Analysis

In all experiments, moisture levels were kept similar on both sides of the testing apparatus. Frogs respond differently to odors of larger and smaller frogs (Lee and Waldman 2002), so we matched sizes of frogs used in every test. This species cannot be reliably sexed by external traits, but adult females can be larger than adult males (Bell 1978). Thus, we were unable to break down results by sex. Times subjects spent on either side of the apparatus were summed over both halves of the experiment and, as data were normally distributed, compared by paired *t*-tests. Differences in preferences as a function of distance between home ranges were analyzed by one-way analysis of variance. Statistical analyses were conducted with Minitab 13.30. All statistical inferences were based on two-tailed probabilities.

23.3 Results

23.3.1 Self/Nonself Recognition

Frogs preferred the substrate that they themselves had marked to that marked by a conspecific but only if they had not previously shared a home range with them. The strength of the preference varied depending on the distance between the home ranges of the two individuals (Fig. 23.1). If frogs were collected together, under the same rock, they did not discriminate between the sides ($t_{19}=0.07$, $P=0.94$). Presumably the individuals had become familiar with each other, and their odors, prior to collecting them. However, frogs demonstrated preferences for their own side

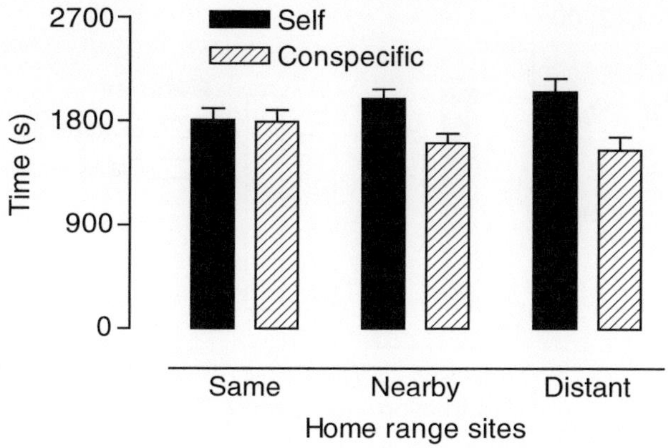

Fig. 23.1 Mean times (+SE) subjects spent on substrate that they had marked themselves and those marked by a conspecific collected in the same home range (under the same rock), nearby home ranges (less than 5 m apart), or distant home ranges (>5 m apart) (modified from Waldman and Bishop 2004)

if the conspecifics were from nearby (t_{19}=2.37, P=0.028) or distant home ranges (t_{19}=2.30, P=0.033). The further apart their home ranges, the stronger were their preferences for their own substrate to that of a conspecific ($F_{2,57}$=3.30, P=0.044). Additional analyses on other behavioral measures confirm these results (Lee and Waldman 2002; Waldman and Bishop 2004).

23.3.2 Self Attraction vs. Conspecific Avoidance

Frogs preferred their own odor to that of an unmarked substrate (t_{19}=2.53, P=0.020) but preferred an unmarked substrate to one marked by a conspecific (t_{19}=2.15, P=0.045) (Fig. 23.2). Thus, individuals recognize and respond both to their own odors and those of conspecifics, moving toward their own marked areas but away from those of others (also see Waldman and Bishop 2004).

23.3.3 Source of Chemical Cues

Urine collected from frogs was sufficient to elicit discrimination (t_{21}=2.11, P=0.047) (Fig. 23.3). In contrast, we did not observe significant preferences for substrate marked with subjects' own feces to those marked by conspecifics (t_{21}=0.96, P=0.35). Discrimination between self and nonself markings was strongest in response to odors collected from swabs of frogs' skin (t_{21}=3.04, P=0.006) (Fig. 23.3). Thus, social discrimination in some contexts may be based on skin secretions or odorants in the urine rather than fecal cues (Waldman and Macfie 2005).

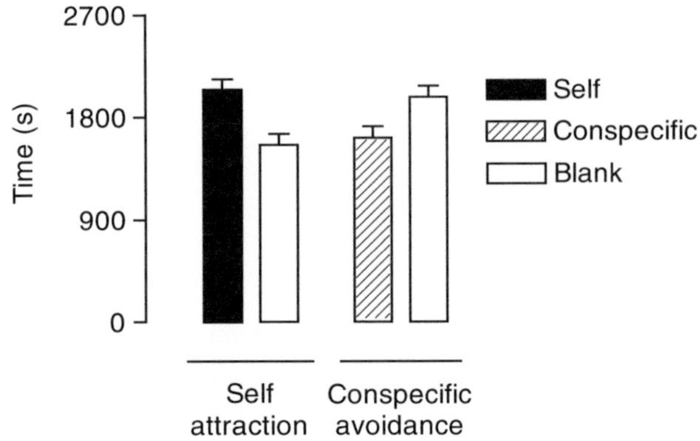

Fig. 23.2 Mean times (+SE) subjects spent on substrate that they had marked themselves and blank substrate (*left*), and on substrate marked by unfamiliar conspecifics (collected >5 m away) and blank substrate (*right*) (modified from Waldman and Bishop 2004)

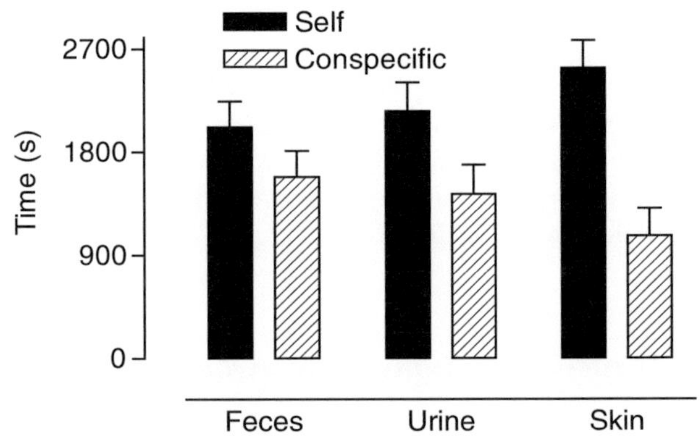

Fig. 23.3 Mean times (+SE) subjects spent on substrates marked with their own urine, feces, or skin secretions and those marked by conspecifics collected from distant home ranges (>5 m apart)

23.4 Discussion

Leiopelma hamiltoni is capable of chemosensory recognition of self, neighbors, and strangers, and maintains the ability to discriminate among these classes even after being held in separate containers for 72 h. Individuals from the same home range appear to tolerate one another and elicit no responses that result in assortative behavior. Individuals from distant home ranges elicit maximal levels of withdrawal behavior. This behavior corresponds with the known dispersal behavior of this species. Individuals travel slowly over small home ranges (<25 m^2) and return to their daytime refuges as morning approaches (Webster 2004). Frogs thus become familiar with the odors of their neighbors and learn to recognize chemical traces that mark their home ranges.

This social discrimination may be useful in a variety of contexts. The ability to discriminate between neighbors from the same or adjacent home ranges and strangers from afar, termed "dear enemy recognition", can be advantageous because it minimizes time and energy expended to maintain exclusive access to resources (Jaeger 1981). Furthermore, chemosignals may function to facilitate cooperation among close kin or selection of non relatives as mates (Madison 1975). The avoidance of close inbreeding, or optimal outbreeding (Bateson 1983), can be particularly important for species such as *L. hamiltoni* that live in highly structured genetic populations (Waldman and McKinnon 1993). The propensity of Hamilton's frogs to travel around fixed home ranges, apparently over their lifetimes, should enable them to recognize and respond appropriately to strangers that may be potential mates or competitors.

The source of the cues used by the frogs for social communication remains to be determined. Skin secretions and urine, and possibly to a lesser extent feces, all seem sufficient to elicit discrimination. In other studies, we found that fecal odors not only were sufficient to elicit discrimination between one's own substrate and those of conspecifics, but effectively conveyed information about individuals' body condition and physiological state. For example, frogs were able to effectively judge the size of conspecifics based on exposure just to their fecal cues, withdrawing from substrates marked by frogs larger than themselves but approaching those marked by subjects smaller than themselves (Lee and Waldman 2002). However, our most recent results suggest that the substrate used in those experiments may have been contaminated with skin secretions or urine.

The skin of *L. hamiltoni* contains mucous glands and two types of granular glands, each of which appears to secrete different peptide mixtures (Melzer et al. 2011). Although thought to be important in predator defense or as antimicrobial peptides that confer disease resistance, the secretions provide a rich source of chemosignals that might be used for social recognition. Further research is needed to analyze the chemical composition and biological properties of these secretions. Aside from information about sex, size, health, and genetic identity, chemical cues may reflect environmental factors, such as the frogs' diet which can vary among home ranges (Bell 1995).

Leiopelma hamiltoni never has been observed to breed in the wild. Because in many aspects of its ecology and behavior, this species resembles *L. archeyi* that lives in parts of New Zealand's North Island, researchers have assumed that, like *L. archeyi*, the frogs breed under rocks or leaf litter, with subsequent paternal care of young (Bell 1978). However, one night while conducting frog surveys on Maud Island, I was surprised to witness a *L. hamiltoni* frog adopting a posture, similar to those used by other species when making advertisement vocalizations, from a crevice in a tree trunk about 3 m above ground (Fig. 23.4). A few hours later, I found the frog amplexed with another less than 50 cm away (Fig. 23.5). These observations raise the possibility that *L. hamiltoni* climbs to find a suitable perch site from which it can more effectively broadcast chemosignals to attract potential mates.

Amphibians worldwide, including New Zealand *Leiopelma* frogs, are declining at precipitous rates as a result of numerous factors including habitat destruction,

Fig. 23.4 *Leiopelma hamiltoni* individual observed on a perch site within a tree crevice about 3 m above ground, possibly broadcasting chemosignals

Fig. 23.5 An amplectant *Leiopelma hamiltoni* pair observed near the perch site from which one had appeared to be signaling immediately before

introduced predators and competitors, chemical contaminants, and emerging infectious diseases. While communication by chemical signaling offers advantages for conveying information about home range boundaries, social status, reproductive condition, and individual identity, chemosignals have inherent properties that also may make frogs more vulnerable to predation (Hamer et al. 2011). Chemical communication systems are especially vulnerable to disruption by anthropogenic change (Park et al. 2001). Pesticides, herbicides, and industrial pollutants, even at low, sublethal concentrations, may have unpredictable effects on the stability of social systems based on communication with chemosignals. Ecotoxicological studies need to be broadened in scope to examine whether chemical noise makes social communication more difficult and interferes with normal reproductive behavior. Knowledge of the underpinnings of how frogs communicate may prove vital to the conservation of threatened and endangered species (Waldman and Tocher 1998).

Acknowledgments I am grateful to two anonymous referees for their comments on the chapter. Much of the research was conducted in collaboration with my students S. F. Lee and K. Macfie, and P. J. Bishop who I invited to the University of Canterbury from the University of the Witwatersrand to work with me on this project. Data analyses were conducted by S. F. Lee, K. Macfie, and me. Many others assisted with the research, including G. Climo, R. Diesel, K. Eggers, L. Ometto, A. Small, K. van de Wolfshaar, and K. Webster. The research was conducted under permits issued by the New Zealand Department of Conservation and the University of Canterbury Animal Ethics Committee. I received a Marsden Fund award (M1069) from the Royal Society of New Zealand for this project. I am grateful to the National Research Foundation of Korea, funded by the government of the Republic of Korea (MSIP) (grants 2010-0002767, 2012R1A1A2044449 and 2012K1A2B1A03000496), and Seoul National University for research support.

References

Asay MJ, Harowicz PG, Su L (2005) Chemically mediated mate recognition in the tailed frog (*Ascaphus truei*). In: Mason RT, LeMaster MP, Müller-Schwarze D (eds) Chemical signals in vertebrates, vol 10. Springer, New York, NY, pp 24–31

Bateson P (1983) Optimal outbreeding. In: Bateson P (ed) Mate choice. Cambridge University Press, Cambridge, pp 257–277

Belanger RM, Corkum LD (2009) A review of aquatic sex pheromones and chemical communication in anurans. J Herpetol 43:184–191

Bell BD (1978) Observations of the ecology and reproduction of the New Zealand leiopelmid frogs. Herpetologica 34:340–354

Bell EA (1995) Habitat use, distribution and population dynamics of the Maud Island frog, *Leiopelma hamiltoni*. MSc thesis, Victoria University of Wellington, Wellington, New Zealand

Bell BD, Pledger SA (2010) How has the remnant population of the threatened frog *Leiopelma pakeka* (Anura: Leiopelmatidae) fared on Maud Island, New Zealand, over the past 25 years? Austral Ecol 35:241–256

Bell BD, Daugherty CH, Hay JM (1998) *Leiopelma pakeka*, n. sp. (Anura: Leiopelmatidae), a cryptic species of frog from Maud Island, New Zealand, and a reassessment of the conservation status of *L. hamiltoni* from Stephens Island. J Roy Soc N Z 28:39–54

Bogert CM (1960) The influence of sound on the behavior of amphibians and reptiles. In: Lanyone WE, Tavolga WN (eds) Animal sounds and communication. American Institute of Biological Sciences, Washington, DC, pp 137–320

Byrne PG, Keogh JS (2007) Terrestrial toadlets use chemosignals to recognise conspecifics, locate mates and strategically adjust calling behaviour. Anim Behav 74:1155–1162

Graves BM, Summers CH, Olmstead KL (1993) Sensory mediation of aggregation among postmetamorphic *Bufo cognatus*. J Herpetol 27:315–319

Hamer R, Lemckert FL, Banks PB (2011) Adult frogs are sensitive to the predation risks of olfactory communication. Biol Lett 7:361–363

Holyoake A, Waldman B, Gemmell NJ (2001) Determining the species status of one of the world's rarest frogs: a conservation dilemma. Anim Conserv 4:29–35

Jaeger RG (1981) Dear enemy recognition and the costs of aggression between salamanders. Am Nat 117:962–974

Lee JSF, Waldman B (2002) Communication by fecal chemosignals in an archaic frog, *Leiopelma hamiltoni*. Copeia 2002:679–686

Lewis ER, Lombard RE (1988) The amphibian ear. In: Fritzsch B, Ryan MJ, Wilczynski W, Hetherington TE, Walkowiak W (eds) The evolution of the amphibian auditory system. Wiley, New York, NY, pp 93–123

Madison DM (1975) Intraspecific odor preferences between salamanders of the same sex: dependence on season and proximity of residence. Can J Zool 53:1356–1361

Melzer S, Clerens S, Bishop PJ (2011) Differential polymorphism in cutaneous glands of archaic *Leiopelma* species. J Morphol 272:1116–1130

Park D, Hempleman SC, Propper CR (2001) Endosulfan exposure disrupts pheromonal systems in the red-spotted newt: a mechanism for subtle effects of environmental chemicals. Environ Health Perspect 109:669–673

Pearl CA, Cervantes M, Chan M, Ho U, Shoji R, Thomas EO (2000) Evidence for a mate-attracting chemosignal in the dwarf African clawed frog *Hymenochirus*. Horm Behav 38:67–74

Poth D, Wollenberg KC, Vences M, Schulz S (2012) Volatile amphibian pheromones: macrolides from mantellid frogs from Madagascar. Angew Chem Int Ed 51:2187–2190

Pyron RA, Wiens JJ (2011) A large-scale phylogeny of Amphibia including over 2800 species, and a revised classification of extant frogs, salamanders, and caeclians. Mol Phylogenet Evol 61:543–583

Roček Z (2000) Mesozoic anurans. In: Heatwole H, Carroll RL (eds) Amphibian biology, vol 4. Surrey Beatty & Sons, Chipping Norton, NSW, pp 1295–1331

Starnberger I, Preininger D, Hödl W (2014a) From uni- to multimodality: towards an integrative view on anuran communication. J Comp Physiol A 200:777–787

Starnberger I, Preininger D, Hödl W (2014b) The anuran vocal sac: a tool for multimodal signalling. Anim Behav 97:281–288

Steiger S, Schmitt T, Schaefer HM (2011) The origin and dynamic evolution of chemical information transfer. Proc R Soc B 278:970–979

Stephenson EM (1951) The anatomy of the head of the New Zealand frog, *Leiopelma*. Trans Zool Soc Lond 27:255–305

Stephenson EM (1955) The head of the frog, *Leiopelma hamiltoni* McCulloch. Proc Zool Soc Lond 124:797–801

Wabnitz PA, Bowie JH, Tyler MJ, Wallace JC, Smith BP (1999) Aquatic sex pheromone from a male tree frog. Nature 401:444–445

Waldman B, Bishop PJ (2004) Chemical communication in an archaic anuran amphibian. Behav Ecol 15:88–93

Waldman B, Macfie K (2005) Chemosignals in the archaic New Zealand frog *Leiopelma hamiltoni*. N Z J Zool 32:231

Waldman B, McKinnon JS (1993) Inbreeding and outbreeding in fishes, amphibians, and reptiles. In: Thornhill NW (ed) The natural history of inbreeding and outbreeding: theoretical and empirical perspectives. University of Chicago Press, Chicago, IL, pp 250–282

Waldman B, Tocher M (1998) Behavioral ecology, genetic diversity, and declining amphibian populations. In: Caro T (ed) Behavioral ecology and conservation biology. Oxford University Press, New York, NY, pp 394–443

Webster JT (2004) Individual identification, disease monitoring and home range of *Leiopelma hamiltoni*. MSc thesis, University of Canterbury, Christchurch, New Zealand

Willaert B, Bossuyt F, Janssenswillen S, Adriaens D, Baggerman G, Matthijs S, Pauwels E, Proost P, Raepsaet A, Schoofs L, Stegen G, Treer D, Van Hoorebeke L, Vandebergh W, Van Bocxlaer I (2013) Frog nuptial pads secrete mating season-specific proteins related to salamander pheromones. J Exp Biol 216:4139–4143

Woodley SK (2010) Pheromonal communication in amphibians. J Comp Physiol A 196:713–727

Woodley SK (2014) Chemical signaling in amphibians. In: Mucignat-Caretta C (ed) Neurobiology of chemical communication. CRC Press, Boca Raton, FL, pp 255–284

Part VI
Studies of Domestic and Zoo-Housed Animals

Chapter 24
Chemical Signals in Giant Panda Urine (*Ailuropoda melanoleuca*)

Martin Dehnhard, Thomas B. Hildebrandt, Cathleen Meerheim, Iain Valentine, and Frank Göritz

24.1 Introduction

The rarest member of the bear family, the Giant panda (*Ailuropoda melanoleuca*) lives mainly in rainy bamboo forests high in the mountains of western China at elevations between 1500 and 3000 m, where they almost entirely feed on the leaves, stems, and shoots of various bamboo species. Restricted and degraded habitat, through logging and agricultural operations, is the greatest threat to giant pandas. Since the mid-1970s three range-wide surveys have been conducted and results from the most recent survey in 2004 indicated a total wild population of 1600 individuals ~40 % more than were thought to exist in the 1980s (http://wwf.panda.org). The Giant panda is a conservation reliant endangered species and establishing self-sustaining panda bear populations in breeding stations and zoos are long-term aims to create healthy, self-sustaining wild panda populations through the release of captive pandas to strengthen small isolated populations of wild pandas. However, a slow reproductive rate in captivity (Hu and Wei 1990) limits the reproductive success of pandas in zoos.

According to Zhang et al. (2004) the most common reasons for mating failure were motivation in males and/or the inability to achieve intromission, even if almost all female pandas displayed strong behavioral estrus. A husbandry reducing stress and establishing a more natural environment are among the factors associated with increased reproduction (Zhang et al. 2004). A 2013 report shows a total number of registered pandas in captivity of 332 individuals held in China and at institutions in

M. Dehnhard (✉) • T.B. Hildebrandt • C. Meerheim • F. Göritz
Leibniz Institute for Zoo and Wildlife Research, PF 601103, 10252 Berlin, Germany
e-mail: dehnhard@izw-berlin.de

I. Valentine
Edinburgh Zoo, Royal Zoological Society of Scotland, Edinburgh, Scotland

B.A. Schulte et al. (eds.), *Chemical Signals in Vertebrates 13*,
DOI 10.1007/978-3-319-22026-0_24

Europe and North America in which only 25 cubs were born in 2011 (International Studbooks for Rare Species of Wild Animals in Captivity 2013).

Unfortunately, panda breeding management is hampered by difficulties inherent in the maintenance and breeding of this species in zoos. This can be attributed, in part, to the unique reproductive life history patterns of Giant pandas. Reproductive biology of pandas is complicated by three main traits, namely experiencing only one single estrus with spontaneous ovulation in the spring (seasonality), delayed implantation and the occurrence of pseudo-pregnancy (Spady et al. 2007).

In the wild, Giant pandas are usually solitary, except during the breeding season between March and May where females may breed with multiple males. Behavioral and non-behavioral cues regulate many aspects of social interactions related to reproduction in mammals. Among the non-behavioral cues, vocal (Charlton et al. 2010) and chemical [pheromones (Karlson and Luscher 1959)] stimuli release specific behavioral or endocrine reactions in the recipient. Mechanisms of pheromone release into the environment are manifold and secretion with urine represents one mechanism. Giant pandas are known to rely heavily on chemical signals released with anogenital gland secretions (AGS) and urine to coordinate their reproductive and social activities (Schaller et al. 1985; Swaisgood et al. 1999; Nie et al. 2012a).

Scent markingin pandas using both sources is higher during the mating season than the non-mating season (Nie et al. 2012a). Male pandas mark their territory by urinating on trees. It has been assumed that they urinate as high as possible so that rival males know how large they are (Swaisgood et al. 2004). Females often mark with urine and AGS, especially around the time of sexual receptivity (Kleiman et al. 1979; Kleiman 1983; Schaller et al. 1985). AGS contain volatile components which consist of a series of short- and long-chain fatty acids (Zhang et al. 2008; Hagey and Macdonald 2003). Female's urine also appears to function as a chemical signal (Swaisgood et al. 2000) and high concentrations of short chain fatty acids have been demonstrated to appear 6–8 days prior to sexual receptivity (Dehnhard et al. 2004). It has been shown that male pandas displayed a significant investigatory preference for estrous over non-estrous female urine, as indicated by sniffing and flehmen (Swaisgood et al. 2002). Thus female scent may play an important role in coordination of mating, e.g., to attract males, to signal estrus, and to act as an aphrodisiac.

Chemical analyses using mass spectrometry identified 951 chemical components from scent glands, urine, and vaginal secretions of pandas collected during the spring breeding season (Hagey and Macdonald 2003). The scent marks of the two genders contained a similar composition of chemicals but varied particularly in the concentration of short chain fatty acids making up more than one half of the gland secretion in males and approximately one eighth in females (Hagey and Macdonald 2003). A study of Zhang et al. (2008) solely found 39 different substances in anogenital gland secretions of pandas collected in December. Short chain fatty acids were not confirmed, however, distinct amounts of long chain fatty acids (C10–C24) were detected. The divergence in compounds among the two studies might be explained by differences in the analytical systems and the season during which samples were collected.

Chemical analyses of urinary compounds are only limited. A study of Yuliang et al. (2012) found 56 substances, 38 of them were putatively identified, whereby fatty acids were not detected. This contrasts to our previous findings demonstrating an increase in fatty acids during spring when ovarian steroid secretion increases (Dehnhard et al. 2004).

The endocrine events associated with the peri-ovulatory period have been described by measuring the corresponding urinary steroid hormone metabolites (McGeehan et al. 2002). The excretions of estrogens by female giant pandas increased approximately 8 days prior to observed matings, were at maximum during the proceptive period, and decreased to basal levels during the period of receptivity (Bonney et al. 1982; Hodges et al. 1984). In captivity, successful reproduction depends on the ability of animal managers to detect estrus for the accurate timing of mate pairing and artificial insemination. Thus, daily estrogen measurements aiming to detect the estrogen peak as well as behavioral and morphological indices are used for reproductive management.

However, a long-lasting continuous estrogen increase which may also include intermittent declines (Bonney et al. 1982) might complicate a precise estrus prognosis because this would require anticipation of the further course; for example, whether an attained high value is followed by a decrease (and thus reflects preovulatory peak concentrations) or by an additional increase closer to ovulation. Thus, urinary pheromones or volatiles with concentrations that are elevated coincident with the preovulatory LH surge might be excellent indicators of ovulation to detect the relatively brief receptive period in female pandas (McGeehan et al. 2002).

We hypothesized that female panda urine might possess estrous-related signals and that urinary volatiles, which might serve as pheromones in panda communication, could be used to evaluate ovarian function. Pheromones may represent more informative indicators of behavioral and physiological conditions than measurements of circulating or excreted hormones, which do not contain communicator functions. Using a combined approach of hormonal analysis and urinary volatile monitoring we aimed to detect reproductive stage-related substances. We selected three different methods for volatile analyses: (1) solid-phase microextraction [SPME; (Arthur and Pawliszyn 1990)], to reproduce data that had been published earlier based on one female panda (Dehnhard et al. 2004), (2) static headspace (SHS) injecting air saturated with volatiles from above the urine, and (3) stir bar sorptive extraction (SBSE) using a TWISTER stir bar adsorbing volatile as well as nonvolatile urinary compounds. Finally, we aimed to identify estrous-related substances and to assess their reliability for monitoring ovarian function.

24.2 Material and Methods

24.2.1 Animal and Sample Collection

One adult female and one adult male (both 11 years) giant pandas maintained in the zoo of Edinburgh were involved in the study. Urine samples from the female were collected on a daily basis between January 21st and October 15th, 2013 and February 28th and September 25th, 2014, respectively. Urine samples from the male were collected irregularly between February 13th and May 21st, 2014. All samples were frozen immediately after collection, stored at –20 °C and shipped to Berlin by overnight express, delivered next day, and kept at –20 °C until analyses.

24.2.2 Determination of Female and Male Reproductive Status

Conjugated steroids were hydrolyzed, extracted, and measured with two enzyme immunoassays (EIA) being specific to either total estrogens or pregnanediol. Hormone determinations were carried out as described in a previous paper (Meyer et al. 1997). Testosterone determinations were carried out following hydrolysis (Meyer et al. 1997) as described in a previous paper (Kretzschmar et al. 2004).

In both years artificial inseminations (AI) were scheduled according to the estrogen courses and performed at the beginning of the post peak decrease indicating the period of receptivity (Bonney et al. 1982; Hodges et al. 1984), additionally confirmed by the detection of the ovulatory LH peak in blood applying a semiquantitative LH assay (WITNESS® LH test).

24.2.3 Solid-Phase Microextraction (SPME)

Solid-phase microextractionuses a fiber coated with an adsorbent that can extract organic compounds from the headspace above urine. Extracted compounds are desorbed upon exposure of the SPME fiber in the heated injector port of a gas chromatograph (GC). SPME was carried out with a CTC Combi Pal system autoinjector at 70 °C for 60 min using a fiber coated with a 85 μm Carboxen/PDMS layer. Sampling was done in the headspace above the surface of 5 ml diluted, Acetate-buffered urine (containing 2 ml urine; 2.5 ml water; 0.5 ml 2 M acetate-buffer, pH 4.8; 1.83 g NaCl, and 2 μg undecanoic acid as internal standard) in 20 ml headspace vials (Shimadzu, Duisburg, Germany).

24.2.4 Static Headspace Analyses (SHS)

Headspace vials (20 ml) were filled with 2 ml diluted, acetate-buffered urine (containing 1 ml urine; 1.0 ml 0.5 M acetate-buffer, pH 4.8 and 2 μg camphor as internal standard). SHS was carried out at an incubation temperature of 70 °C for 30 min using a CTC Combi Pal system autoinjector. Thereafter a volume of 2000 μl from the air above the urine surface was injected into the gas chromatograph (GC) using a syringe temperature of 85 °C.

24.2.5 Stir Bar Sorptive Extraction (SBSE)

SBSE was done in 20 ml headspace vials (Shimadzu, Duisburg, Germany) containing 1 ml urine, 1 ml 0.5 M acetate-buffer (pH 4.8), and 2 μg undecanoic acid as internal standard. Polydimethylsiloxane (PDMS) coated stir-bars (Twisters) for sorptive extraction (Gerstel GmbH, Mulheim an der Ruhr, Germany) were conditioned according to Jakubowska et al. (2009). They consisted of a 10 mm long glass-encapsulated magnetic stir bar with an external PDMS coating of 0.5 mm thickness. The stir bar was inserted into the headspace vial filled with acetate-buffered urine and stirred at 1000 rpm for 60 min at room temperature. After extraction, the stir bar was removed from the sample using a magnetic tweezer and excess of water was dabbed away with a lint-free tissue.

Due to the criteria of higher recoveries of urinary compounds including decanoic acid SBSE was chosen as standard method to continue urine analyses (see Table 24.1). In addition, SBSE also allows enrichment of nonvolatile urinary compounds also adsorbing on the PDMS phase which are undetectable by SPME (Benet et al. 2015). SBSE is also more concerned with ghost peaks or carryover peaks than SPME because it uses 50 μl of sorption media compared to 0.61 μL of sorption medium as typical of 100 μm PDMS fibers (Rodil and Moeder 2008).

Table 24.1 Comparison of extraction efficiency between solid phase microextraction (SPME) and stir bar sorptive extraction (SBSE) based on ten major urinary compounds naturally occurring in panda urine

No	Substance	Retention time	SPME area × 10^6	SBSE area × 10^6
1	Unknown01[a]	9.11	0.54	5.28
2	Methyl salicylate[b]	9.16	1.16	41.97
3	4-vinylphenol[b]	9.36	n. d.	14.93
4	1-(2,4,6-trimethylphenyl)-ethanone[c]	10.63	1.19	15.38
5	Decanoic acid[b]	11.40	0.20	4.93
6	4-(1,1-dimethylethyl)-2-methyl-phenol[c]	12.18	n. d.	41.53
7	Unknown03[a]	13.12	1.74	5.50
8	Unknown04[a]	13.72	0.25	4.44
9	1-(2,6,6-trimethyl-1-cyclohexen-1-yl)-2-butene-1-one[c]	14.60	n. d.	167.10
10	Civetone[c]	21.45	0.15	2.45

n. d. = not detected

[a]Unknown: compounds that did not match >80 % with any substance (NIST MS search)

[b]Compounds verified with authentic standards

[c]Compounds that did match >80 % with substances from library (NIST MS search)

24.2.6 GC-MS Analysis

Sample analysis was performed on an Agilent 7890A Gas Chromatograph (Agilent Technologies, Böblingen, Germany) interfaced with an Agilent 5975C mass selective detector. For SPME we used the rear injector with the splitless injection mode. For SHS as well as SBSE the split/splitless inlet of the front injector was replaced with a Gerstel thermo desorption unit (TDU) which mounts directly onto a Gerstel cooled injection system (CIS). The GC was equipped with an Agilent HP-5MS (5 % phenyl-, 95 % methylsiloxane) fused silica capillary column (30 m × 0.25 mm i.d. × 0.25 μm). The initial oven temperature was held at 45 °C for 2 min, increased to 105 °C at 15 °C/min, then increased to 175 °C at 10 °C/min, and finally increased to 255 °C at 4 °C/min with a final temperature hold of 2 min. Ultrapure helium at a flow rate of 1.1 ml/min was used as the carrier gas. The TDU was operated in the solvent vent mode.

When conducting SHS the TDU was maintained at 250 °C and volatiles were swept into the CIS. The CIS inlet was operated using an initial temperature of 0 °C cryo-focussing the volatiles and ramped up to 300 °C at 720 °C/min with a final temperature hold of 2 min.

When operating SBSE an initial TDU temperature was held at 20 °C for 30 s, and then increased to 200 °C at 60 °C/min with a final temperature hold of 5 min. The TDU thermally desorbs substances adsorbed on the PDMS layer and sweeps these analytes into the CIS as well. The CIS inlet was also operated in the solvent vent mode using an initial temperature of 25 °C ramped up to 320 °C at 720 °C/min with a final temperature hold of 2 min. The purge flow was set at 3.0 ml/min.

Both, the desorption (TDU to CIS) and the injection (CIS to column) were performed in splitless mode at a helium flow of 70 mL/min. A liner filled with glass wool was installed in the CIS.

The CIS and TDU parameters were set using the Gerstel software installed on the GC-MS computer system, which consisted of a Hewlett-Packard computer with Windows 7 and Agilent MSD Chemstation software.

The MS was operated in the EI mode with the electron voltage set at auto-tune value. MS acquisition was performed in TIC (total ion chromatogram). The chemstation data analysis software was used for peak integration and library searches. Spectra of unknown urinary components were compared with spectra of known components stored in the NIST library containing well over 100,000 spectra. Compounds that match (>80 %: methyl salicylate, 4-vinylphenol, decanoic acid) were verified with authentic standards using both, retention time and spectra. Compounds that did not match (<80 %) with substances from library were regarded as unknowns.

24.2.7 Data Analyses

Urinary steroid metabolite concentrations were standardized to creatinine (Cr) levels to control for water content and are given as ng hormone/mg creatinine (crea). For urinary estrogen profiles, an iterative process was used to calculate basal concentrations (Herrick et al. 2010). Briefly, the mean of all samples for the female was calculated and samples with concentrations greater than three standard deviations (mean + 3SD) above this mean were removed. This iterative process was repeated until there were no samples with concentrations greater than the mean + 3SD. The mean of the remaining values was considered as the individual basal concentration. An increase above basal was defined as the day when urinary estrogen concentrations exceeded basal concentrations + 3SD for at least three consecutive samples.

All hormone and volatile profiles shown in the figures refer to the period of receptivity when artificial insemination (AI) was carried out (day 0, see Sect. 24.2.2).

24.3 Results

24.3.1 Seasonal Profile of Reproductive Hormones

Figure 24.1 shows the course of urinary steroid metabolites in 2013. During January to March basal excretion of estradiol remained essentially constant at approximately 2.6 ng/mg creatinine (crea). On the 8th of April (day −13) estrogens increased above

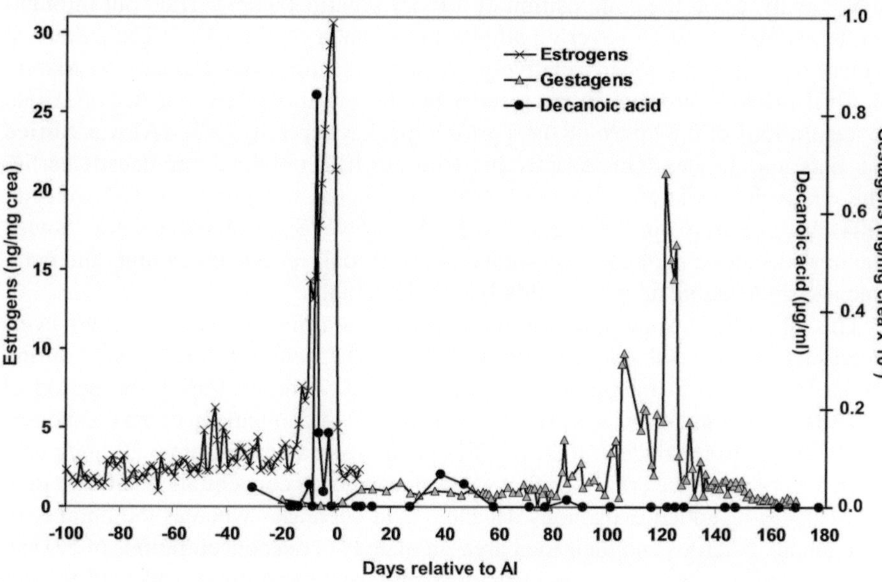

Fig. 24.1 Profiles of urinary estrogens, gestagens and the stir bar sorptive extraction (SBSE) profile of decanoic acid throughout the year 2013. Hormone concentrations were standardized to creatinine (crea) levels to control for water content and are given as ng hormone/mg creatinine. Day 0 is defined as the day of artificial insemination (AI) corresponding to peak receptivity (confirmed by the detection of the ovulatory LH peak in blood)

basal reaching maximal concentrations of 32.8 ng/mg crea 2 days prior to artificial insemination (AI: April 21, equivalent to peak receptivity). Urinary estrogen concentrations fell to baseline levels by 23rd of April and remained low thereafter. Following AI urinary gestagen analyses (pregnanediol) were performed to monitor luteal activity. The slight increase in gestagens mid-July (day 85) might reflect the period of delayed implantation, the distinct increase above basal + 3SD on the 6th of August (282 ng/mg crea, day 105) probably indicates implantation and thus the beginning of pregnancy, which was terminated early September due to an abortion (as evidenced by an ultrasound investigation in the following year).

A similar profile was obtained in 2014 (data not shown) with estrogen concentrations increasing above basal on the 31th of March (day −13) and achieving maximal concentrations between the 9th and 11th of April and dropping to basal levels by the 13th of April, the date when AI was carried out (data not shown). Similar to 2013 luteal activity increased above basal on the 12th of July (day 90) end ended in early September without delivery. Thus, we suspected this period of higher gestagen activity signaled a pseudo-pregnancy.

24.3.2 SPME: Is the Increase in Urinary Fatty Acids Unique Among Individual Pandas?

SPME analyses on the composition of urinary volatiles were carried out throughout the entire year 2013 covering all phases of panda reproduction. The pattern of urinary volatiles remained relatively constant during from January to March. In April urinary fatty acids increased whereby decanoic acid reached maximal concentrations of 0.84 μg/ml 8 days prior to peak receptivity (when AI was carried out, data not shown). This exactly fits with our data obtained one decade earlier where fatty acids likewise peaked 8 days prior to peak receptivity (Dehnhard et al. 2004). In contrast to our 2004 report on pandas, however, this female panda showed maximal decanoic acid concentrations that were only one-tenth as high and octanoic and dodecanoic acid were only barely detectable.

This might be attributed to extreme differences in urinary creatinine (crea) levels increasing from basal concentrations of 1.06 ± 0.75 mg/ml (mean ± SD) during January to March to maximal concentration of 10.7 mg/ml during the period of intense estrogen secretion in April (data not shown). A comparison of area responses in relation to urine concentration by diluting an urine sample from 6.20 mg/g crea down to 0.14 mg/g crea revealed that when crea exceeds concentrations of 2.48 μg/g the recovery of added exogenous decanoic acid decreases whereas the number of endogenous volatiles and their total area stagnated at crea concentrations of 5.0 and 6.2 μg/g. This might indicate an inadequate buffering capacity as well as fiber saturation. Therefore we applied stir bar sorptive extraction (SBSE) using PDMS coated stir-bars (Twister).

24.3.3 Stir Bar Sorptive Extraction (SBSE)

24.3.3.1 Comparison of the Sensitivity of SBSE and SPME

To comparethe extraction efficiency between SBSE and SPME we used ten endogenous substances some of them identified by computer MS library research and confirmed with those of the authentic standards. Table 24.1 shows a distinct higher recovery when using SBSE. Substances undetectable (n. d.) when applying SPME became clearly detectable following SBSE (no 3, 6, and 9), whereas other substances were extracted with 3- (no 7) up to 25-fold higher recoveries (decanoic acid) when applying SBSE. Therefore SBSE was chosen as standard method to continue urine analyses.

24.3.3.2 Identification of Urinary Volatiles Related to Estrus

Fatty Acids

To increase the buffering capacity and to account for often limited amounts of urine, 1 ml of urine was buffered with an equal amount of 0.5 M acetate buffer adjusting the pH to ~4.8 irrespective diverging urine concentrations. The analysis of the composition of urinary volatiles throughout the samples collected in 2013 reveals dramatic changes during the period of estrogen secretion in April. Again an increase in decanoic acid during that period was confirmed peaking 5 days following the urinary estrogen increase (day −13, see Sect. 24.3.1) and therefore 8 days prior to the day of peak receptivity (Fig. 24.1: day 0). Decanoic acid levels decreased thereafter until the end of the year except a weak intermediate increase at the end of May (day 37). A similar profile was seen in the same set of samples when SPME was used (see Sect. 24.3.2).

The composition of urinary volatiles in samples collected 2014 changed dramatically during the period of estrogen secretion in April. This time the increase in decanoic acid reached peak levels 7 days following the estrogen increase (day −13) and 6 days prior to the day of peak receptivity (data not shown). Decanoic acid levels decreased thereafter until the end of the collection period except two weak intermediate increases on days 34 and 85 (May 17th, July 7th).

Other (Unknown) Estrus Related Substances

Similar to decanoic acid an unknown substance appeared in panda urine during the phase of intense estrogen secretion in 2013 that reached peak levels on day −3 and remained at low levels until the end of the year (Fig. 24.2). The mass spectra of the substance that has yet to be identified is characterized by an intense mass fragment at *m/z* 127 (see Fig. 24.2 insert). A similar seasonal pattern was also obtained in

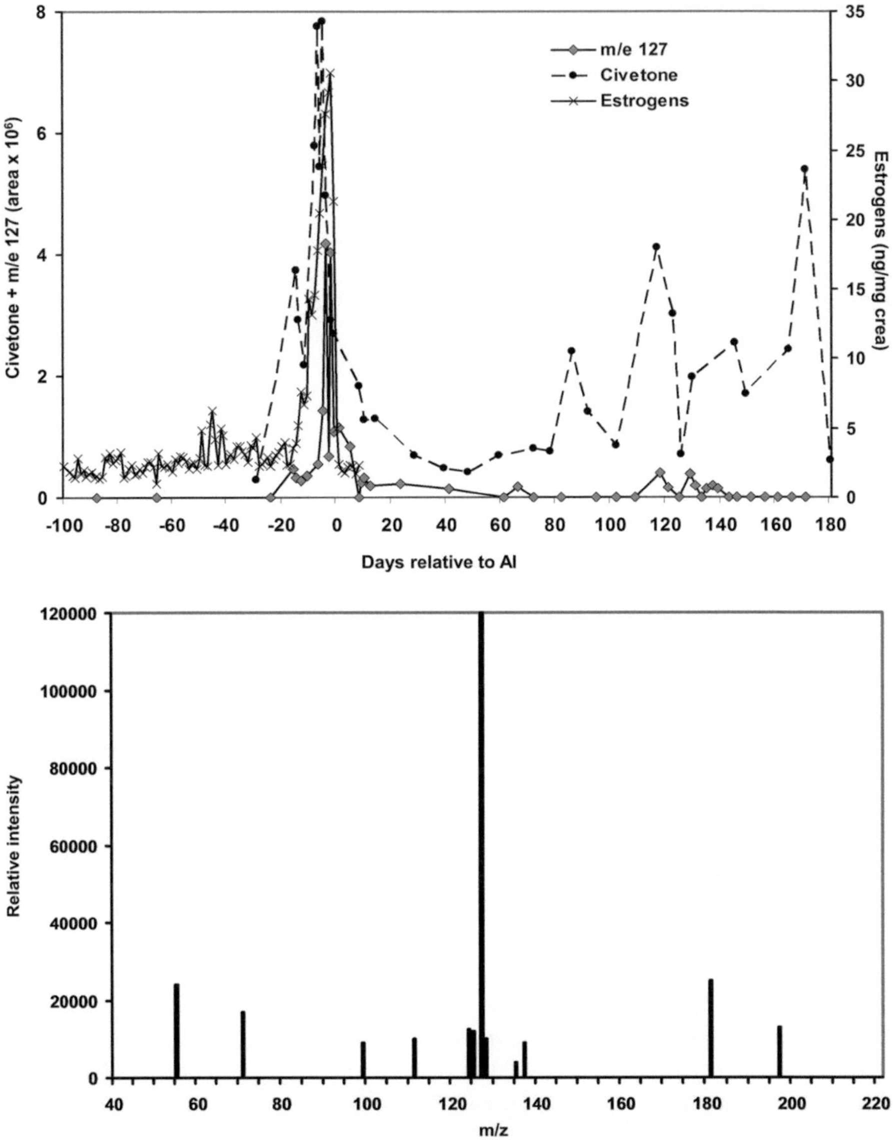

Fig. 24.2 SBSE profiles of an unknown substance with a distinct mass fragment of 127 (see *insert*) and civetone in relation to the urinary estrogen profile during the phase of intensive estrogen secretion and thereafter throughout the year 2013. Day 0 corresponds to the day of AI

2014 when the unidentified substance reached maximal concentrations on day −2. In contrast to decanoic acid, which has been demonstrated in the Berlin panda Yan Yan during the two consecutive years 2002 and 2003, we were unable to detect this unknown substance in GCMS files from SPME analyses of 2002 and 2003 based

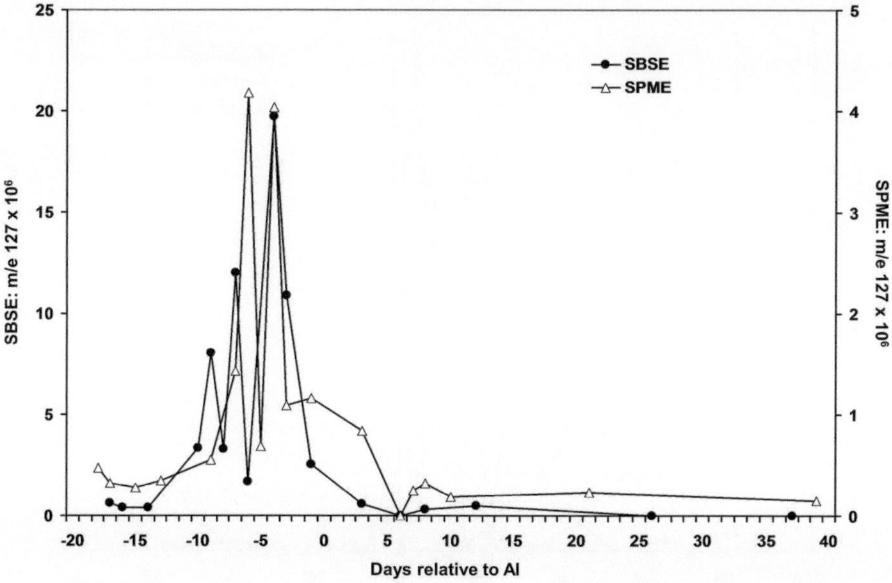

Fig. 24.3 Comparative SBSE and solid phase microextraction (SPME) profiles of urinary decanoic acid throughout the year 2013

on retrospective evaluations. Thus we cannot exclude the *m/z* 127 substance as an individual phenomenon, as it was clearly detectable in the samples from 2013 when SPME was applied (see Fig. 24.3). In addition, *m/z* 127 and its estrogen dependency was also confirmed in the samples collected in 2014 (data not shown) when applying SBSE. Regarding *m/z* 127 SBSE was about five-times more efficient than SPME when comparing samples from 2013 (Fig. 24.3).

In addition to the appearance of decanoic acid and the *m/z* 127 substance several other compounds became detectable during the episode of intensive estrogen secretion in April such as civetone (see Fig. 24.2) and 1-(4-methylphenyl)-ethanone ($C_9H_{10}O$: 93 % match with substance from NIST MS library, not shown). Despite low concentrations in the subsequent months, both substances increased again in mid-July (civetone) and September (methylphenyl-ethanone), respectively, remaining at high levels until the end of the year as shown for civetone (Fig. 24.2) following patterns unrelated to urinary gestagens.

24.3.3.3 Substances with Temporal Relationship to Gestagenes

During the episode of increased luteal activity in August 2013 the pattern of urinary volatiles changed again whereby two substances α-terpineol and endo-borneol (match >80 % with substances from NIST MS library) reached peak levels during the episode of highest gestagen production between August 19th and 26th (not shown).

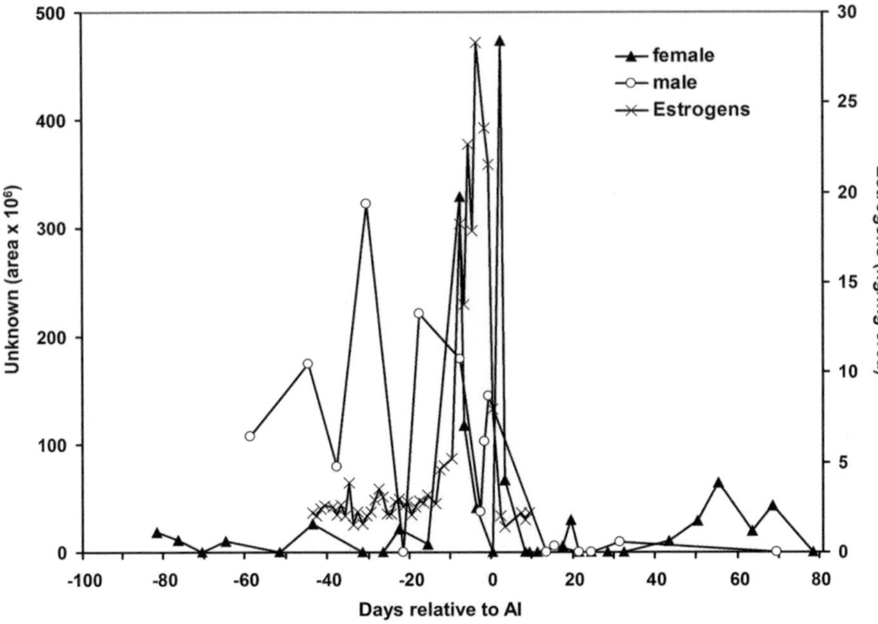

Fig. 24.4 Results of static headspace analyses (SHS) of urine samples of a female and a male panda during 2014 showing the substance at *m/z* 135/192 in comparison. In addition the profile of urinary estrogens is shown

A different result was obtained in samples from the subsequent year when both compounds increased late in the year but were not associated with a high level of luteal activity.

24.3.4 Static Headspace Analyses (SHS)

Samples were analyzed as described in material and methods. Best outcome was reached with incubation at 70 °C for 30 min and a pH-adjustment at 4.8. Adding of salt and enzyme (β-glucuronidase from Helix pomatia) did not improve the analysis (data not shown).

Through SHS only a few compounds were detectable in the urine. One reoccurring substance was an unidentified compound with the *m/z* of 135/192 and a retention time at 11.74 min. The substance reached peak levels on the 15th of April, 2 days following artificial insemination (Fig. 24.4). The substance was found in urine samples from 2013 as well, where its concentration clearly peaked on the 16th of April (here 6 days prior to A.I., data not shown).

In addition, urine samples from the male panda Yang Guang in Edinburgh collected in 2014 were analyzed with SHS. Contrary to our expectations, urine from the male panda contained this substance as well. Peak levels for Yang Guang were reached on March 13th (Fig. 24.4).

The substance was found in urine samples from both animals until the end of June, when SHS analyses were terminated in favor of SBSE. However, in both animals concentrations dropped to lower levels during April and remained low until the end of June.

24.4 Discussion

The Giant panda (*Ailuropoda melanoleuca*) is an endangered species native to southwestern China living in fragmented habitats and being vulnerable to continued human threats, especially habitat loss. By 2004, there were estimated to be 1600 pandas alive in the wild (IUCN red list). Thus ex situ breeding programs play an important role in the study and recovery of endangered species. One of the many challenges to such programs is establishing a mating protocol that simulates, as closely as possible, what normally occurs under natural conditions. Because females are mono-estrous with a single receptive period of 1–3 days each spring (Kleiman 1983; Schaller et al. 1985) successful reproduction in captivity depends on the ability of animal managers to detect estrus for the accurate management of pairing and artificial insemination. The most useful approach is a skilled combination of monitoring urinary estrogen hormone profiles and careful observations of reproductive behaviors (Zhang et al. 2004).

Giant pandas are a solitary, seasonally mono-estrous species, and as such they certainly rely on chemical signals to coordinate mating efforts (Schaller et al. 1985; Swaisgood et al. 2002). Males displayed a significant investigatory preference for estrous over non-estrous female urine (Swaisgood et al. 2002).

Our systematic SBSE investigations of the composition of urinary volatiles in the female panda throughout two consecutive years revealed a characteristic increase of urinary decanoic acid associated with the seasonal estrogen increase which triggers estrus behavior. The appearance of decanoic acid was also demonstrated in our earlier study where we investigated urine from another female panda during two consecutive years (Dehnhard et al. 2004). Even with a sample size of only two animals, the estrous related appearance of decanoic acid in consecutive years implies that this substance might serve as a pheromone to attract males to females which should be emitted in time before ovulation. In nature, females approaching estrus may attract up to five males that travel long distances and then assemble in the female's home range to compete for copulation opportunities (Schaller et al. 1985; Nie et al. 2012a, b). In captivity it has been shown that 71.0 % of the panda cubs were produced from matings occurring on the day of or the day after the urinary estrogen peak (Huang et al. 2012). In our study AI was performed on the day following the estrogen peak and ovulation was confirmed in 2013 and 2014 by semiquantitative detection of the LH surge in blood plasma when AI was carried out (F. Göritz, personal communication). The occurrence of the decanoic acid peak 6 and 8 days prior to peak receptivity is similar to our earlier findings (Dehnhard et al. 2004) and might indicate a role of attracting males. This is corroborated with the observation that first scent marking activity usually occurs 6 days prior to the estrogen maximum (McGeehan et al. 2002).

The origin of decanoic acid is unknown. However, our lab identified decanoic acid in four samples of bladder urine obtained between 1999 and 2003. In addition, decanoic acid has not been found in anogenital gland secretion (AGS) of giant pandas (Zhang et al. 2008) which might rule out the possibility of an extrarenal source from accessory glands of the reproductive tract. However, long chain fatty acids from tridecanoic acid to tetracosanoic (C24) acid had been shown to occur in AGS (Zhang et al. 2008). So far decanoic acid has been found in temporal gland secretions of Asian elephant bulls (*Elephas maximus*) during musth (Rasmussen et al. 1990) and in scents on cheeks and foreheads of the tiger (*Panthera tigris*) emitted during facial marking behavior (Soini et al. 2012). In both species decanoic acid was part of a cocktail of chemical compounds and a pheromonal action of the substance was not investigated.

In the panda comparisons of urinary volatiles can also be complicated by the appearance of a metabolic by-product due to changes in bamboo quality. Profiling of volatile compounds of bamboo from Taiwan has shown changes in urinary constituents and their relative contents when comparing spring and winter bamboo shoots. This particularly concerns the concentrations of methyl salicylate that differed markedly between spring and winter (Chung et al. 2012). This substance has been clearly demonstrated recently in panda urine (Table 24.1) but also in our former analyses. Thus when searching for reproductive state related volatiles that may have semiochemical functions it is important to differentiate between hormone and nutrition caused changes in the profile of substance. Thus, feeding the first spring bamboo might interfere with the appearance of estrus related volatiles. In addition, the increase in urine concentration from approximately 1.06 + 0.75 mg/ml and 0.57 + 0.36 mg/ml creatinine up to 10.7 mg/ml and 9.5 mg/ml, respectively, in 2013 and 2014 during the follicular phase might also impair the recovery of volatiles. Thus an adequate buffering capacity should be chosen to exclude an effect of pH on the recovery of decanoic acid.

When applying static headspace analysis (SHS) we assumed that we would obtain a pattern reflecting similar to what the panda could smell. However, only a few compounds of the urine became detectable. Among them was an unknown substance that reached peak levels in April which were not closely related to estrogen secretion as they occur 6 days before (2013) and 2 days after the day of AI (2014). At approximately the same time, this substance was also detected in urine samples from the male panda coinciding with high urinary testosterone concentrations, but became undetectable when urinary testosterone levels decreased to zero (data not shown). The results from both animals might indicate a relationship to reproductive activities, however, additional analyses and substance identification are needed until potential physiological aspects of the substance can be discussed.

In addition, with SBSE we were able to demonstrate the presence of an unknown substance with a marked ion of 127. The substance shows a clear estrogen dependency increasing during the follicular phase in two consecutive years. Unfortunately our finding is limited to one animal and reanalyses of GCMS data collected 10 years ago did not reveal evidence for this substance in the panda investigated in our previous paper (Dehnhard et al. 2004).

In this paper we also confirmed the presence of civeton in panda urine, a substance which was absent in urines of 20 other mammal species we investigated earlier (Dehnhard et al. 2004). Urinary civetone did not reflect a cycle that can be clearly attributed to reproductive activity. The first distinct increase appears during the estrogenic phase in April but increases were also measured late in September and October. Currently, we have no evidence for a signal function of civetone. Its continuous but variable presence in female panda urine and its absence in other investigated species might imply a biological function. Civetone belongs to the macrocyclic musk compounds. In the wild the powerful musky odor of panda urine had been detected from as far as 5 m (Schaller et al. 1985). The exocrine odor glands of the musk deer (*Moschus moschiferus*) also produce an intensive musk odor which is used by female musk deer for tail marking which attracts male partners over great distances during the rutting season (Meng et al. 2008). There is a distinct possibility that civetone in panda urine might serve as an ingredient for territorial marking in females.

We obtained no evidence for volatile substances that might depend on luteal activity and, thus, may be used as pregnancy indicators as previously described in the Asian elephant (Dehnhard et al. 2003). In pandas, pregnancy diagnosis is complicated by the occurrence of pseudo-pregnancies; which are characterized by a similar duration of luteal gestagen production after infertile matings but also infertile AI. Thus a "true" pregnancy indicator as recently described for felids (Dehnhard et al. 2012) would be of great diagnostic value. However, a comparison between samples from this study between 2013 (when the animal was pregnant but aborted) and 2014 (when the animal was pseudo-pregnant) is inappropriate because of SBSE methodological improvements between the 2 years.

In conclusion, the role of decanoic acid as a pheromone remains to be investigated and studies on their behavioral relevance are required. Is has been shown that environmental enrichment can enhance welfare of captive animals through provisioning of stimuli needed for optimal physiological and psychological well-being which has gained popularity in the last few decades (Rafacz and Santymire 2014). If decanoic acid is of biological relevance for male pandas it could be used as olfactory stimuli to enhance their reproductive behavior and, therefore may have implications for captive-breeding programs.

Acknowledgements We thank the keepers from Edinburgh Zoo for collecting urine samples from the two pandas.

References

Arthur CL, Pawliszyn J (1990) Solid-phase microextraction with thermal-desorption using fused-silica optical fibers. Anal Chem 62:2145–2148

Benet I, Guardia MD, Ibanez C, Sola J, Arnau J, Roura E (2015) Analysis of SPME or SBSE extracted volatile compounds from cooked cured pork ham differing in intramuscular fat profiles. LWT Food Sci Technol 60:393–399

Bonney RC, Wood DJ, Kleiman DG (1982) Endocrine correlates of behavioural oestrus in the female giant panda (*Ailuropoda melaneleuca*) and associated hormonal changes in the male. J Reprod Fertil 64:209–215

Charlton BD, Keating JL, Rengui L, Huang Y, Swaisgood RR (2010) Female giant panda (*Ailuropoda melanoleuca*) chirps advertise the caller's fertile phase. Proc Biol Sci 277: 1101–1106

Chung MJ, Cheng SS, Lin CY, Chang ST (2012) Profiling of volatile compounds of Phyllostachys pubescens shoots in Taiwan. Food Chem 134:1732–1737

Dehnhard M, Hatt JM, Eulenberger K, Ochs A, Strauss G (2003) Headspace solid-phase microextraction (SPME) and gas chromatography-mass spectrometry (GC-MS) for the determination of 5alpha-androst-2-en-17-one and -17beta-ol in the female Asian elephant: application for reproductive monitoring and prediction of parturition. J Steroid Biochem Mol Biol 84:383–391

Dehnhard M, Hildebrandt T, Knauf T, Ochs A, Ringleb J, Göritz F (2004) Chemical signals in Giant panda urine (*Ailuropoda melanoleuca*). In: Chemical Signals in Vertebrates, vol 10., pp 110–117

Dehnhard M, Finkenwirth C, Crosier A, Penfold L, Ringleb J, Jewgenow K (2012) Using PGFM (13,14-dihydro-15-keto-prostaglandin F2alpha) as a non-invasive pregnancy marker for felids. Theriogenology 77:1088–1099

Hagey L, Macdonald E (2003) Chemical cues identify gender and individuality in giant pandas (*Ailuropoda melanoleuca*). J Chem Ecol 29:1479–1488

Herrick JR, Bond JB, Campbell M, Levens G, Moore T, Benson K, D'Agostino J, West G, Okeson DM, Coke R, Portacio SC, Leiske K, Kreider C, Polumbo PJ, Swanson WF (2010) Fecal endocrine profiles and ejaculate traits in black-footed cats (*Felis nigripes*) and sand cats (*Felis margarita*). Gen Comp Endocrinol 165:204–214

Hodges JK, Bevan DJ, Celma M, Hearn JP, Jones DM, Kleiman DG, Knight JA, Moore HDM (1984) Aspects of the reproductive endocrinology of the female giant panda (*Ailuropoda melanoleuca*) in captivity with special reference to the detection of ovulation and pregnancy. J Zool 203:253–267

Hu JC, Wei F (1990) Development and progress of breeding and rearing giant pandas in captivity within China. In: Hu J, Wei F, Yuan C, Wu Y (eds) Research and progress in biology of the giant panda. Sichuan Publishing, Sichuan, pp 322–325

Huang Y, Zhang H, Li D, Zhang G, Wei R, Huang Z, Zhou Y, Zhou Q, Liu Y, Wildt DE, Hull V (2012) Relationship of the estrogen surge and multiple mates to cub paternity in the giant panda (*Ailuropoda melanoleuca*): implications for optimal timing of copulation or artificial insemination. Biol Reprod 87:112

International Studbooks for Rare Species of Wild Animals in Captivity (2013). International Zoo Yearbook 47:420

Jakubowska N, Henkelmann B, Schramm KW, Namiesnik J (2009) Optimization of a novel procedure for determination of VOCs in water and human urine samples based on SBSE coupled with TD-GC-HRMS. J Chromatogr Sci 47:689–693

Karlson P, Luscher M (1959) Pheromones': a new term for a class of biologically active substances. Nature 183:55–56

Kleiman DG (1983) Ethology and reproduction of captive giant pandas (*Ailuropoda, Melanoleuca*). Z Tierpsychol 62:1–46

Kleiman DG, Karesh WB, Chu PR (1979) Behavioural changes associated with oestrus in the giant panda (*Ailuropoda melanoleuca*) with comments on female proceptive behaviour. In: International Zoo Yearbook, vol 19., pp 217–223

Kretzschmar P, Ganslosser U, Dehnhard M (2004) Relationship between androgens, environmental factors and reproductive behavior in male white rhinoceros (*Ceratotherium simum simum*). Horm Behav 45:1–9

McGeehan L, Li XB, Jackintell L, Huang SQ, Wang AP, Czekala NM (2002) Hormonal and behavioral correlates of estrus in captive giant pandas. Zoo Biol 21:449–466

Meng XX, Perkins GC, Yang QS, Feng ZJ, Meng ZB, Xu HF (2008) Relationship between estrus cycles and behavioral durations of captive female alpine musk deer. Integr Zool 3:143–148

Meyer HH, Rohleder M, Streich WJ, Goltenboth R, Ochs A (1997) Sex steroid profiles and ovarian activities of the female panda Yan Yan in the Berlin Zoo. Berl Munch Tierarztl Wochenschr 110:143–147

Nie YG, Swaisgood RR, Zhang ZJ, Hu YB, Ma YS, Wei FW (2012a) Giant panda scent-marking strategies in the wild: role of season, sex and marking surface. Anim Behav 84:39–44

Nie YG, Zhang ZJ, Swaisgood RR, Wei FW (2012b) Effects of season and social interaction on fecal testosterone metabolites in wild male giant pandas: implications for energetics and mating strategies. Eur J Wildl Res 58:235–241

Rafacz ML, Santymire RM (2014) Using odor cues to elicit a behavioral and hormonal response in zoo-housed African wild dogs. Zoo Biol 33:144–149

Rasmussen LE, Hess DL, Haight JD (1990) Chemical analysis of temporal gland secretions collected from an Asian bull elephant during a four-month musth episode. J Chem Ecol 16:2167–2181

Rodil R, Moeder M (2008) Stir bar sorptive extraction coupled to thermodesorption-gas chromatography-mass spectrometry for the determination of insect repelling substances in water samples. J Chromatogr A 1178:9–16

Schaller GB, Jinchu H, Pan W, Zhu J (1985) The giant pandas of Wolong. University of Chicago Press, Chicago, IL, p 298

Soini HA, Linville SU, Wiesler D, Posto AL, Williams DR, Novotny MV (2012) Investigation of scents on cheeks and foreheads of large felines in connection to the facial marking behavior. J Chem Ecol 38:145–156

Spady TJ, Lindburg DG, Durrant BS (2007) Evolution of reproductive seasonality in bears. Mammal Rev 37:21–53

Swaisgood RR, Lindburg DG, Zhou X (1999) Giant pandas discriminate individual differences in conspecific scent. Anim Behav 57:1045–1053

Swaisgood RR, Lindburg DG, Zhou X, Owen MA (2000) The effects of sex, reproductive condition and context on discrimination of conspecific odours by giant pandas. Anim Behav 60:227–237

Swaisgood RR, Lindburg DG, Zhang H (2002) Discrimination of oestrous status in giant pandas (*Ailuropoda melanoleuca*) via chemical cues in urine. J Zool 257:381–386

Swaisgood RR, Lindburg D, White AM, Zhang H, Zhou X (2004) Chemical communication in giant pandas: experimentation and application. In: Lindburg D, Baragona K (eds) Giant pandas: biology and conservation. University of California Press, Berkeley, CA, pp 106–120

Yuliang L, Qiuhong X, Dingzhen L, Zhigang C, Zhihe Z, Jingchao L, Rongping W, Desheng L, Rong H (2012) The application of HS-SPME-GC/MS in detecting chemical components in giant panda (*Ailuropoda melanoleuca*) urine. Acta Theriologica Sinica 32:48–57

Zhang GQ, Swaisgood RR, Zhang HM (2004) Evaluation of behavioral factors influencing reproductive success and failure in captive giant pandas. Zoo Biol 23:15–31

Zhang JX, Liu D, Sun L, Wei R, Zhang G, Wu H, Zhang H, Zhao C (2008) Potential chemosignals in the anogenital gland secretion of giant pandas, *Ailuropoda melanoleuca*, associated with sex and individual identity. J Chem Ecol 34:398–407

Chapter 25
Semiochemical Communication in Dogs in the Context of Reproductive Behavior

Michał Dzięcioł, Antoni Szumny, Wojciech Niżański, Marcin Wrzosek, Ewa Stańczyk, and Joanna Polikowska

25.1 Semiochemical Communication

Semiochemical signaling is one of the oldest means of communication used by organisms of all taxa. This kind of communication plays an important role in initiating and modifying many types of behavior, including sexual behavior (Archunan 2009; Dzięcioł et al. 2012b; Stevens et al. 1982; Pageat and Gaultier 2003; Raymer et al. 1986; Tod et al. 2005; Vázquez and Orihuela 2001). Taking into account very strong impact of sex pheromones on the behavior as well as on physiology of signals' recipient (releaser and primer effect), the understanding of mechanism of action of this group of semiochemical substances could be beneficial for better understanding of whole semiochemical communication mechanisms. In this context, considering the length of estrus period, strength of the expression of sexual behavior and level of domestication allowing for close, direct observation of undisturbed natural reactions and reflexes, dog seems to be a good experimental model. The aim of this study was the analysis of the chosen aspects of the semiochemical communication in animals in the context of sex pheromones on the model of domestic dogs (*Canis familiaris*).

M. Dzięcioł (✉) • W. Niżański • E. Stańczyk
Department of Reproduction and Clinic of Farm Animals, Wrocław University of Environmental and Life Sciences, 50-366 Wrocław, Poland
e-mail: michaldzieciol@wp.pl

A. Szumny • J. Polikowska
Department of Chemistry, Wroclaw University of Environmental and Life Sciences, 50-375 Wrocław, Poland

M. Wrzosek
Centre for Experimental Diagnostics and Biomedical Innovations, University of Environmental and Life Sciences, 50-366 Wrocław, Poland

B.A. Schulte et al. (eds.), *Chemical Signals in Vertebrates 13*,
DOI 10.1007/978-3-319-22026-0_25

25.2 What Do We Know About Canine Sex Pheromones?

The rapid development of sophisticated, modern, analytic methods of identification of chemical compounds (GC/MS) in the second half of the twentieth century held great promise of fast and easy identification of active substances able to modify animal behavior. The search for sex pheromones in dogs was performed mostly in the 1970s and 1980s; however, the results obtained by different researchers present conflicting opinions. In the study by Goodwin et al. (1979), methyl *p-hydroxybenzoate* (methyl paraben, MP) was postulated to be the main sex pheromone in dogs. When this compound (MP) was applied to the vulva of bitches that were not in estrus, male dogs behaved toward the females as though they were in estrus. As the most often-cited article in the area of canine sex pheromones, this study seemed to settle the question of the identity of the primary canine sex pheromone (Kustritz 2005; Kutzler 2005; Pageat and Gaultier 2003; Person 1985; Santos et al. 2013; Wani et al. 2013). However, some authors exploring the issue of canine sex attractants did not share the opinion that methyl paraben was the estrous pheromone. Kruse and Howard (1983) stated that methyl *p*-hydroxybenzoate cannot be considered as a key sexual attractant for male dogs. The study by Schultz et al. (1985) as well as results obtained by Dzięcioł et al. (2014c) did not confirm the presence of methyl paraben in samples of urine collected from bitches during estrus. Santos et al. (2013) obtained better quality semen from dogs after stimulation with methyl paraben during electroejaculation procedures, but these results could not confirm its effectiveness because these males also received an injection of prostaglandins, which could have significantly influenced the results (Estienne and Harper 2000; Kozink et al. 2002). Moreover, Santos et al. (2013) clearly stated that in their study they did not observe any excitement of the males during contact with methyl paraben.

Even though there have been many critical opinions over the usefulness of methyl paraben as a sex pheromone in dogs, there still exists a common belief that this substance is effective and can be useful in veterinary practice as well as in breeding. This feeling is supported by the presence of a commercially available product called Eau D'Estrus (Synbiotics, USA) containing methyl paraben as a major component. However, similar to the observations of Santos et al. (2013), we did not note any kind of interest of male dogs either toward bitches marked with MP or during contact with samples of cotton swabs soaked in MP (this method of stimulation was also suggested by the producer of artificial pheromones) (Dzięcioł et al. 2011). Taking into account the common presence of methyl paraben in our environment (cosmetics, food preservatives, etc.), previous reports presenting MP as a compound of canine urine could result from contamination: in the studies by Goodwin et al. (1979) and also by Kruse and Howard (1983), the samples of urine examined were collected from the floor. This idea could be correct considering that other compounds identified during the Goodwin et al. (1979) studies nowadays are usually identified as artifacts comprising very popular industrial plasticizers.

25.3 Sex Pheromones in Dogs: Suspected Structure

In animal reproduction, female attractiveness to a male, which is essential for proper conduct of the subsequent sexual act, is determined mostly by the presence of specific semiochemical substances emitted into the environment (Kustritz 2005; Mertens 2006).

According to current knowledge, semiochemicals usually act not as a single substance, but rather as mixtures of a few compounds that are active together. Also, while a few different types of chemical substances are usually considered as candidate pheromones, however, volatile compounds are the predominant type in most reports. Evaluation of male dogs' behavior as a response to the message seems to be important for proper analysis of chemical signaling of bitches during estrus.

During analysis of stud dogs' behavior, a few phases of breeding behavior could be distinguished. The first step is approaching the bitch, which serves as a reaction to volatile substances, because this signal is recognizable from a great distance. Although this signal is present from the first to the last day of estrus, during the first few days the females, however attractive, do not allow the males to copulate. From the biological point of view, this mechanism could be beneficial because during this time a number of males are able to receive the signal and close in on the females. This is also probably the time for competition among the male dogs and when the strongest male can dominate other competitors.

The second step during the process of recognizing the status of the female after approaching is sniffing from close proximity of the different parts of her body (mostly the ano-genital region, the tarsal region, and muzzle areas). It appears important that this phase usually involves intensive licking behavior. During the act of licking, a temporary immobilization is usually observed and the ano-vaginal region is the predominant place of interest. Moreover, males like to sniff and lick objects (grass, etc.) marked by the bitch with urine, a type of behavior that is often observed in other species (cattle, elephants, and others) (Rasmussen et al. 1982; Marchlewska-Koj 2011). The final decision (whether to mate or not) is usually made after completing both of the aforementioned behaviors. This behavioral pattern might suggest a complex characteristic of the message sent by the female. Thus, a nonvolatile property of one or more of the substances collected by males during licking should be strongly considered. Such substances could be transported straight into the vomeronasal entry area. This route of semiochemical detection also is suspected because of a specific behavioral characteristic observed in dogs—the equivalent of the Flehmen response—called tonguing, which has been frequently observed after the male dogs lick a urine spot (Case 2010).

The complex character of semiochemical signals emitted by the female in estrus could also be suspected considering the suggestion of some authors, and what has also been strongly confirmed by many breeders, that some male dogs are "able to determine not just if the bitch is in estrus, but if she is at the optimal time for breeding" (Kustritz 2005). In our previous study, the behavior of experienced stud dogs toward females in different stages of estrus (confirmed by vaginal cytology, blood progesterone levels) was evaluated. We observed that, while there was increased interest of the males (sniffing, licking the ano-vaginal region of the bitches), the mating decision in most cases was not made until females were at the optimal time for mating (Dzięcioł et al. 2012a). The fact that males were able to distinguish between specific phases of estrus seems to confirm the complex character of the chemical message sent by females. Taking into account the duration of the canine heat period, with proestrus and estrus lasting approximately 9 days each and dynamic changes in hormone levels occurring, the idea mentioned above appears to be consistent with the observations of other authors who found that changes in the levels of reproductive hormones could lead to changes in the composition of the semiochemical signal (Raymer et al. 1986).

It is worth noting that the above-described kinds of skills are present only in some dogs, and usually in the older, more experienced males that are kept in kennels in close contact with the females. This observation could indicate that learning is important in the development of such skills. A learning process could also be indicated in the context of acquired skills of dogs, mice, and rats that are able to detect the chemical signals of estrus of a different species, i.e., ruminants (Jezierski 1992). However, the question if learned skills are involved in detection of the most fertile period in bitches needs further study.

25.4 The Meaning of Fully Recognizable Signals

The role of bacterial flora in the production of semiochemical substances has been confirmed in both animals and humans (Dravnieks et al. 1970; Michael et al. 1974, 1975; Albone et al. 1978; Bird and Gower 1982; Jackman and Noble 1983; Gower et al. 1986, 1994; Kohl et al. 2001; Konopski and Koberda 2003).

Changes in the number of bacteria in the environment of the bitches vagina are linked closely with the ovarian cycle (Allen and Dagnall 1982; Baba et al. 1983; Watts et al. 1996; Grundy et al. 2002; Janowski et al. 2006; Zduńczyk et al. 2006; Ulgen et al. 2010). During anestrus, which usually lasts several (5–11) months, the vagina remains almost sterile. The number of bacteria starts to increase at the beginning of estrus and persists throughout proestrus and estrus. At the end of estrus, a sudden decrease in cornified cells, characteristic for the fertile period, can be noted. This process is combined with an increased number of leukocytes that eliminate bacteria from the bitch's vagina. At the same time, a sudden decrease or even absence of attractiveness of the females toward the males is observed. These findings show the correlation between the presence of bacteria in the canine vagina and the emission of a semiochemical signal into the environment.

In our previous study, we observed reduced attractiveness of bitches in estrus treated with antibiotics toward experienced stud dogs (Dzięcioł et al. 2013). Moreover, we proved that this observation was not caused by the presence of the antibiotic carrier (Myglyol 480), because no aversion behavior by the males was observed during their contact with this compound. However, while the presence of metabolites of administered drugs could potentially be involved in the altered "smell" of the females, it is also very probable that for the production of the final semiochemical signal (sex pheromones) the presence of the physiological vaginal flora is necessary (Dzięcioł et al. 2013). Similar results were also observed in different species by Ungerfeld and Silva (2005) as well by Merkx et al. (1988).

Marking females in estrus with specific chemicals (repellents) is the popular method for reducing their attractiveness. The low effectiveness of this method prohibits its classification as a recommended contraceptive method. However, interesting notes were made during observation of the behavior of males toward females wearing collars containing so-called appeasing pheromones, which are described as analogs of the substances produced in the skin glands of the mammary gland of the females just a few days after parturition (Pageat and Gaultier 2003; Dzięcioł et al. 2014a). According to preliminary data, the simultaneous presence of two different kind of semiochemicals, sex pheromones (estrus) and maternal pheromones (released from the collar) characteristic for lactation, reduced the mating attempts by some experienced stud dogs (Dzięcioł et al. 2014a). The underlying mechanism of this possible interaction is still unknown and needs further research.

25.5 How to Investigate Sex Pheromones in Dogs?

The classical scheme of research dedicated to pheromone identification in vertebrates includes: collection of the material, chemical evaluation of the samples (usually by GC/MS and similar methods), identification of the putative pheromones, synthesis of the substances suspected of the semiochemical activity, and biological, behavioral tests confirming their activity. However, this ideal schedule in practice seems to be very difficult to perform. There are several reasons for this. Aside from the fact that the detection limits of the analytical equipment used remain coarser than the detection capability of the animals' receptors, the complex structure of the semiochemical signal causes problems in creating an artificial equivalent of the natural signal. Furthermore, in vertebrates the critical part of analysis of the semiochemical interaction is the last step: evaluation of the reaction of the recipient to the presented putative pheromone. Many methods have been described, but the predominant one still seems to be analysis of the behavior of the individual receiving the proffered signal. Unfortunately, in many cases this method is far from perfect, since different temperaments, degrees of reactivity, as well as the influence of more or less predictable additional factors, can all significantly influence visible behavior, and the skill of the observer can also affect the final results.

Methods based on evaluation of the bioelectrical activity of the neural system are to be used in insects for detection of semiochemical events (Konopski and Koberda 2003). However, in mammals, these methods carry limitations due to ethical objections. Also atraumatic EEG (Electroencephalography) examination did not support its usefulness in this kind of study, because of difficulties with interpretation of the results obtained (pers. obs.). The numerous artifacts connected with measurement of muscle activity can potentially be eliminated by pharmacological sedation (immobilization); however, the use of sedative drugs usually significantly reduces, or in stronger sedation almost eliminates signals that might be evaluated. Moreover, knowledge about the neural pathway connecting the VNO (vomeronasal organ) mainly with the brain's subcortical structures raise doubts whether evaluation of cortical electrical activity could be relevant and sufficient for this kind of analysis (Konopski and Koberda 2003; Marchlewska-Koj 2011).

The usefulness of evaluation of the heart rate (HR) in the context of sexual arousal and response of the organism toward sexual pheromones were evaluated in animals and humans (Dzięcioł et al. 2012b). However, even though it was proved that semiochemicals could stimulate HR, this method seems to have limited utility because of its reduced specificity (non-specific stimuli such as any kind of stress could be also the reason for increased HR) (Galosy et al. 1979).

Looking for a more specific test measuring the reactivity of the recipient to semiochemical stimuli, perfusion of the vessels directing blood into the corpora cavernosa was evaluated (Dzięcioł et al. 2012b). It was proved previously that non-contact stimulation from females in estrus evokes penile erection and this reaction is specific for estrous signals (Sachs et al. 1994; Sachs 1997). While in dogs the penis has a fibro-cavernous structure, the erection is strictly due to the increased perfusion of this organ. In our previous study, we showed that evaluation of the blood flow in this vessel could be a specific and sensitive method for evaluating the reaction to the semiochemical stimulation, since the presentation of natural sex pheromones significantly increased blood flow in the vessels sending blood into the corpora cavernosa (Dzięcioł et al. 2012b). While due to the small size of the vessels examined, the traditional method of blood flow evaluation (pulsed Doppler) is difficult to perform, the color Doppler technique seems to be satisfactory when combined with objective post-examination, digital evaluation of the results using PIXEL FLUX software (Chameleon, Germany), dedicated for dynamic tissue perfusion measurement (Dzięcioł et al. 2014b). Unfortunately, this method also seems to have limitations because it requires trained dogs, familiar with the long-lasting ultrasound examination, because the stress connected with this kind of procedure in untrained dogs could easily inhibit natural arousal, and a surge of the stress hormones (adrenaline, corticosteroids) can modify blood flow in the organs of the animal (e.g., central sequestration of the blood).

Another new and promising method for evaluating the effect of sex pheromones on the physiology of the recipient of a specific signal is the modern technique of brain activity evaluation (PET, positron emission tomography, and fMRI, functional magnetic resonance imaging). In humans, gender-specific chemical compounds presented to individuals of the opposite sex caused increased activity in particular parts of the brain (Savic et al. 2005). In dogs, an evaluation of brain activity was performed in the context of voice-sensitive recognition (Attila et al. 2014). Investigations into the response of the canine brain toward the natural sex pheromones are in progress and we can expect that implementation of this technique will be beneficial for studies on canine sex pheromones.

25.6 Conclusion

Despite the many published reports and studies investigating the canine sex pheromones, there are still many unanswered questions and further studies are required. While most of the efforts attempting to identify the active semiochemical compounds in canine urine were performed at least 30 years ago and frequently present opposing conclusions, renewed attempts using modern, more sensitive methods of detection should be considered. Moreover according to current knowledge and taking into account the results of analysis of male dog behavior, such attempts should probably focus not only on the volatile but also on nonvolatile compounds. In this context, special attention should be given to the role of proteins.

Identification of the complex structure of sex pheromones in dogs, taking into account its possible practical applications, could be beneficial for the breeders and owners. It could contribute to creation of a better environment for coexistence of animals and humans. Moreover preparing model of experiment allowing for sex pheromones identification could be also very useful in future, in managing the reproduction of captive canines including endangered species.

Considering wider aspect of studies concerned to identification of sex pheromones in animals it is worth to notice that except the possible different sources of semiochemicals, the pathway dedicated to identification of putative pheromones seems to be similar in various species (GC/MS and similar methods). On the other hand, choosing the methods of evaluation of the recipient reaction greater account should be taken on species specificity (e.g., different type of penis combined with different mechanism of erection—dog and bull; presence or absence of typical Flehmen response). In context of species specificity probably also the cooperation of the additional (A.O.B.) and the main olfactory bulb (M.O.B.) in estrus detection should also be considered.

Acknowledgments Presented results were obtained during studies supported by Statutory Research and Development Activity Funds assigned to the Faculty of Veterinary Medicine, Wroclaw University of Environment and Life Sciences.

References

Albone ES, Gosden PE, Ware GC, MacDonald DW, Hough NG (1978) Bacterial action and chemical signalling in the red fox. Proc Am Chem Soc 67:78–91

Allen W, Dagnall GJR (1982) Some observations on the aerobic bacterial flora of the genital tract of the dog and bitch. J Small Anim Pract 23:325–335

Archunan G (2009) Vertebrate pheromones and their biological importance. J Exp Zool India 12:227–239

Attila A, Gacsi M, Kis T, Miklosi A (2014) Voice-sensitive regions in the dog and human brain are revealed by comparative fMRI. Curr Biol 4:574–578

Baba E, Hata H, Fukata T, Arakava A (1983) Vaginal and uterine microflora of adult dogs. Am J Vet Res 44:606–609

Bird S, Gower DB (1982) Axillary androstenone, cholesterol and squalene in men: preliminary evidence for androstenone being a product of bacterial action. J Steroid Biochem Mol Biol 17:517–522

Case LP (2010) Social behavior & communication in dogs and cats. In: Garza D (ed) Canine and feline behaviour and training: a complete guide to understanding our two best friends. Delmar, New York, NY, pp 48–49

Dravnieks A, Krotoszynski BK, Keith L, Bush IM (1970) Odor threshold and gas -chromatographic assays of vaginal odors: changes with nitrofurazone treatment. J Pharm Sci 59:495–501

Dzięcioł M, Niżański W, Ochota M, Kozdrowski R, Twardoń J (2011) The usefulness of the synthetic bitch sex pheromone for stimulation of reproductive reflexes in dogs (*Canis familiaris*). Reprod Domest Anim 46(Suppl 3):101

Dzięcioł M, Niżański W, Ochota M, Kozdrowski R, Stańczyk E (2012a) Observation on possibility to identify by the stud dogs the signs of the fertile period in bitches. JAVA 11:962–967

Dzięcioł M, Stańczyk E, Noszczyk-Nowak A, Niżański W, Ochota M, Kozdrowski R (2012b) Influence of bitches sex pheromones on the heart rate and other chosen parameters of blood flow in stud dogs (*Canis familiaris*). Res Vet Sci 93:1241–1247

Dzięcioł M, Niżański W, Stańczyk E, Kozdrowski R, Najder-Kozdrowska L (2013) The influence of antibiotic treatment of bitches in oestrus on their attractiveness to males during mating. Pol J Vet Sci 16:509–516

Dzięcioł M, De Porter TL, Niżański W (2014a) Can dog appeasing pheromones (D.A.P.®) influence the reproductive attractiveness of a female dog? ICSE-CSiV 8-12 July, Urbana-Champaign, USA

Dzięcioł M, Scholbach T, Stańczyk E, Ostrowska J, Kinda W, Woźniak M, Atamaniuk W, Skrzypczak P, Niżański W, Wieczorek A, Scholbach J, Kiełbowicz Z (2014b) Dynamic tissue perfusion measurement in the reproductive organs of the female and male dog. Bull Vet Inst Pulawy 58:149–155

Dzięcioł M, Politowicz J, Szumny A, Nižanski W (2014c) Methyl paraben as sex pheromone in canine urine-is the question still open? Pol J Vet Sci 17:601–605

Estienne MJ, Harper AF (2000) PGF2alpha facilitates the training of sexually active boars for semen collection. Theriogenology 54:1087–1092

Galosy RA, Clarke LK, Mitchell JH (1979) Cardiac changes during behavioral stress in dogs. Am J Physiol Heart Circ Physiol 236:750–758

Goodwin M, Gooding KM, Regnier F (1979) Sex pheromone in the dog. Science 203:559–561

Gower DB, Nixon A, Jackman PJH, Malet AI (1986) Transformation of steroids by axillary coryneform bacteria. Int Cosm Sci 8:149–158

Gower DB, Holland KT, Mallet AI, Rennie PJ, Watkins WJ (1994) Comparison of 16androstene steroid concentrations in sterile apocrine sweat and axillary secretions: interconversions of 16-androstenes by the axillary microflora-mechanism for axillary odour production in man. J Ster Biochem Molec Biol 48:409–418

Grundy SA, Feldman E, Davidson A (2002) Evaluation of infertility in the bitch. Clin Tech Small Anim Pract 17:108–115

Jackman PJ, Noble WC (1983) Normal axillary skin microflora in various populations. Clin Exp Dermatol 8:259–268

Janowski T, Zduńczyk S, Jurczak A (2006) Bacterial flora of the genital tract in healthy bitches during the oestrous cycle and in bitches with vaginal discharge. Reprod Dom Anim 41 (suppl 1):17

Jezierski T (1992) The effectiveness of estrus detection in cows by a trained dog. Anim Sci Pap Rep 10:57–66

Kohl J, Atzmueller M, Fink B, Grammer K (2001) Human pheromones: integrating neuroendocrinology and ethology. Neuro Endocrinol Lett 22:309–321

Konopski L, Koberda M (2003) Feromony człowieka. Środki komunikacji chemicznej między ludźmi. Scholar, Warszawa, pp 129–167

Kozink DM, Estienne MJ, Harper AF, Knight JW (2002) The effect of lutalyse on the training of sexually inexperienced boars for semen collection. Theriogenology 58:1039–1045
Kruse SM, Howard WE (1983) Canid sex attractant studies. J Chem Ecol 9:1503–1510
Kustritz MVR (2005) Reproductive behavior in small animals. Theriogenology 64:734–746
Kutzler MA (2005) Semen collection in the dog. Theriogenology 64:747–754
Marchlewska-Koj A (2011) Rozród – feromony sygnalizujące. In: Grzesiak A (ed) Feromony ssaków – ich rola w fizjologii i zachowaniu. Polska Akademia Umiejętności, Kraków, pp 61–63
Merkx J, van der Werff ten Slob AK, Bosch JJ (1988) Vaginal bacterial flora partially determines sexual attractivity of female rats. Physiol Behav 44:147–149
Mertens PA (2006) Reproductive and sexual behavioral problems in dogs. Theriogenology 66:606–609
Michael RP, Bonsall RW, Warner P (1974) Human vaginal secretion: volatile fatty acid content. Science 186:1217–1219
Michael RP, Bonsall RW, Kutner M (1975) Volatile fatty acids "Copulins", in human vaginal secretions. Psychoneuroendocrinology 1:153–163
Pageat P, Gaultier E (2003) Current research in canine and feline pheromones. Vet Clin Small Anim Pract 33:187–211
Person JR (1985) Mounting evidence of paraben sensitivity in dogs. Arch Dermatol 121:1107
Rasmussen LE, Schmidt MJ, Henneous R, Groves D, Daves GD Jr (1982) Asian bull elephants: flehmen like response to extractable compounds in female elephant estrous urine. Science 217:159–160
Raymer JD, Wiesler D, Novotny M et al (1986) Chemical scent constituents in urine of wolf (*Canis lupus*) and their dependence on reproductive hormones. J Chem Ecol 12:297–314
Sachs BD (1997) Erection evoked in male rats by airborne scent from estrus females. Physiol Behav 62:921–924
Sachs BD, Akasofu K, Citron JH, Daniels SB, Natoli JH (1994) Noncontact stimulation from estrus females evokes penile erection in rats. Physiol Behav 55:1073–1079
Santos IP, Ramos CL, Ramos JL, Oliveira RF, Cunha IC (2013) Efficient association between PGF2α and methyl 4-hydroxybenzoate sex pheromone prior to electroejaculation in dogs. Reprod Domest Anim 48:160–164
Savic I, Berglund H, Lindstrom P (2005) Brain response to putative pheromones in homosexual men. Proc Natl Acad Sci U S A 102:7356–7361
Schultz TH, Kruse SM, Flath RA (1985) Some volatile constituents of female dog urine. J Chem Ecol 11:169–175
Stevens K, Perry GC, Long S (1982) Effect of ewe urine and vaginal secretions on ram investigative behavior. J Chem Ecol 8:23–29
Tod E, Brander D, Waran N (2005) Efficacy of dog appeasing pheromone in reducing stress and fear related behavior in shelter dogs. Appl Anim Behav Sci 93:295–308
Ulgen G, Kaan O, Aytekin G, Mihriban U (2010) Vaginal, cervical and uterine bacterial flora at the different stages of the reproductive cycle in ovariohysterectomized bitches. JAVA 9:478–481
Ungerfeld R, Silva L (2005) The presence of normal vaginal flora is necessary for normal sexual attractiveness of estrous ewes. Appl Anim Behav Sci 93:245–250
Vázquez R, Orihuela A (2001) Effect of vaginal mucus and urine from ewes in estrus on plasma testosterone levels and weight gain of feedlot rams. Small Rum Res 42:173–177
Wani AA, Dhindsa SS, Shafi TA, Chowdhary SRA, Kumar B (2013) The role of pheromones in animal reproduction – a review. Prog Res 8:14–18
Watts JR, Wright PJ, Whithear KC (1996) Uterine, cervical and vaginal microflora of the normal bitch. J Small Anim Pract 37:54–60
Zduńczyk S, Janowski T, Borkowska I (2006) Vaginal and uterine bacterial flora in bitches with physiological and inflammatory conditions. Med Wet 62:1116–1119

Chapter 26
Hormonal and Behavioral Responses to Odor Cues in Zoo-Housed African Painted Dogs (*Lycaon pictus*)

Michelle L. Rafacz, Matthew R. Heintz, and Rachel M. Santymire

26.1 Introduction

African painted dogs (APD; *Lycaon pictus*), the largest canid in Africa, are specialist predators of medium-sized antelope (Creel and Creel 1995; Woodroffe and Sillero-Zubiri 2012). In the wild, APDs always live in packs containing a potential breeding pair, and they disperse as single sex groups of either sex. In captivity, however, they are often kept in single-sex groups to prevent reproduction.

Environmental enrichment is a tool that can be used to improve the welfare of zoo-housed animals because it can increase the diversity of natural species-specific behavior by promoting interaction with the environment (Shepherdson 1998; Clark and King 2008). Olfactory enrichment has primarily been studied in zoo-housed carnivores, especially large felids (for a review see Clark and King 2008). Only a few studies have investigated the effect of olfactory enrichment on activity levels in APDs (Packard et al. 2010; Price 2010; Rafacz and Santymire 2014). Recently, we investigated the effect of fecal odors from a predator/competitor (lion; *Panthera leo*), and two prey species (gazelle; *Gazella granti*), and (cattle; *Bos taurus*) on both behavior and stress physiology of a pair of older male APDs using non-invasive measurements of fecal glucocorticoid metabolites (FGM) and unobtrusive behavioral observations (Rafacz and Santymire 2014). The gazelle odor elicited an increase in activity in both dominant and subordinate males, but only the subordinate individual had a decrease in FGM concentrations following presentation of fecal odors. Our results suggested that exposure to odors from natural prey may be used as enrichment, and that changes in stress may depend on

M.L. Rafacz (✉)
Science and Mathematics Department, Columbia College Chicago, Chicago, IL 60605, USA
e-mail: mrafacz@colum.edu

M.R. Heintz • R.M. Santymire
Lincoln Park Zoo, Conservation and Science Department, Chicago, IL 60614, USA

B.A. Schulte et al. (eds.), *Chemical Signals in Vertebrates 13*,
DOI 10.1007/978-3-319-22026-0_26

social rank (Rafacz and Santymire 2014). Thus, we used methods identical to those used in Rafacz and Santymire (2014) to investigate whether feces from the lion, Grant's gazelle and cattle affects the activity and FGM of a younger, all-female group of zoo-housed APDs.

26.2 Materials and Methods

26.2.1 Subjects

Four zoo-born female APDs (3-year-old sisters) were housed together at Lincoln Park Zoo (LPZ; Chicago, IL) in an outdoor exhibit space (approximately 225 m^2) with several structures. The females had access to both the exhibit and an off-exhibit indoor holding area at most times, depending on the weather and the needs for husbandry. They were individually fed a meat-based diet with water provided *ad libitum*. We distinguished individuals by distinctive markings and assigned social rank from keeper judgments. Female APD-1 and APD-2 were considered dominant, with APD-1 as the highest ranked, and APD-3 and APD-4 were subordinate, with APD-4 being the lowest ranked. Research was approved through the LPZ Research Committee and complied with all legal requirements for research use in the United States.

26.2.2 Odor Cues and Behavioral Measurements

Behavioral data were collected using instantaneous scan-sampling with a 1-min fixed interval and all-occurrence methods (Altmann 1974). An ethogram modified from Rafacz and Santymire (2014) was used. Behaviors were divided into two categories: inactive comprised sleeping either alone or together and active included locomote, forage, other, self-groom, dig, waste elimination, scent-mark, and inspect-feces (Rafacz and Santymire 2014). Observations were conducted by co-author MH during continuous 60-min sessions (11:00–12:00 h), followed by 30-min follow-up sessions (14:00–14:30 h) 2 or 3 times per week.

Feces were collected from the LPZ's lions, gazelle, and cattle to serve as odor cues for the APDs. The lion feces served as the "predator/competitor" odor, and the gazelle and cattle feces served as prey odors. Feces were thawed and homogenized, approximately 200 g was removed for presentation, and each odor was presented over 3 consecutive days behind a chain-linked fence where the APDs could smell but not access it. A new 200 g portion of the fecal sample was used on each day. The order in which the odors were presented was random, resulting in the gazelle odor being presented first, followed by lion and then cattle. Odors were presented to the APDs at 11:00 h, at the start of the 60-min session (11:00–12:00 h), remained on exhibit until the end of the 30-min follow-up session (14:00–14:30 h), and were then removed (Freeman 2005).

Behavioral data were collected for 1 week prior to the experimental portion of this study, and a 2-week period of behavioral observations without an odor separated the presentation of each odor cue, resulting in a design of: 1 week without odor, gazelle odor, 2 weeks without odor, lion odor, 2 weeks with no odor, cattle odor, and 1 week without odor. The pre-cue period for each odor cue was defined as 1 week prior to presentation of the odor. The during-cue period included the 3 days of odor presentation, and the post-cue period was defined as the first week of the 2-week period separating odors for lion and cattle odor cues, or for the gazelle cue, the 1-week follow up period that followed presentation of all odor cues.

26.2.3 Fecal Sample Collection and Enzyme Immunoassay Analysis

Fecal samples were also collected 2–3 times per week over the entire study. The pre-, during-, and post-cue time periods were the same as the schedule for behavioral data collection. Samples were typically collected at the end of the day, after odor removal and were stored at −20 °C until processing. A fecal marker (Gordon Food Service, Tampa, FL) and non-digestible food items (corn and sunflower seeds) were added to each female's rations to identify feces of each individuals (Fuller et al. 2011). The FGM concentrations were measured using a corticosterone enzyme immunoassay (EIA) modified from a previously published method (Munro and Stabenfeldt 1984; Santymire and Armstrong 2009). Briefly, wet feces were homogenized and 0.50 g (±0.02 g) were weighed into 16 × 125 mm glass tubes and mixed with 5.0 mL of 90 % ethanol:distilled water. The tubes were capped, shaken for 30 min (Glascol Mixer, Terre Haute, IN), and centrifuged at 1500 rpm for 20 min. The supernatant was poured off into labeled tubes and the pellet was suspended in an additional 5 mL of 90 % ethanol:distilled water by vortexing for 30 s. Suspensions were centrifuged as above for 15 min, the second supernatant was combined with the first, and then dried down on a warm bath (60 °C) using an air stream directed over the extracts until completely dry. The dried extracts were reconstituted in 1 mL of methanol, ultrasonicated for 20 min and then diluted 1:60 (sample to phosphate buffered saline, 0.2 M NaH_2PO_4, 0.2 M Na_2HPO_4, NaCl) and stored in the freezer until analyzed.

Horseradish peroxidase (HRP) ligands and polyclonal antiserum (CJM006) were provided by C. Munro (University of California, Davis, CA). Antiserum and HRP to corticosterone were used at dilutions of 1:6000 and 1:20,000, respectively. Antiserum cross-reactivities were: deoxycorticosterone, 14.25 %; tetrahydrocorticosterone, 0.9 %; 11-deoxycortisol, 0.03 %; prednisone, 0.01 %; prednisolone, 0.07 %; cortisol, 0.23 %; cortisone, <0.01 %; progesterone, 2.65 %; testosterone, 0.64 %; and estradiol 17β, <0.01 % (Santymire and Armstrong 2009). An adrenocorticotrophic (ACTH) challenge has validated the stress analysis (Rafacz and Santymire 2014).

26.2.4 Data Analysis

All statistical analyses were performed using *Sigma Stat* Version 3.0 (SPSS Inc., Chicago, IL, USA) at a $P<0.05$ level of significance with exact P-values provided. Normality was tested with a Kolmogorov–Smirnov test and equal variance with the Levene median test. If data were not normal, a non-parametric test was used. Results are reported as the mean ± standard error (SEM). A one-way Kruskal–Wallis ANOVA on ranks with a Dunn's post-doc comparison was used to evaluate the influence of time period (pre-, during-, and post-cue presentation) on the percentage of time spent in active behaviors. A two-way ANOVA was used to test for differences in means among individuals and among cues for each time period separately with a Student–Newman–Keuls post-doc comparison. A RM ANOVA was used to test for differences in FGMs across the three time periods for each individual and each cue. A Kruskal–Wallis ANOVA on ranks with a Dunn's post-doc comparison was used to compare FGM values among individuals. A three-way ANOVA was used to analyze the influence of individual, time period, and cue species on FGMs.

26.3 Results

26.3.1 Behavior

All APDs (APD1-4) were more active ($H_2=7.544$; $P=0.023$) when odors were present than the pre- and post-cue periods (Fig. 26.1). For the pre-odor cue time periods, percentage of time spent in active behaviors did not differ ($F_{2,11}=4.388$; $P=0.067$) between cues. For all cues, APD-4 was the most active individual both before odor presentation ($q=5.084$; $P=0.043$; Fig. 26.2) and when the odor was present (APD-1 $q=12.298$, $P<0.001$; APD-2 $q=10.967$, $P<0.001$; APD-3 $q=6.808$; $P=0.003$; Fig. 26.3a). APD-1 was more active during all cues than APD-2 ($q=4.159$; $P=0.026$) and APD-3 ($q=5.49$; $P=0.019$) (Fig. 26.3a). When combined, all APDs spent a greater percentage of time ($F_{2,11}=6.477$; $P=0.032$) in active behaviors when the lion odor was present versus the cattle odor (Fig. 26.3b). There were no differences ($F_{3,11}=1.514$; $P=0.304$) among individuals in percentage of time spent in active behaviors following presentation of all odors; however, all APDs were more active ($F_{2,11}=11.385$; $P=0.009$) after the lion (19.84 ± 3.33 %) and gazelle (14.44 ± 4.82 %) odor cues than after the cattle (2.50 ± 1.00 %) cue.

26.3.2 Fecal Hormone Metabolites

In ascending order of overall mean FGM concentrations, the APDs ranked APD-4, APD-2, APD-1, and APD-3, which did not reflect social rank (Fig. 26.4). Although FGM concentrations were unaffected by odor exposure for the lion ($F_{3,5}=0.637$;

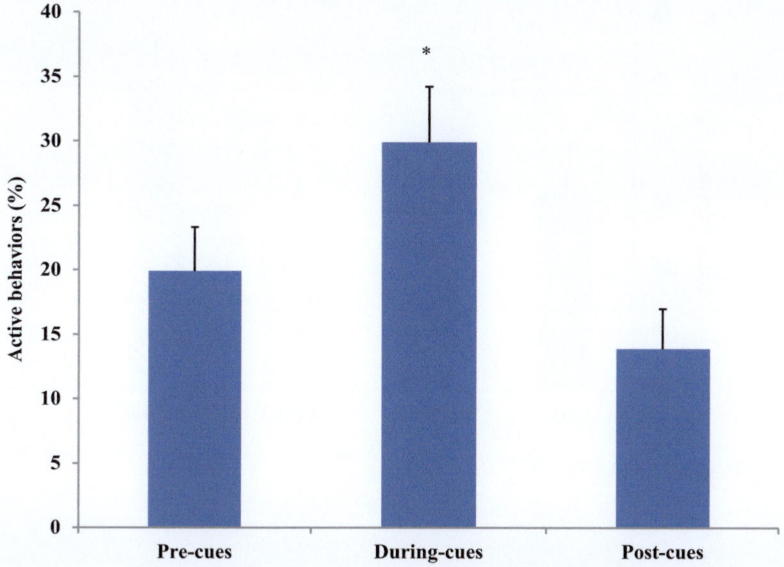

Fig. 26.1 The total percentage of time spent in active behaviors for all four female African painted dogs (APDs) at the three time periods for all three odor cues combined (lion, cattle, and gazelle). Asterisks indicate a difference ($P < 0.05$) among the time periods

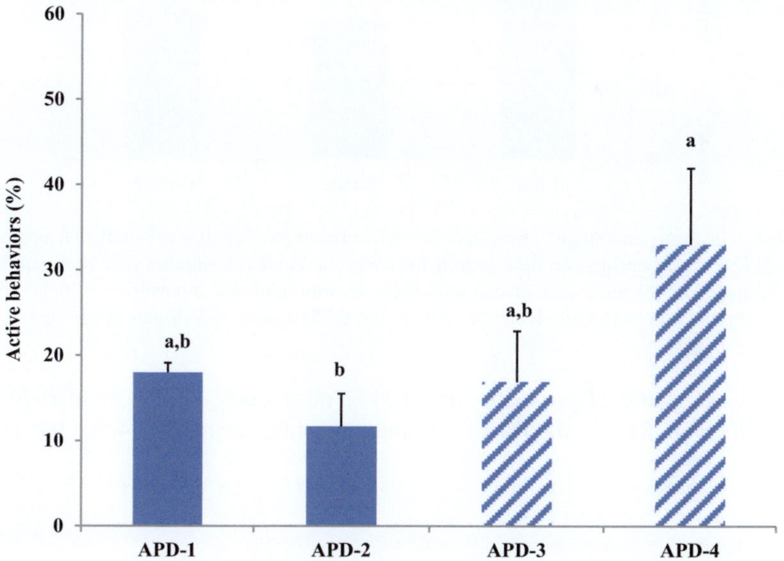

Fig. 26.2 The total percentage of time spent in active behaviors for all four female African painted dogs (APDs) during all the pre-cue time periods. Solid and striped bars indicate dominant and subordinate females, respectively. Different superscripts indicate differences ($P < 0.05$) among individuals

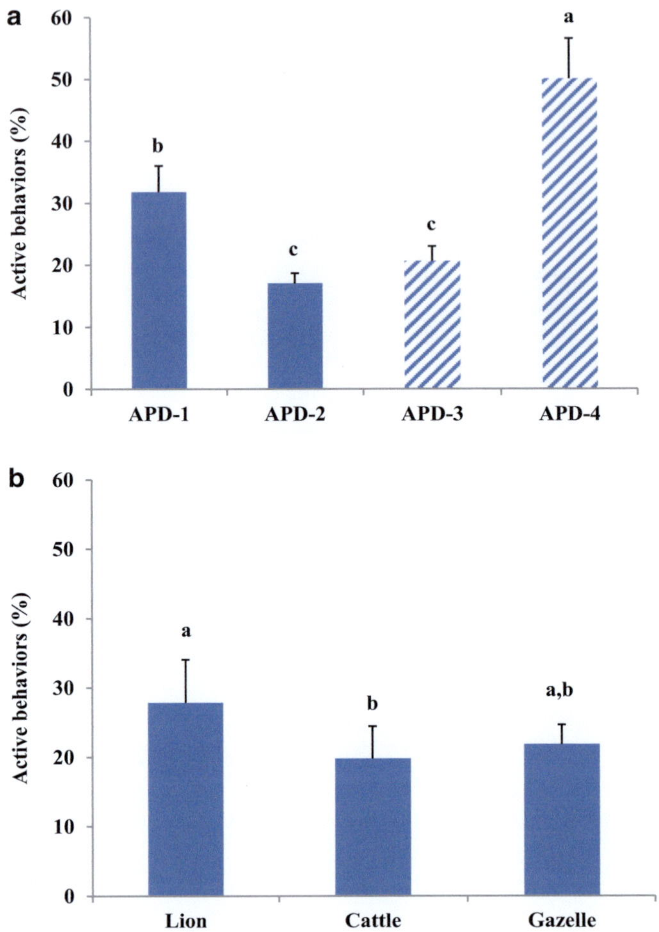

Fig. 26.3 The total percentage of time spent in active behaviors for all four female African painted dogs (APDs) in the during-cue time period for each cue (a) and separated into each cue (cattle, lion, and gazelle); (b) Solid and striped bars indicate dominant and subordinate females, respectively. Different superscripts indicate difference (P < 0.05) among individuals or species

$P = 0.701$) or cattle ($F_{3,7} = 1.066$; $P = 0.517$) odor cues, APD-1 tended to have higher ($F_{3,9} = 6.781$; $P = 0.057$) FGM concentrations than APD-4 for the gazelle odor cue.

26.4 Discussion

The present study is one of only few that have investigated how scent cues from a competitor and prey species elicit both behavioral and hormonal changes. Using completely non-invasive behavioral and hormonal monitoring techniques, our

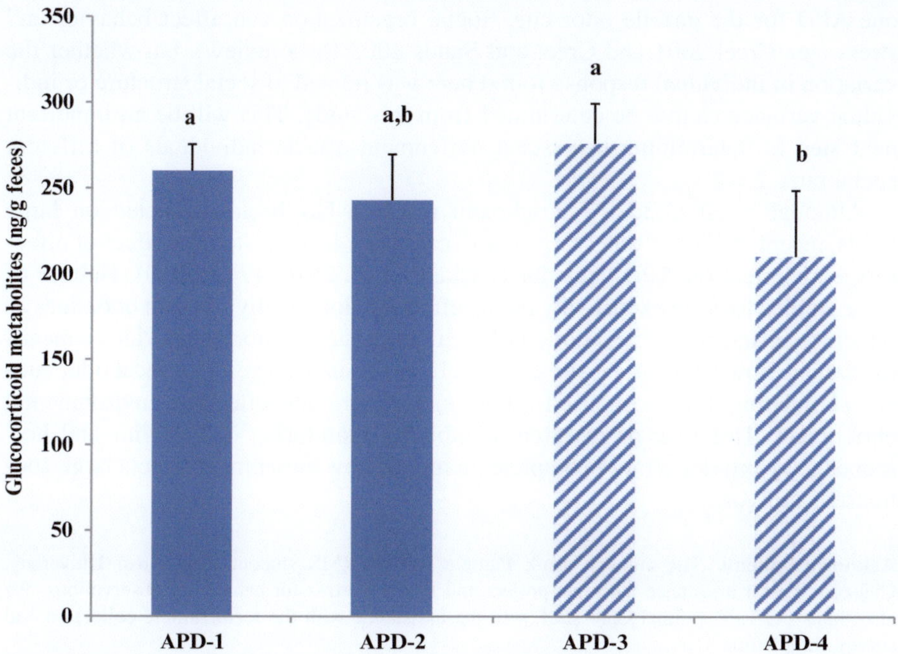

Fig. 26.4 Overall fecal glucocorticoid metabolites (FGM; ng/g feces) for the four female African painted dogs (APDs). Solid and striped bars indicate dominant and subordinate females, respectively. Different superscripts indicate differences ($P < 0.05$) among individuals

research examined whether fecal odor cues influenced behavior, stress physiology, and animal welfare. Other research has investigated the response of prey species to predator scents. For example, Buchanan-Smith et al. (1993) found that tamarins are able to detect the scent of a predator and exhibit stress-induced behavior. Similarly, Monclus et al. (2006) found that rabbits detect predator scent and respond with an increase in stress hormones.

In a previous study, we presented odor cues from predator and prey species to two older male APDs and determined that zoo-housed carnivores become more active when the prey odor was present (Rafacz and Santymire 2014). Similarly, every female APD in the present study spent a higher percentage of time in active behaviors during all cues than in the pre- or post-cue time periods, but overall activity levels were similar among all odor cue exposures. The subordinate APD in the previous study had lower FGM concentrations following presentation of all fecal odors suggestive of an enrichment effect of the cues. However, the stress response was variable among individuals both pre- and during-cues (Rafacz and Santymire 2014). Additionally, the dominant APD had lower overall FGM concentrations than the subordinate APD, and presentation of the prey odor cue (gazelle) coincided with increased FGM concentrations for the subordinate APD (Rafacz and Santymire 2014). In the present study, FGM concentrations did not vary across individual APDs, but there was a trend for higher concentrations in one APD for the gazelle odor cue. Social organization can affect behavior and stress (see Creel 2001 and Creel and Sands 2003 for a review), but whether the variation in individual response found here was related to social structure or individual variance cannot be determined from this study. This will be an important next step in determining how scent enrichment affects individuals of different social rank.

Although most olfactory enrichment research has been conducted on large felids, recent studies of other large carnivores have investigated the effect of olfactory enrichment on APD behavior (Packard et al. 2010; Price 2010). Neither of these studies focused exclusively on the effect of biologically relevant odor cues as olfactory enrichment. The addition of fecal hormone monitoring provides a means to examine how odors affect stress. Overall, our results suggest that fecal odor cues are stimulating for zoo-housed APDs and so provide effective environmental enrichment. This type of research, combining monitoring of behavior and hormones, may provide a more complete picture of how enrichment affects large zoo-housed carnivores.

Acknowledgments The authors thank Pamela Roberts (MS student from Rush University, Chicago, IL) for assistance with this project and Emily Portraz for behavioral observations. We also thank the LPZ Animal Care staff with the assistance with the fecal sample collection and survey completion.

References

Altmann J (1974) Observational study of behavior: sampling methods. Behaviour 49:227–267

Buchanan-Smith HM, Anderson DA, Ryan CW (1993) Response of cotton-top tamarins (*Saguinus oedipus*) to faecal scents of predators and non-predators. Anim Welfare 2:17–32

Clark F, King AJ (2008) A critical review of zoo-based olfactory enrichment. In: Hurst J, Beynon R, Roberts SC, Wyatt T (eds) Chemical signals in vertebrates, vol 11. Springer, New York, NY, pp 391–398

Creel S (2001) Social dominance and stress hormones. Trends Ecol Evol 16:491–497

Creel S, Creel NM (1995) Communal hunting and pack size in African wild dogs, *Lycaon pictus*. Anim Behav 50:1325–1339

Creel S, Sands JL (2003) Is social stress a consequence of subordination or a cost of dominance? In: de Waal F, Tyack P (eds) Animal social complexity. Harvard University Press, Cambridge, MA, pp 153–179

Freeman E (2005) Behavioral and socio-environmental factors associated with ovarian acyclicity in African elephants. Ph.D. dissertation, George Mason University Fairfax, Virginia, USA

Fuller G, Margulis SW, Santymire R (2011) The effectiveness of indigestible markers for identifying individual animal feces and their prevalence of use in North American zoos. Zoo Biol 30:379–398

Monclus R, Rodel H, Palme R, Von Holst D, Miguel FJ (2006) Non-invasive measurement of the physiological stress response of wild rabbits to the odour of a predator. Chemoecology 16:25–29

Munro CJ, Stabenfeldt GH (1984) Development of a microtitre plate enzyme immunoassay for the determination of progesterone. J Endocrinol 101:41–49

Packard JM, Turner SJ, Shepard S (2010) Behavioral enrichment for African wild dogs (*Lycaon pictus*): response to stimuli used in an ongoing program at the Houston Zoo. Biodiversity Stewardship Report No. BS10-2 Biodiversity Stewardship Lab, TAMUS, College Station, Texas

Price LJ (2010) A preliminary study of the effects of environmental enrichment on the behavior of captive African wild dogs (*Lycaon pictus*). Biosci Horiz 3:32–140

Rafacz M, Santymire R (2014) Using odor cues to elicit a behavioral and hormonal response in zoo-housed African wild dogs. Zoo Biol 33:144–149

Santymire RM, Armstrong DA (2009) Development of a field-friendly technique for fecal steroid extraction and storage using the African wild dog (*Lycaon pictus*). Zoo Biol 28:1–14

Shepherdson DJ (1998) Second nature: environmental enrichment for captive animals. Smithsonian Press, Washington, DC, p 376

Woodroffe R, Sillero-Zubiri C (2012) *Lycaon pictus*. The IUCN Red List of Threatened Species. Version 2014.2. www.iucnredlist.org. Accessed on 19 July 2014

Chapter 27
Responses to Domestic Cat Chemical Signals in the House Mouse Are Modulated by Early Olfactory Experience

Vera V. Voznessenskaya, Ilya G. Kvasha, Artyom B. Klinov, and Tatiana K. Laktionova

27.1 Introduction

Predator–prey relationships provide an excellent model for the study of interspecies chemical communication. Chemosensory detection is a very important aspect for predator avoidance strategy for many mammals including the house mouse *Mus musculus*. Odors from carnivores may elicit fear-induced stereotypic behaviors, change activity patterns, and feeding rate; affect neuroendocrine system, reproductive behavior, and reproductive output in potential prey (Harvell 1990; Kats and Dill 1998; Apfelbach et al. 2005; Hayes et al. 2006; Müller-Schwarze 2006; Hayes 2008). The ability of predator odors or carnivore signals to produce profound effects on the behavior of prey in general and especially on the reproductive behavior and neuroendocrine system is associated with natural predators which suggests that there is an evolutionary link between signaling predator and potential prey. This observation suggests that prey species may be able to distinguish between potential predator species on a chemosensory basis. A number of studies support this observation (see review Apfelbach et al. 2005). One of the most specialized predators towards house mouse is domestic cat *Felis catus*. There is a long history of coexistence in the same environments that led to the development of mutual adaptations. Thus, these two species provide a perfect model for the study of innate responses to predator odors and raise a question about the multicompound nature of the kairomones as well as about existence of species-specific predator chemical cues (Voznessenskaya 2014). However, none of the known molecules may produce a full set of defensive behaviors.

V.V. Voznessenskaya (✉) • I.G. Kvasha • A.B. Klinov • T.K. Laktionova
A. N. Severtsov Institute of Ecology and Evolution Russian Academy of Sciences,
33 Leninski prospect, 119071 Moscow, Russia
e-mail: veravoznessenskaya@gmail.com

B.A. Schulte et al. (eds.), *Chemical Signals in Vertebrates 13*,
DOI 10.1007/978-3-319-22026-0_27

Felinine is a unique sulfur-containing amino acid found in the urine of domestic cats (Rutherfurd et al. 2002). Sulfur-containing volatile compounds, 3-mercapto-3-methyl-1-butanol, 3-mercapto-3-methylbutyl formate, 3-methyl-3-methylthio-1-butanol, and 3-methyl-3-(2-methyl-disulfanyl)-1-butanol, are identified as species-specific odorants and candidates of felinine derivatives from the cat urine. The levels of these compounds were found to be sex- and age-dependent (Miyazaki et al. 2006a, b). These cat-specific volatile compounds may represent pheromones used as territorial markers for conspecific recognition or reproductive purposes by mature cats (Miyazaki et al. 2008). Our data indicate that species-specific compounds may be used also by mice to recognize potential predators and their physiological status. These compounds may affect reproductive output in the house mouse (Voznessenskaya 2014).

Neonatal exposure to odorants may influence the function of the olfactory system of an animal, and it produces changes in its responses to these odorants later in its life. Our earlier studies showed that regardless of the odorant (conspecific or heterospecific urine, androstenone), early exposure to it resulted in an increase in a rodent's sensitivity to that stimulus (Voznessenskaya et al. 1995). Interestingly, exposure to the odorant during 2 weeks after the pups opened their eyes appeared to induce the greatest level of sensitization, suggesting a sensitive period to such stimuli (Voznessenskaya et al. 1999). Thus, the specific aim of our current study was to examine whether early olfactory experience of mice with chemosignals of cats during "critical" period for sensitization to odors may modulate behavioral or neuroendocrine responses to the target cues later in adulthood.

27.2 Methods and Materials

27.2.1 Test Subjects

Test subjects were C57BL/6 J mice. Inbred C57BL/6 J mice are known for high investigatory activity and low level of anxiety (Voznessenskaya and Poletaeva 1987; Crawley et al. 1997). Experimental rooms were illuminated on 12:12 h light:dark schedule, and maintained at 20–22 °C. Animals were kept in standard plastic cages. Within each experiment all animals received the same, standard, laboratory chow (Laboratorkorm, Moscow) and water ad libitum. Virgin females, 3 months of age, in proestrus/estrus, as determined by vaginal cytology, were chosen for the mating experiments. Sexually experienced males were used as sires. The morning after pairing, the females were checked for successful mating, as indicated by the presence of a vaginal plug. Successfully mated females were then housed singly. Females with offspring were kept in separate cages. For each litter the exact date of birth was recorded. All litters were reduced at birth to six pups per dam. We tried to group litters so that the pups were similar in age. On completion of the exposures, the cages were changed and dams were provided with clean nest material.

Pups were weaned from mothers at 30 days of age. At weaning, the weight of the pups was 16–18 g. At the time of testing, when the pups were 9 weeks, they weighed between 22 and 24 g.

27.2.2 Urine Collection

Urine collected from sexually mature male and female domestic cats was used as a source of chemical cues from a natural sympatric predator. Cats were maintained on meat diet for 14 days before urine collection. Freshly voided urine from several animals was mixed, aliquoted in 1-ml amounts, and frozen and stored at −22 °C. Once defrosted, each 1 ml-aliquot of urine was used only once. L-feinine (US Biologicals, USA) was used in concentration 0.05 %, which is comparable with concentration in intact cat urine stored at 0–2 °C (Hendriks et al. 1995).

27.2.3 Neonatal Exposure of Mouse Pups to Stimulus

All nursing female mice with offspring were assigned into three experimental groups. One group received exposures to tap water (control); another group was exposed to cat urine, the third group was exposed to L-felinine. When the pups in each group were between 14 and 28 days of age (Voznessenskaya et al. 1999) they were exposed five times to target odors, once every 3 days. All exposures took place in environmental chambers with controlled incoming and outcoming air flow (ASP 130, Flu France).

Exposures consisted of pups being presented with a sterile cotton swab containing 50 μ of cat urine, 50 μ L-felinine (0.05 % in water), or 50 μ tap water. Cotton swabs were placed directly into home cage once in 3 days and they were used by females as nest material providing an instant exposure to target odors.

27.2.4 Assay for Corticosterone

Upon completion of the final 15-min exposure to predator urine (25 μl), blood samples (20 μl) from the orbital area were drawn with heparinized Natelson blood-collecting tubes (Fisher Scientific, USA). Samples were centrifuged, the plasma separated, and frozen at −22 °C until subsequent analysis. Plasma corticosterone was assayed (in duplicate) by enzyme immunoassay method (EIA K014, Arbor Assays, USA). Before analysis, plasma samples were brought to the room temperature and treated with the supplied in the kit the dissociation reagent to yield the total corticosterone concentration in plasma. Final plasma dilutions were ≥1:100. Moderate to severely hemolyzed samples were excluded from analysis.

Concentration of corticosterone was measured with spectrophotometer SpectraMax 340PC 384 (Molecular Devices, USA) at 450 and 670 nm. We used SpectraMax Software (http://www.moleculardevices.com/pages/software/softmax.html) to calculate concentrations of corticosterone.

27.2.5 Open Field Test

We used one of the modifications of a standard open field test using a hole board. The hole-board apparatus consisted of a plastic nontransparent cylinder measuring 40 cm in diameter. The walls were 50 cm high (Voznessenskaya and Poletaeva 1987). The arena was raised 3 cm above the ground on a plastic stand. The hole board contained 17 cone-shaped holes (3 mm in diameter) that were located symmetrically in two rows: eight holes each and one hole in the center of the board. Plastic Petri dishes with cat urine (0.15 ml), L-felinine (0.15 ml) or tap water (0.15 ml) on filter paper were placed under each of the holes. In control experiments water samples (0.15 ml) were placed under all the holes.

All testing was performed in a separate room; experimental animals were brought to the room 1 day prior to testing. All experiments were carried out during late evening hours when locomotor activity of C57BL/6 J mice is highest (Voznesenskaya and Poletaeva 1983). Mice were transported to the hole board using a plastic tube to reduce handling stress (Hurst and West 2010).

During 15 min trial, we recorded on a spreadsheet and videotaped the number of rearing (the mouse is stationary on its back paws while raising its forepaws off the ground and extending the body vertically); we considered this to be a measure of vertical investigatory activity (Walsh and Cummins 1976). We also recorded the number of holes that the mouse investigated; this was a measure of horizontal investigatory activity. We recorded the number of urinations and defecations by the mouse as a measure of its emotionality. In addition, we recorded the total amount of time that the mouse sat without moving (freezing behavior) as a measure of passive-avoidance behavior. Lastly, we recorded the frequency and duration of self-grooming as an indicator of conflict between exploratory behavior and passive-avoidance behavior (Stone et al. 1995; Stanford 2007). Statistical analysis was performed using parametric and nonparametric tests with STATISTICA 7.0 software.

27.3 Results and Discussion

Exploration of new environments may provide an animal with new information about potential food resources, availability of cover, or potential sexual partner. At the same time, entering a new environment or approaching to a novel stimulus, an animal might also increase its risk of predation, aggression from conspecifics, or

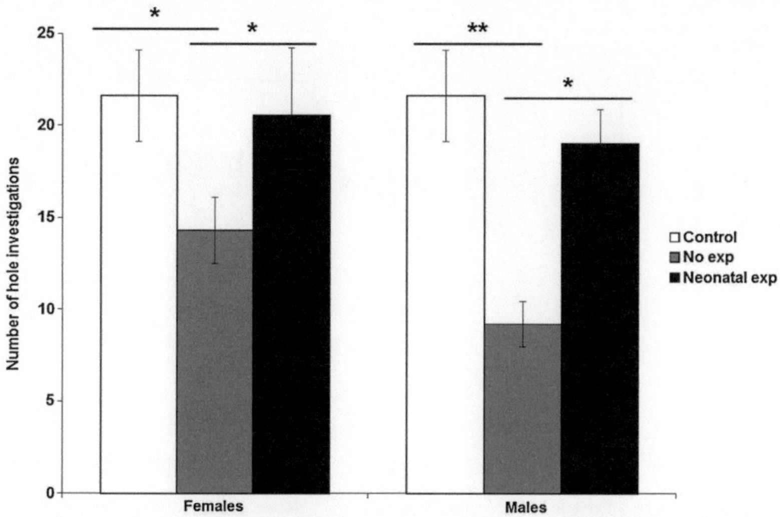

Fig. 27.1 The influence of neonatal exposures to cat urine on investigatory activity (hole investigations) in adult mice in "Open Field" test. Control group ($n = 13$)—no neonatal exposures; no cat urine samples presentations during testing. "No exp." group ($n = 20$)—exposed to tap water neonatally during days 14–28; cat urine samples presented during testing in "open field." "Neonatal exp." group ($n = 22$)—exposed to cat urine neonatally during days 14–28; cat urine samples presented during testing in "open field." $^{*}p < 0.05$; $^{**}p < 0.01$, *t*-test

other hazards (Brown and Nemes 2008). The situation when an animal investigates or avoids novelty has been described as the outcome of an approach–avoidance conflict (Montgomery 1955; Montgomery and Monkman 1955) or as a balance between neophilic and neophobic tendencies (Greenberg 2003).

Placement of an animal into open arena provokes a fear of new environments on the one hand and an intention to investigate these new environments, on another and the open field test represents a dynamic balance between these two behavioral tendencies (Hughes 2007). Presentation of cat urine (under the central hole) or L-FELININE significantly suppressed both measures of investigatory activity: rearing (vertical) and hole investigations (horizontal) in male and female mice (Figs. 27.1 and 27.2). In addition, the number of defecations increased and the total time the mice spent grooming decreased (Figs. 27.3 and 27.4). It was shown by a number of investigations that extended periods of grooming correlate with elevated emotional stress (Stone et al. 1995). Elevated level of defecations favors this observation. Under the influence of domestic cat chemical signals balance between exploratory behavior and passive-avoidance behavior shifted in favor the latter.

Mice responded to predator scents in a gender specific manner. In presence of cat urine female mice investigated holes more often than males (*t*-test $p < 0.01$, $n = 10$), had higher defecation rates (*t*-test $p < 0.05$, $n = 10$), more acts of freezing behavior than did males (Mann–Whitney test, $p < 0.01$, $n = 10$). Likewise, females exposed to

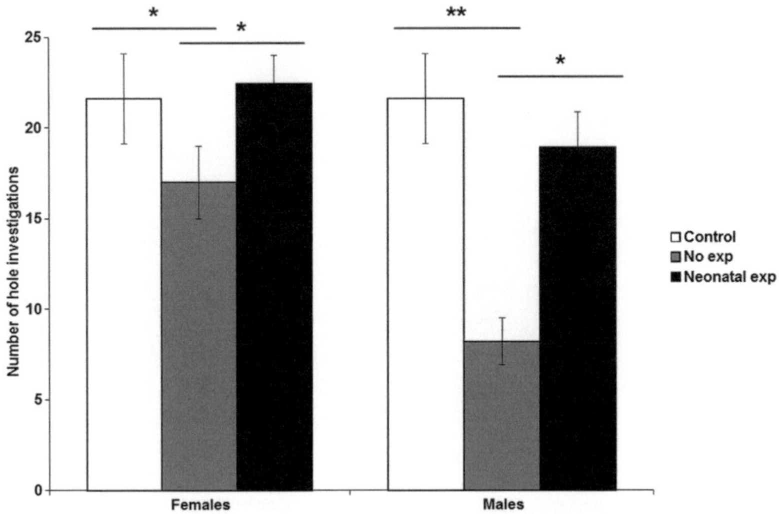

Fig. 27.2 The influence of neonatal exposures to L-felinine (0.05 %) on investigatory activity (hole investigations) in adult mice in "Open Field" test. Control group ($n=13$)—no neonatal exposures; no L-felinine samples presentations during testing. "No exp." group ($n=20$)—exposed to tap water neonatally during days 14–28; L-felinine samples presented during testing in "open field." "Neonatal exp." group ($n=24$)—exposed to L-felinine neonatally during days 14–28; L-felinine samples presented during testing in "open field." $^{*}p<0.05$; $^{**}p<0.01$, *t*-test

L-felinine investigated more holes (*t*-test, $p<0.05$, $n=10$), reared more often (*t*-test, $p<0.01$, $n=10$), and groomed more frequently than did males (Mann–Whitney test, $p<0.05$, $n=10$). Thus, we found a more profound decrease in investigatory activity in males then in females in response to exposure to domestic cat chemical signals.

Exposure of mice pups to cat urine orL-felinine when they were 14–28 days of age significantly affected their behavior as adults when they were 9 weeks of age relative to that of control mice. Female mice exposed to cat urine during this 2-week period showed significant decreases in the number of freezing acts (Mann–Whitney test, $p<0.01$, $n=11$) and number of defecations (*t*-test $p<0.01$, $n=11$) in response to presentation of cat urine samples during the hole board test. Also female mice in the experimental groups spent less time on grooming (*t*-test, $p<0.01$, $n=11$) and freezing (Mann–Whitney test, $p<0.01$, $n=11$) than did control females exposed during critical period to tap water (Fig. 27.3). Likewise, male mice exposed to cat urine spent less time on grooming (*t*-test, $p<0.01$, $n=11$) and freezing (Mann–Whitney test, $p<0.01$, $n=11$) as well as showed fewer acts of grooming (Mann–Whitney test, $p<0.01$, $n=11$) and fewer acts of freezing (Mann–Whitney test, $p<0.01$, $n=11$) relative to those mice that were exposed to the tap water control (Fig. 27.3). Both males and females of this experimental group showed higher investigatory activity relative to control group, exhibiting more rearing (*t*-test, $p<0.05$, $n=22$) and more hole investigations (*t*-test, $p<0.0\ 5$, $n=22$; Fig. 27.1)

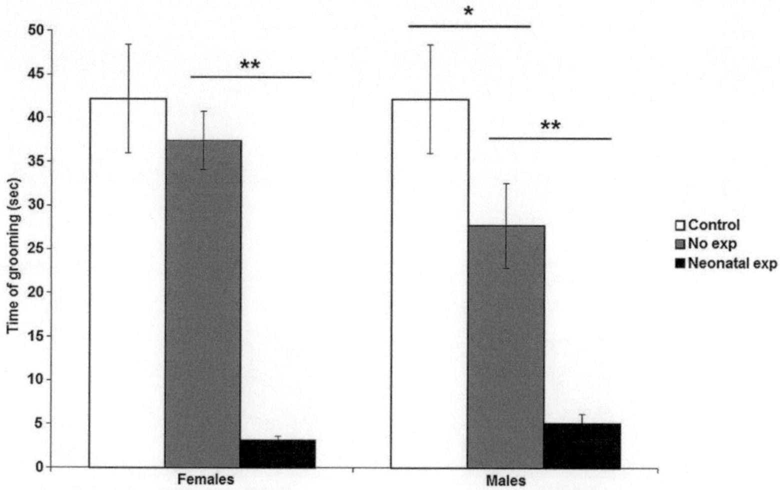

Fig. 27.3 The influence of neonatal exposures to cat urine on grooming duration in adult mice in "Open Field" test. Control group ($n=13$)—no neonatal exposures; no cat urine samples presentations during testing. "No exp." group ($n=20$)—exposed to tap water neonatally during days 14–28; cat urine samples presented during testing in "open field." "Neonatal exp." group ($n=22$)—exposed to cat urine neonatally during days 14–28; cat urine samples presented during testing in "open field." $^{*}p<0.05$; $^{**}p<0.01$, *t*-test

than did controls. Thus, early olfactory experience with cat urine significantly reduced fear-induced behavior towards its chemosignals in adulthood for male and female mice.

Females exposed to L-felinine during days 14–28 after birth, later, at age of 9 weeks more often investigated holes (Fig. 27.2, *t*-test, $p<0.05$, $n=13$) and performed more vertical rearings (*t*-test, $p<0.05$, $n=13$) in the hole board test (with L-felinine present) compared to that of controls exposed to tap water. The total time of grooming by females was shorter compared to that of the control (*t*-test, $p<0.01$, $n=13$; Fig. 27.4). We also observed less individual acts of freezing (Mann–Whitney test, $p<0.01$, $n=13$) and total time of freezing had become shorter in duration (Mann–Whitney test, $p<0.01$, $n=13$) relative to that of the control group. The number of defecations significantly decreased (*t*-test, $p<0.01$, $n=13$). Males exposed to L-felinine when they were 14–28 days of age also showed higher investigatory activity in the hole board test compared to that of the controls (Fig. 27.2). In addition, male mice spent less time grooming than did controls (Fig. 27.4). Thus, early olfactory experience with L-felinine produced changes in behavior of adult mice which were very similar with those caused by exposures to cat urine. The balance between investigatory activity and passive-avoidance behavior towards predator chemical signals changed in favor of exploratory behavior.

In spite of clear changes at the behavioral level, we did not observe significant changes in corticosterone response in mice that were exposed to cat urine 14–28 days

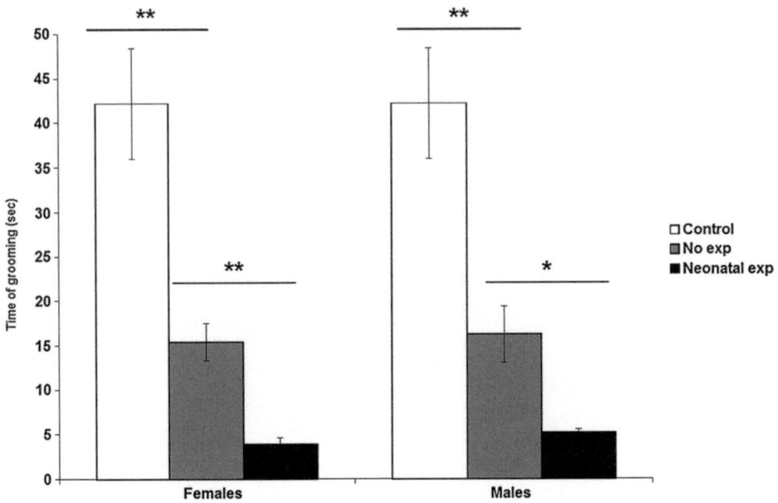

Fig. 27.4 The influence of neonatal exposures to L-felinine (0.05 %) on grooming duration in adult mice in "Open Field" test. Control group ($n = 13$)—no neonatal exposures; no L-felinine samples presentations during testing. "No exp." group ($n = 20$)—exposed to tap water neonatally during days 14–28; L-felinine samples presented during testing in "open field." "Neonatal exp." group ($n = 24$)—exposed to L-felinine neonatally during days 14–28; L-felinine samples presented during testing in "open field." $^{*}p<0.05$; $^{**}p<0.01$, *t*-test

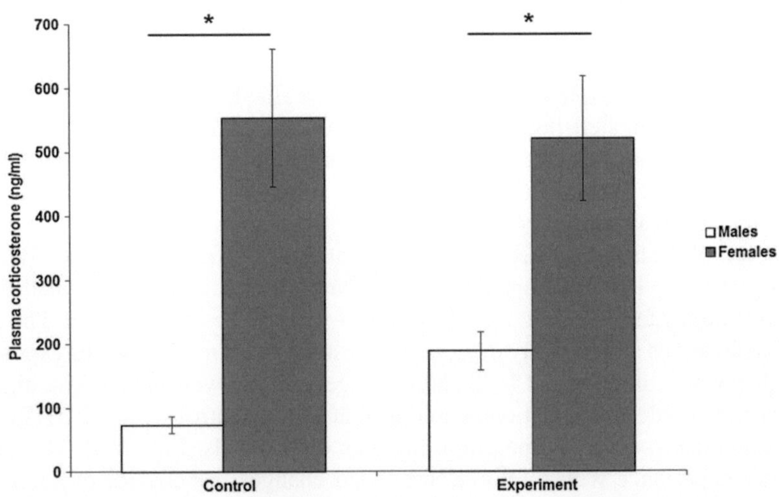

Fig. 27.5 The influence of neonatal exposures to cat urine on plasma corticosterone in response to cat urine presentations in adult mice. Control group ($n = 10$)—exposed to tap water neonatally during days 14–28. Experimental group ($n = 10$)—exposed to cat urine neonatally during days 14–28. $^{*}p<0.05$, Mann–Whitney test

after birth (Fig. 27.5). We observed sexual differences in response to the 15-min presentation of cat urine, but at the same time corticosterone elevation was significant ($p<0.05$, $n=10$) relative to basal level. The data obtained indicate an innate nature of corticosterone response for mice exposed to cat odors. There have been a number of studies investigating the effects of cat odor exposure on the glucocorticoid and ACTH production of the potential prey (Blanchard et al. 1998; Sullivan and Gratton 1998; Cohen et al. 2000; Figueiredo et al. 2003; Voznessenskaya et al. 2003; Papes et al. 2010). In one study, File et al. (1993) showed that a cloth that had been rubbed on a cat cause an increase in circulating corticosterone in rats. However, with repeated exposure to the cat odor the rats habituated and did not show an increase in corticosterone (File et al. 1993). We also observed habituation to cat odor at the level of plasma corticosterone response in Wistar rats (Voznessenskaya et al. 2003). In the present study, however, mice, unlike rats, did not habituate to repeated exposures of the cat odor. It may imply a hardwired processing of a cat odor as a potential pheromone in the house mouse. It seems likely that a profound aversive effect of predator odors and lack of habituation under repeated exposures might only exist "if predator and prey have a long evolutionary history in parallel so that a prey species becomes genetically pre-disposed to avoid the odors of sympatric predators" (Stoddart 1980) and there is an extremely high risk of fatal outcome for the direct interactions of predator and prey (Apfelbach et al. 2005). Domestic cats are more specialized predator towards house mice (Blanchard et al. 2001a; Apfelbach et al. 2005). In contrast to mice, rats are less vulnerable to such predators as cats (Apfelbach et al. 2005). In addition, rats show lower stress-reactivity to a wide range of stressors including predator odors (Zangrossi and File 1994; Blanchard et al. 2001a, b).

27.4 Conclusions

Taken as a whole, our data suggest that although odors of a cat and a chemosignal from a cat can induce innate responses, environmental influences are still important. Mice appear to habituate to cat odors at the level of the behavior, if they were neonatally exposed to target chemosignals during critical for odor sensitization period. At the same time neuroendocrine response of house mice to cat odor has a hardwired nature and more profound in females than in males. Early olfactory experience with cat odors produced dissociation in responses to these odors later in the life at the behavioral level and at the hormonal level. Habituation to cat odors at the behavioral level is the major limitation of using these odors as natural repellents while hardwired nature of corticosterone response makes cat chemosignals potential reproductive disrupters in mice.

Acknowledgments Research was supported by Russian Foundation for Basic Research # 14-04-05011 to V.V.V.

References

Apfelbach R, Blanchard CD, Blanchard RJ, Hayes RA, McGregor IS (2005) The effects of predator odors in mammalian prey species: a review of field and laboratory studies. Neurosci Biobehav Rev 29:1123–1144

Blanchard RJ, Nikulina JN, Sakai RR, McKittrick C, McEwen B, Blanchard DC (1998) Behavioral and endocrine change following chronic predatory stress. Physiol Behav 63:561–569

Brown GR, Nemes C (2008) The exploratory behaviour of rats in the hole-board apparatus: is head-dipping a valid measure of neophilia? Behav Proc 78:442–448

Blanchard DC, Griebel G, Blanchard RJ (2001a) Mouse defensive behaviors: pharmacological and behavioral assays for anxiety and panic. Neurosci Biobehav Rev 25:205–218

Blanchard RJ, McKittrick CR, Blanchard DC (2001b) Animal models of social stress: effects on behavior and brain neurochemical systems. Physiol Behav 73:261–271

Cohen H, Benjamin J, Kaplan Z, Kotler M (2000) Administration of high-dose ketoconazole, an inhibitor of steroid synthesis, prevents posttraumatic anxiety in an animal model. Eur Neuropsychopharmacol 10:429–435

Crawley JN, Belknap JK, Collins A, Crabbe JC, Frankel W, Henderson N, Hitzemann RJ, Maxson SC, Miner LL, Silva AJ, Wehner JM, Wynshaw-Boris A, Paylor R (1997) Behavioral phenotypes of inbred mouse strains: implications and recommendations for molecular studies. Psychopharmacology 132:107–124

Figueiredo HF, Bodie BL, Tauchi M, Dolgas CM, Herman JP (2003) Stress integration after acute and chronic predator stress: differential activation of central stress circuitry and sensitization of the hypothalamo–pituitary–adrenocortical axis. Endocrinology 144:5249–5258

File SE, Zangrossi H Jr, Sanders FL, Mabbutt PS (1993) Dissociation between behavioral and corticosterone responses on repeated exposures to cat odor. Physiol Behav 54:1109–1111

Greenberg R (2003) The role of neophobia and neophilia in the development of innovative behaviour of birds. In: Reader SM, Laland KN (eds) Animal innovation. Cambridge University Press, Cambridge, pp 175–196

Harvell CD (1990) The ecology and evolution of inducible defenses. Q Rev Biol 65:323–340

Hayes RA (2008) Seasonal responses to predator fecal odours in Australian native rodents vary between species. In: Hurst JL, Beynon RJ, Roberts SC, Wyatt TD (eds) Chemical signals in vertebrates 11. Springer, New York, pp 379–387

Hayes RA, Nahrung HF, Wilson JC (2006) The response of native Australian rodents to predator odours varies seasonally: a by-product of life-history variations? Anim Behav 71:1307–1314

Hendriks WH, Moughan PJ, Tarttelin MF, Woolhouse AD (1995) Felinine: a urinary amino acid of Felidae. Comp Biochem Physiol B Biochem Mol Biol 112:581–588

Hughes RN (2007) Neotic preferences in laboratory rodents: issues, assessment and substrates. Neurosci Biobehav Rev 31:441–464

Hurst JL, West RS (2010) Taming anxiety in laboratory mice. Nat Methods 7:825–826

Kats LB, Dill LM (1998) The scent of death: chemosensory assessment of predation risk by prey animals. Ecoscience 5:361–394

Miyazaki M, Yamashita T, Hosokawa M, Taira H, Suzuki A (2006a) Species-, sex-, and age-dependent urinary excretion of cauxin, a mammalian carboxylesterase family. Comp Biochem Physiol B Biochem Mol Biol 145:270–277

Miyazaki M, Yamashita T, Suzuki Y, Soeta S, Taira H, Suzuki A (2006b) A major urinary protein of the domestic cat regulates the production of felinine, a putative pheromone precursor. Chem Biol 13:1071–1079

Miyazaki M, Yamashita T, Taira H, Suzuki A (2008) The biological function of cauxin, a major urinary protein of the domestic cat (*Felis catus*). In: Hurst JL, Beynon RJ, Roberts SC, Wyatt TD (eds) Chemical signals in vertebrates 11. Springer, New York, pp 51–60

Montgomery KC (1955) The relation between fear induced by novel stimulation and exploratory behavior. J Comp Physiol Psychol 48:254–260

Montgomery KC, Monkman JA (1955) The relation between fear and exploratory behavior. J Comp Physiol Psychol 48:132–136
Müller-Schwarze D (2006) Chemical ecology of vertebrates. Cambridge University Press, New York
Papes F, Logan DW, Stowers L (2010) The vomeronasal organ mediates interspecies defensive behaviors through detection of protein pheromone homologs. Cell 141:692–703
Rutherfurd KJ, Rutherfurd SM, Moughan PJ, Hendriks WH (2002) Isolation and characterization of a felinine-containing peptide from the blood of the domestic cat (Felis catus) J Biol Chem 277:114–119 catus) J Biol Chem 277:114–119
Stanford SC (2007) The open field test: reinventing the wheel. J Psychopharm 21:134–135
Stoddart M (1980) The ecology of vertebrate olfaction. Chapman & Hall, London
Stone A, Manavalan S, Zhang Y, Quartermain D (1995) Beta-adrenoceptor blockade mimics effects of stress on motor activity in mice. Neuropsychopharmacology 12:65–71
Sullivan RM, Gratton A (1998) Relationships between stress-induced increases in medial prefrontal cortical dopamine and plasma corticosterone levels in rats: role of cerebral laterality. Neuroscience 83:81–91
Walsh RN, Cummins RA (1976) The open-field test: a critical review. Psychol Bull 83:482–504
Voznessenskaya VV (2014) Influence of cat odor on reproductive behavior and physiology in the house mouse: (*Mus musculus*). In: Mucignat-Caretta C (ed) Neurobiology of chemical communication. CRC Press, Boca Raton, pp 389–405
Voznessenskaya VV, Feoktistova NY, Wysocki CJ (1999) Is there a period during neonatal development for maximal imprinting an odor? In: Johnston RE, Müller-Schwarze D, Sorensen PW (eds) Advances in chemical signals in vertebrates 8. Kluwer, New York, pp 617–621
Voznesenskaya VV, Poletaeva II (1983) The effect of $ACTH_{4-10}$ on the diurnal cycle of locomotor activity in mice. Zh Vyssh Nerv Deiat Im I P Pavlova 33:960–961
Voznessenskaya VV, Poletaeva II (1987) Orienting-investigatory reaction of mice with different genotypes under injection of $ACTH_{4-10}$. Zh Vyssh Nervn Deiat im I P Pavlova 37:174–176
Voznessenskaya VV, Naidenko SV, Feoktistova NY, Krivomazov GJ, Miller LA, Clark L (2003) Predator odors as reproductive inhibitors for Norway rats. In: Singleton GR, Hinds LA, Krebs CJ, Spratt DM (eds) Rats, mice and people: rodent biology and management, vol 96, ACIAR monograph. ACIAR, Canberra, pp 131–136
Voznessenskaya VV, Parfyonova VM, Wysocki CJ (1995) Induced olfactory sensitivity in rodents: a general phenomenon. Adv Biosci 93:399–406
Zangrossi JH, File SE (1994) Habituation and generalization of phobic responses to cat odor. Brain Res Bull 33:189–194

Part VII
Field and Analytical Approaches

Chapter 28
Does Deconvolution Help to Disentangle the Complexities of Mammal Odors?

Peter Apps

28.1 Introduction

Two of the challenges which make mammal chemical signals difficult to characterize are the very low concentrations at which they exert their effects (e.g. Laska et al. 2005; Dorries et al. 1995; Leinders-Zufall et al. 2000) and the complexity of the chemical mixtures in which they are embedded (Albone 1984 pp. 6, 9; Burger 2005; Willse et al. 2005).

The concentrations of mammal signals are actually well within the working range of modern analytical instruments. The lower limits of detection (LODs) for mammalian noses range down to attomolar (Moulton and Marshall 1976; Laska et al. 2005), but it is unlikely that signaling compounds are present only at the lower end of this range. It would be a surprise to find any signal being sent at its detection limit; animals can be expected to shout rather than whisper, and the concentration of mouse urine is 1,000,000 times higher than the detection threshold for vomeronasal organ (VNO) neurons (He et al. 2010). Even allowing for short-term increases in volatile emissions (e.g. Goodwin et al. 2012) the rates of emission from scent marks trend downwards, and must be much higher when they are fresh and readily detectable than after they have aged and are difficult for animals to detect. Also, scents are detected at a distance and their concentration near the source must be higher than on the fringes of their active space. The GC–MS systems that are routinely applied to analyze mammal odors in the search for semiochemicals have lower LODs of around 0.1 ng, even for the intractable compounds that occur in mammal odors (Apps and Mmualefe 2012). With samples in the 0.1–10 g range this provides LODs in the nano- to picomolar range, which is well below the concentrations of active

P. Apps (✉)
Paul G. Allen Family Foundation Laboratory for Wildlife Chemistry,
Botswana Predator Conservation Trust, Private Bag 13, Maun, Botswana
e-mail: peterjapps@gmail.com

B.A. Schulte et al. (eds.), *Chemical Signals in Vertebrates 13*,
DOI 10.1007/978-3-319-22026-0_28

components that have been found in mammal chemical signals. The concentration of (Z)-7-dodecen-1-yl acetate in the urine of female Asian elephants (*Elephas maximus*) reaches 0.146 mM just before ovulation (Rasmussen et al. 1997). Dehnhard et al. (2013) found steroid pheromones in boar (*Sus scrofa*) saliva at low μg/ml. Male red fox (*Vulpes vulpes*) urine has active constituents at μg/ml (Whitten et al. 1980).

Mammal odors are typical of natural mixtures in which a multitude of compounds with a disordered distribution of chromatographic retention times generate numerous peak overlaps and co-elutions, and no single column is able to separate all their components (Davis and Giddings 1983; Davis 1999; Schure and Davis 2011). Because peaks overlap, when real samples are analyzed the LODs are limited by chemical noise from the sample, not the signal:noise limit of the detector. Even if the signals that are embedded within mammal odors are simple (Apps 2013) there is a high probability that the peaks from a chemical signal will be obscured or overlapped by peaks from the matrix. The problems of complexity and dilution compound one another; because the number of components increases as their concentration decreases (Enke and Nagels 2011), and large peaks obscure many of the smaller ones; the impact of overlap on the analysis of the smaller peaks on the chromatogram is worse than on the larger ones (Fig. 28.1).

Separation can be enhanced by running a sample on columns with different selectivities by multidimensional chromatography (Giddings 1987, 1995). Targeted "heart cuts" transfer the effluent from one column to the other over a selected time period (often 10–30 s long) and carry out the second dimension separation on a normal analytical column (Deans 1968; MacNamara et al. 2003; Apps 2006). Comprehensive GC×GC transfers all of the effluent from one column to the other in very short slices, and carries out the second dimension separation on a short, narrow-bore column (Marriott et al. 2003). Nevertheless, mammal odors often contain large quantities of free acids which will overload one or the other of the "orthogonal" columns in a GC×GC column set. Heartcutting is more robust than GC×GC but heartcutting the whole of a 60 min separation into 30 s slices with a 30 min separation in the second dimension would take 8 days per sample. Only one study of mammal odor (Nielsen et al. 2006) has used heartcutting.

Because two-dimensional separations are impractical, we need to do the best we can with the data from one dimension. Our search for signaling compounds in complex mammal odors is a special case of the metabolomics and bioinformatics approach to finding biologically significant compounds against a background of chemical noise (Likic 2009; Baker 2014), where post-run GC–MS data processing by deconvolution is a well-established way of extracting hidden information from separations of complex mixtures (e.g. Aggio et al. 2011; Du and Zeisel 2013; Kessler et al. 2013; Luedemann 2008; O'Callaghan et al. 2012). Despite its potential value in mammal semiochemistry, the only reports of deconvolution of mammal odors are by Willse et al. (2005), Kwak et al. (2008, 2012) and Chap. 31 this volume. How well deconvolution performs has previously been tested with defined mixtures (Colby 1992; Lu et al. 2008; O'Callaghan et al. 2012) which almost certainly underrepresent the challenges of real mammal odors. To explore how well deconvolution works with a mammal odor, the GC–MS data from analysis of a set

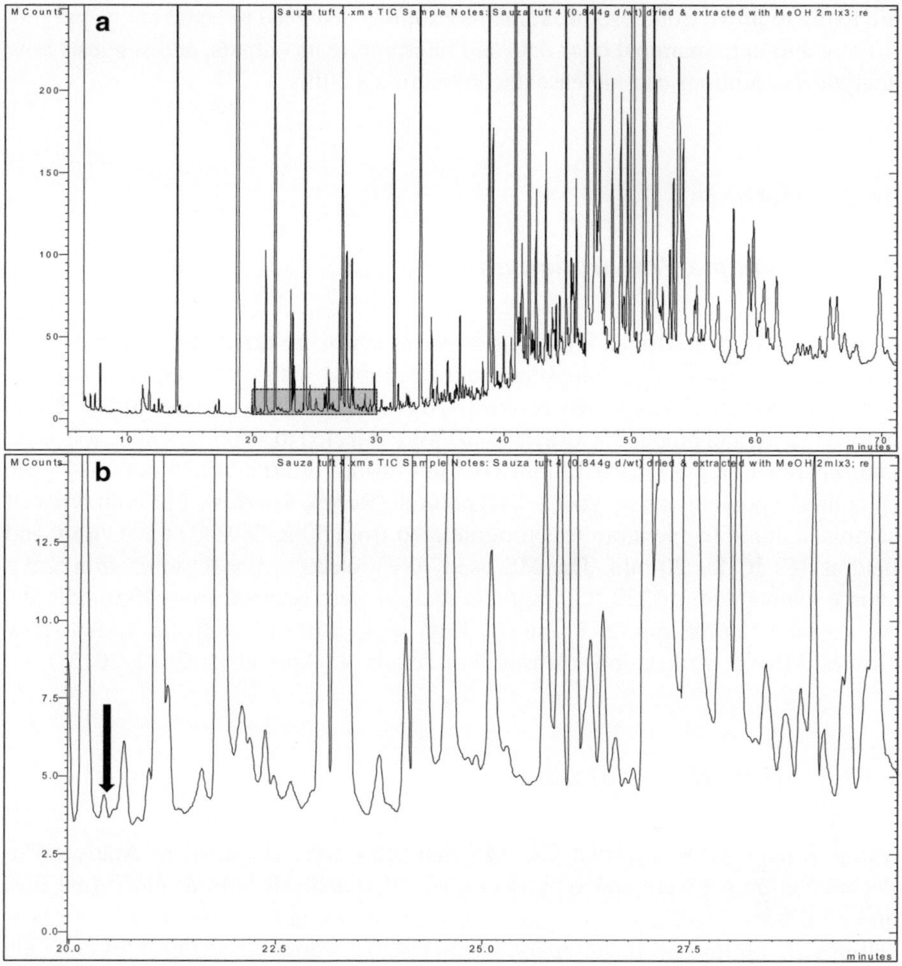

Fig. 28.1 The complexity problem in mammal odors. The small *grey rectangle* from the TIC of an African wild dog preputial tuft (**a**) is expanded in the lower trace (**b**). Only one of the peaks (*arrowed*) in (**b**) is not overlapped by others

of 59 African wild dog urine extracts, and four extracts of the odorous secretion adhering to the hairs of African wild dog preputial tufts was processed using a commercially available deconvolution software package: AnalyzerPro (SpectralWorks, Runcorn, UK http://www.spectralworks.com/default.html). The urine sample set was deliberately chosen to present a "torture test" with the widely variable concentrations and high background noise that are typically generated with raw extracts of mammal secretions and excretions.

This paper deals with general features of the application of deconvolution to mammal odors rather than the specifics of any particular method or software. It highlights features of the chromatographic behavior of real mammal odors that are likely to prove troublesome, some precautions that need to be taken in applying deconvolution to mammal odor data and interpreting its outputs, and suggests how analytical conditions can increase deconvolution's utility.

28.2 Methods

28.2.1 Sample Preparation and Analysis

Samples of voided African wild dog urine on soil were dried and extracted with methanol, and tufts of preputial hair were extracted with methanol. The extracts were analyzed on a Varian (now Bruker) 450 GC coupled to a Varian 320 MS. Urine extracts (2 μl) and preputial extracts (1 μl) were injected to a splitless inlet at 240 °C and separated on a 30 m × 0.32 mm × 0.5 μm (urine) or 30 m × 0.25 mm × 0.25 μm (preputials) polyethylene glycol (PEG) column (Restek RtxWax). For both types of sample, column temperature was programmed from 50 to 240 °C at 5 °C/min, and held at 240 °C for 20 min. The MS used 70 eV electron impact ionization and a source temperature of 220 °C, fragment masses were scanned from 33 to 450 Da, with a scan rate of approx 2 scans/s. Data were captured with the Varian (now Bruker) MSWS workstation software. For details see Apps et al. (2012, 2013).

28.2.2 Data Deconvolution

Varian/Bruker MSWS format GC–MS data files were imported to AnalyzerPro deconvolution software and subject to qualitative analysis both as individual files and sequences. I will not go into the details of the operation of the software or how settings are optimized. In summary, fragment ions whose elution profiles coincide are assigned to a component, the list of components from a sample can be transferred to a target list, and the components that are found in other samples can be compared to those on the list with respect to how similar their spectra are to the targets' and whether they elute within a set retention time window of a target. Components from multiple samples can be added to an existing target list so that a list of all the components in a set of samples can be compiled, and the occurrence and peak area of each component that occurs in each sample can be tabulated. Ideally, each compound has a unique target and is always assigned to the same target in every sample.

The detection of components can be optimized by adjusting software parameters. Reject masses is a list of the masses of spectral fragments that will not be used to detect components. The retention time window is the part of the chromatogram that will be processed, and can be used to set different parameters for different parts of the chromatogram. Minimum masses is the smallest number of fragments that must coincide for a component to be detected. The area and height thresholds are minimum values for lower detection limits. They apply to individual fragment peaks, not to the sum of the fragments in a component. The width threshold sets the minimum width for a component peak, and is used to exclude noise that changes at a higher rate than the signal. Resolution depends on the number of data points across a chromatographic peak, the fewer the points the lower the setting, and scan windows compensates for spectral skew from scanning spectrometers. Smoothing sets the order of a Gaussian smooth of the raw data and is conveniently adjusted to the same value as the optimum smooth in the mass spectrometer's data processing software.

AnalyzerPro's ability to find peaks hidden under baseline noise or overlapped by other peaks was checked by examining whether the ions that it assigned to a component also created a peak in manually extracted single ion traces in the MSWS software.

Comparisons between samples depend on three parameters; relative intensity threshold, retention window, and spectral match. Relative intensity threshold is the lowest fragment abundance relative to the spectral base peak that will be included in the component's spectrum. The retention window is the maximum absolute difference in retention time for two components to be assigned to the same target. The spectral matches are the minimum values for forward and reverse match for components to be assigned to the same target.

Whether the detection of components was consistent from sample to sample, and which features of the analytical data caused problems for the deconvolution were investigated by detailed inspection of deconvolution results from sets of samples, and their comparison against sets of up to 4 MSWS data traces (four is the maximum that can realistically be compared at a time).

28.3 Results

28.3.1 Detecting Trace Components and Separating Overlapped Peaks

With settings optimized for peak detection AnalyzerPro found about 50 components per minute in the densest part of the preputial gland chromatograms, with an average separation between peak apices of 3.4 scans (Fig. 28.2). The smallest peaks that could be detected were in the 4–800 pg range (Fig. 28.3), at the signal:noise limit for the MS. If the settings were more sensitive than the optimum, false components were detected when the noise generated by baseline drift and background fluctuated on a time scale similar to the peak width (Fig. 28.4). The number of false components was reduced by excluding fragment ions that contributed to noise but that were not important to the detection of genuine components. Low mass fragments at m/z 33–44 are so widespread that they are hardly any use for discriminating components from background noise or other peaks, and they are the main contributors to baseline drift and fluctuating noise. Excluding them is more effective

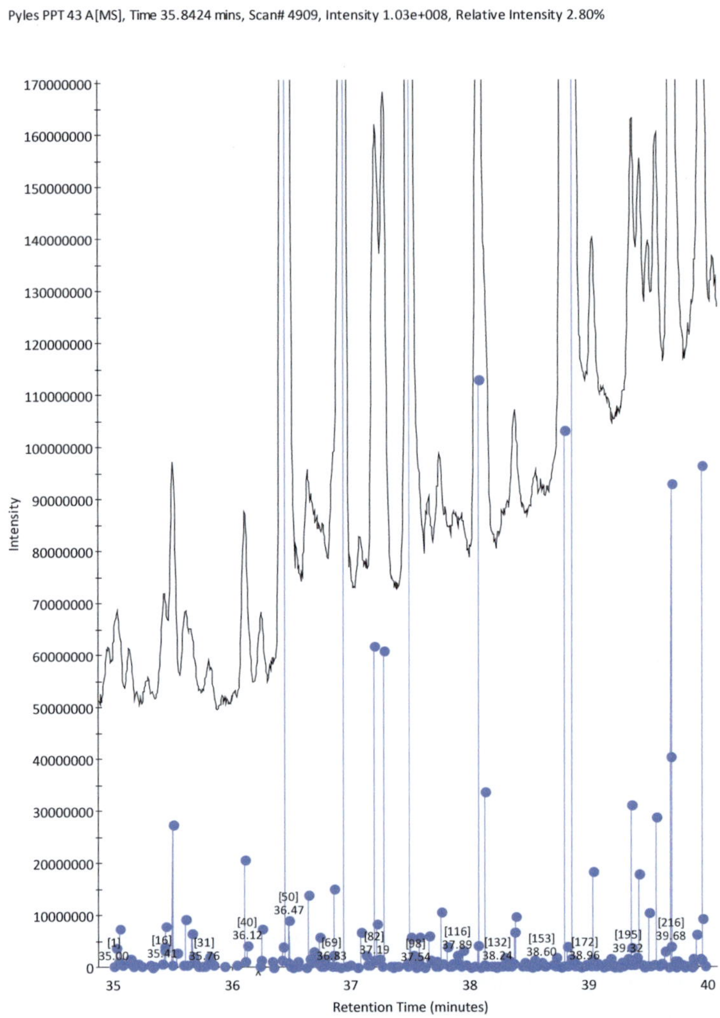

Fig. 28.2 With settings selected to find the maximum number of peaks, deconvolution finds 235 components (shown by *blue markers*) in a 5-min section of the data from an African wild dog preputial tuft extract

for reducing the number of false components than increasing the minimum area threshold (Fig. 28.4 and Table 28.1). In methanol extracts of urine, the m/z 129 ion from cyanuric acid, which is a gas-phase polymerization product of urea, was a major source of fluctuating noise and was excluded from the component detection. The m/z 129 ion is also prominent in spectra of free acids, but its exclusion did not affect their detection because they have numerous other stronger ions in their spectra. Although excluded ions and ions that are below the minimum area threshold are not used to find components, they are still in the original mass spectra and are available

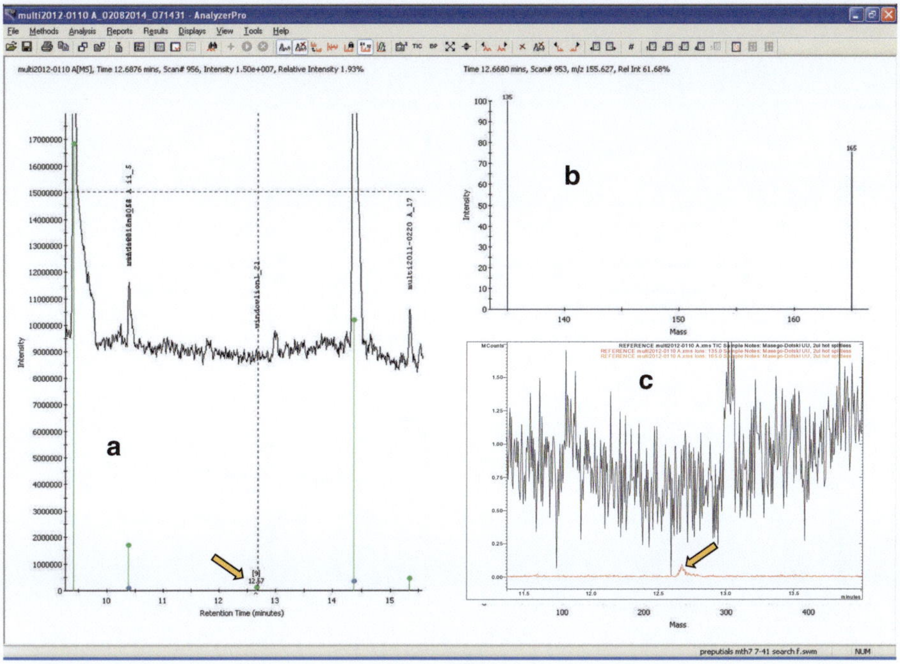

Fig. 28.3 An example of a trace component detected under the TIC baseline. The *arrowed* component at 12.67 min (**a**) has ions at m/z 135 and 165 (**b**) which are confirmed by extracted ions from the original data file (*arrowed*, **c**). The peak area corresponds to approximately 4 pg

to identify compounds from fragment patterns, so that the matches returned from library searches of a component's spectrum are not affected by the deconvolution settings that were used to find it.

False detections from fluctuating noise and baseline drift were also reduced when a larger number of different ions was required for component detection, because a larger number of noise ions had to fluctuate in synchrony to generate a false component (Fig. 28.4 and Table 28.1). The disadvantage of requiring more fragment ions in a component was that it raised the detection limit by making the minimum area and signal:noise thresholds apply to weaker ions in the component's spectrum. For compounds whose spectra are dominated by a few strong ions, raising the minimum number of ions to include any of the weakest ions increases the detection limit sharply. For example in 4-methylphenol, a very common component of mammal odors, the three most abundant fragments have m/z 108 (100 %), 107 (97 %) and 77 (23 %), and increasing the minimum number of ions from two to three increased the LOD by a factor of four. Wild dog urine has a family of peaks (possibly C_4-substituted hydantoins) with a very dominant m/z 100 fragment and another family (possibly dimethylamines) with a very dominant m/z 58 fragment and little else in their spectra, and these were not detected when the minimum number of ions was set above one.

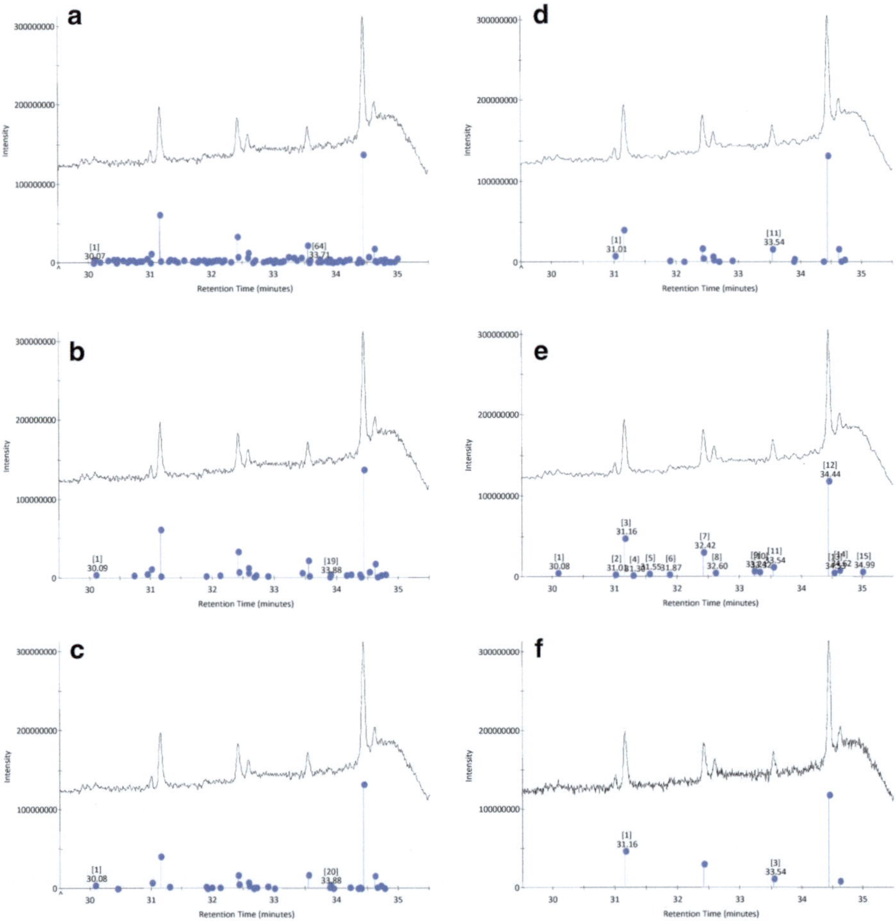

Fig. 28.4 The effect of deconvolution settings on how many components are detected and how many are due to noise. (**a**) With no ions excluded, minimum number of ions 1 and minimum area 1,000,000. (**b**) With no ions excluded, minimum number of ions 2, minimum area 1,000,000. (**c**) With m/z 33–44 ions excluded, minimum number of ions 1, minimum area 1,000,000. (**d**) With m/z 33–44 ions excluded, minimum number of ions 2, minimum area 1,000,000. (**e**) No ions excluded, minimum number of ions 1, minimum area 10,000,000. (**f**) No ions excluded, minimum number of ions 2, minimum area 10,000,000. See also Table 28.1

When GC conditions are optimized, peak width stays the same throughout a temperature programmed run, but in many cases it is necessary to hold the column temperature at a maximum determined by the thermal stability of the stationary phase. Wild dog volatile peaks that eluted after the end of the temperature ramp were progressively wider than those that eluted while the column temperature was increasing. To maintain detectability required a progressive increase in settings for the number of points in the Gaussian smooth from 7 to 15, in the peak width from 0.04 to 0.1 min and in scan window from 1 to 3. In addition the larger molecules that

Table 28.1 The effect on component detections and false detections of changing deconvolution settings

	Settings			Number of components	
Figure 28.4	Excluded ions (m/z)	Minimum ions	Minimum area (10^6 counts)	Real peaks	From noise
a	None	1	1	32	64
b	None	2	1	30	10
c	33–44	1	1	31	1
d	33–44	2	1	18	0
e	None	1	10	9	6
f	None	2	10	5	0

Chromatograms are shown in Fig. 28.4

elute later provide more fragments and so the minimum ion setting was increased in order to reduce the adverse effects of baseline drift and fluctuating noise. If possible, temperature ramps need to continue to the end of a separation in order to maintain the same peak widths throughout.

28.3.2 *Comparing Samples*

The usual strategy in mammalian semiochemistry is to compare the chemical compositions of different classes of samples to find those features that correlate with biology, usually those that differ between classes more than within classes (Willse 2005; Wyatt 2014, pp. 57–58). If inter-sample comparisons of deconvoluted data are to be meaningful, the deconvolution must be consistent from sample to sample, but this aspect has been neglected hitherto. All deconvolution procedures compare peaks between samples on the basis of how similar their spectra and retention times are, anything that changes spectra or shifts retentions between samples can cause peaks that are from the same compound to be assigned to different components in different samples.

28.3.2.1 Retention

Retention windows had to be wide enough to accommodate retention shifts, but narrow enough that different peaks with similar spectra did not elute within the same window. Consistent deconvolution of the wild dog data was adversely affected by changes in retention times. If the retention of a compound in sample B differed from its retention in A by more than the width of the retention window, the peak in B was not matched to the target that had been assigned to the compound in A. Then, after the unmatched components of sample B were added to the target list, the list contained two targets for the same compound at two different retention times.

If other samples had the same compound within the retention window of both these targets, their peaks were matched to both of them. These redundant targets were easily recognized in the tabulated data since the two targets had identical areas, and the sum of the numbers of matched targets and unmatched targets was larger than the number of components in the sample.

Deconvolution was not adversely affected by random retention shifts due to fluctuations in inlet or oven temperature, gas flow, or sample volume, because the shifts were typically less than 1 s which was smaller than the half-height width of the peaks and narrower than retention windows of 6 s which generated the required selectivity and consistent deconvolution for multiple samples. In contrast, retention drift due to column aging and inlet contamination caused significant difficulties. Over the 59 urine samples that were included in the data set there was a 12 s range of retention times for nearly all the peaks. This is the baseline width of the peaks and so a 12 s retention window made deconvolution entirely dependent on differences in spectra between adjacent peaks.

28.3.2.2 Spectral Match

Although a given compound's mass spectrum is constant (with a given instrument and conditions), the fragments in the spectrum of a chromatographic peak that are extracted by deconvolution depend on the signal:noise ratio and which fragments discriminate the peak from its overlapping neighbors. More stringent match thresholds increase specificity but mean that differences in the spectrum of the same compound in different samples lead to missed matches, the creation of multiple redundant targets and the assignment of single peaks to multiple targets.

Large Differences in Peak Size Produce Spectra with Low Match Scores

The concentrations of the African wild dog urine extracts spanned four orders of magnitude (Fig. 28.5); three orders more than the concentration range examined by Lu et al. (2008) in simulated metabolomics samples. As a component peak got smaller, the number of fragments that were above the signal:noise and minimum area thresholds also got smaller until, at the limit, only the most abundant fragment was detected [compare Tagfinder (Luedemann et al. 2008) which assumes constancy of spectra]. The sparse spectra from small peaks gave poor spectral matches to the rich spectra from large peaks, and vice versa (Fig. 28.6), leading to the same compound at widely different concentrations being assigned to different targets. To avoid redundant targets being assigned to peaks of widely different sizes from the same compound in different urine samples, it was necessary to set the forward and reverse match parameters to 350 and 450 respectively—much lower than would be regarded as a match in an MS library search. This reduced selectivity by increasing the rate at which different peaks in different samples were assigned to one target.

The impact of peak size on spectral match was reduced by reordering file processing sequences to start with samples that had peak sizes close to the mean,

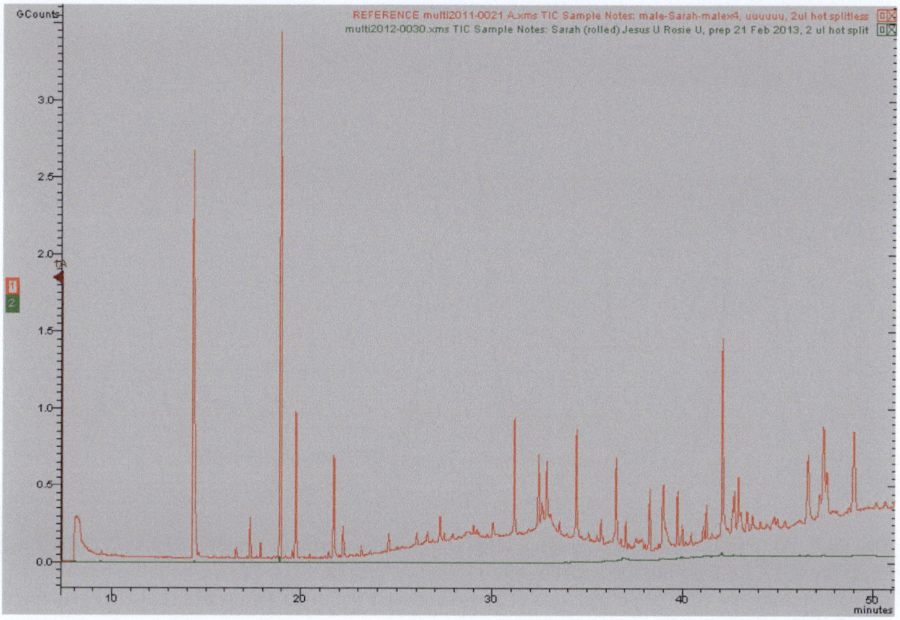

Fig. 28.5 TICs from the highest (*red*) and lowest (*green*) sample concentrations among the African wild dog urine samples

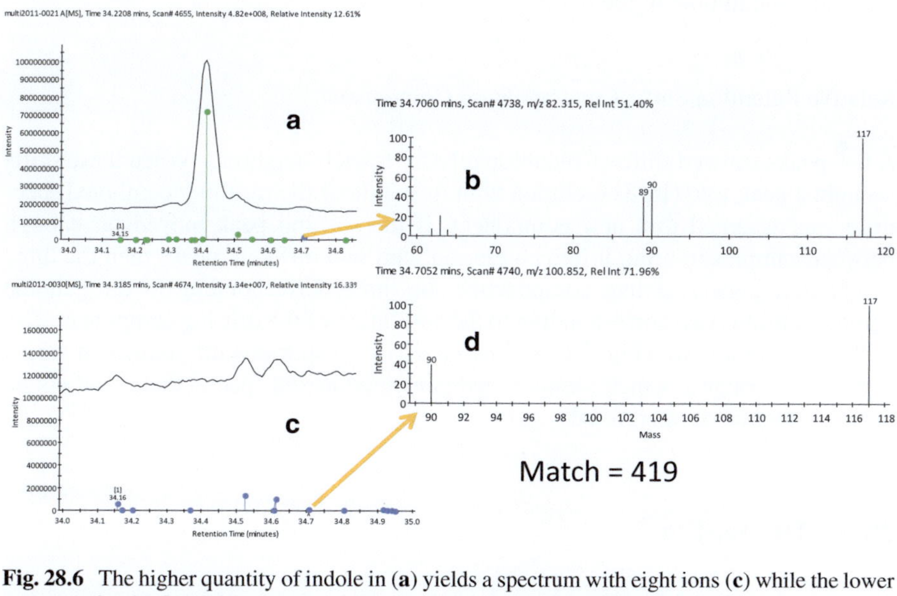

Fig. 28.6 The higher quantity of indole in (**a**) yields a spectrum with eight ions (**c**) while the lower concentration in (**b**) yields a spectrum with only two ions (**d**). Although the two spectra are from the same compound, they have a forward match score of only 419

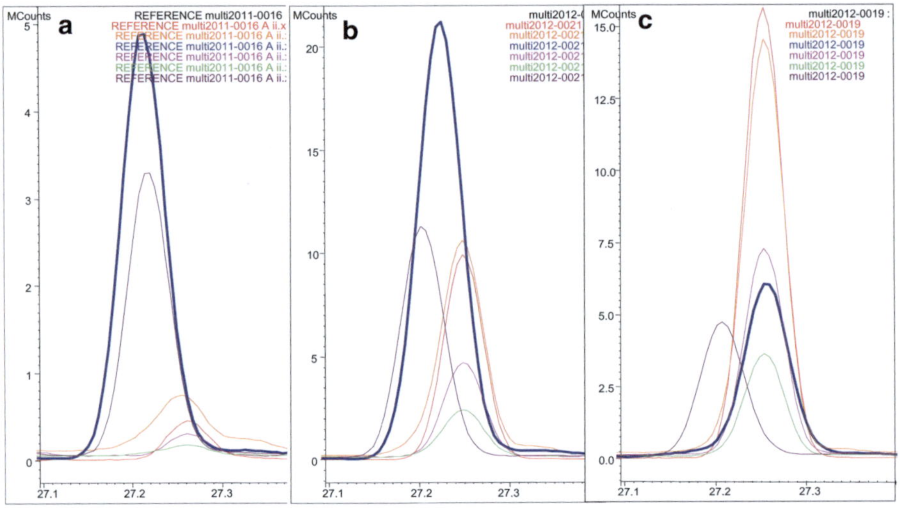

Fig. 28.7 Among urine extract samples relative retention shifts in the m/z 85 ion (thicker blue trace) moved it from co-elution with the m/z 71 ion (*purple*) (**a**) to co-elution with the m/z 102, 58, 75, and 70 ions (**c**) via a retention time midway between the other two peaks (**b**). This generated mixed spectra in (**a**) and (**c**) even after deconvolution

and then working outwards to the extremes. Similarly, the effect of retention differences was reduced by beginning the sequence with samples whose retention times were near the middle of the range.

Relative Retention Shifts Generate False Components

A few peaks showed shifts in retention relative to their neighbors. When these shifts brought a peak into close co-elution with its neighbor the resulting combined spectrum was assigned to a new component. If the shifting peak moved far enough between samples to bring it into co-elution with two different peaks then the three compounds generated five components; the three corresponding to the genuine compounds plus two corresponding to the co-elution of the drifting component with each of the other two (Fig. 28.7). This gave rise to spurious differences in which compounds occur in which samples, and generated mixed spectra that could not be library searched or interpreted.

28.4 Discussion

In contrast to the false detections and missed targets reported by Lu et al. (2008) when comparing AnalyzerPro, ChromaTOF, and AMDIS, deconvolution by AnalyzerPro with the appropriate settings did not produce new artifacts; all of the components that were detected were present in the raw GC–MS data—although some of them were analytical artifacts rather than characteristics of the sample. In other words, false components in the deconvoluted data were the result of analytical noise, not the deconvolution, and false or missed target matches were due to spectral skew, large differences in peak size, and retention shifts. The impact of these analytical shortcomings will not be specific to a particular deconvolution method, and although their effects can be reduced by painstaking optimization of deconvolution settings and manual cleanup of the data, they really need to be addressed by improving the analysis.

28.4.1 Detection

AnalyzerPro's ability to find hidden peaks was as good as or better than an experienced human, and much faster (minutes rather than hours) per chromatogram. AMDIS software provides the same benefits (Meyer et al. 2010), but will output results only for predefined targets. The ability of deconvolution to find tiny, hidden peaks presents a temptation to use settings that maximize the number of components (Fig. 28.2), but these results will not be robust to the inevitable variation in retention time and peak area that arise from analysis of real samples. To get useful comparisons between samples it is necessary to trade sensitivity for robustness. The compromises that are required to optimize deconvolution are equivalent to those that are required with conventional data processing, and will be familiar to all chromatographers. Highly sensitive settings lead to noise being seen as peaks, and insensitive settings lead to small peaks being missed. Low detection limits and very high resolution have to be traded-off against reproducibility and robustness.

Background noise generates many false components that have only one fragment ion, but there are also genuine peaks with only one strong fragment. Excluding the noise requires an increase in the minimum area and signal:noise parameters, or an increase in the minimum number of ions required to detect a component. Which to use for a particular analysis will depend on whether the lower limit of detection or the detection of the maximum number of compounds is given priority.

28.4.2 Resolution and Peak Capacity

Under optimum conditions the peaks from 0.25 or 0.32 mm diameter capillary columns are 8–11 s wide at baseline, and cover about 20 MS scans with the ion traps or quadrupoles that are the workhorses of mammal semiochemistry. Clean separation for Gaussian peaks needs the peak maxima to be 18 scans apart, and at 10 scans apart the valley is half the peak height (or more if one peak is much bigger than the other). Deconvolution can separate peaks that are one scan apart, increasing effective resolution and the number of separated peaks by a factor of 18. In a typical 60 min run, with a baseline width of 10 s, the maximum number of peaks that can be baseline resolved is 360. At a scan rate of 2 s^{-1} the maximum number of deconvoluted components that can be separated by one scan is 7200, and with a more robust two-scan separation it is 3600. The number of components that AnalyzerPro deconvoluted in the dense regions of the preputial secretion chromatograms corresponds to an empirical total peak capacity of 3000 in a 60 min run on a 30 m column, a remarkably high efficiency, and evidence that the complexity of these samples (and probably other mammal odors) is close to the upper limit of what can be separated on a 30 m capillary column.

28.4.3 Inter-sample Comparisons

In contrast to the detailed examinations of the process of deconvolution that have been published (Luedemann 2008; Likic 2009; Aggio et al. 2011; O'Callaghan et al. 2012; Du and Zeisel 2013; Kessler et al. 2013), its reproducibility when applied to data from different samples has been neglected.

28.4.3.1 Retention Issues

Deconvolutions that detect components separated by only one scan are obviously vulnerable to one-scan shifts in peak maxima. A scanning MS takes discontinuous signal measurements from the smooth elution profiles of chromatographic peaks, and if the apex of the chromatographic peak is close to the midpoint between two scans, which scan has the higher signal and is the peak apex in the MS data can be changed by very small changes in retention, and by noise. In addition, one-scan differences in the apparent retention of different fragments from one chromatographic peak can be generated by the time-skewed detection of different fragment masses by scanning MS instruments. Consequently the settings for smoothing, resolution, and scan windows need to be adjusted to give at least two scans between adjacent components.

Retentions of polar compounds can be shifted by differences in the water content of the injected sample, and retentions of acidic and basic compounds by differences in pH. Very careful attention to uniform sample preparation will ameliorate these problems, but retention shifts can also be intrinsic to the properties of the volatiles themselves. If sample concentrations vary widely, then relative retention shifts can be generated by interactions among components; for example high concentrations of free acids will retard the elution of trace amines. The very high resolution achievable by deconvolution means that concentration overloading leads to detectable shifts in retention before the peak shape is affected.

The retention window needs to span the range of a peak's retention times in the sample set, but also needs to be narrower than the difference in retention between any two peaks with similar spectra. It may be impossible to meet both these requirements, for example; 3- and 4-methylcresols whose spectra are very similar, elute within 6 s of each other on 30 m PEG columns. The complex mixtures of free acids from about C_{14} upwards that occur in extracts of mammal secretions generate multiple overlapping peaks whose similar spectra make their deconvolution dependent on narrow retention windows. AnalyzerPro retention windows and spectrum match values in target lists can be set individually and can be tailored to particular challenges. For example, the spectral match criteria for the cresols can be set higher to distinguish between the closely eluting isomers.

When batches of dirty samples are run, progressive retention drift of several seconds due to column ageing and inlet contamination is inevitable and gets worse as more samples are run. Consequently, larger sample sets require wider retention windows, which lead to more multiple matches and mis-assignments. It is probably better to run small numbers of samples chosen for informative comparisons rather than large numbers in the hope that patterns will emerge. Inlet contamination and column deterioration also affect column selectivity and cause changes in relative retention times, and contribute to background noise and baseline drift. Frequent inlet maintenance and the use of replaceable uncoated guard columns are sensible precautions. Although sample clean-up is a good way of reducing contamination and noise, its selectivity prejudges the characteristics of the signal.

Peak alignment is an alternative to retention windows to accommodate differences in retention time between samples, but retention shifts that move a peak close to the retention time of a different peak with a similar spectrum will also confound chromatogram alignment (Willse et al. 2005) since there is no independent way to discriminate which of the two compounds generated either of the peaks. Aligning retention times becomes even more of a problem when large differences in peak size between samples reduce the similarity of the spectra for a given compound in different samples, and it cannot address changes in relative retention times.

28.4.3.2 Spectral Distortion

The impact of four order differences in peak size on spectral match is particularly troublesome, since if the quantity of sample is uniform (as it was for the wild dog urine samples) the size of the peaks is a feature of the sample, and possibly of the signal. Lu et al. (2008) varied the concentration of their test mixture by only one order of magnitude, and still found problems with all three of the deconvolution programs that they tested. It may be practical to standardize the concentrations of extracts by diluting the concentrated ones and concentrating the dilute ones, and then calculate back to the original volumes.

Retention shifts and spectral mismatches have the same overall effects: multiple redundant targets are generated for some single compounds and some single peaks are assigned to more than one target. Although this preserves the original data content (peaks are detected as components which are assigned to a target with a label, spectrum, retention, and area), it degrades specificity because different compounds in different samples can be matched to a single target. In general the similarity between samples will be underestimated because there are many ways to distort a spectrum or shift a retention, but only one way to generate a good match. On the other hand, setting the spectral match thresholds low enough to accommodate the effects of peak size on spectra could lead to different compounds in different samples being assigned to the same target, and thus an overestimate of similarity.

Deconvolution settings that provide consistent inter-sample comparisons are a compromise; setting very strict criteria for spectrum matches so that different compounds in different samples are not assigned to the same component even if they have similar retention times, or setting very narrow retention windows so that closely eluting peaks with similar spectra do not fall into one another's retention windows both generate false differences between samples, and redundant targets. If the range of retention times in a set of chromatograms is wider than the separation of peaks with similar spectra the data have to be interrogated manually no matter what data processing is employed, and peaks drifting into co-elution with others can be detected only by inspection of the data.

28.4.4 General

The better the GC–MS analysis, the better deconvolution works on the data. Deconvolution's ability to find very small peaks is enhanced by low and consistent background noise. Retention times need to be as repeatable as possible, and removing extraneous influences on relative retention times improves the reproducibility of deconvolution.

As well as the extra data it generates, deconvolution is also a quick way of finding discrepancies and anomalies in the data, although this still requires an experienced eye for unexpected features in the data. An additional benefit of post-run processing that is not tied to a particular instrument platform is that it provides comparability (in the strict sense) between data from different instruments and different laboratories.

If the African wild dog samples are typical of mammal odors, their complexity and analytical intractability mean that in studies of mammal chemical communication deconvolution will have to be supplemented by manual clean-up of the data (see also e.g. Luedemann et al. 2008; Kessler et al. 2013). Although automated processing eliminates most of the tedious repetitive tasks of manual processing, interrogation and clean-up of the data is still dependent on the analyst's skill and experience. Mammal odors are so complex and intractable that deconvolutions that are robust enough to run as the automated "workflows" used in metabolomics will need to be so insensitive and coarsely resolved that they will miss the trace components and close co-elutions whose discovery in mammal odors is the reason to be using deconvolution in the first place.

Many of the problems posed by the African wild dog urine samples could have been avoided by a different choice of study animal; Figs. 2 and 5 in Willse et al. (2005) show a much closer quantitative similarity between samples from inbred laboratory mice than I found from the urine of free-ranging wild dogs.

28.5 Summary

Deconvolution is a fast and powerful way of finding obscured and overlapped peaks in GC–MS data from mammal odors.

Optimizing deconvolution imposes the same compromises as conventional peak integration between limit of detection, resolution, and reproducibility, but total data yield is much higher, and the lower limit of detection is much lower.

Because deconvolution can resolve peaks that elute very close together, retention repeatability is more critical when data is deconvoluted than when it is integrated conventionally, and due to the lower LODs in deconvoluted data, clean baselines are more beneficial when data is deconvoluted than when it is integrated conventionally.

The variability and complexity of mammal odors require manual interrogation and cleanup of deconvoluted data.

Acknowledgments I am very grateful to John Moncur of SpectralWorks who provided AnalyzerPro free of charge, the Paul G. Allen Family Foundation which funded the research by a grant to the Botswana Predator Conservation Trust, Tico McNutt, director of the Trust, Neil Jordan and Geoff Gillfillan who collected most of the samples and Lesego Mmualefe who analyzed many of them. Permission to conduct research in Botswana was granted by the Department of Wildlife and National Parks under permit number EWT 3/3/8 XXIV.

References

Aggio R, Villas-Bôas SG, Ruggiero K (2011) Metab: an R package for high-throughput analysis of metabolomics data generated by GC-MS. Bioinformatics 27:2316–2318

Albone ES (1984) Mammalian semiochemistry. The investigation of chemical signals between mammals. Wiley, Chichester

Apps PJ (2006) Low cost, robust, in-house hardware for heart cutting two-dimensional gas chromatography. J Sep Sci 29:2338–2349

Apps PJ (2013) Are mammal olfactory signals hiding right under our noses? Naturwissenschaften 100:487–506

Apps P, Mmualefe L (2012) A versatile cryo-focussing flow switching gas chromatography inlet for trace analysis of intractable compounds. J Chromatogr A 1257:171–188

Apps P, Mmualefe L, McNutt JW (2012) Identification of volatiles from the secretions and excretions of African wild dogs (*Lycaon pictus*). J Chem Ecol 38:1450–1461

Apps P, Mmualefe L, McNutt JW (2013) A reverse engineering approach to identifying which compounds to bioassay for signaling activity in the scent marks of African wild dogs (*Lycaon pictus*). In: East ML, Dehnhard M (eds) Chemical signals in vertebrates XII. Springer, Berlin, pp 417–432

Baker TC (2014) It's still simple: signal plus response equals communication. J Chem Ecol 40:310

Burger B (2005) Mammalian semiochemicals. Top Curr Chem 240:231–278

Colby BN (1992) Spectral deconvolution for overlapping GC/MS components. J Am Soc Mass Spectrom 3:558–562

Davis J (1999) Prediction of numbers of singlets, doublets, and triplets in poorly resolved separations by statistical-overlap theory. J Chromatogr A 83:37–49

Davis JM, Giddings JC (1983) Statistical theory of component overlap in multicomponent chromatograms. Anal Chem 55:418–424

Deans DR (1968) A new technique for heart cutting in gas chromatography. Chromatographia 1:18–22

Dehnhard M, Rohrmann H, Kauffold J (2013) Measurement of 16-androstenes (5α-androst-16-en-3-one, 5α-androst-16-en-3α-ol, 5α-androst-16-en-3ß-ol) in saliva of German landrace and Göttingen minipig boars. In: East ML, Dehnhard M (eds) Chemical signals in vertebrates XII. Springer, Berlin, pp 381–390

Dorries KM, Adkins-Regan E, Halpern BP (1995) Olfactory sensitivity to the pheromone, androstenone, is sexually dimorphic in the pig. Physiol Behav 57:255–259

Du X, Zeisel SH (2013) Spectral deconvolution for gas chromatography mass spectrometry- based metabolomics: current status and future perspectives. Comput Struct Biotechnol J 4:1–10

Enke CG, Nagels LJ (2011) Undetected components in natural mixtures: how many? What concentrations? Do they account for chemical noise? What is needed to detect them? Anal Chem 83:2539–2546

Giddings JC (1987) Concepts and comparisons in multidimensional separations. J High Res Chromatogr Chromatogr Comm 10:319–323

Giddings JC (1995) Sample dimensionality: a predictor of order-disorder in component peak distribution in multidimensional separation. J Chromatogr A 703:3–15

Goodwin TE, Broederdorf LJ, Burkert BA, Hirwa IH, Mark BD, Waldrip ZW, Kopper RA, Sutherland MV, Freeman EW, Hollister-Smith JA, Schulte BA (2012) Chemical signals of elephant musth: temporal aspects of microbially-mediated modifications. J Chem Ecol 38:81–87

He J, Ma L, Kim S, Schwartz J, Santilli M, Wood C, Durnin MH, Yu CR (2010) Distinct signals conveyed by pheromone concentrations to the mouse vomeronasal organ. J Neurosci 30:7473–7483

Kessler N, Neuweger H, Bonte A, Langenkämper G, Niehaus K, Nattkemper TW, Goesmann A (2013) MeltDB 2.0 – advances of the metabolomics software system. Bioinformatics 29:2452–2459

Kwak J, Willse A, Matsumura K, Opiekun MC, Yi W, Preti G, Yamazaki K, Beauchamp GK (2008) Genetically-based olfactory signatures persist despite dietary variation. PLoS One 3:e3591

Kwak J, Grigsby CC, Rizki MM, Preti G, Köksal M, Josue J, Yamazaki K, Beauchamp GK (2012) Differential binding between volatile ligands and major urinary proteins due to genetic variation in mice. Physiol Behav 107:112–120

Laska M, Fendt M, Wieser A, Endres T, Hernandez Salazar LT, Apfelbach R (2005) Detecting danger—or just another odorant? Olfactory sensitivity for the fox odor component 2,4,5-trimethylthiazoline in four species of mammals. Physiol Behav 84:211–215

Leinders-Zufall T, Lane AP, Puche AC, Ma W, Novotny MV, Shipley MT, Zufall F (2000) Ultrasensitive pheromone detection by mammalian vomeronasal neurons. Nature 405:792–796

Likic V (2009) Extraction of pure components from overlapped signals in gas chromatography-mass spectrometry (GC-MS). BioData Min 2:6. doi:10.1186/1756-0381-2-6

Lu H, Dunn WB, Shen H, Kell DB, Liang Y (2008) Comparative evaluation of software for deconvolution of metabolomics data based on GC-TOF-MS. Trends Anal Chem 27:215–227

Luedemann A, Strassburg K, Erban A, Kopka J (2008) TagFinder for the quantitative analysis of gas chromatography-mass spectrometry (GC-MS)-based metabolite profiling experiments. Bioinformatics 24:732–737

MacNamara K, Leardi R, Hoffmann AI (2003) Developments in 2D GC with heartcutting. LC-GC Eur 16:14–22

Marriott PJ, Morrison PD, Shellie RA, Dunn MS, Sari E, Ryan D (2003) Multidimensional and comprehensive two-dimensional gas chromatography. LC-GC Eur 16:23–31

Meyer M, Peters FT, Maurer HH (2010) Automated mass spectral deconvolution and identification system for GC-MS screening for drugs, poisons, and metabolites in urine. Clin Chem 56:575–584

Moulton DG, Marshall DA (1976) The performance of dogs in detecting *alpha*-ionone in the vapor phase. J Comp Physiol A 110:287–306

Nielsen L, Eaton D, Wright D, Schmidt-French B (2006) Characteristic odors of *Tadarida brasiliensis mexicana* Chiroptera: Molossidae. J Cave Karst Stud 68:27–31

O'Callaghan S, De Souza DP, Isaac A, Wang Q, Hodkinson L, Olshansky M, Erwin T, Appelbe B, Tull DL, Roessner U, Bacic A, McConville MJ, Likić VA (2012) PyMS: a Python toolkit for processing of gas chromatography-mass spectrometry (GC-MS) data. Application and comparative study of selected tools. BMC Bioinf 13:115–133

Rasmussen L, Lee TD, Zhang A, Roelofs WL, Daves GD (1997) Purification, identification, concentration and bioactivity of (*Z*)-7-dodecen-1-yl acetate: sex pheromone of the female Asian elephant, *Elephas maximus*. Chem Senses 22:417–437

Schure MR, Davis JM (2011) The statistical overlap theory of chromatography using power law (fractal) statistics. J Chromatogr A 1218:9297–9306

Whitten WK, Wilson MC, Wilson SR, Jorgenson JW, Novotny M, Carmack M (1980) Induction of marking behavior in wild red foxes (*Vulpes vulpes* L.) by synthetic urinary constituents. J Chem Ecol 6:49–55

Willse A, Belcher AM, Preti G, Wahl JH, Thresher M, Yang P, Yamazaki K, Beauchamp GK (2005) Identification of major histocompatibility complex-regulated body odorants by statistical analysis of a comparative gas chromatography/mass spectrometry experiment. Anal Chem 77:2348–2361

Wyatt TD (2014) Pheromones and animal behaviour chemical signals and signatures. Cambridge University Press, Cambridge, pp 57–58

Chapter 29
Guidelines for Collecting and Extracting Avian Odors in a Remote Field: Case Study of a Subantarctic Seabird

Marianne Gabirot, Jérôme Mardon, Sylvie Campagna, Nigel West, Francesco Bonadonna, and Sandra M. Saunders

29.1 Introduction

For many animals, chemical signals or "chemosignals" (i.e. pheromones, scent marks, body odors) are an important feature of social behaviors. Chemical cues vary enormously in their complexity. It might be one special molecule or a bouquet of components that elicit a particular biological response (Wyatt 2010). Bird chemosignaling is, however, a relatively unexplored field of research: avian olfactory

M. Gabirot (✉)
Organisms and Environment Research Division, Cardiff School of Biosciences, Cardiff University, Cardiff, UK

CEFE UMR 5175, CNRS – Université de Montpellier – Université Paul-Valéry Montpellier – EPHE, 1919 route de Mende, F-34293 Montpellier, Cedex 5, France
e-mail: GabirotM@cardiff.ac.uk

J. Mardon
CEFE UMR 5175, CNRS – Université de Montpellier – Université Paul-Valéry Montpellier – EPHE, 1919 route de Mende, F-34293 Montpellier, Cedex 5, France

Atmospheric and Environmental Chemistry Research (AECR) Group, School of Chemistry and Biochemistry, The University of Western Australia, Crawley, WA, 6009, Australia

S. Campagna
Université de Nimes, F-30021 Nimes Cedex 1, France

N. West
ChemCentre, Resources and Chemistry Precinct, Curtin University, Bentley, WA, 6102, Australia

F. Bonadonna
CEFE UMR 5175, CNRS – Université de Montpellier – Université Paul-Valéry Montpellier – EPHE, 1919 route de Mende, F-34293 Montpellier, Cedex 5, France

S.M. Saunders
Atmospheric and Environmental Chemistry Research (AECR) Group, School of Chemistry and Biochemistry, The University of Western Australia, Crawley, WA, 6009, Australia

B.A. Schulte et al. (eds.), *Chemical Signals in Vertebrates 13*,
DOI 10.1007/978-3-319-22026-0_29

capabilities and avian volatile emissions (essentially limited to plumage scent) have long been overlooked (for review see Hagelin and Jones 2007; Campagna et al. 2012; Soini et al. 2013; Caro et al. 2014). In addition, all the studies have focused on the analysis of either uropygial secretions or feather compounds. The uropygial gland, located at the dorsal base of the tail, is indeed the only sebaceous gland of birds and as such, is often considered as the key source of avian chemical substances. Yet to date no published study has focused on the bouquet of airborne volatiles surrounding a bird, although the odors in a bird's headspace are essentially what is being used in the communication. Some partial results were provided in two studies that collected the volatiles emitted by a crested auklet (Douglas 2006), a chicken or a pigeon (Syed and Leal 2009) enclosed in an aeration chamber. In both studies, the authors have targeted only one (octanal in Douglas 2006) or a few specific compounds (Syed and Leal 2009) so that the whole body odor was not determined. As well as the relative infancy of the field of avian chemosignals, major reasons for the sparse data are undoubtedly the absence of an appropriate methodological framework and experimental difficulties associated with collecting such samples.

To facilitate deciphering the complex scent of birds especially in remote field conditions, we provide in this paper a methodological comparison of various combinations of sampling and analytical techniques that have been tested or recently developed in the field and laboratory. However, we do not aspire to suggest the best method to achieve sampling and analysis of birds' airborne potential signals, because, as the reader will see, there is not one, but probably a combination of some. We used the particular case study of blue petrels (*Halobaena caerulea*), a Subantarctic procellariiform seabird known for its musky smell and the associated olfactory capabilities (Bonadonna and Nevitt 2004; Bonadonna et al. 2007; Mardon and Bonadonna 2009; Bonadonna and Mardon 2010, 2013; Mardon et al. 2010, 2011). Previous works have showed that volatile and semivolatile molecules contained in preen oil vary between species (Bonadonna and Mardon 2010) and sexes (Mardon et al. 2010). While further eco-behavioral evidence is needed to confirm the "signaling" function of their odors, blue petrels provide a particularly appropriate candidate for the investigation of avian social chemosignals. Moreover, blue petrels breed on small oceanic islands around Antarctica that constitute a complicated and far-off environment to collect and to analyze chemical samples. Consequently this model is appropriate to discuss the merits of different methods for chemical sampling in remote areas.

The sampling approaches involved uropygial secretions, feather lipids, and airborne volatiles emitted by birds. Three different extraction methods were also compared: solvent extraction (SE), solid-phase microextraction (SPME), and thermal desorption (TD). In all cases, extracted materials were chemically analyzed with a gas chromatograph coupled to a mass spectrometer (GC–MS). We considered the following approaches: (1) analysis by SE of uropygial secretion samples, (2) analysis by SE of feather lipids, (3) analysis by direct solid-phase TD of feather lipids, (4) analysis by SPME of airborne volatiles collected on a cotton swab, (5) analysis by direct solid-phase TD of airborne volatiles collected on a cotton swab and

(6) analysis by TD of airborne volatiles emitted by a bird and collected on a Tenax® TA polymer. The outcomes of the different combinations of sampling and extraction methods are analyzed in terms of chromatographic quality, number and properties of resolved analytes. A comparison of the various options currently available to ecologists for post-processing their chromatographic data (i.e. peak detection, peak integration, identification, statistics) is not, however, within the scope of the present paper. Although designed to target avian chemosignals, some of the presented protocols should be readily adaptable to other animal models.

29.2 Methods

All samples were collected during four austral summers (November to January 2006–2007, 2007–2008, 2008–2009, and 2011–2012) as part of the French Subantarctic program ETHOTAAF no. 354, on Ile Verte (49°51′S, 70°05′E), from the Kerguelen Archipelagos in the southern Indian Ocean.

29.2.1 Chemical Analysis

All lab procedures reported in the methods detailed below, whether extraction or chromatographic protocols, were optimized through a variety of preliminary tests.

29.2.1.1 Analysis of Uropygial Secretions by Solvent Extraction (Method 1)

Field Sampling

Secretionsamples were obtained from 16 blue petrels using a protocol adapted from Burger et al. (2004) and Mardon et al. (2010). Uropygial gland contents were collected by gently squeezing the area around the gland, while wearing clean nitrile gloves, until a small amount of waxy secretion was discharged. The secretion was then caught with a 100 μl glass capillary. The end of the capillary containing the secretion was inserted and broken into a small opaque chromatographic vial (Sigma-Aldrich®, Castle Hill, Australia). Finally the vial was stored at −4 °C in the field during 2–3 weeks and at −20 °C in the laboratory until extraction.

Sample Preparation and Extraction

Uropygial samples were solvent extracted in 400 μl of a 1:3 mix of dichloromethane (distilled High Performance Liquid Chromatography grade, ≥99.9 %, Sigma-Aldrich®, Australia) and *n*-hexane (distilled Analytical Reagent grade, ≥99.0 %,

Fluka®, Castle Hill, Australia), poured directly into the field vial containing the capillary tube. The vial was then resealed and left to stand 7 min in a beaker of ice to keep the extracting temperature as low as possible. The extraction mixture in the vial was subsequently transferred into a second clean chromatographic vial, passing through a Pasteur pipette filled with a glass wool plug to filter out impurities (dust, feather debris). In parallel, with the same protocol several blank vials with solvent only (as controls) were prepared. Finally, all samples were spiked with 10 μl of an internal standard solution of 2-bromophenol (purum, ≥99.0 %; Fluka®, Australia) in methanol (Analytical Reagent grade, ≥99.6 %; Sigma-Aldrich®, Australia) at 504 ng/μl for indicative quantification purposes (5040 ng injected in each vial).

Chromatographic Analysis

The analyses were carried out at the University of Western Australia using a GC–MS (Shimadzu QP2010; Shimadzu Corp., Rydalmere Vic., Australia) equipped with an autosampler (Shimadzu AOC-20i+s; Shimadzu Corp., Australia) and a generalist Rtx®-5MS capillary column (L=30.0 m; Diameter=0.25 mm; Film thickness=0.10 μm; Restek®, Bellefonte PA, USA). The injection port temperature was set at 250 °C and helium was used as carrier gas at a constant linear velocity of 35 cm/s. A volume of 1 μl of solvent extracts was injected, in splitless mode, and cold-trapped at 40 °C on the column tip for 3 min. Samples were subsequently separated using a temperature program of 8 °C/min from 40 to 150 °C, then 6 °C/min from 150 to 200 °C and then 2 °C/min from 200 to 280 °C (hold 15 min). The interface temperature was held at 280 °C and the ion source temperature at 200 °C. The mass spectrometer (MS) was used in scan mode (scan speed=625; scan interval=0.5 s) with an electron source voltage of 70 eV and over the mass range of 45–350 amu.

29.2.1.2 Analysis of Feathers by Solvent Extraction (Method 2)

Field Sampling

Feather samples were obtainedfrom the same 16 blue petrels as above (Method 1). Wearing clean nitrile gloves, we cut between 60 and 100 mg of feathers from the ventral duvet of the birds with clean steel scissors rinsed with methanol (Lab Reagent grade, Sigma-Aldrich, Australia). Feathers were then packed in aluminum foil, placed in a sealed plastic bag and stored at −4 °C in the field during 2–3 weeks and at −20 °C in laboratory until chemical analysis.

Sample Preparation and Extraction

Feather samples were solvent extracted with the same solvent mix as in Method 1. Sixty milligrams of feathers were placed in a 50 ml conical flask, together with 4 ml of the solvent mix and a magnetic stirrer. The flask was then sealed with a hermetic stopper, placed in a beaker of ice to minimize volatilization of lighter compounds, and the content left to macerate on a magnetic stirring apparatus for 2.5 h. After the maceration, the extract was transferred into a 4 ml vial through a Pasteur pipette filled with a glass wool plug. To concentrate samples, we used a low-pressure drying approach (Burger et al. 2004). The extract volume was reduced to 400 μl by attaching the vial to a low-pressure liquid nitrogen cold finger manifold (the vacuum provided by a rotary vacuum pump Genevac® type GRS2; Ipswich, UK). With the same protocol, we prepared several blank vials with only solvent (i.e. control). Finally, the 400 μl concentrated extract was transferred into a second clean chromatography vial and spiked with 10 μl of the same internal standard solution of 2-bromophenol in methanol as in Method 1.

Chromatographic Analysis

GC–MS conditions for the analysis of the feather lipid extracts were identical to the ones detailed in Method 1.

29.2.1.3 Analysis of Feathers by Direct Solid-Phase Thermal Desorption (Method 3)

Usually, Thermal Desorption consists of exposing samples to a flow of hot dry inert gas so that chemical compounds present within the sample are progressively volatilized. Released analytes eventually transfer, via the gas flow, to the GC–MS where they are generally retrapped (e.g. cold-trap) before being injected into the chromatograph without any solvent dilution. Although the retrieving efficiency of TD is lower than through solvent extraction (Baltussen et al. 2002), the absence of a dilution effect generally makes it more sensitive than the latter. The novel use of TD tested here was partly inspired by the work of another team investigating mosquitoes-chicken chemical interactions (Santos et al. 2005).

Field Sampling

Feathers were collected as stated above in Method 2 from 22 birds during the field campaigns 2011–2012, and were conserved in nalophan® (polyethylene terephthalate) and in aluminum foil at −4 °C in the field during 2–3 weeks and then at −20 °C in the laboratory until analysis. This method of conservation by nalophan® with aluminum was preferred rather than Method 3 (only aluminum). It leads to a lower contamination of samples by aluminum and plastic bags (Hudson et al. 2009).

Sample Preparation

Using clean nitrile gloves, 10 mg of feathers from each sample were first placed in a "Loose Fit" Teflon® insert [Liner polytetrafluoroethylene (PTFE); Markes International Limited, Rhondda Cynon Taff, UK] and then dropped into a clean empty TD stainless steel tube (OD=6 mm; L=88 mm; Perkin-Elmer France, Courtaboeuf, France). At the top of the insert, a silanized treated glass wool plug (Perkin Elmer USA) was added to avoid any loss of feathers. Empty tubes (with insert and without feathers) were used as a control. All samples were spiked with 1 μl of an internal standard solution of biphenyl (99.5 % Sigma-Aldrich®, France) dissolved at a concentration of 0.1 mg/ml in dichloromethane/*n*-hexane, 1:3 (v/v) (Sigma-Aldrich®, France) (100 ng injected in each TD tube). In this case, no extraction was achieved and a direct desorption was realized as described below in the chromatographic analysis section.

Chromatographic Analysis

Chromatographic analyses were carried out in the PACE-Labex CEFE-CNRS (Montpellier, France), using a GC–MS (Shimadzu QP2010; Shimadzu Corp.) equipped with a TD autosampler (Shimadzu AOC-20i+s; Shimadzu Corp.). As the combustion point of feathers is around 230 °C (J. Mardon, personal data), samples were thermally desorbed for 10 min at 180 °C to avoid combustion products collecting during analysis. Volatilized analytes were sent to a secondary Tenax® TA trap held at −10 °C which was then desorbed in turn, by rapidly heating it from −10 to 250 °C within a few seconds. From there, samples were injected with a split of 10:1 into a DMS-5 capillary column (L=60.0 m; Diameter=0.25 mm; Film thickness=0.25 μm; Supelco, Bellafonte, PA, USA). The GC oven temperature was programmed at 30 °C (hold 5 min), then 2 °C/min–100 °C, 5 °C/min–270 °C 2 °C/min to 100 °C and then 5 °C/min to 270 °C (hold 5 min). The transfer line was set at 250 °C, the ion source at 200 °C. Standard positive electron ionization (70 eV) was used with the scanning rate of 0.3 scans/s over the mass range of 35–350 amu.

29.2.1.4 Analysis of Cotton Swab by Solid-Phase Microextraction (Method 4)

In previous studies, a specialized cotton swab was successfully used to collect avian odors as indicated by an effective discrimination of odors in a Y-maze with birds (Mardon and Bonadonna 2009) and by mice in an olfactometer procedure (Celerier et al. 2011). This special cotton swab is produced in Hungary (Prada et al. 2010) and also employed by the French scientific police service in an "odorology protocol" (Service Central d'Identité Judiciaire of Lyon police, France, validated by INTERPOL).

Field Sampling

Ten birds were gently rubbed (back, under wings, rump) with the special cotton swab for 3–4 min (experimenter wearing disposable nitrile gloves) during field campaigns 2011–2012. Field controls were made by exposing the cotton swab to the same conditions as samples but in open air only. Each cotton swab was then sealed in an opaque glass jar, kept at −4 °C in the field during 2–3 weeks and at −20 °C in laboratory until analysis.

Sample Preparation and Extraction

A static headspace solid-phase microextraction (SPME) was performed as used in other similar studies, such as in the analysis of otter scent (Kean et al. 2011) or gray catbirds (Shaw et al. 2011). A SPME fiber [StableFlexTM 65 μm Polydimethylsiloxane (PDMS)–Divinylbenzene (DVB), Supelco, Bellefonte, PA, USA] was exposed to the headspace over the cotton swab for 20 min at ambient temperature to collect volatile organic compounds eluting from samples. Then the fiber was retracted and the adsorbed chemicals were analyzed by GC–MS. Fibers were conditioned according to manufacturer's recommendations and reconditioned for 10 min in a gas chromatography (GC) injection port at 250 °C between each sample. An analysis of a fiber not exposed to any sample was conducted at least every sixth sample to detect any contamination or deterioration of the fiber, and fibers were replaced when damaged. For this method, there was no internal standard spiked.

Chromatographic Analysis

Following exposure, fibers were immediately analyzed at the PACE-Labex CEFE-CNRS (Montpellier, France), using a GC–MS (Shimadzu GCMS 2010 QP plus). Fibers were injected manually and desorbed for 5 min at 250 °C in the injection port fitted with an SPME liner (Supelco) in direct mode. Samples were analyzed on a Zebron Zb-5MSi capillary column (L=30 m; Diameter=0.25 mm; Film thickness=0.25 μm; Phenomenex Inc., CA, USA) with helium as carrier gas at constant pressure (7.2 psi). The oven was held initially at 40 was held initially at 40 °C for 5 min, then heated at 5 °C/min to 200 °C, followed by heating at 16 °C/min to 240 °C, and held at 240 °C for 4 min. Temperature of transfer line, and ion source was set to 250, and 200 °C, respectively. Standard positive electron ionization (70 eV) was used with the scanning rate of 0.3 scans/s over the mass range of 35–350 amu.

29.2.1.5 Analysis of Cotton Swab by Thermal Desorption (Method 5)

Field Sampling

Bird odors were collected as described above in Method 4 on the same individuals.

Sample Preparation

Using nitrile gloves and cleaned scissors, 2 cm^2 of cotton swab from each sample were packed in TD stainless steel tubes (OD = 6 mm; L = 88 mm; Perkin-Elmer France, Courtaboeuf, France). All samples were spiked with 1 µl of an internal standard solution of biphenyl (99.5 % Sigma-Aldrich®, France) dissolved at 0.1 mg/ml in dichloromethane/n-hexane 1:3 (v/v) (Sigma-Aldrich®, France) (100 ng injected in each TD tube).

Chromatographic Analysis

The chromatographic analyses were carried out at the PACE-Labex CEFE-CNRS (Montpellier, France), using a GC–MS (Shimadzu QP2010; Shimadzu Corp.) equipped with a TD autosampler (Shimadzu AOC-20i+s; Shimadzu Corp.). To avoid the combustion process of swab during analysis, samples were thermally desorbed for 10 min at 80 °C. Volatilized analytes were sent to a secondary Tenax® TA trap held at −10 °C which was then desorbed in turn, by rapidly heating it from −10 to 250 °C within a few seconds. From there, samples were injected with a split of 10:1 into an RXi-624Sil MS capillary column (L = 30.0 m; Diameter = 0.25 mm; Film thickness = 1.40 µm; Restek Corporation, Bellafonte, PA, USA). The GC oven temperature was programmed as 30 °C (hold 4 min), then 4 °C/min to 100°C, 3 °C/min to 180°C, and 6 °C/min to 270 °C (hold 3 min). The transfer line was set at 250 °C, the ion source at 200 °C. Standard positive electron ionization (70 eV) was used with the scanning rate of 0.3 scans/s over the mass range of 35–350 amu.

29.2.1.6 Analysis of Airborne Volatiles by Thermal Desorption (Method 6)

Paradoxically, few studies on vertebrates' chemosignaling have so far focused directly on the airborne volatiles emitted by an animal. This is possibly due to experimental difficulties associated with collecting such samples, particularly in the field. We present a method inspired by laboratory studies of plant, insect, or rodent semiochemicals (Moritz and Crewe 1988; Dicke et al. 1990; Röck et al. 2006), allowing the quantitative sampling of a known volume of air passing through a chamber containing the animal. Similar studies have been also carried out on birds (Douglas 2006; Syed and Leal 2009). In both studies, the volatiles were trapped on

SuperQ polymer (Douglas 2006; Syed and Leal 2009) or Tenax® (Douglas 2006). However, the collection traps were not thermodesorbed, but were eluted with dichloromethane or methanol.

Field Sampling

The sampling apparatus (Fig. 29.1) consisted of an activated charcoal trap (400 cc in-line gas purifier, filled with a charcoal refill kit; Grace/Alltech®, Deerfield IL, USA) connected with Teflon® PTFE tubing (OD=6 mm, Clean Air Engineering Inc., Palatine IL, USA) to a chamber formed by an hermetic stainless steel autoclave (modified SEB® pressure cooker model Clipso Ovale, size 44×24×24 cm). A constant laminar air flow was created through the apparatus using a portable air sampling pump (Escort® ELF Pump; MSA France, Châtillon-sur-Chalaronne, France) located at the end of the circuit. The air, filtered through the activated charcoal filter, was drawn through the chamber where a bird sat, and then out of the chamber through a TD tube. All junctions to and from the chamber were made with Teflon® PTFE connectors (Grace/Alltech®, USA).

Stainless steel TD tubes (OD=6 mm, L=88 mm; Perkin-Elmer®) packed with a single bed of 150 mg of Tenax® TA (mesh size 35/60) sealed with glass wool, were used.

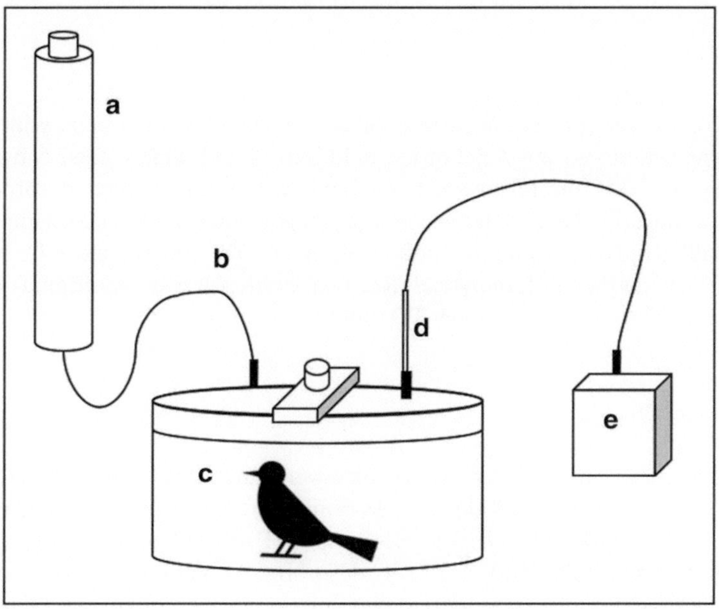

Fig. 29.1 Experimental apparatus for the collection of airborne compounds emitted by birds. (**a**) activated charcoal air filter; (**b**) Teflon® tubing connectors; (**c**) air chamber; (**d**) adsorbent tube; (**e**) air sampling pump

Among the various adsorbents commercially available for the trapping of volatile compounds, Tenax® TA (mesh size: 35/60; Grace/Alltech®, USA) was selected in this study. Tenax® TA, a commonly used organic adsorbent, possesses the most wide-ranging affinities in terms of chemical class and size and is hydrophobic (Scientific Instrument Services Inc. 1995). This makes it appropriate for the study of complex mixtures of volatiles such as scents, particularly in areas where there may be a significant amount of moisture. Air flow was set at 500 ml min^{-1} (Guillot et al. 2000). A high flow rate, within the range of optimal adsorption efficiency, was chosen considering the volume of air required (dependent on the concentration of the volatiles) and the level of disturbance for the animal (dependent on sampling time). Air sampling was carried out for 30 min, corresponding to 15 L of air sampled.

The above protocol was used to collect the "odors" from 20 blue petrels during the austral summer 2007–2008 and 2008–2009. Several field and apparatus blank samples were also collected to provide information regarding background noise in subsequent analyses. Birds were captured from their burrow, transported to the apparatus in a cotton bag and placed in the chamber. Animals were then transported back to their burrow. Interestingly, birds seemed more stressed during handling phases (capture or ringing) than while in the apparatus chamber, during which they often sat quietly in the dark (F. Bonadonna, personal observations). Between two consecutive samples, the chamber was cleaned with water then methanol (Lab Reagent grade; Sigma-Aldrich®, Australia), and allowed to dry. Activated charcoal from the air filter was renewed every 10–15 samples.

Sample Preparation

After the air sampling, the adsorbent tube was sealed at both ends with Teflon® PTFE caps and stored at −4 °C in the field during 3–4 weeks, then conserved in laboratory at −20 °C until analysis. In the laboratory, before a direct desorption was realized (as described below in the chromatographic analysis part), all samples were spiked with 10 µl of a 10 ng/µl solution of toluene-D8 (99.6 % atom D; Aldrich®, Australia) in methanol (Analytical Reagent grade, ≥99.6 %; Sigma-Aldrich®, Australia) (100 ng injected in each TD tube).

Chromatographic Analysis

Chromatographic analyses of these samples were carried out at the ChemCentre of Perth, using a Varian Saturn 2000 GC–MS (ion-trap) (Varian Inc Australia, Mulgrave Vic., Australia), equipped with a TD autosampler (Perkin Elmer ATD 400 Automatic Thermal Desorption System). A 270 °C desorbing temperature (below the 350 °C upper limit of Tenax® TA) was used to maximize analyte recovery and minimize the background contribution from the thermal degradation of Tenax® TA. Volatilized analytes were sent to a secondary Tenax® TA trap held at −10 °C which was then desorbed in turn, by rapidly heating it from −10 to 270 °C within a few seconds.

From there, samples were injected with a split of 10:1 into the GC capillary column (DB-5MS; $L=30.0$ m; Diameter = 0.25 mm; Film thickness = 0.25 μm; Agilent technologies®, Forrest Hill, Vic., Australia). The GC oven temperature was programmed as 30 °C (hold 1 min), then 5 °C/min to 100 °C, 10 °C/min to 200 °C, 15 °C/min to 290 °C (hold 4 min). The transfer line was set at 170 °C, the ion trap at 150 °C. Standard positive electron ionization (70 eV) was used with the scanning rate of 2 scans/s over the mass range of 45–270 amu.

29.2.2 Chemical Data Processing

Chemical data processing was carried out with the GCMS Solution software v2.40© (Shimadzu Corp., France). In all analyses, background noise was first removed from the data by subtracting the baseline signal and peaks of compounds obtained from blank samples run regularly within our sample batches. Blanks were designed to account for potential noise from the sampling procedure, the extraction protocol, and the instrument in different methods.

Qualitative identification of all analytes of interest was determined by cross-checking the best matches obtained from the NIST Mass Spectral Search Program v2.0© (Faircom Corp.; Columbia MO, USA) with the calculated Retention Index (RI) of the analytes. Calculated RIs were obtained by calibrating the GCMS Solution software with the retention times of various linear alkanes between C_{10} and C_{40} ($n=15$), run under identical chromatographic conditions. For quantitative analysis, the peak area of the internal standard was used to account for variations in the GC–MS instrument response when standardization was made possible. For method 6, calibration curves were constructed from a suite of standards of various masses and chemical classes, using air-monitoring National Association of Testing Authorities, Australia (NATA) accredited methods.

29.3 Results

29.3.1 Method 1 (Solvent Extraction of Uropygial Secretions)

Chemical profiles showed in particular a 30 min section (25–55 min) containing an abundance of analytes (more than 200 compounds), essentially fatty acid esters and alcohols (Table 29.1 and Fig. 29.2a). The sample signal was strong compared to background noise (instrument, sampling) and peak chromatography was good overall. A slower ramp than the one used could provide even better separation of fatty contents although this would considerably extend the total analysis time. Compounds detected and tentatively identified ranged from lower semi-volatiles (octanoic acid, tetradecane) to large fatty acid esters (nonadecanoic acid, eicosyl ester). Data analysis led to the identification of clear sociochemical information within the uropygial

Table 29.1 Advantages and disadvantages of the different methods tested

	Method 1	Method 2	Method 3	Method 4	Method 5	Method 6	Alternative method (not tested here) Stir bar sorptive extraction
Sample type	Uropygial secretions	Feathers	Feathers	Cotton swab	Cotton swab	Airborne volatiles	Liquid–solid–volatile phase
Extraction method	SE	SE	Direct TD	SPME	Direct TD	TD	SBSE
N analytes detected	266	321	89	27	18	22	–
Carbon chain size	C_8–C_{39}	C_8–C_{39}	C_2^a–C_{19}	C_6–C_{15}	C_8–C_{14}	C_2^a–C_{13}	–
Class of analytes detected	Aldehydes, alkanes, alcohols, alkenes, amides, carboxylic acids, acid esters	Aldehydes, alkanes, alcohols, alkenes, amides, aromatic hydrocarbons, carboxylic acids, acid esters	Aldehydes, alkanes, alcohols, aromatic hydrocarbons, carboxylic acids, acid esters, Furans, sulfides	Aldehydes, alkanes, alcohols, carboxylic acids, acid esters, sulfides	Aldehydes, alkanes, alcohols, carboxylic acids, acid esters, sulfides	Aldehydes, alkanes, alcohols, carboxylic acids, acid esters, furans, sulfides	–
Storage container	Glass vial	Aluminum foil and plastic bag	Nalophan® bag	Glass jar	Glass jar	TD Tenax TA tubes	–
Storage quality	✓✓✓ But heavy and fragile	✓ Lightweight but only for "solid" samples	✓✓ Lightweight but only for "solid" samples"	✓✓✓ But heavy and fragile	✓✓✓ But heavy and fragile	✓✓ Some contamination seems to occur during airplane travel	–
Fast sampling	✓✓	✓✓	✓✓	✓✓✓	✓✓✓	✓	–

Minimal manipulation to bird	✓	✓	✓	✓✓	✓✓	✓✓	✓✓ (when used directly on feathers)
Minimal analytical steps	✓ (pre-concentration step is required)	✓ (pre-concentration step is required)	✓✓✓	✓✓	✓✓✓	✓✓	✓✓
Data close to bird's "true" odor signal[b]	✓	✓	✓✓	✓✓	✓✓	✓✓✓	✓✓
Extraction efficiency (*estimated*)	✓✓✓		✓✓	✓	✓	✓✓✓	✓✓
General disadvantages	Solvent can impact some reactive/fragile compounds such as amines Analysis of chromatogram is complex. Highly abundant analytes may be "masking" less abundant compounds		Low temperatures of desorption may reduce the efficiency of recovery	Fragility of the coating of SPME fibers can lower the reproducibility	Low temperatures of desorption may reduce the efficiency of recovery	Desorption at high temperatures can breakdown some analytes (mainly the polar ones) Low signal/noise ratio due to high level of contamination	Diversity of coating is limited

SE Solvent extraction, *TD* Thermal desorption, *SPME* Solid-phase microextraction. *SBSE* Stir bar sorptive extraction was not tested in this study but is added in the table as a matter of comparison (✓: Fair; ✓✓: Good; ✓✓✓: Very good)

[a]C_2 is Dimethyl disulfide

[b]Similar to what we expect may contain the actual compounds involved in a chemosignal that a bird emits

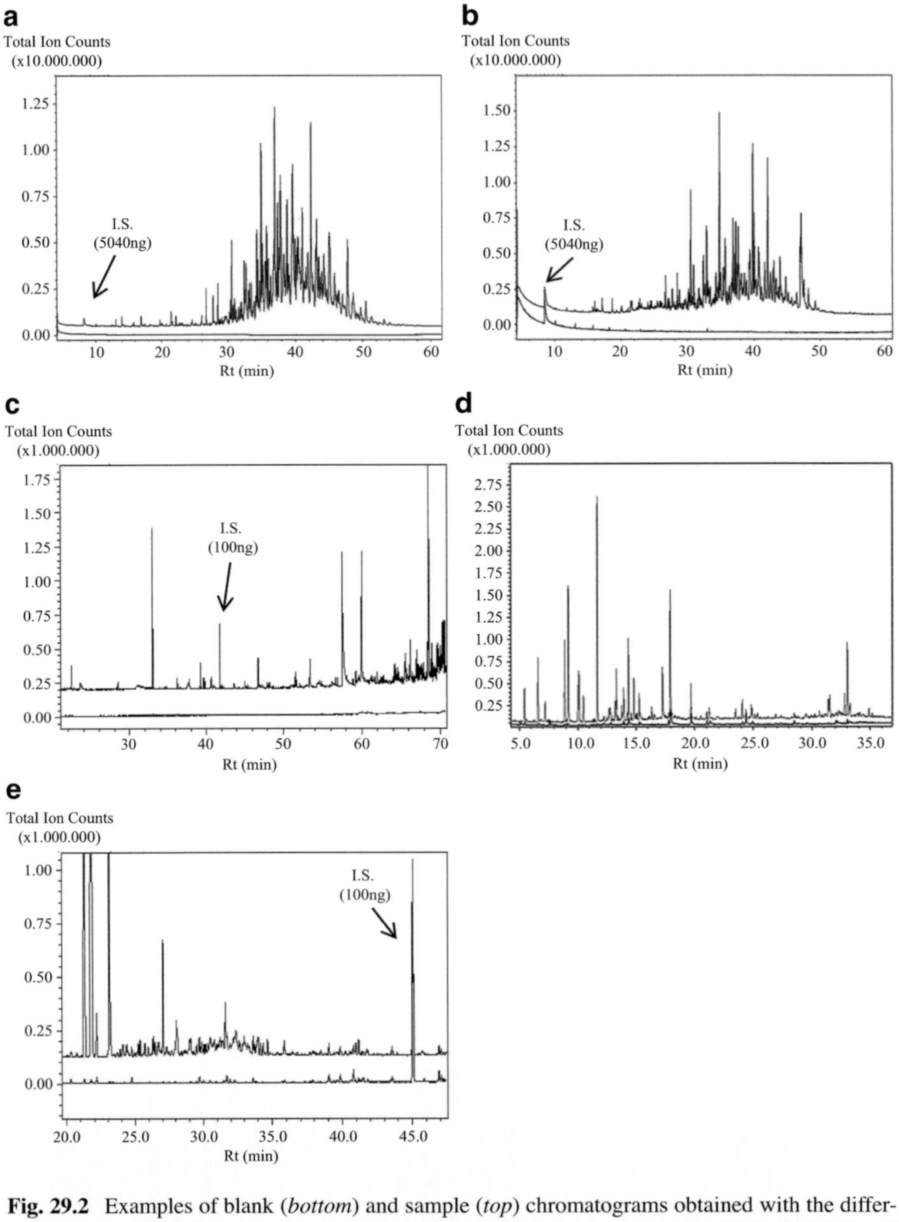

Fig. 29.2 Examples of blank (*bottom*) and sample (*top*) chromatograms obtained with the different methods. (**a**) Method 1 (analysis of uropygial secretions by solvent extraction); (**b**) Method 2 (analysis of feathers by solvent extraction); (**c**) Method 3 (analysis of feathers by thermal desorption); (**d**) Method 4 (analysis of cotton swab by SPME); (**e**) Method 5 (analysis of cotton swab by thermal desorption). The peak of the internal standard (I.S.) and the quantity injected in the sample are indicated when present

secretions of blue petrels including species-specific, sex-specific, and individually specific badges (details reported in Mardon et al. 2010). Importantly, chemical compounds associated with these chemical badges were large fatty acid esters and alcohols of relatively low volatility, which are unlikely to be the final carriers of the olfactory information. Method 1 therefore did not lead to the identification of the final olfactory signals but provided instead chemical correlations or precursors of the possible signals. If this method is still useful to compare individuals, genders, or species (the actual signal is probably contained in the chromatogram obtained), nonvolatile compounds, contamination, ejection of parasites, bacteria, etc., take part in this chromatogram, perhaps masking the actual signal that the individual may use.

29.3.2 Method 2 (Solvent Extraction of Feathers)

The sample signal again largely dominated the background noise and peak chromatography was satisfactory (Fig. 29.2b). The lower edge of detection sensitivity was also similar to Method 1. Although the quantity of chemical material retrieved was slightly less, more analytes were detected in this analysis than in the previous one (more than 300 components found, see Table 29.1). This is explained by the appearance of new compounds, essentially semi-volatile compounds, and cyclic hydrocarbons, on the bird feathers due to different processes (detailed in Mardon et al. 2011). The feather chemical profiles were nevertheless analogous to the uropygial profile with a similar 30 min section (25–55 min) containing abundant analytes. Furthermore, 95 % of the secretion analytes were present on the feathers while 79 % of the feather analytes originated from the preen secretions; thus confirming, in petrels at least, the uropygial origin of most feather lipids (Mardon et al. 2011). Data analysis of the blue petrel feather chemicals led to the identification of sociochemical badges similar to the ones found within the uropygial secretions (detailed in Mardon et al. 2011). The chemical nature of these feather-borne badges was consistent with that previously elucidated within uropygial contents. In the same way as in method 1, molecules coupled with this method were analytes with low volatility and not necessary the final product of the bird's odor.

29.3.3 Method 3 (Thermal Desorption of Feathers)

A good chromatographic signal from the feathers was obtained (Fig. 29.2c). Peak chromatography was good overall with little co-elution, although most carboxylic acid peaks tailed significantly. The lower edge of detection sensitivity for this method was found to be about five times lower than for the SE technique. The TD approach permitted the detection of a low number of analytes (less than 100). Components found were more volatile and of a much smaller size than in Method 1 and 2 (e.g. heptanal) (see Table 29.1). Interestingly, some of these early eluting compounds included sulfides (dimethyl disulfide), furans (2-furanmethanol), and imidazole. Several contaminant components (benzene derived) were also detected, that might be due to bird's exposure to pollutants or during the sampling procedure. The remainder of the analytes detected were from chemical classes that were already encountered (acids, esters, alcohols, and alkanes). Preliminary data analysis of these samples suggest that the volatiles desorbed directly from the feathers may still carry a significant badge of the species, i.e. a chemical signature of blue petrels, but no clear-cut sex-specific badge (M. Gabirot, personnal data).

29.3.4 Methods 4 and 5 (Cotton Swab by Solid-Phase Microextraction and by Thermal Desorption)

The results obtained with these two methods were very similar. A satisfactory chromatographic signal was obtained in both cases (Fig. 29.2d, e) with a sample signal that largely dominated the background noise. Peak chromatography was good overall with little co-elution, although most peaks tailed significantly. There was a lower detection threshold of sensibility for these two methods. The number of peaks was also significantly lower than in method 1, 2, and 3 (less than 30 compounds; see Table 29.1). The main compounds were detected between RT=20 and RT=35 min for the SPME method and in a 21 min section (RT=20–41 min) for the TD method. In one section of the chromatogram, peaks were too close together (mostly between RT=26 and RT=29 min), thus a slower ramp than the one used could provide better separation of compounds present. In the same way as the analysis of feathers by thermal desorption, analysis of the swab by SPME or TD produced more volatile and semi-volatile compounds compared to methods using solvent extraction. Compounds released from the swab were tentatively identified and ranged from volatiles (e.g. octanal, hexanal, heptanal, nonanal) and semi volatiles (e.g. hexanone, undecane, methyl nonane, and undecane) to heavier compounds such as fatty acid ester or methyl ketone or methyl alkanes. Once again, atypical compounds such as sulfides and molecules that were present probably due to pollution (benzene derived) were detected.

29.3.5 Method 6 (Thermal Desorption of Airborne Volatiles)

This method produced chromatograms that were dominated by background noise (data not shown). The contamination found in this method might be due to a problem during the volatiles sampling or/and storage and travel to the laboratory. This contamination suggests for example that activated charcoal filters may not be sufficient to purify ambient air to a satisfactory level for the screening of subtle volatiles signals. The noisy background made the discrimination of the bird-associated chemicals significantly harder, resulting in fewer analytes detected in this TD-analysis (Table 29.1). The smaller early eluting analytes identified in Method 3, in particular furans and sulfides were also found with this method. Most of the analytes detected were from the more usual acids, alkanes, alcohols, and esters classes. Data analysis did not provide clear-cut evidence of sociochemical information but the high background noise prevents any definitive conclusion at this stage of our research.

29.4 Discussion

We provide different approaches to study the odor of birds, and discuss the advantages and limitations of each method for field chemo-ecologists. We took into consideration the methodological difficulties of analyzing vertebrate scents, coupled with the logistical challenges of working with animals in remote field locations. Studying wild populations far away from laboratory facilities implies additional constraints, including that sampling might not be under completely controlled conditions (e.g. cold conservation of samples).

29.4.1 Sampling Considerations

Different samples were collected (uropygial secretions, feather lipids and/or volatiles, bird odor) as they might carry partly or totally the compounds involved in the bird scent. We also used different sorbent materials for volatile collection such as Tenax® TA polymer to trap the molecules from the air surrounding the bird's headspace. Another sorbent material was also tested for the first time i.e. a cotton swab that was rubbed directly on the animal. These collection techniques were chosen because they offered significant advantages in the field such as their ability to sample quickly, logistical simplicity, and the relatively high yield of information. As the compounds present in the scent can have vastly different physical properties, any particular method of sampling will certainly fail to collect all of them. Therefore, a combination of different approaches is recommended until the key-components of the scent are identified. For example, in the field one can easily collect feathers and cottons swabs rubbed on birds. Successively, in the lab, one may use complementary methods such as direct thermodesorption of feathers (Method 3), or exposing samples to SPME fibers or Stir Bars with different polymers sensitive complementary classes of compounds. However, there are significant issues that need to be addressed before starting any sampling of animal material.

The rules of thumb can be summarized in three points.

1. Avoid stress of the animals. The collection of uropygial and feathers involves a significant amount of handling which can be stressful (Le Maho et al. 1992). Stress might imply modification of physiology and might have consequences in variation of animal scents (e.g. in rats, Valenta and Rigby 1968).

2. Avoid as well as possible contamination of the samples. In our study some compounds were already present in the samples, such as atmospheric pollutants (sulfides and benzene derivatives) deposited on the feathers (Mardon et al. 2011). These can be corrected by blanks, considering that the level of contamination is relatively minor and does not affect the clarity of the biogenic emissions (Mardon et al. 2011). It is worth remembering, however, that contamination can occur during the process at any stage from initial sample collection, sample transportation, sample preparation, to chemical analysis. For example GC–MS analysis of tubes that made a round trip to the field and back to the laboratory without being opened, showed some contamination (Gabirot et al. unpublished results). A possible explanation may be that variation in temperature and/or pressure during the flight may have affected the efficiency of sealing caps fitted with PTFE ferrules.
3. Store and cool samples to limit degradation and compounds loss. Glass vials are usually reliable and better containers than plastic ones, if correctly sealed with screw caps or lined stoppers which can provide a gas tight seal (e.g. screw caps Teflon-faced or screw caps with PTFE faced silicone septa) and stored at cold temperature (Hudson et al. 2009). For example a deficient sealing, often due to dirty closure, and/or to septa expansion and contraction during temperature fluctuations associated with sample cooling/freezing, has an average of 41–46 % lower volatile organic compound concentrations than the samples correctly sealed (Hewitt et al. 1995). The loss of components can also be minimized by limiting the number of analytical steps (as pre-concentration in solvent extraction, weighing of feathers, etc., Drea et al. 2013). Finally, maintaining samples at cold temperatures (at least 4 °C) during storage and transportation minimizes vapor loss of highly volatile compounds and limits uncontrolled microbial degradation.

In spite of problems given by point (1) and (2), relative comparisons between groups are still possible to find putative candidate compounds that can be successively used in a finer-grained quantitative chemical analysis or behavioral assays. However, point (3) may be more critical. Apart from degrading the samples, bacterial communities can be individual specific and prevent relative comparisons between groups (i.e. individual, gender, species, etc.) by generating different odorous metabolites (Archie and Theis 2011) that contribute to the final scent (Ezenwa and Williams 2014), although this hypothesis has not been directly tested in birds.

29.4.2 Novel Techniques for Sampling Odors

Our work presents new methods for sampling avian odors focusing on the volatile nature of scent samples that might contain the cues involved in chemical communication. Firstly, we developed a direct solid-phase thermal desorption method in which the compounds are removed from the natural trap in the feathers without any further step (see Method 3). In this approach, feathers are considered in place of the adsorbent material such as graphite or organic polymer (e.g. Tenax® TA) that is traditionally used. Moreover, we introduced the use of nalophan® material, a usual component of food bags which is odor and taste free as well as impermeable to water and fat (Hudson et al. 2009). Nalophan®, which is already employed in plant volatile collection (Noble et al. 2001) and air sampling (Van Harreveld 2003; Nagata 2010), appears in our case as a very convenient storage bag for "solid" samples (e.g. feathers). It is lightweight and does not release noticeable intrinsic components. Other polymer bags such as Teldar, Teflon, and PTFE sample bags might be suitable as well, but were not tested here (Zarra et al. 2012; Guillot 2012). We also presented another unusual trapping material: a cotton swab, employed by the French scientific police and produced in Hungary (Prada et al. 2010). Other materials often used in forensic science exist such as Dukal, Kings Cotton, or Johnson and Johnson pads (Hudson-Holness and Furton 2010). Cotton swabs were rubbed on the bird and then directly introduced in a TD tube (see Method 5) or its headspace was collected and concentrated on a SPME fiber (see Method 4) before thermal desorption. During the analysis of samples, the releasing capacity of the cotton swab was limited to only 20–30 components that were detected either with direct thermal desorption (Fig. 29.2e) or by sampling the headspace (Fig. 29.2d). However, this number is comparable with the number of compounds (around 20) extracted from human scent using various sorbent pads (Prada et al. 2010). Finally, the SPME fibers were used only in the laboratory because their complex handling requirements (due to their friability and the fragility of the coating), and the difficulty conserving all volatiles into the fiber absorbent during the long travel times, made them unsuitable for field use.

A novel apparatus to access the airborne excretions of birds was also explored in our study (see Method 6). We put the bird in a chamber formed by a hermetic stainless steel autoclave with an activated charcoal trap connected to a Tenax® TA tube. Similar studies have been also carried on crested auklet (Douglas 2006) and chicken or pigeon (Syed and Leal 2009). In both studies, the collection traps were not thermodesorbed as we did but eluted with a solvent (dichloromethane or methanol). We also tested the solvent extraction of the trap in a previous work (data not published) and it proved inadequate when working at low levels of volatiles trapped on the adsorbent. Indeed, no matter the combination of solvent quantity, sonication time, evaporating technique (cold finger manifold or stream of purified nitrogen), level of pre-concentration or GC–MS parameters tried on the samples, we were unable to reach a satisfactory level of sensitivity for this analysis. Although minor peaks were detected, indicating the verge of sensitivity, most chemical profiles were not usable. Indeed, the chromatograms provided by Syed and Leal (2009) displayed much more compounds when a SPME fiber was used rather than with solvent extraction of the SuperQ trap. In that case, solvent desorption is not a particularly sensitive technique as sample dilution seems to reduce the sensitivity of this approach. The combination of adsorbent trapping and solvent extraction thus appears not suited for the profiling of whole animal scents. The approach may nevertheless prove useful for the analysis of a priori identified targets by using restricted search options (e.g. the specific ion mode of mass spectrometers).

Sampling of airborne volatiles with the Method 6 is logistically more complex than the other techniques described above. It also requires a longer time to ensure sufficient levels of airborne volatiles can be collected on the adsorbent traps. However, the stress induced by this method appears reduced as animals were universally much calmer during sampling than while being handled. Although some bird-emitted airborne compounds were successfully collected, the apparatus led to significant environmental noise from unfiltered anthropogenic ambient volatile organic compounds (data not shown). Further methodological refinements, including the use of pure-air canisters and a glass chamber, are currently being tested to explore this method further.

Each of these new approaches is suitable for collecting samples in remote areas where the handling of animals is limited and benefit by not requiring time-consuming chemical extraction steps that can be a source of contamination and/or losses of volatile analytes. Finally, the choice of the sorbent material is crucial as the chemical composition of the materials plays a great role in governing their trapping and releasing capabilities (Hudson-Holness and Furton 2010). Further standardization and validation are still necessary to improve these methods, mainly regarding the extraction step.

29.4.3 Comparison of Extraction Methods

At least two classic approaches appear to extract the chemical molecules from a sample: solvent extraction or thermal desorption of molecules trapped onto an adsorbent material (e.g. SPME fiber or Tenax® TA polymer). Solvent extracting approaches are based on the chemical affinity between sample materials and solvents. As such, they require reasonable chemical skills in order to tune protocols to samples (Burger et al. 2004). They can lead to good chromatographic data provided there is sufficient chemical material extracted. Solvent extracts give the most convenient method of sample handling (i.e. easy storage). Some downsides of these techniques are, however, critical to consider when studying animal chemosignals (Table 29.1). Because there is no universal solvent (or solvent mix) able to achieve a comprehensive recovery of polar and nonpolar mixtures, solvent extraction inevitably results in a partial qualitative loss of chemical mixtures. Nevertheless, the solvent extraction usually keeps a more complete profile but is consequently more complex which may complicate analysis. Solvents can also directly impact the most reactive/fragile fraction of these organic mixtures (e.g. amines). The addition of the solvent itself induces an important dilution effect which is undesired when examining subtle olfactory chemosignals. Although generally chosen for their low boiling points, solvents result in large chromatographic peaks that affect the exploration of the most volatile fraction of chromatograms. In addition, many times during extraction procedures, it may be necessary to concentrate the sample by evaporating a fraction of the solvent using a gas flow such as nitrogen. This process ensures that relatively low-abundance compounds will be concentrated to detectable levels.

Finally as previously mentioned, contaminants can be introduced during extraction and throughout sample preparation. An alternative to the use of organic solvents is extraction with supercritical fluids such as supercritical carbon dioxide. This presents the advantage of using a totally volatile solvent but specialized equipment is required, and this could be problematic in the field.

In one of our studies, we introduced a new extraction method: the direct solid-phase thermal desorption on two different samples (feathers and cotton swab). A small amount of the sample was placed directly in the thermal desorption tube (cotton swab) or in a PTFE insert (feather). Desorption of the compounds trapped in the adsorbent materials was achieved using heat and a flow of inert gas. In this straightforward and cost effective procedure, no tedious and lengthy sample extractions are needed. This approach combines advantages from using a relatively abundant material (i.e. feather lipids) with the advantages of using thermal desorption (see above), while avoiding most environmental and adsorbent noise (Table 29.1). However some limitations must be considered. Direct thermal desorption is only appropriate where the small sample size introduced in the tube is representative of the sample as a whole in order to achieve quantitative analysis. The extraction can only take place at a temperature of desorption below the decomposition point of the sample, which can impair the efficiency of release. However a complete extraction of volatiles during a qualitative study is not required until a representative profile of the scent is obtained. Some refinements of the procedure are actually tested in order to evaluate its repeatability and reproducibility.

As illustrated by our results, thermal desorption provides a higher sensitivity than solvent extraction due to the absence of dilution and the quantitative transfer of desorbed material to the analytical instrument. The absence of solvent facilitates the examination of early eluting compounds and is a key aspect for chemosignal research. For example, the detection of dimethyl disulfide in both the desorbed feather lipids and airborne volatiles is intriguing because blue petrel chicks and adults respond to the odor of related compound (dimethyl sulfide) (Nevitt et al. 1995; Bonadonna et al. 2006). The presence of dimethyl disulfide could arise from the environment or from bacterial metabolism. A range of bacteria and fungi from the soil (e.g. *Lactobacillus spp.*, *Clostridium spp.*, *Citrobacter spp.*, *Pseudomonas spp.*, and *Flavobacterium spp.*) producing this molecule (Effmert et al. 2012) could be present in the burrow of the birds. However, more works is required to ascertain the origin of the dimethyl disulfide detected in our samples, a biogenic origin and behavioral function might be probable (Singer et al. 1976).

Several issues have to be addressed before the combination of adsorbent trapping and thermal desorption leads to the types of unequivocal results obtained with solvent extraction of feathers and secretions. First, the high temperatures needed to achieve enough recoveries from the adsorbent may result in the breakdown of the trapped analytes, particularly for polar analytes such as alcohols, carboxylic acids, and so on (Baltussen et al. 2002). Second, there is a clear need for better generalist adsorbents that could reduce noise while increasing the sample signal. Indeed, Tenax® TA (and most other adsorbents) releases significant amounts of breakdown products in the GC–MS and its strong affinity for polar analytes leads to poor recovery rates of these (Baltussen et al. 2002). This is particularly problematic in chemical ecology because polar compounds are very often involved in biogenic chemosignals (Soini et al. 2005). While awaiting the development of better adsorbents, Tenax® TA remains the most logical option for studies like ours, as it offers quantitative trapping and efficient release for volatile and semivolatile organic compounds ranging in volatility from n-hexane to *n*-C_{26}. The combination of different polymers such as Tenax® TA with Carbograph or SulfiCarb could also be a good alternative. For example SulfiCarb offered quantitative trapping and efficient release for compounds ranging in volatility from C_3 to *n*-C_6. In the same way, the use of a SPME fiber coated with PDMS can be optimized. For example, some recent studies examining volatiles emitted by birds have used a Carboxen-PDMS coating fiber (Supelco, Bellefont, PA, USA) which led to the detection of C_2 compounds (Shaw et al. 2011).

29.4.4 Final Considerations

The choice of a particular sampling or extraction technique is generally influenced by the theoretical framework of a study. For instance, the scarcity of studies examining avian chemical emissions from the perspective of chemical communication has led to a methodological bias towards solvent extraction of uropygial secretions and tissues. This is critical to consider when examining the resolution and relevance of data obtained so far. Avian secretions and feather lipids typically contain a complex mix of both large waxy nonvolatile and smaller potentially volatile compounds (Jacob 1978). To screen these whole mixtures in search of potential olfactory chemosignals presents some analytical shortcomings. First, the relative abundances of volatile and waxy contents may be very dissimilar. As a result, the levels of analytical sensitivity required to explore the full range of chemical fractions has proven to be extremely different. On a similar note, highly abundant analytes may be "masking" the presence of other less abundant compounds (for instance because of co-elution, or detector saturation). Second, the often very large number of compounds detected in avian chemical substances can compel biologists to select a priori the compounds, or class of compounds, which could be relevant (Soini et al. 2007). Such a decision, apart from requiring good chemical expertise, is not entirely satisfactory as the nature of the biologically active signals remain in most cases completely unknown. Finally, screening secretions or feather lipids for volatile signals might overlook the various degradation processes (oxidation, enzymatic breakdown, and photolysis) which may exogenously convert secreted lipid precursors into their biologically active forms (Wisthaler and Weschler 2010; Mardon et al. 2011). This is, in essence, why we have recently designed and started optimizing alternative techniques as presented here. There are, in addition, other recent and newly emerging alternative methods for the study of animal chemosignals, which are not considered in our comparison, due to the limitations of available resources and equipment in our study. For example, stir-bar sorptive extraction (SBSE; marketed by Gerstel, Linthicum, MD) is a newer technique that is more versatile and more sensitive than SPME (Baltussen et al. 2002). In SBSE, the sorptive polymer is coated on a small stir bar, which may be immersed in the liquid of interest or adapted for headspace analysis (Soini et al. 2005). SBSE has recently been successfully applied to biological media and avian endogenous chemicals (Soini et al. 2005, 2007, 2013; Whittaker et al. 2013; Tuttle et al. 2014) and is gradually replacing SPME. Compared with other microextraction techniques, SBSE provides a higher extraction efficiency and better reproducibility owing to the much greater amount of the extraction phase (for review, see: He et al. 2014). On the other hand, the implementation of SBSE is quite expensive and applicable coatings are still limited. A high-sensitivity sample enrichment probe has also been developed (Mitra et al. 2013) which can be implemented on any gas chromatograph at practically no cost.

In conclusion, ecologists working in the field of animal chemical communication still await the development of satisfactory, sensitive, and logistically reasonable techniques. The diverse and extreme physical properties of compounds forming the animal scent provide unique analytical challenges. The presence of highly volatile compounds increases the difficulty of developing an appropriate procedure (Rowan 2011). Despite that, we have presented here several techniques specifically used for the study of avian chemical emissions, and have compared their relative properties, advantages, and inconveniencies (Table 29.1). The limits of classical techniques, such as solvent extraction, are illustrated and new alternative techniques, such as adsorbent trapping or solid-phase thermal desorption, were introduced. Although targeting seabirds specifically, these methods have broad applicability to the study of vertebrate olfactory communication, in remote areas or otherwise. This study contributes to the general methodological effort dedicated to these challenges, and there is little doubt that exciting discoveries are around the corner.

Acknowledgements This work was supported by the Institut Polaire Français Paul-Emile Victor (IPEV) through the program ETHOTAAF no. 354 to FB, and the Agence Nationale de la Recherche Française (AMBO ANR-08-BLAN-0117-01 to FB). All aspects of the study were performed according to guidelines established by the IPEV and the CNRS for the ethical treatment of animals and complied with current French regulations. The authors wish to thank Tony Reader and Gavin Flemmati from the BBCS School of UWA, for their expertise on GC–MS techniques, and Heidi Woods and Michael Grocott from the ChemCentre, for technical assistance. Special thanks are also due to Aurélie Colomb from LaMP for her help on improving the parameters and the chromatographic outputs of Method 3. We finally thank Eleanor Kean to read over our manuscript. We would also like to thank the anonymous reviewers for all their constructive and helpful suggestions.

References

Archie EA, Theis KR (2011) Animal behavior meets microbial ecology. Anim Behav 82: 425–436

Baltussen E, Cramers CA, Sandra PJF (2002) Sorptive sample preparation – a review. Anal Bioanal Chem 373:3–22

Bonadonna F, Mardon J (2010) One house two families: petrel squatters get a sniff of low-cost breeding opportunities. Ethology 116:176–182

Bonadonna F, Mardon J (2013) Besides colours and songs, odour is the new black of avian communication. In: East ML, Dehnhard M (eds) Chemical signals in vertebrates 12. Springer, New York, pp 325–339

Bonadonna F, Nevitt GA (2004) Partner-specific odor recognition in an Antarctic seabird. Science 306:835

Bonadonna F, Caro S, Jouventin P, Nevitt GA (2006) Evidence that blue petrel, *Halobaena Caerulea*, fledglings can detect and orient to dimethyl sulfide. J Exp Biol 209:2165–2169

Bonadonna F, Miguel E, Grosbois V, Jouventin P, Bessiere JM (2007) Individual odor recognition in birds: an endogenous olfactory signature on petrels' feathers? J Chem Ecol 33:1819–1829

Burger BV, Reiter B, Borzyk O, Du Plessis MA (2004) Avian exocrine secretions. I. Chemical characterization of the volatile fraction of the uropygial secretion of the green woodhoopoe, *Phoeniculus purpureus*. J Chem Ecol 30:1603–1611

Campagna S, Mardon J, Celerier A, Bonadonna F (2012) Potential semiochemical molecules from birds: a practical and comprehensive compilation of the last 20 years studies. Chem Senses 37:3–25

Caro SP, Balthazart J, Bonadonna F (2014) The perfume of reproduction in birds: chemosignaling in avian social life. Horm Behav. doi:10.1016/j.yhbeh.2014.06.001

Celerier A, Bon C, Malapert A, Palmas P, Bonadonna F (2011) Chemical kin label in seabirds. Biol Lett 7:807–810

Dicke M, Vanbeek TA, Posthumus MA, Bendom N, Vanbokhoven H, Degroot AE (1990) Isolation and identification of volatile kairomone that affects acarine predator-prey interactions – involvement of host plant in its production. J Chem Ecol 16:381–396

Douglas HD (2006) Measurement of chemical emissions in crested auklets (*Aethia cristatella*). J Chem Ecol 32:2559–2567

Drea CM, Boulet M, Del Barco-Trillo J, Greene LK, Sacha CR, Goodwin TE, Dubay GR (2013) The "secret" in secretions: methodological considerations in deciphering primate olfactory communication. Am J Primatol 75:621–642

Effmert U, Kalderas J, Warnke R, Piechulla B (2012) Volatile mediated interactions between bacteria and fungi in the soil. J Chem Ecol 38:665–703

Ezenwa VO, Williams AE (2014) Microbes and animal olfactory communication: where do we go from here? Bioessays 36:847–854

Guillot JM (2012) Odour measurement: focus on main remaining limits due to sampling. Chem Eng Trans 30:295–300

Guillot JM, Fernandez B, Cloirec PL (2000) Advantages and limits of adsorption sampling for physico-chemical measurements of odorous compounds. Analusis 28:180–187

Hagelin JC, Jones IL (2007) Bird odors and other chemical substances: defense mechanism or overlooked mode of intraspecific communication? Auk 124:741–761

He M, Chen BB, Hu B (2014) Recent developments in stir bar sorptive extraction. Anal Bioanal Chem 406:2001–2026

Hewitt AD, Jenkins TF, Grant CL (1995) Collection, handling, and storage: keys to improved data quality for volatile organic compounds in soil. Am Environ Lab 2:25–28

Hudson DT, Curran AM, Furton KG (2009) The stability of collected human scent under various environmental conditions. J Forensic Sci 54:1270–1277

Hudson-Holness DT, Furton KG (2010) Comparison between human scent compounds collected on cotton and cotton blend materials for SPME-GC/MS analysis. J Forensic Res 1:101. doi:10.4172/2157-7145.1000101

Jacob J (1978) Uropygial gland secretions and feather waxes. In: Florkin M, Scheer BT (eds) Chemical zoology. Academic, New York, pp 165–211

Kean EF, Müller CT, Chadwick EA (2011) Otter scent signals age, sex and reproductive status. Chem Senses 36:555–564

Le Maho Y, Karmann H, Briot D, Handrich Y, Robin JP, Mioskowski E, Cherel Y, Farni J (1992) Stress in birds due to routine handling and a technique to avoid it. Am J Physiol Regul Integr Comp Physiol 263:775–781

Mardon J, Bonadonna F (2009) Atypical homing or self-odour avoidance? Blue petrels (*Halobaena caerulea*) are attracted to their mate's odour but avoid their own. Behav Ecol Sociobiol 63:537–542

Mardon J, Saunders SM, Anderson MJ, Couchoux C, Bonadonna F (2010) Species, gender and identity: cracking petrels' socio-chemical code. Chem Senses 35:309–321

Mardon J, Saunders SM, Bonadonna F (2011) From preen secretions to plumage: the chemical trajectory of petrels' social scent. J Avian Biol 42:29–38

Mitra S, Burger BV, Poddar-Sarkar M (2013) Headspace volatile oxylipins of Eastern Himalayan moss *Cyathophorella adiantume* by sample enrichment probe. Lipids 48:997–1004

Moritz RFA, Crewe RM (1988) Chemical signals of queens in kin recognition of honeybees, *Apis mellifera* L. J Comp Physiol A 164:83–89

Nagata Y (2010) Measurement of odor threshold by triangle bag method. Environmental Sanitation Center, Kawasaki, pp 118–127

Nevitt GA, Veit RR, Kareiva P (1995) Dimethyl sulphide as a foraging cue for Antarctic procellariiform seabirds. Nature 376:680–682

Noble R, Hobbs PJ, Dobrovin-Pennington A, Misselbrook TH, Mead A (2001) Olfactory response to mushroom composting emissions as a function of chemical concentration. J Environ Qual 30:760–767

Prada PA, Curran AM, Furton KG (2010) Comparison of extraction methods for the removal of volatile organic compounds (VOCs) present in sorbents used for human scent evidence collection. Anal Methods 2:470–478

Röck F, Mueller S, Weimar U, Rammensee HG, Overath P (2006) Comparative analysis of volatile constituents from mice and their urine. J Chem Ecol 32:1333–1346

Rowan DD (2011) Volatile metabolites. Metabolites 1:41–63

Santos SM, Bernier UR, Allan SA, Quinn BP, Booth MM, Kline DL, Barnard DR, Yost RA (2005) Identification of candidate attractant compounds from chicken feathers for the mosquito vector of the West Nile virus by GC/MS. In: Proceedings of the 53rd ASMS conference on mass spectrometry and allied topics, San Antonio

Scientific Instrument Services Inc. (1995) Adsorbent resins guide (updated: 2009). http://www.sisweb.com/index/referenc/resins.htm

Shaw CL, Rutter JE, Austin AM, Garvin MC, Whelan RJ (2011) Volatile and semivolatile compounds in gray catbird uropygial secretions vary with and between breeding and wintering grounds. J Chem Ecol 37:329–339

Singer AG, Agosta WC, O'Connell RJ, Pfaffmann C, Bowen DV, Field FH (1976) Dimethyl disulfide: an attractant pheromone in hamster vaginal secretion. Science 191:948–950

Soini HA, Bruce KE, Wiesler D, David F, Sandra P, Novotny MV (2005) Stir bar sorptive extraction: a new quantitative and comprehensive sampling technique for determination of chemical signal profiles from biological media. J Chem Ecol 31:377–392

Soini HA, Schrock SE, Bruce KE, Wiesler D, Ketterson ED, Novotny MV (2007) Seasonal variation in volatile compound profiles of preen gland secretions of the dark-eyed junco (*Junco hyemalis*). J Chem Ecol 33:183–198

Soini HA, Whittaker DJ, Wiesler D, Ketterson ED, Novotny MV (2013) Chemosignaling diversity in songbirds: chromatographic profiling of preen oil volatiles in different species. J Chromatogr A 1317:186–192

Syed Z, Leal WS (2009) Acute olfactory response of Culex mosquitoes to a human- and bird-derived attractant. Proc Natl Acad Sci U S A 106:18803–18808

Tuttle EM, Sebastian PJ, Posto AL, Soini AH, Novotny MV, Gonser RA (2014) Variation in preen oil composition pertaining to season, sex, and genotype in the polymorphic white-throated sparrow. J Chem Ecol 40:1025–1038

Valenta JG, Rigby MK (1968) Discrimination of the odor of stressed rats. Science 161:599–601

Van Harreveld AP (2003) Odor concentration decay and stability in gas sampling bags. J Air Waste Manage Assoc 53:51–60

Whittaker DJ, Gerlach NM, Soini HA, Novotny MV, Ketterson ED (2013) Bird odour predicts reproductive success. Anim Behav 86:697–703
Wisthaler A, Weschler CJ (2010) Reactions of ozone with human skin lipids: sources of carbonyls, dicarbonyls, and hydroxycarbonyls in indoor air. Proc Natl Acad Sci U S A 107:6568–6575
Wyatt TD (2010) Pheromones and signature mixtures: defining species-wide signals and variable cues for identity in both invertebrates and vertebrates. J Comp Physiol A 196:685–700
Zarra T, Reiser M, Naddeo V, Belgiorno V, Kranert M (2012) A comparative and critical evaluation of different sampling materials in the measurement of odour concentration by dynamic olfactometry. Chem Eng Trans 30:307–312

Chapter 30
Pair-Specific Scents in African Wild Dogs, *Lycaon pictus*, and an Example of a Potential Method to Identify Signals Within Complex Mixtures

Neil R. Jordan, Peter J. Apps, Krystyna A. Golabek, and J. Weldon McNutt

30.1 Introduction

Olfactory communication plays a major role in the lives of mammals (Ewer 1968), and scent is pivotal in modulating an extremely wide range of behaviors across an equally wide range of taxa. Chemical signals, and an animal's ability to extract this information from scent-marks, are fundamental components of parental (e.g. Poindron et al. 1988), sexual (e.g. Rasmussen et al. 1997), and territorial behavior (e.g. Müller and Manser 2007). A fundamental component of territorial scent marking depends upon discriminating differences between signals from "self" and "other" at the very least, although more complex discrimination such as neighbor–stranger (e.g. banded mongoose, *Mungos mungo*, Müller and Manser 2007) and group-specific discrimination have been described in some species (e.g. raccoon

N.R. Jordan (✉)
Botswana Predator Conservation Trust, Private Bag 13, Maun, Botswana

Centre for Ecosystem Science, School of Biological, Earth and Environmental Sciences, University of New South Wales, Sydney, NSW 2052, Australia

Taronga Conservation Society Australia, Taronga Western Plains Zoo, Wildlife Reproduction Centre, Obley Road, Dubbo, NSW 2830, Australia
e-mail: neil.jordan@unsw.edu.au

P.J. Apps
Paul G. Allen Family Foundation Laboratory for Wildlife Chemistry, Botswana Predator Conservation Trust, Private Bag 13, Maun, Botswana

K.A. Golabek
Botswana Predator Conservation Trust, Private Bag 13, Maun, Botswana

Department of Zoology, Wildlife Conservation Research Unit, University of Oxford, Oxford, UK

J.W. McNutt
Botswana Predator Conservation Trust, Private Bag 13, Maun, Botswana

B.A. Schulte et al. (eds.), *Chemical Signals in Vertebrates 13*,
DOI 10.1007/978-3-319-22026-0_30

dog, *Nyctereutes procyonoides*, Yamamoto 1984; European badger, *Meles meles*, Davies et al. 1988) suggesting that these abilities may be common among mammals. For animals to recognize or discriminate between scents there must be consistent differences between important categories of each unit (e.g. sex and pack). Although unit-specific discrimination has been described in many species (e.g. sex discrimination in domestic dogs, *Canis familiaris*, Dunbar 1977; individual discrimination in Indian mongooses, *Herpestes auropunctatus*, Gorman 1976), in mammalian semiochemistry in particular, the specific distinguishing compound or compounds typically remain unknown (Apps 2013). As mammals commonly use urine and feces in communication, distinguishing active signals from metabolic waste products constitutes a significant part of the challenge in this process.

The most commonly accepted mechanism of territorial defense using scent is by "scent matching" (Gosling 1982). Intruders match the scents of encountered individuals with scents encountered in the environment, and subsequently modulate their behavior in the presence of the identified territory owner(s) due to asymmetry in the costs/benefits of escalating a challenge. In order for such scent matching to occur, individuals must be able to discriminate between scents and identify territory residents as such. In group-living species, territoriality via scent-matching is likely to be facilitated through group-specific odors deployed as scent marks communicating territory residence.

Long-term ranging behaviors of a subpopulation of African wild dogs (*Lycaon pictus*) in northern Botswana followed in this study strongly illustrate characteristic territorial residence, including the use of exclusive areas (Pomilia et al. 2015). Furthermore, observations of daily movements by resident packs indicate that residence may be communicated using scent marks (e.g. Jordan et al. 2013). As part of ongoing research to manipulate the movement and territorial behavior of African wild dogs, we investigated whether the chemical compositions of wild dog scent-marks are pack-specific. We used gas chromatography and mass spectrometry (GC–MS), deconvolution (Chap. 28) and multivariate statistics to separate and identify any pack-specific components that may contribute to a signal of territory residence. Previous work on the scent-marking behavior of African wild dogs showed that the dominant urine overmarks (DUOs) of wild dog pairs—where one member of a pair deposits urine on top of the urine of its partner—are a likely source of semiochemical information (Jordan et al. 2013). Distinct postures were almost uniquely associated with DUO deposition and increased the likelihood that these deposits would be investigated by other dogs (Jordan et al. 2013). Among all scent types, dominant urine was of greatest interest to other wild dogs. Additionally, the dominant pair in a pack may remain stable over many years, whereas the composition of the rest of the pack changes from year to year, as subdominants disperse or die (McNutt 1996). Consequently, DUOs provide the focus for our search for a pack-specific semiochemical signal. If DUOs contain a signal advertising residence, then we would expect the DUOs from each pair/pack to be distinguishable from others (i.e. a pair-specific signature), since residents must be capable of distinguishing their own scents from those of their neighbors.

Searching through complex multicomponent mixtures for signals presents a challenge in analytical semiochemistry. Focusing our investigation on territorial scent marking in African wild dogs in this study, we attempt to address the following questions: Are the chemical compositions of DUOs of wild dogs pair-specific, and which particular components might impart pair-specificity? In attempting to answer these questions we utilize established statistical techniques in a novel way. We used a multivariate (factorial) approach as a statistical filter to remove components that varied within pairs while retaining components that differed between pairs. This approach uses Principal Components Analysis, and particularly its utility in locating and retaining variables (chemical components) which account for the highest proportion of inter-sample variance. Our approach may be broadly applicable across signaling modalities, as finding signals amongst the noise of multicomponent mixtures is a common problem in animal communication.

30.2 Materials and Methods

30.2.1 Study Site and Population

Data were collected between July 2011 and September 2012 from six packs of free-ranging African wild dogs in Botswana. The study area (approx. 2600 km^2; 19°31′S, 23°37′E; elevation approx. 950 m) is bordered by the Okavango Delta and includes the Moremi Game Reserve and surrounding Wildlife Management Areas. Further details can be found in McNutt (1996). A pack was defined as a group containing at least one adult male and female which form a potential reproductive unit (cf. Malcolm 1979; McNutt 1996). Mean adult (>1 year) pack size (at the study mid-point) was 7 ± 4.29 (mean ± SD; range; 3–14). Packs were located by radiotracking from the air and from a vehicle, with one to six individuals in each pack fitted with GPS radiocollars (<350 g) with a VHF tracking pinger (Vectronic Aerospace GmbH, Berlin, Germany; Royal Veterinary College, London, UK,) or VHF radiocollars (Sirtrack, Havelock West, New Zealand; <180 g) following procedures described elsewhere (Osofsky et al. 1996). Dominant pairs were easily identified in each pack by their mutual overmarking or "tandem marking" behavior, where one individual places its urine directly on top of (sensu Johnston et al. 1994) the urine of its opposite-sex partner (Jordan et al. 2014). This behavior is associated with sniffing and the adoption of specific leg postures, and was not observed in subdominant wild dogs (Jordan et al. 2013). During this study, breeding attempts were only observed between dominant pairs.

30.2.2 Sample Collection

Packs were observed from a vehicle while resting (at distances of 3–40 m) and traveling (20–200 m). We remained at the DUO deposition site until the dogs had moved away sufficiently to allow the sample to be collected on foot beside the vehicle.

Samples used in this study were collected 9.04 ± 7.33 (Mean ± SD; range 2–26) min post-deposition using clean metal spoons to scoop patches of urine-soaked soil into cleaned glass jars with lids lined with aluminum foil. Jars were double rinsed with boiling distilled water. Caps (80 °C), jars, and liners (220 °C) were baked for at least 24 h at in a forced convection oven, assembled while still warm, and were not opened until the sample was put in them. Samples from the same pair were never collected from the same location, and all samples were frozen (ca. −5 °C) on the day of collection and until analysis. Samples were frozen for 355 ± 17 (Mean ± SD) days (range: 168–606) before analysis.

30.2.3 Chemical Analysis

Samples of urine on soil were spread in stainless steel dishes in a fume cupboard and samples were stirred intermittently to get even drying. Room temperature is air-conditioned to 22–25 °C. Once the sample was dry enough to be free-flowing (which usually took between 2 h and overnight) it was sieved through a stainless steel mesh with holes of about 1 mm to remove fragments of vegetation, and then a weighed subsample was taken from it. The dried, sieved remainder was returned to the sample's original jar, and subsamples were weighed accurately (0.1 mg resolution) into glass Pasteur pipettes with a small pledget of silanized glass wool (Sigma-Aldrich) in the stem. Pipettes were prepared by inserting the glass wool, rinsing with 2 ml of methanol and then baking overnight (at least) at 220 °C. Approximately 2.5 g of sample filled the body of the pipette to about 10 mm below the open end. The pipettes were supported vertically with their tips in 2 ml autosampler vials that had been rinsed with methanol and dried at room temperature. Methanol (Signal-Aldrich Chromasolve Gradient grade) was percolated though the sample under gravity and the first 1 ml (to a mark on the vials) was collected. GC–MS analyses of 2 μL of injected extract were carried out on a Varian 450 GC gas chromatograph with a Varian 1079 programmable inlet, interfaced to a Varian 320 MS single quadrupole mass spectrometer. The inlet was fitted with a fritted liner (LJ-18-24-1) and the column was a 30 m × 0.32 mm × 0.5 μm polyethylene glycol (PEG) (Restek Rtx-Wax #12349), held at 50 °C for 3 min and programmed at 5 °C/min–240 °C. The carrier gas was helium with a constant volume flow of 2 mL/min. Electron impact mass spectra were acquired with electron energy of 70 eV, and compounds were tentatively identified by searches of the NIST 05 mass spectral library. For further details see Apps et al. (2012). Varian Bruker MSWS mass spectral data files from 59 samples (selected from a larger set to provide inter-pack comparisons) were imported to the AnalyzerPro deconvolution software package (SpectralWorks) and processed using the Matrix Analyzer routine with the following settings up to 40 min retention: reject masses 33–45, minimum masses 2, area threshold 800,000, height threshold 1, signal:noise 1, width threshold 0.04 min, resolution high, scan windows 1, smoothing 5, forward fit 550, reverse fit 550, retention window 0.25 min. After 40 min retention m/z 129 was added to the excluded ions, minimum masses was increased to four, area threshold increased to 1,000,000, smoothing increased to 15, and fits to 650. A known series of methyl-imidazolidinedione isomers, which have a single strong ion at m/z 114, and groups of peaks with shifting retentions and similar spectra were deconvoluted using ion and retention settings specific to the peaks (see also Apps, this volume).

30.2.4 Statistical Analysis

Data were exported to Microsoft Excel for manual cleanup, and then imported to R (R Core Team 2013) for all statistical processing. Statistical analyses were applied to 24 DUOs (four from each of six packs). Raw peak areas were converted to percentages of the total peak area on the chromatogram. As the dataset contained many more components than samples, we first reduced the number of variables using a Principal Component Analysis (PCA). PCA cannot deal with 0 values, and almost all components had peak areas below the minimum threshold in at least one sample, so we assigned an arbitrarily low percentage abundance (0.00001) to all zero values. Following this, all abundance data were log-transformed (using a natural log function) as PCA would otherwise over-inflate the importance of variables (components) with higher values (abundances). As the resultant (post-transformation) variances entered into the PCA reflect proportional changes and not absolute changes, this removes any bias towards components with higher abundances, and gives equal weight to components at low measured abundances; an important consideration in chemical signaling.

30.2.4.1 Are African Wild Dog Duos Pair-Specific?

To determine whether African wild dog DUOs had pair-specific qualities, we conducted Discriminant Function Analyses (DFA) on the dataset of 24 samples (4/pack). Post-hoc "bootstrapping" analyses were conducted to determine the probability that a cross-validated correct assignment value from the DFA was achieved by chance, and followed the methods of Müller and Manser (2008).

30.2.4.2 Which Components Code for Pair?

To determine which components may code for pair, we processed the data to remove intra-pair variation and retain inter-pair variation using PCA; particularly its utility in identifying and retaining maximum variation between samples. PCA is a dimension-reduction tool that can be used to reduce a large set of variables to a small set that still contains most of the variation of the original set. If several samples from each pack were subjected to PCA, then the analysis would retain the maximum variation across the whole data set, therefore resulting in the retention of both intra- and inter-pair variation. Instead, by conducting PCA with one sample from each pair (six samples total) we were able to maximize the retention of inter-pair variation while excluding intra-pack variation from the resulting principal components (PCs). We conducted PCA using a covariance matrix on 20 different subsets of data (each subset contained a randomly selected single sample from each of the six packs).

From these outputs we identified, by ranking and examining the derived loadings (percentage variance in that PC accounted for by the factor/chemical component, averaged for each component in each PC over the 20 trials), which individual chemical components were consistently associated with this (inter-pair) variance. As the loadings describe the contribution of each of the individual chemical components to each PC, we were then able to estimate the importance of each of the chemical components to inter-pair variation. Loadings were ranked by absolute value (i.e. loadings with high negative values were also included). We identified the component with the highest loading for PC1, and then calculated the proportion of that loading value that the loading for each other component represented (again using absolute values). We identified and extracted any components whose loadings represented >85 % of the loading value of the component with the highest loading. This resulted in six components being extracted for PC1. The initial 85 % cut-off was chosen as a starting point because, with our particular dataset, it would locate a reasonable number of components that could be tested in combination in the field given the constraints of our study system. The relevance of this cut-off was checked by comparing the success of post-hoc DFA classification analyses for combinations which either included additional components, or excluding each of the original components in turn. In this case we could not improve upon the initial cut-off selected.

Since PC1 represented only 31.12 % of the estimated inter-pair variation, we also extracted the highest loading components from PC2 to PC5, as this explained >99 % of the overall variation. The number of components we extracted from these PCs was dependent upon the percentage of variation that each PC accounted for in the inter-pair variation (see Table 30.1). As six components were extracted from PC1 and PC1 represented 31.12 % of the variation, the top four components were extracted from PC2, as this PC represented 22 % of the variation (and so contributed 22 % of the extracted components). The number of components extracted from each PC was proportional to the variation that each PC explained.

Eleven of the 19 components had been previously identified (Apps et al. 2012, shown bold in Table 30.2), and nine of these were considered unlikely to be signals

Table 30.1 Percentage variance explained by five principal components across 20 trials

Principal component	PC1	PC2	PC3	PC4	PC5	Total
Mean % variance explained	31.12	22.44	19.03	15.78	11.63	
Max. % variance explained	39.76	25.49	20.94	17.84	15.45	
Min. % variation explained	24.78	20.31	16.22	10.49	8.34	
N (highest loading) components extracted	6	4	4	3	2	19

Table 30.2 Nineteen components potentially contributing to a signal of pair-specificity

Ret. order	Identity (confirmed in bold)	% detected ($n=24$)	Likely signal?	Comment
1	**2-methylpropanoic acid**	62.5	No	Widespread in mammals
2	**Quinazoline**	50	Yes	Wild dog specific
3	Unknown	58.33	Yes	
4	***N,N*-dimethylurea**	54.17	No	Widespread in mammals
5	**2-piperidinone**	75	No	Ubiquitous urine metabolite
6	**Glycerine**	66.67	No	Ubiquitous
7	m/z 85/86 unknown compound 1	41.67	Yes	
8	m/z 85/86 unknown compound 2	70.83	Yes	
9	**Methyl tridecanoate**	20.83	No	Probably from methanol + tridecanoic acid
10	Unknown	50	Unknown	
11	**Tetradecanoic acid**	45.83	No	Ubiquitous
12	**1-methyl-2,4-imidazolidinedione**	79.17	Unknown	Carnivore-specific. 3-methyl- and 5-methyl-isomers also present
13	Unknown	4.17	Unknown	
14	Pentadecen-1-ol	29.17	Unknown	Isomer unknown
15	**Nonacosane**	58.3	Unknown	
16	**Hexadecanoic acid**	75	No	Ubiquitous
17	**Hexadecenoic acid**	29.17	No	Ubiquitous-isomer unknown
18	2-aminocarbonyl-1-methylimidazole	37.5	Unknown	
19	**Octadecanoic acid**	54.17	No	Ubiquitous

of pair because they are very widespread in mammal odors (Burger 2005). To test whether the remaining 10 extracted components allowed classification of pairs in the absence of any other chemical information, log-transformed percentage abundance data were extracted for these components. Because multiple components were extracted from each PC, the abundance of each extracted component was likely to be correlated with the abundance of at least one other extracted component. As DFA requires the input variables to be uncorrelated, we ran the dataset ($n=24$, four samples from six pairs) through PCA before assessing the classification of samples using DFA. Using PCA reduced the dimensions of the dataset to produce uncorrelated/orthogonal input variables (PCs) thus removing the common problem of multicollinearity in tests of this nature (Field 2005). Bootstrapping was used to assess whether statistical classification of samples to pair was significantly better than that expected by random assignment.

It is important to point out that our method allows for the possibility of both single components and ratios to be the signal. It also allows for the possibility of binary coding; i.e. that the presence or absence of each component (or combination of components) is what differs among pairs. We explore this possibility further by conducting an additional DFA on PCs derived from presence absence data. In each of the 24 samples, any of the 10 components that were detected at any abundance were assigned a score of "1," and undetected components were assigned a score of "0."

30.3 Results

Deconvolution located 990 components in the 24 samples. Only one component, benzoic acid (Apps et al. 2012) was present in all 24 samples. 27 (2.73 %) components were detected in 75 % or more samples, 66 (6.67 %) components were detected in 50 % or more samples and 135 (13.64 %) components were detected in at least 25 % of samples. The majority of components (703; 71.01 %) were each detected from only one pair, with a large majority of these (669/703, 95.16 %) being found in only one sample. Of the remaining 287 components that were detected in more than one sample, 34 were located exclusively in samples from a single pair, but not in all the samples from a pack. Twenty-six of these components (76.47 %) were found in only two samples from the same pair, and the remaining eight (23.53 %) were found in three of four samples from the same pair. Only one pair had components in ¾ of the samples, and no component found exclusively in one pair was found in all samples from that pair (None of these components were extracted on the basis of their loadings below).

30.3.1 Are African Wild Dog DUOs Pair-Specific?

Fifteen PCs explaining 85.6 % of the measured variance contained in 990 chemical components in 24 samples (four samples/pack from six packs) were input into a DFA which correctly classified 21/24 (87.5 %) of DUOs to the pair that produced them (Table 30.3). This is significantly better than the 16.67 % correct classification

Table 30.3 A comparison of correct classification of 24 African wild dog dominant urine overmarks (DUOs) to six pairs with that achieved by Discriminant Function Analysis using five Principal Components derived from: (1) log-transformed percentage abundance data on all 990 located chemical components; (2) log-transformed percentage abundance data on 10 extracted chemical components; and (3) presence/absence data on 10 extracted chemical components

% correct classification results		
Log-transformed % abundances (990 components)	Log-transformed % abundances (10 components)	Presence/absence (10 components)
87.5	79.17	79.17
$p<0.0001$	$p<0.0001$	$p<0.0001$

p values are derived from a bootstrapping analysis and assess whether the classification obtained is relative to that expected at random (16.67 % in all cases) expected by chance (bootstrapping; $p < 0.0001$), and is evidence that the chemical composition of samples is pair-specific.

30.3.2 Which Components Code for Pair?

With intra-pair variability excluded by analyzing one sample from each pack, in 20 trials, 5 PCs were sufficient to explain 100 % of measured inter-pair variation (Table 30.1). Nineteen of 990 components possibly contributed to a signal of pair specificity (Table 30.2). Of these, nine were unlikely to be signals, due either to their ubiquity (#1, #4, #6, #11, #16, #17, #19) or their known derivation in excretory pathways (#5, #9), and were excluded from further analyses. Of the 10 remaining possible signaling components, quinazoline (#2) has, to our knowledge, so far only been identified in the scent-marks of African wild dogs (Apps et al. 2012) and 1-methyl-2,4-imidazolidinedione (#12) appears to be carnivore-specific and has also been found in the urine of lions, leopards, and spotted hyenas, but not in that of elephants, buffalo, impala, or zebra (Apps, unpublished). Pentadecen-1-ol and 2-aminocarbonyl-1-methylimidazole (#18) were tentatively identified from mass spectra, but the other six components of interest (#3, #7, #8, #10, #13, and #15) are unknowns. Some components are at picogram levels and their identification will present formidable problems.

Seven Principal Components (PCs) encompassing 96.02 % of the variance contained within log-transformed percentage abundance data from 10 components only were sufficient to correctly classify 19/24 (79.1 %) of DUOs to the pair that produced them (Table 30.3). This classification is significantly better than the 16.67 % expected by chance alone (bootstrapping; $p < 0.0001$).

Eight PCs containing 98.39 % of the variation contained within a binary coded (presence/absence) data frame were sufficient to correctly classify 19/24 (79.1 %) of DUOs to the pair that produced them, which is the same percentage correct classification as for analyses including percentage abundance data (Table 30.3). Again this classification is significantly better than the 16.67 % expected by chance alone (bootstrapping; $p < 0.0001$), and may suggest that presence/absence, rather than absolute or relative abundance, may be a sufficient signal.

30.4 Discussion

We assessed the chemical composition of scent marks from African wild dogs using multivariate statistics and showed that DUOs have a high degree of pair-specificity in their chemical compositions. From 990 components, we statistically located 10 candidate components based on inter-pack variation that could contribute to a signal of pair-specificity. Information on the presence or absence of these 10 components in combination was sufficient to statistically classify a significantly high proportion of samples to the correct dominant pair/pack. Overall our results show that African wild dog DUOs contain statistically identifiable characteristics to enable pair-specific discrimination, and that binary-based multicomponent discrimination could be sufficient for pack-discrimination by scent in African wild dogs.

A degree of pack-specificity has previously been shown in the scents of some other species. For example, analytical studies based on gas chromatography showed that sub-caudal secretions from European badgers have inter-group differences in their chemical profile (Gorman et al. 1984), and that the castoreum secretions of Eurasian beavers (Schulte 1998) and North American beavers (Sun and Müller-Schwarze 1998) are family-specific. In general, we would expect territoriality and/or ownership of resources to be advertised by unit-specific signals; in solitary territorial species an individually specific signal may be expected (e.g. Indian mongoose, Gorman 1976), whereas in group-living species territory ownership should be advertised with a group-specific signal (e.g. European badger, Gorman et al. 1984). A high degree of pair-specificity in the DUOs of African wild dogs may be evidence that DUOs play an important role in communicating residence and facilitating territoriality in this species. In common with many cooperative breeders, the dominant breeding pair has most to gain by repelling intruders from their range (e.g. Ethiopian wolves, Sillero-Zubiri and Macdonald 1998), and indeed subdominants may benefit from occasional incursions by intruders, because these can provide dispersal and breeding opportunities (e.g. meerkats, Jordan 2007). The dominant pair in a pack also remains stable over many years, whereas the composition of the rest of the pack changes from year to year as subdominants disperse or die (McNutt 1996). DUOs therefore represent the most likely source of a territorial residence signal, and that we find a specific signature in the combined urine-based scent marks of dominant African wild dog pairs is perhaps unsurprising. However, it should be noted that we have not yet determined whether there are group-wide consistencies in scent, as only DUOs have been compared in this way.

We have shown that 10 components could contribute to a pair-specific signal on the basis of some combination or combinations of their presence and absence. This suggests that African wild dogs may need only to detect the presence of these chemicals (in combination) in order to discriminate between the scents of different packs, which is a similar potential mechanism as shown in the ant (*Formica fusa*), where nine different positional isomers encode the colony signal (Martin et al. 2008). However it is important to keep in mind that, although wild dog scents have pair-specific compositions, this obviously does not demonstrate that wild dogs can discriminate between the scents of different packs, or that if they can, they use that information. Further field work is required to test whether pack-discrimination by scent occurs in African wild dogs, and scent presentation experiments can then test whether we have located the signaling components.

Whether African wild dogs use the pair-specific components in encountered scents to detect and identify intruders is not yet known, and it is theoretically possible that retaining a cognitive map of where (and perhaps when) you scent marked, and comparing this to scents that you encounter could facilitate the detection of intruders on a territory (cf. Bekoff 2001) without the need for a pack/pair-specific signal. Indeed some species have been suggested to retain the locations of several 100 features of interest throughout their range (e.g. refuge holes in meerkats, Manser and Bell 2004). However, although some components appear to trigger spatial and scent fingerprint memory (e.g. darcin in mouse scent-marks, Roberts et al. 2012), it seems that such "cognitive recall" is an unlikely mechanism of intruder detection, particularly in species with large territories. Given the prevalence of unit-specific signals (e.g. group-specificity in European badgers scents, Gorman et al. 1984) even across modalities (e.g. group-specificity in green woodhoopoe calls, Radford 2005), and the discrimination of these (e.g. partner-specific scent-discrimination in Antarctic prions, *Pachiptila desolata*, Bonadonna and Nevitt 2004; group discrimination in green woodhoopoe, *Phoeniculus purpureus*, Radford 2005), it is expected that detecting differences between self (or group companion) and other scents (e.g. European badger, Palphramand and White 2007), or the scents of different groups (e.g. colony scent-discrimination in big brown bat, *Eptesicus fuscus*, Bloss et al. 2002), is a more likely mechanism for territorial advertisement.

Although we do not test whether pair-specificity results in pair discrimination, the evidence presented here suggests that a multicomponent signal could code for pair in wild dogs. First there were 10 components whose combined abundance or presence/absence reliably discriminated between pairs, and data from each component suggested that any single component alone would be insufficient to distinguish between pairs. Additionally, although we found multiple chemical components that were specific to a pair, most of these were present in only one, two, or three samples (of four) from that pair. Thus single compound signaling of pair did not occur in our dataset, but multicomponent signaling—where the presence of one component and/or another or more in the scent may be used to discriminate between packs—may have been possible.

Although multivariate statistical approaches have often been employed to sift through complex chemical mixtures in search of category-specific information (e.g. Jordan et al. 2010; Safi and Kerth 2003), the validity of these approaches and therefore the results are rarely questioned, and opinions differ as to whether they are fit for the purpose. Martin and Drijfhout (2009) formerly questioned the validity of a multivariate approach, particularly with respect to the analysis of cuticular hydrocarbon profiles of hymenopteran insects. As our specific research problem presents a similar challenge, it is worth considering their concerns, three of which are particularly relevant in this context: (1) compounds have a high chance of being correlated (and should not therefore not be subjected to classification analyses that require independent input variables); (2) minor compounds would have a disproportionately large effect on the analysis; (3) PCA retains all of the variation in a system, not just the variation in the system that the researchers are interested in. In this paper we have attempted to deal with all of these potential pitfalls by specific data processing.

First, while it is true that some chemical compounds within a scent sample have a high chance of being correlated and therefore not being independent of one another (e.g. Martin and Drijfhout 2009), this does not preclude appropriately processed data being used in multivariate statistical analyses for classification (in this case discriminant function analysis). As potential correlation or multicollinearity problems may lead to erroneous results if multivariate methods are used (Field 2005), we ran potentially correlated data (the abundance of located chemical components) through a PCA which groups correlated variables into PCs which are orthogonal to each other. We then used the resulting principal components for classification. This ensured that our input variables (now principal components) for DFA were independent, but retained the majority of the variation found in the original dataset of potentially correlated variables. Using a PCA in this way is common in studies of scent (e.g. Jordan et al. 2010; Safi and Kerth 2003) and sound (e.g. Golabek and Radford 2013; Radford 2005), where multiple potentially correlated variables are measured, although authors rarely state that producing independent data was a reason for doing so.

Second, the suggestion that minor compounds have a disproportionately large effect on the analysis is important, particularly given that some components we detected were in abundances at picogram levels. This issue derives from the fact that small fluctuations in the abundance of components present at low abundances would result in relatively large changes in their percentage abundance compared to similar raw changes in components present at higher abundances. In such cases, variance—which PCA uses to assign importance—will be larger for variable components at lower abundances and their importance may then be overinflated. With the methods and specific data-processing we have employed in this paper however, this is only a problem if proportional changes are related to abundance (i.e. if low abundance components are more variable in terms of their proportional shifts between samples), which we have no a priori reason to suspect. It is also important to realize that, for example, while components at relatively low and relatively high abundance respectively and displaying the same proportional changes across two samples (e.g. 1, 10, 100 in one sample and 2, 20, 200 in another) will produce different variances, therefore overinflating the importance of the more abundant component in the sample, log-transforming these data (as we did) ensures that the variance is equal for the two series and therefore reflects proportional changes regardless of relative abundances. As the resultant (post-transformation) variances entered into the PCA reflect proportional changes and not absolute changes, this removes any bias towards components with higher abundances, and gives equal weight to components irrespective of their measured abundances. This is an important consideration in chemical signaling generally, and such log transformations are commonplace in problems of this nature (e.g. Jordan et al. 2010; Safi and Kerth 2003). This is primarily because it is rarely possible to use the absolute abundance of a peak, as the total amount of eluted components may differ between samples based on sampling method (e.g. the ratio of sand to urine) and sample dilution introduced both naturally (related to individual hydration levels) and analytically. However, while raw abundances will differ in different dilutions of the same mixture, the ratios of components within them will be constant.

Finally, Martin and Drijfhout (2009) suggest that a multivariate approach may only be valid in the search for potential signaling components if all other variation can be removed. This is a problem since PCA in particular attempts to maximize the retention of ALL variation between samples, not only the variation (in this case *between* pairs) that the researcher is interested in retaining. For example, even where individually specific signals occur (e.g. Jordan et al. 2011), scent samples from the same individuals are highly variable due to intrinsic and extrinsic factors. In attempting to locate components potential signaling pair identity, including multiple samples from the same pair in a PCA is unhelpful since much within pair variation ("noise" in this context) will also be retained. The key is to attempt to remove as much variation that is unrelated to the factor (pair) of interest and retain the variation describing the factor. In this chapter we have attempted to remove, or at least reduce, the variation that is not attributable to inter-pair differences as much as possible while retaining that variation describing inter-pair differences. We have done so by first running a PCA using only one sample from each pair, repeating this 20 times using different combinations of samples each time, and then using average loadings to locate components of interest (i.e. those describing inter-sample, and therefore inter-pair, variation). This, we believe, is a novel, appropriate and effective method of removing variation in factors other than that of interest for analyses. It may also be important to note that, for confidence in locating the correct components, multiple trials are required using different exemplars from members of each category of interest.

Despite addressing all of the above issues, and successfully reducing the number of components (variables) by nearly two orders of magnitude, we must acknowledge that our final dataset was imperfect for the classification analyses (DFA) we used to evaluate the success of our extraction method. As Mitteroecker and Bookstein (2011) point out, if the sample:variable ratio is less than 5:1, grouping can occur purely as a statistical artifact, and this ratio was 2.4:1 in this study. It is important to stress however that this issue only affects the confirmatory classification analyses at the end of the study, which we have merely used in an attempt to evaluate the efficacy of the described method in locating the components of interest. Further confirmation is therefore required to fully demonstrate the utility of this method in identifying the components contributing to a pair-specific signal in this study, and will take the form of field discrimination experiments ("bioassay"). However, identifying (not just locating) the components of interest is required before the optimal bioassays can be completed, and remains a priority work in progress. It is also important to stress that while in our particular case study we must interpret the final DFA classification results with caution, this issue does not undermine the method described to locate potential signaling components. Sample size is, as always, an important consideration for other studies wishing to use the method. We therefore contend then that the method of deriving these results is sound, and may fruitfully be employed generally where datasets fulfil the above criteria.

Identifying potential signals within complex chemical mixtures is invariably a multivariate problem, and as such it seems appropriate to adopt a multivariate approach to tackling it. While a multivariate *statistical* approach is most commonly used, it is also theoretically possible to take a multivariate *experimental* approach in the field. However, when working with most mammalian secretions, this is clearly impractical as a first step. In this study we located 990 chemical components in our sample, which would be logistically impossible to test individually, let alone in combinations. A multivariate statistical approach is therefore necessary as a first step, but data input and inference must be undertaken with care to ensure the validity of the results. Having located a manageable number of components (10) using multivariate statistics, we are now in a position to test the relevance of these components (if any) in signaling pair-identity in African wild dogs. Following identification of all of the located components of interest, a carefully designed bioassay-based approach will be the acid test of our method for identifying potential signals in complex mixtures.

In summary, we have used multivariate statistics—and PCA in particular—to filter through data from African wild dog urinary scent samples containing almost 1000 components and reduce the number of candidate signal components by two orders of magnitude. Following this data reduction, experimental tests to determine whether the pair-specific chemical differences identified in these scents are used by the animals themselves in pair discrimination and territory defense are now feasible in the field. Identifying components signaling territory ownership is a critical stage in our research towards developing a synthetic scent-mark which mimics the natural territorial signals and responses to them by African wild dogs. Identifying the specific communication components (signals) is a common problem in the study of animal communication and chemical ecology in particular, and it is suggested that ours or a similar approach may be successfully employed as a first filter in attempts to locate components with communicative value from complex calls or chemical mixtures in particular.

Acknowledgments We thank the Paul G. Allen Family Foundation for funding this work. SpectralWorks (www.spectralworks.com/analyzerpro.html) kindly donated the AnalyzerPro software. All work was carried out under a permit from the Botswana Ministry of Environment Wildlife and Tourism, Department of Wildlife and National Parks. Reena H. Walker kindly assisted with data extraction. Briana Abrahms, Geoffrey Gilfillan and Jessica Vitale assisted in collecting the samples used in this study.

References

Apps PJ (2013) Are mammal olfactory signals hiding right under our noses? Naturwissenschaften 100:487–506

Apps P, Mmualefe L, McNutt JW (2012) Identification of volatiles from the secretions and excretions of African wild dogs (*Lycaon pictus*). J Chem Ecol 38:1450–1461

Bekoff M (2001) Observations of scent-marking and discriminating self from others by a domestic dog (*Canis familiaris*): tales of displaced yellow snow. Behav Processes 55:75–79

Bloss J, Acree TE, Bloss JM, Hood WR, Kunz TH (2002) Potential use of chemical cues for colony-mate recognition in the big brown bat, *Eptesicus fuscus*. J Chem Ecol 28:819–834

Bonadonna F, Nevitt GA (2004) Partner-specific odor recognition in an Antarctic seabird. Science 306:835

Burger BV (2005) Mammalian semiochemicals. In: Schulz S (ed) The chemistry of pheromones and other semiochemicals II (Topics in current chemistry 240). Springer, Heidelberg, pp 231–278

Davies JM, Lachno DR, Roper TJ (1988) The anal gland secretion of the European badger (*Meles meles*) and its role in social communication. J Zool 216:455–463
Dunbar IF (1977) Olfactory preferences in dogs: the response of male and female beagles to conspecific odors. Behav Biol 20:471–481
Ewer RF (1968) Ethology of mammals. Logos Press, London
Field A (2005) Discovering statistics using SPSS, 2nd edn. Sage, London
Golabek KA, Radford AN (2013) Chorus-call classification in the southern pied babbler: multiple call types given in overlapping contexts. Behaviour 150:691–712
Gorman ML (1976) A mechanism for individual recognition by odour in *Herpestes auropunctatus* (Carnivora: Viverridae). Anim Behav 24:141–146
Gorman ML, Kruuk H, Leitch A (1984) Social functions of the sub-caudal scent gland secretion of the European badger *Meles meles* (Carnivora: Mustelidae). J Zool 204:549–559
Gosling LM (1982) A reassessment of the function of scent marking in territories. Z Tierpsychol 60:89–118
Johnston RE, Chaing G, Tung C (1994) The information in scent over-marks of golden hamsters. Anim Behav 48:323–330
Jordan NR (2007) Scent marking investment is determined by sex and breeding status in meerkats. Anim Behav 74:531–540
Jordan NR, Mwanguhya F, Kyabulima S, Rüedi P, Cant MA (2010) Scent marking within and between groups of wild banded mongooses. J Zool 280:72–83
Jordan NR, Mwanguhya F, Furrer RD, Kyabulima S, Rüedi P, Cant MA (2011) Scent marking in wild banded mongooses: 2. Intrasexual overmarking and competition between males. Anim Behav 81:43–50
Jordan NR, Golabek KA, Apps PJ, Gilfillan GD, McNutt JW (2013) Scent-mark identification and scent-marking behavior in African wild dogs (*Lycaon pictus*). Ethology 119:1–9
Jordan NR, Apps PJ, Golabek KA, McNutt JW (2014) Top marks for top dogs: tandem marking and pair-bond advertisement in African wild dogs (*Lycaon pictus*). Anim Behav 88:211–217
Malcolm JR (1979) Social organization and communal rearing in African wild dogs. Dissertation, Harvard University, Cambridge
Manser MB, Bell MB (2004) Spatial representation of shelter locations in meerkats, *Suricata suricatta*. Anim Behav 68:151–157
Martin SJ, Drijfhout FP (2009) How reliable is the analysis of complex cuticular hydrocarbon profiles by multivariate statistical methods? J Chem Ecol 35:375–382
Martin SJ, Helanterä H, Drijfhout FP (2008) Colony-specific hydrocarbons identify nest mates in two species of Formica ant. J Chem Ecol 34:1072–1080
McNutt JW (1996) Sex-biased dispersal in African wild dogs, *Lycaon pictus*. Anim Behav 52:1067–1077
Mitteroecker P, Bookstein F (2011) Linear discrimination, ordination, and the visualization of selection gradients in modern morphometrics. Evol Biol 38:100–114
Müller CA, Manser MB (2007) 'Nasty neighbours' rather than 'dear enemies' in a social carnivore. Proc R Soc B 274:959–965
Müller CA, Manser MB (2008) The information banded mongooses extract from heterospecific alarms. Anim Behav 75:897–904
Osofsky SA, McNutt JW, Hirsch KJ (1996) Immobilization of free-ranging African wild dogs (*Lycaon pictus*) using a ketamine/xylazine/atropine combination. J Zoo Wildl Med 27: 528–532
Palphramand KL, White PCL (2007) Badgers, *Meles meles*, discriminate between neighbor, alien and self scent. Anim Behav 74:429–436
Poindron P, Levy F, Krehbiel D (1988) Genital, olfactory, and endocrine interactions in the development of maternal behavior in the parturient ewe. Psychoneuroendocrinology 13:99–125
Pomilia MA, McNutt JW, Jordan NR (2015) Ecological predictors of African wild dog ranging patterns in northern Botswana. J Mamm DOI:10.1093/jmammal/gyv130.
R Core Team (2013) R: a language and environment for statistical computing. R Foundation for Statistical Computing, Vienna. http://www.R-project.org

Radford AN (2005) Group-specific vocal signatures and neighbor-stranger discrimination in the cooperatively breeding green woodhoopoe. Anim Behav 70:1227–1234

Rasmussen LEL, Lee TD, Zhang A, Roelofs WL, Daves GD (1997) Purification, identification, concentration and bioactivity of (*Z*)-7-dodecen-1-yl acetate: sex pheromone of the female Asian elephant, *Elephas maximus*. Chem Senses 22:417–437

Roberts SC, Davidson AJ, McLean L, Beynon RJ, Hurst JL (2012) Pheromonal induction of spatial learning in mice. Science 338:1462–1465

Safi K, Kerth G (2003) Secretions of the interaural gland contain information about individuality and colony membership in Bechstein's bat. Anim Behav 65:363–369

Schulte BA (1998) Scent marking and responses to male castor fluid by beavers. J Mamm 79:191–203

Sillero-Zubiri C, Macdonald DW (1998) Scent-marking and territorial behavior of Ethiopian wolves *Canis simensis*. J Zool 245:351–361

Sun L, Müller-Schwarze D (1998) Anal gland secretion codes for family membership in the beaver. Behav Ecol Sociobiol 44:199–208

Yamamoto I (1984) Latrine utilization and feces recognition in the raccoon dog, *Nyctereutes procyonoides*. J Ethol 2:47–54

Chapter 31
Automated Headspace Solid-Phase Microextraction of Urinary VOCs from Eleven Maned Wolves (*Chrysocyon brachyurus*): A Recursive Workflow for GC–MS Analysis

Marieke E. Kester, Elizabeth W. Freeman, Nucharin Songsasen, and Thomas B. Huff

31.1 Introduction

31.1.1 Maned Wolf Biology

The maned wolf (*Chrysocyon brachyurus*) is native to the tropical grassland habitats of South America (Dietz 1985). The species is currently listed as "Near Threatened" (IUCN 2015) with an estimated wild population of around 20,000 (Songsasen and Rodden 2010). Like most canids, maned wolves are monoestrous: females cycle only once per year (Asa and Valdespino 1998; Sillero-Zubiri et al. 2004). The onset of the breeding season appears to be a response to decreasing day length (Maia and Gouveia 2002; Rodden et al. 2007; Valdespino 2007), though several occurrences of breeding outside the typical season have been recorded (see below). For this species, breeding

M.E. Kester
Department of Environmental Science and Policy, George Mason University, Fairfax, VA 22030, USA

Center for Species Survival, Smithsonian Conservation Biology Institute, National Zoological Park, Front Royal, VA 22630, USA

E.W. Freeman
New Century College, George Mason University, Fairfax, VA 22030, USA

Center for Species Survival, Smithsonian Conservation Biology Institute, National Zoological Park, Front Royal, VA 22630, USA

N. Songsasen
Center for Species Survival, Smithsonian Conservation Biology Institute, National Zoological Park, Front Royal, VA 22630, USA

T.B. Huff (✉)
Potomac Environmental Research and Education Center, George Mason University, Fairfax, VA 22030, USA
e-mail: thuff@gmu.edu

B.A. Schulte et al. (eds.), *Chemical Signals in Vertebrates 13*,
DOI 10.1007/978-3-319-22026-0_31

typically takes place from September through February in the northern hemisphere and from March through June in the southern hemisphere (Kleiman 1972; Rodden et al. 1996, 2007; Maia and Gouveia 2002; Songsasen et al. 2014).

Maned wolves are induced ovulators; females enter estrus and/or ovulate only in the presence of a male (Songsasen et al. 2006; Johnson et al. 2014). Although not all canids have been investigated for this trait, the only other known induced ovulating canid is the Channel Island fox (*Urocyon littoralis*) (Asa et al. 2007). The domestic dog (*Canis familiaris*) (Concannon 2011), gray wolf (*Canis lupus*) (Seal et al. 1979; Asa et al. 2006), coyote (*Canis latrans*) (Carlson and Gese 2008), African wild dog (*Lycaon pictus*) (Monfort et al. 1997), bush dog (*Speothos venaticus*) (DeMatteo et al. 2006), red fox (*Vulpes vulpes*) (Möller 1973), red wolf (*Canis rufus*) (Walker et al. 2002) and arctic fox (*Alopex lagopus*) (Mondain-Monval et al. 1977) are spontaneous ovulators.

Female maned wolves not paired with a male show baseline progestagen levels through the entire breeding season, demonstrating a lack of ovulation (Songsasen et al. 2006; Reiter 2012; Johnson et al. 2014). Treating females with a gonadotropin releasing hormone (GnRH) agonist can successfully induce estrus and ovulation in paired females, but not in females housed singly (Johnson et al. 2014). However, singleton females treated with the GnRH agonist followed by an exogenous luteinizing hormone treatment ovulate successfully (Johnson et al. 2014).

31.1.2 Scent Marking

Chemical signaling through scent-marking is central to social recognition, territoriality, reproduction, and alarm signaling across Mammalia (Albone 1984; Wyatt 2014). Scent marks continue to emit olfactory signals after the depositor has left, which makes scent-marking especially important for solitary animals that rarely come into direct contact with conspecifics (Hagey and Macdonald 2003; Wyatt 2014; Delbarco-Trillo et al. 2011; Kean et al. 2011; Harris et al. 2012).

Urine is considered to be a more important source of scent signaling than feces for canids (Gese and Ruff 1997; Parker 2010; Jordan et al. 2013). In general, canids urinate more frequently than they defecate, and they assume specialized postures for urination that are not necessary for urine elimination but are thought instead to provide a visual signal to accompany the urine scent mark (Asa et al. 1984; Pal 2003; Wirant et al. 2007). African wild dogs respond at higher rates to urine than to feces (Parker 2010; Jordan et al. 2013). Urine scent-marking (but not defecation) increases in frequency during the breeding season in several canid species, including the maned wolf (Hradecký 1985; Asa et al. 1990; Rodden et al. 1996; Gese and Ruff 1997; Pal 2003; Parker 2010; Jordan et al. 2013). In maned wolves, the frequency of female urine scent-marking is highest during proestrus and predicts reproductive success, as defined by the birth of pups (Rodden et al. 1996). Therefore, it is likely that maned wolves release signals in their urine leading up to breeding to communicate about reproduction eventually prompting the female to enter estrus and/or to ovulate.

Recent anecdotal evidence from individuals housed that the Smithsonian Conservation Biology Institute supports the role of a chemical signal in reproduction. A female maned wolf ovulated when housed singly while sharing a fence line with a male. The female had visual access to the male as well as the ability to contact his urine scent marks (Johnson et al. 2014). However, other females housed within sight of a male but with no contact with his urine failed to ovulate. These findings indicate that chemical signaling may be responsible for inducing estrus and/or ovulation in maned wolves.

31.1.3 Olfactory Stimulation of Estrus and/or Ovulation in Mammals

In laboratory mice (*Mus musculus*) estruscan be accelerated and ovulation can be induced by exposing group-housed females to male chemosignals (Marsden and Bronson 1964; Bronson and Whitten 1968; Whitten et al. 1968; Jemiolo et al. 1985, 1986; Marchlewska-Koj et al. 1990; Ma et al. 1999; Morè 2006). When major urinary proteins from male mice were applied to the noses of estrous females, the number of eggs ovulated was similar to the number ovulated when females were exposed to whole male urine, suggesting that major urinary proteins elicit the ovulation response in female mice (Morè 2006). A similar effect of male urine has also been observed for deer mice (*Peromyscus maniculatus*) (Bronson and Marsden 1964) and rats (*Rattus norvegicus*) (Johns et al. 1978). Female rats allowed contact with male urine ovulated at the same rate as females exposed to bedding from the male's enclosure (Johns et al. 1978), demonstrating that male urine is the bioactive substance.

Olfactory effects of males on estrus and ovulation have also been documented in prairie voles (*Microtus ochrogaster*) (Carter et al. 1980), Siberian hamsters (*Phodopus sungorus*) (Dodge et al. 2002), sheep (*Ovis aries*) (Cohen-Tannoudji et al. 1989, 1994; Gelez et al. 2004), and goats (*Capra aegagrus hircus*) (Iwata et al. 2000; Murata et al. 2009, 2014; Bedos et al. 2010). Furthermore, an olfactory cue prompting ovulation has been hypothesized for the dromedary camel (*Camelus dromedarius*) (Adams and Ratto 2013).

In carnivores, the so-called male effect is studied far less. There does seem to be a male effect in the bush dog, the closest living relative to the maned wolf, where the presence of an adult male decreases the inter-estrus interval of females (DeMatteo et al. 2006). However, to our knowledge, the compounds and mechanisms responsible for this effect in carnivores remain completely uninvestigated.

31.1.4 Maned Wolf Urinary Volatile Organic Compounds

Becausemammalian excretions are exceptionally complex (Burger 2005; Apps 2013), characterizing all the constituents that make up maned wolves' urine is a formidable task. Previously, only two studies of maned wolf urine have been published.

They examined volatile organic compounds (VOCs) as likely candidates for signaling. In one study, Goodwin et al. (2012) found that sulfur-containing hemiterpenoids, hemiterpenoid alcohols, and pyrazines were the main components contributing to this species' odiferous urine. This study also supported the idea that different canid species have unique urinary VOCs; a compound found to be abundant in other canids, 3-methyl-1-methylthiobut-3-ene, was not found in maned wolves; whereas, an isomer tentatively identified as 3-methyl-1-methylthiobut-2-ene was abundant. The only other analysis identified several of the pyrazine compounds (Childs-Sanford 2005) also found by Goodwin et al. (2012).

Analysis of VOCs in urine samples typically employs gas chromatography–mass spectrometry (GC–MS) to identify compounds and measure their abundances. Even under the best conditions, chromatograms are crowded with peaks that can overlap or co-elute in such a way that some poorly resolved compounds are hidden among those with higher abundance. Peaks for compounds with high abundance can be distorted, and peaks for compounds of low abundance may not be sufficiently resolved or distinguished from background noise. When peaks overlap, the mass spectrum at any given retention time (RT) may contain ions from one, two, or several compounds, thus making identification difficult and time-consuming. Preliminary analysis (unpublished) of several maned wolf urine samples revealed around 800 peaks in each chromatogram after spectral deconvolution. Even in a relatively small-scale experiment consisting of around 100 samples, the number of peaks quickly explodes to an intractable 80,000. With a rate of 5 min per peak working 10 h per day, checking all peak alignments manually would require 667 days.

In this study, a data-analysis software bundle was used to facilitate the analysis of GC–MS data from maned wolf urine samples. The strength of this data analysis method is that any sort of differential analysis can be conducted to generate a list of candidate semiochemical compounds to be tested for biological relevance in behavioral bioassays. For these analyses we focused on searching specifically for two lists of compounds: those that differ between sexes and those that are common across maned wolf urine samples. Identifying compounds that differ between the sexes is a solid first step toward semiochemical candidate identification while generating a list of compounds present across the entire data set will be useful to create a synthetic maned wolf urine mixture to be used as a control or a vehicle for bioassay studies.

31.2 Methods

31.2.1 Animals

Eleven maned wolves housed at institutions within the United States were used in this study (Table 31.1). Monthly urine samples (4–15 mL) were collected when individuals urinated on a clean stainless steel pan placed in their enclosure. Each morning the pan was cleaned and the back of the pan (not the collecting side)

Table 31.1 Maned wolves that supplied urine samples

Institution	Studbook#[a]	Sex	Age (years)
Connecticut's Beardsley Zoo, CT	3231	F	3
Connecticut's Beardsley Zoo, CT	3232	F	3
Fossil Rim Wildlife Center, TX	2845	F	7
Smithsonian Conservation Biology Institute, VA	2810	M	7
Smithsonian Conservation Biology Institute, VA	2844	M	7
Smithsonian Conservation Biology Institute, VA	2926	F	7
Smithsonian Conservation Biology Institute, VA	2954	M	7
Smithsonian Conservation Biology Institute, VA	3120	M	6
Smithsonian Conservation Biology Institute, VA	3184	F	3
White Oak Conservation Center, FL	2660	M	9
White Oak Conservation Center, FL	2945	F	7

[a]Association of Zoos and Aquariums reference number of individual pedigree and demographic history

was sprayed with around 0.5 mL of maned wolf urine, prompting individuals to urinate on the front of the pan. The samples were collected and frozen at −20 °C within 8 h of elimination, a period of time dictated by the logistics at the four participating institutions. This project was reviewed and approved by the Institutional Animal Care and Use Committee (IACUC) or its equivalent at each participating institution.

31.2.2 Head Space Solid-Phase Microextraction-Gas Chromatography–Mass Spectrometry

Urine samples were defrosted, vortexed, and 1.0 mL aliquots were dispensed into 10 mL glass headspace vials with metal twist caps and silicone/PTFE-layered septa (La-Pha-Pack, Langerwehe, Germany). Prior to use vials were rinsed with ultrapure water, rinsed with methanol, and baked at 425 °C overnight. Samples were saturated with high-purity sodium chloride to decrease the solubility of VOCs (Mills and Walker 2000). Samples were run in triplicate on a 7890A-5975C gas chromatograph–mass spectrometer (Agilent Technologies, Santa Clara, CA) fitted with a CombiPAL robotic sampling preparation and injection system (CTC Leap Technologies, Carrboro, NC). The CombiPAL was configured with a solid-phase microextraction (SPME) adapter and a sample heater-agitator to automate the entire sample extraction and chromatography process. The instrument and autosampler system were controlled using MSD Chemstation software ver E.02.02 (Agilent Technologies, Santa Clara, CA).

Prior to running each sample batch, a 1 cm 50/30 μm divinyl benzene-carboxen-poly(dimethylsiloxane) (DVB/CAR/PDMS) stable-flex SPME fiber (Sigma-Aldrich, St. Louis, MO) was cleaned and preconditioned in a spare split-splitless inlet according to manufacturer recommendations of 270 °C for 60 min. Following each run, the SPME fiber was cleaned for 20 min in this extra inlet prior to being re-used with another sample. VOCs in the sample were equilibrated between the headspace and urine at 500 rpm for 30 min at 37 °C. Heating and agitation continued for 45 min while the SPME fiber was exposed to the headspace. The 50/30 μm DVB/CAR/PDMS has been shown to extract the widest array of compound classes with the best sensitivity for commercial SPME fibers (Risticevic and Pawliszyn 2013).

Following extraction, compounds were thermally desorbed from the fiber in a temperature-programmable multimode GC inlet with a 0.75 mm ID SPME liner (Restek Corporation, Bellefonte, PA). The initial inlet temperature was 50 °C for 0.12 min followed by a ramp to 240 °C at 600 °C min^{-1}. The column was a 0.25 mm ID by 30 m long RTX-VMS with a 1.0 μm film (Restek Corporation, Bellefonte, PA). The carrier gas was helium, and the initial GC oven temperature was 50 °C for 3 min, followed by a ramp of 7 °C min^{-1} to 240 °C. Total run time was 45 min. The GC was equipped with Agilent's backflush module, which reversed the column flow for 5 min following the 45 min GC run to eliminate sample-to-sample carry over and to maintain inertness of the flow path.

Retention-time locking was used to maintain consistent peak retention times, which facilitates identifications and comparisons of analytes across large sample sets.

The transfer line temperature was held at 290 °C. The 5975C mass spectrometer operated with an electron energy of 70 eV in the full scan mode with a range from 40 to 350 m/z at a rate of 4.51 scans s^{-1}. The ion source was at 300 °C. The quadrupole was at 180 °C.

31.2.3 Data Analysis

Urine sample data were analyzed using MassHunter Workstation software (Agilent) including MassHunter Qualitative Analysis ver B.06.00 and MassHunter Quantitative Analysis ver B.07.00 for chromatographic data processing and Mass Profiler Professional ver 12.6.1 for visualization and statistical analysis. Subsequently these programs will be referred to as Qual, Quant, and MPP respectively. This software suite was originally developed for metabolomics and proteomics (Gu et al. 2011; Álvarez-Sánchez et al. 2012; Kim et al. 2014) but has applications in several fields: pharmaceutical impurity testing (West et al. 2010), food science (Bondia-Pons et al. 2014), as well as environmental toxicity studies (Hindle et al. 2013).

Qual analyzed all the samples as a batch with parameters selected in a Qual method. The program began the analysis by creating an ion chromatogram for every nominal ion. It integrated each ion chromatogram and created a peak list that was put through a deconvolution algorithm (a chromatographic covariance test to create compounds from related ions that eluted at the same time) (Fig. 31.1). Thus, for each compound there was an associated deconvoluted spectrum that contained far fewer ions from noise and adjacent peaks. The retention time window size factor was set to 80 and the extraction window was set to ±0.3 AMU. Ions 73, 207, and

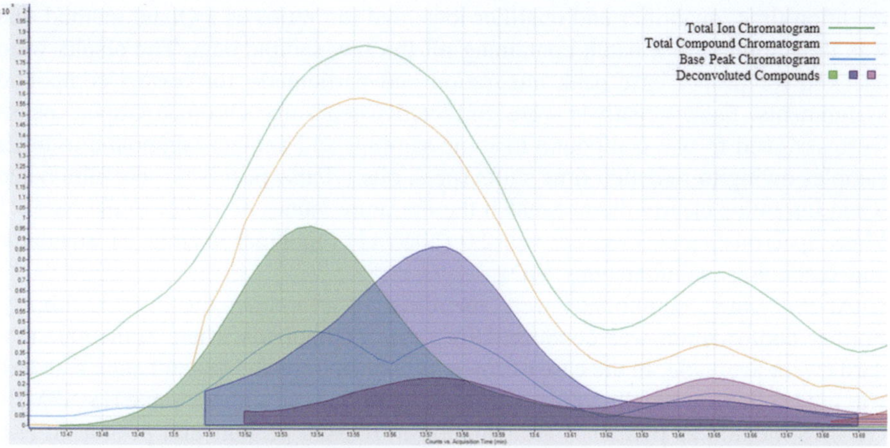

Fig. 31.1 The deconvolution algorithm is able to separate overlapping, co-eluting compounds on the basis of individual ion chromatograms with high covariance. Total ion chromatogram shown in *green*, total compound chromatogram shown in *orange*, and base peak chromatogram shown in *blue*. Three overlapping, co-eluting compounds shown shaded

281 m/z were excluded since they represent GC column bleed. For each data file, the compounds were exported into a .CEF file for evaluation in MPP.

MPP was then used to perform differential analyses to determine relationships among sample groups or experimental parameters. The software aligned each compound across the entire sample batch using an alignment algorithm that matched similar spectra at the same retention time. A retention time tolerance of ±0.20 min was used along with a spectral match factor of 0.30 and a delta m/z window of 0.20. The relative abundance values were then $\log_2$ transformed. Relative abundances of zero were set $= 1 \times 10^{-5}$ so that the $\log_2$ value $= -16.61$ rather than being undefined (van den Berg et al. 2006).

Once compounds were aligned, the differential analyses could be performed across all samples with greater confidence. Compounds that potentially differed by sex were determined using a moderated *t*-test (Smyth 2004) on the $\log_2$ transformed relative abundances. Comparatively loose criteria were used in this step to minimize false negatives that could be present due to retention time shifts. Compounds with a corrected p-value < 0.05 and a fold change difference between the sexes of >2.0 were retained. Fold change between conditions *A* and *B* is calculated on the natural scale as condition *A*/condition *B*. Once on the $\log_2$ scale, $\log_2(\text{fold change}) = \log_2(\text{condition } A) - \log_2(\text{condition } B)$.

To identify compounds that were found most commonly, the aligned peaks were filtered by frequency. Peaks initially found in >80 % of data files were retained to minimize false negatives. The resulting two compound lists were far shorter than the total list of compounds across all samples and were exported as .CEF files for screening in Quant.

Each .CEF file from an MPP analysis formed a compound library for use in Quant to be used in a targeted analysis. The library contained the compounds identified by retention time, quantifier and qualifier ions, and ion ratios as a Quant method that was then applied to a batch containing all the raw sample data. The results were checked for proper base peak integration and particularly for correct compound assignments. False negatives were corrected when some chromatographic peaks were assigned to one compound in some samples and another compound in other samples owing to retention time shifts (pers. obs.). False positives were eliminated when ion ratios clearly did not match the literature values or values in the NIST11 Mass Spectral Search Program ver. 2.0 (NIST11) (National Institute of Standards and Technology, Gaithersburg, MD).

At this point, compounds were identified by searching the NIST11 library. Further confirmation was conducted for those compounds where authentic standards were available. In the case of some pyrazines for which an authentic standard was not available, a small sample of peanut butter was run with the same HS-SPME–GC–MS method as pyrazines are common in roasted peanuts and peanut oil (Ku et al. 1998; Liu et al. 2011). The manually screened data were then exported into a .CEF file for final analysis.

Once again, MPP was used to perform visualizations and differential analyses using the $\log_2$ transformed relative abundances. Principal components analysis (PCA) allowed visualization of the similarities and differences between samples. PCA is a dimension reduction technique to visualize groupings of samples based on a combination of variables, in this case compounds (Wold et al. 1987). Each principal component is chosen to maximize the amount of variance it explains while being orthogonal to the other principal components. Finally, a moderated t-test and a fold change analysis offered insight into statistical significance for compounds that differed by sex. A compound was considered significantly different between groups if it met two criteria: a corrected p-value of <0.001 and a fold change between groups of >3.0. Numerical results are reported as mean ± standard deviation.

31.3 Results and Discussion

A total of 103 urine samples from 11 wolves were analyzed, covering 6–12 months of 2013 with an average of 9.36 ± 2.06 months per individual. One sample with low volume (Female #3184, 9/11/2013) was run in duplicate rather than triplicate, resulting in a total of 308 data files.

31.3.1 Qualitative Analysis, Peak Alignment, and Preliminary Analysis

Overall, 1682 compounds were aligned across the 308 data files. On average, samples showed 126.75 ± 24.92 compounds, meaning the vast majority of compounds were only found in one sample. The average relative standard deviation between replicates of each sample was 6.46 %.

After compound alignment, the $\log_2$ of the relative abundance for each compound was reported for each of the samples. Based on these relative abundances, two lists were created: compounds that differed by sex with a $p < 0.05$ and a fold change between the sexes of >2.0 (63 compounds) and compounds occurring in >80 % of samples (39 compounds). Each list described the compounds with chromatographic and mass-spectral parameters rather than with a compound name.

31.3.2 Compounds That Differed by Sex

Forthe analysis of compounds differing between sexes, false positives were removed and false negatives were added manually in Quant. The resulting candidate compound list contained 27 compounds.

The candidate list of 27 compounds generated from Quant was then imported back into MPP for a second round of statistical analysis since the preliminary statistical analysis occurred prior to manual checks of the integration and compound assignment verification (removal of false positives and addition of false negatives).

After averaging across replicates, PCA was used to visualize the grouping of the samples for the 27 compounds identified in Quant as differing by sex (Fig. 31.2). In this analysis, three principal components explained 43.03 % of variance between samples. Because principal component 1 (PC1) mainly described the variation attributable to the difference between the sample from Female 3184 on October 29, 2013 and the rest of the samples (Fig. 31.2a), the PCA was re-run excluding this data point. Results remained stable with good separation between the sexes using three principal components. In both cases, principal component 2 (PC2) seemed to relate to sex with male wolves showing mainly positive scores and females showing negative scores (Fig. 31.2a, b). Principal component 3 (PC3) mainly described the variation between individuals (Fig. 31.2b). The loadings plots (Fig. 31.2c, d) show the contribution of each compound to the three principal components.

There were seven compounds that showed significantly higher relative abundances in males (Table 31.2): ψ-diosphenol, [fold change (FC) = 3292.22], 1,3-di-tert-butylbenzene (FC = 603.86), 2,4-dimethyl-1-heptene (FC = 85.22), 2-nonen-4-one (FC = 43.70), 2-acetyl-6-methyl pyrazine (FC = 16.63), 6-methyl-3,5-heptadiene-2-one (FC = 6.56), and 2-ethenyl-6-methyl pyrazine (FC = 5.53). The compound ψ-diosphenol (synonym: 2-hydroxy-6-methyl-3-(1-methylethyl)-2-cyclohexen-1-one) is a terpenoid found in the essential oil of the buchu plant (*Agathosma betulina*), endemic to southern Africa (Fluck et al. 1961). 2,4-Dimethyl-1-heptene is a biomarker for bacterial infections, particularly for *Pseudomonas aeruginosa* (Sohrabi et al. 2014) and has also been identified as a byproduct of high-temperature pyrolysis of polypropylene (De Amorim et al. 1982). As far as we are aware, the above two compounds have not yet been reported in mammalian excretions.

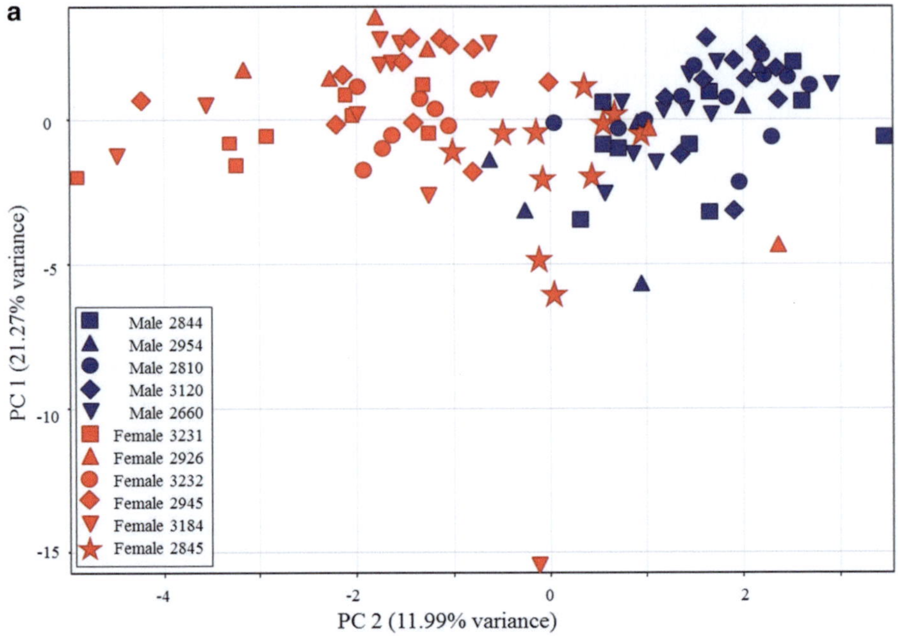

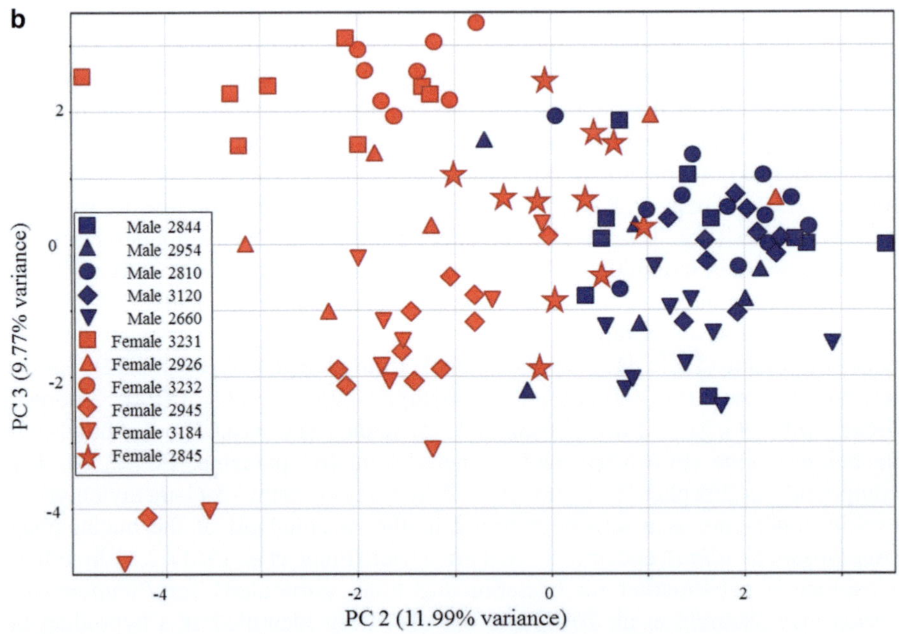

Fig. 31.2 (**a**) Principal component scores of 103 samples on principal components 1 and 2 using 27 compounds found to vary by sex. Samples are colored by sex and shaped by individual wolf. (**b**) Component loadings of 27 compounds on principal components 1 and 2. (**c**) Principal component scores of 103 samples on principal components 2 and 3 using 27 compounds found to vary by sex. (**d**) Component loadings of 27 compounds on principal components 2 and 3

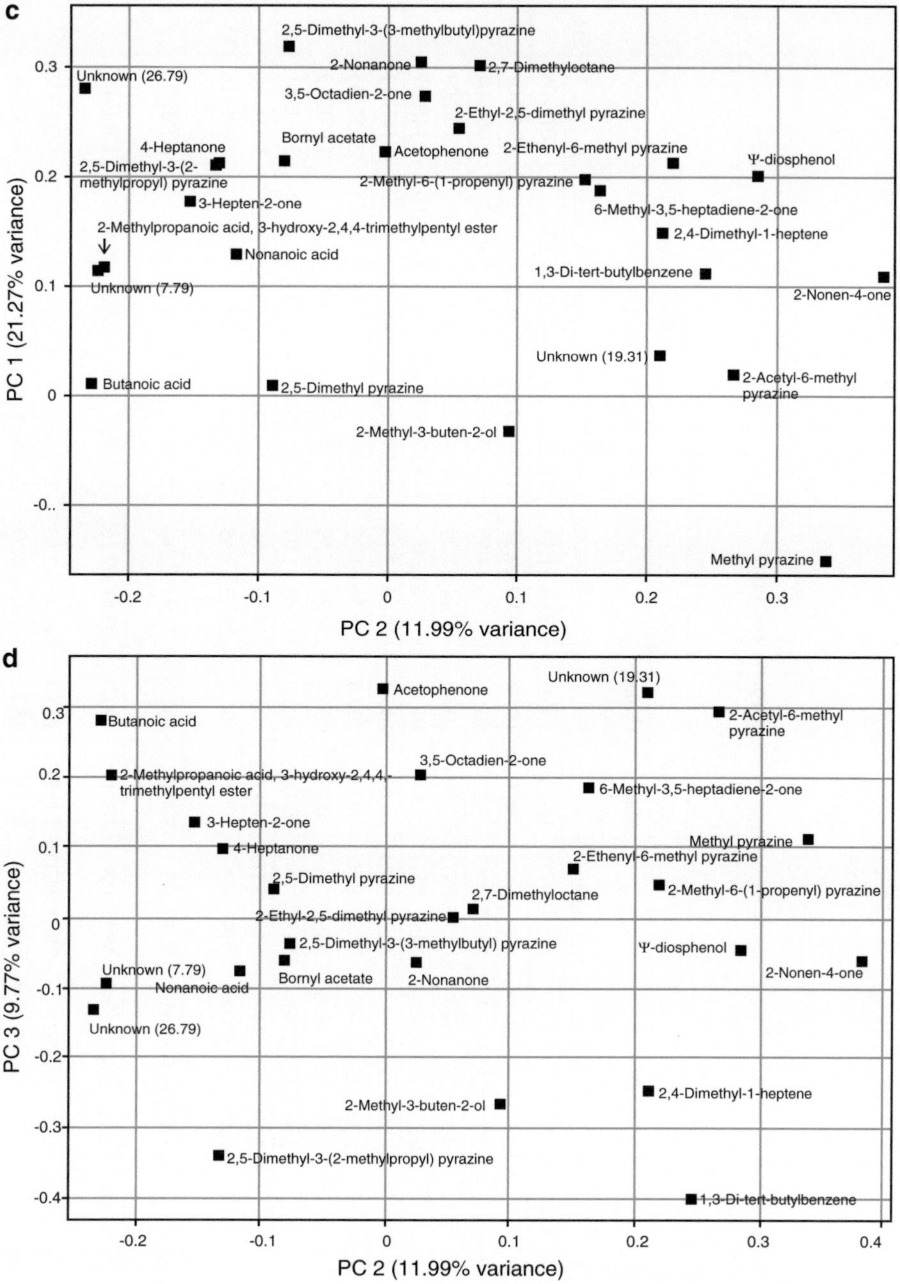

Fig. 31.2 (continued)

Table 31.2 Compounds that differed in relative abundance between sexes in maned wolf urine

				Percentage samples where present		Log_2 (relative abundance) (mean ± SD)		Fold change[b]			
Compound	RT	CAS no.	ID method[a]	Male (N=48) (%)	Female (N=55) (%)	Male (N=48)	Female (N=55)	Log_2	Natural	p	Sig.[c]
2-Methyl-3-buten-2-ol[d]	4.20	115-18-4	NIST, STD	100.0	100.0	22.68 ± 1.13	22.08 ± 1.10	0.60	1.51	4.01E-05	
Unknown (7.79)	7.79			100.0	100.0	13.90 ± 0.87	14.48 ± 0.86	−0.58	−1.50	1.94E-07	
2,4-Dimethyl-1-heptene	8.22	19549-87-2	NIST	68.8	50.9	1.45 ± 14.55	−4.96 ± 13.11	6.41	85.22	5.96E-04	***
Methyl pyrazine[d]	9.50	109-08-0	NIST, STD	100.0	100.0	17.21 ± 1.24	16.24 ± 1.37	0.97	1.96	2.10E-09	
Butanoic acid[e]	10.35	107-92-6	NIST, STD	0.0	16.4	−16.61 ± 0	−10.77 ± 13.27	−5.84	−57.10	2.14E-07	***
4-Heptanone	11.09	123-19-3	NIST, STD	100.0	100.0	20.30 ± 2.13	22.04 ± 1.59	−1.75	−3.35	3.14E-12	***
2,5-Dimethyl pyrazine[d,e]	11.73	123-32-0	NIST, STD	100.0	100.0	25.58 ± 1.33	25.96 ± 0.37	−0.38	−1.30	2.95E-03	
2,7-Dimethyloctane	12.52	1072-16-8	NIST	100.0	98.2	14.26 ± 1.59	13.28 ± 4.24	0.98	1.97	1.07E-02	
2-Nonanone	12.61	821-55-6	NIST	100.0	100.0	15.63 ± 1.91	15.55 ± 3.87	0.08	1.06	8.41E-01	
3-Hepten-2-one	13.50	1119-44-4	NIST, STD	100.0	100.0	14.76 ± 1.66	16.12 ± 1.51	−1.36	−2.57	1.35E-10	
2-Ethenyl-6-methyl pyrazine[d]	14.47	13925-09-2	NIST, PB	100.0	100.0	21.93 ± 2.65	19.46 ± 2.3	2.47	5.53	2.38E-16	***
2-Ethyl-2,5-dimethyl pyrazine[d]	15.52	13360-65-1	NIST, PB	100.0	100.0	22.81 ± 2.14	22.85 ± 1.37	−0.03	−1.02	8.75E-01	
2-Methyl-6-(1-propenyl) pyrazine	16.31	55138-67-5	NIST, PB	100.0	100.0	22.64 ± 2.16	21.68 ± 2.21	0.96	1.95	1.76E-03	
Acetophenone	16.93	98-86-2	NIST, STD	100.0	100.0	16.43 ± 2.63	17.06 ± 2.39	−0.63	−1.55	4.20E-02	
3,5-Octadien-2-one	17.67	38284-27-4	NIST	97.9	98.2	12.01 ± 6.59	11.84 ± 5.59	0.17	1.13	8.23E-01	

2-Nonen-4-one	17.85	32064-72-5	NIST	100.0	100.0	17.41±2.61	11.96±3.39	5.45	43.70	6.66E-28	***
2-Acetyl-6-methyl pyrazine	17.92	22047-26-3	NIST, PB	100.0	96.4	14.86±1.86	10.80±7.23	4.06	16.63	4.40E-09	***
6-Methyl-3,5-heptadiene-2-one	18.07	1604-28-0	NIST	100.0	100.0	13.66±2.42	10.94±4.73	2.71	6.56	1.19E-05	***
2,5-Dimethyl-3-(2-methylpropyl) pyrazine	18.50	32736-94-0	NIST	100.0	100.0	16.86±1.63	17.25±2.64	−0.39	−1.31	1.43E-01	
1,3-Di-tert-butylbenzene	19.17	1014-60-4	NIST, STD	95.8	69.1	15.60±7.83	6.36±15.8	9.24	603.86	7.31E-10	***
Unknown (19.31)	19.31			97.9	94.5	12.20±4.65	10.91±6.91	1.28	2.44	7.20E-02	
ψ-diosphenol	20.36	54783-36-7	NIST	97.9	74.5	11.70±6.98	0.02±11.74	11.68	3292.22	7.57E-17	***
2,5-Dimethyl-3-(3-methylbutyl) pyrazine	20.61	18433-98-2	NIST	100.0	96.4	16.75±1.84	16.10±7.02	0.65	1.57	2.95E-01	
(−)-Bornyl acetate	21.00	5655-61-8	NIST, STD	93.8	85.5	9.99±8.61	7.95±11.33	2.04	4.11	9.26E-02	
Nonanoic acid	21.41	112-05-0	NIST	14.6	30.9	−14.20±6.48	−9.94±11.38	−4.26	−19.21	6.81E-04	***
2-Methylpropanoic acid, 3-hydroxy-2,4,4-trimethylpentyl ester	23.75	74367-34-3	NIST	100.0	100.0	14.99±2.03	15.94±1.99	−0.95	−1.93	1.01E-04	
Unknown (26.79)	26.79			100.0	100.0	14.56±1.16	15.92±2.69	−1.36	−2.57	4.28E-06	

[a]Identification methods: NIST National Institute of Standards and Technology 2011 library of mass spectra; STD retention time match to authentic standard; PB retention time and mass spectral match to compound in peanut butter (see text)

[b]In the natural scale, fold change = condition *A*/condition *B*. In the $\log_2$ scale, $\log_2(A/B) = \log_2(A) - \log_2(B)$

[c]Significance determined when $p < 0.001$ and |Fold change(raw relative abundance)| > 3

[d]Also found in maned wolf urine by Goodwin et al. (2012)

[e]Also found in maned wolf urine by Childs-Sanford (2005)

Of the compounds shown to have significantly greater relative abundances in male maned wolves, two have been previously identified in mammalian excretions: 2-nonen-4-one is a putative reproductive semiochemical found in the rutting pits of male Alaskan moose (*Alces alces gigas*) (Whittle et al. 2000). Female moose roll in the pits that have been marked extensively with male urine. This compound is also a constituent in the urine of 67 out of 84 European badgers (*Meles meles*) (Service et al. 2001). 2-Ethenyl-6-methyl pyrazine is present in the urine of sexually intact male brown antechinus (*Antechinus stuartii*) and was not identified in females (Toftegaards et al. 1999). It is also a reproductive semiochemical in the papaya fruit fly (*Toxotrypana curvicauda*) (Robledo and Arzuffi 2012).

Three compounds were reported in foods: 1,3-di-tert-butylbenzene is one of the main VOCs produced by the mycelium of *Tuber borchii*, an edible species of truffle (Tirillini et al. 2000). 2-Acetyl-6-methyl pyrazine is a common flavor compound responsible for the roasted aroma of grains, peanuts, and liquors (Buttery et al. 1997, 1999; Fan et al. 2007; Liu et al. 2011). 6-Methyl-3,5-heptadiene-2-one is found in tomato products (Buttery et al. 1990), artichoke leaves (*Cynara scolymus L.*) (Saucier et al. 2014), and also is found in the essential oil of *Hypericum annulatum* (Radulović et al. 2010) and species of *Erodium* (Stojanović-Radić et al. 2010), as well as paprika (*Capsicum annuum* L.) oleoresin (Guadayol et al. 1997).

One compound, butanoic acid, was found only in female maned wolves (Table 31.2), though not in all females tested. Butanoic acid seems to be common in excretions from other members of Canidae. This compound is present in anal gland secretions, feces, urine, and the preputial hair tufts of the African wild dog (Parker 2010; Apps et al. 2012), in the anal gland secretion and feces of black-backed jackal (*Canis mesomelas*) (Apps et al. 2012), and in relatively high abundance in the feces of the Iberian wolf (*Canis lupus signatus*) (Mártin et al. 2010). This compound also is found in tiger (*Panthera tigris*) urine marks (Burger et al. 2008), in cloaca secretions of the Tasmanian short-beaked echidna (*Tachyglossus aculeatus setosus*) (Harris et al. 2012), in the circumanal gland of the marmoset monkey (*Callithrix jacchus*) (Smith et al. 2001), in the buccal gland of the dwarf hamster (*Phodopus sungorus*) (Burger et al. 2001), and in milk of the European rabbit (*Oryctolagus cuniculus*) (Schaal et al. 2003).

Two compounds were significantly more abundant in female maned wolves than males (Table 31.2): nonanoic acid (FC=19.21) and 4-heptanone (FC=3.3). Nonanoic acid is present in a wide variety of mammalian excretions including the urine marks of tigers (Burger et al. 2008) and leopards (*Panthera pardus*) (Poddar-Sarkar and Brahmachary 2004), the urine of *Mus domesticus* and *Mus spicilegus* (Soini et al. 2009), anal gland secretion of the wolverine (Wood et al. 2005) and black-backed jackal (Apps et al. 2012), the preputial hair tuft and urine of the African wild dog (Apps et al. 2012), and the cloaca secretion of the Tasmanian short-beaked echidna (Harris et al. 2012). As far as we are aware, differential analyses based on sex have not yet been done for any of the above species.

4-Heptanone has been previously identified in urine across a wide variety of mammals including the gray wolf (Raymer et al. 1986), wolverine (*Gulo gulo*) (Wood et al. 2009), African wild dog (Parker 2010), and several species of strepsirrhine primate (Delbarco-Trillo et al. 2011). Differences in the abundance of 4-heptanone between the sexes are known for some species. In the cheetah (*Acinonyx jubatus*) this compound is only found in female urine (Burger et al. 2006), while in the ferret (*Mustela furo*) 4-heptanone is more abundant in males than in females (Zhang et al. 2005), and in the rat (*Rattus norvegicus*) this compound is only found in male urine (Zhang et al. 2008). In the red fox, 4-heptanone is identified in both males and females but there is a difference in abundance for males between breeding and non-breeding seasons (Jorgenson et al. 1978). Similarly, 4-heptanone increases in concentration from nonbreeding to breeding season in male white-tailed deer (*Odocoileus virginianus*) (Miller et al. 1998). This compound also occurs in the urine of African elephant (*Loxodonta africana*) females during both surges of the reproductive hormone, luteinizing hormone (Goodwin et al. 2005).

31.3.3 Compounds Found Commonly

Once false positives were removed and false negatives were added via the manual recursion process in Quant, the candidate list of compounds found commonly across samples was pared down from 39 to 25 compounds. The resulting compounds were found in >98 % of the samples (Table 31.3). The compound found at a retention time of 11.4 min was likely 3-methyl-1-methylthiobut-2-ene for the reasons discussed in Goodwin et al. 2012. The ion relative abundance ratios in mass spectra here matched those described in that work. Most of the pyrazines and hemiterpenoids reported in Childs-Sanford (2005) and Goodwin et al. (2012) were also found here (Table 31.3).

To further analyze the 25 compounds found to be common across samples, heat maps were created to visualize the changes in compound abundance over the year for males and females separately. No discernible patterns were detected that would indicate changes in abundance of these compounds between breeding- and non-breeding seasons.

31.4 Conclusions

The data analysis conducted with this multistep software method filtered a nearly unmanageable body of raw GC–MS sample data into a list of compounds that were shown to be of considerable interest to this study. The same data analysis protocols can be used with other differential analyses to provide a better understanding of various chemical signaling mechanisms and their associated semiochemicals in many other vertebrates.

The resulting list of putative semiochemicals includes both compounds that have been previously identified in maned wolves and in other species, and compounds

Table 31.3 Compounds common in maned wolf urine

Compound	RT	CAS no.	ID method[a]	Percentage samples where present ($N=103$) (%)	Log_2 (relative abundance) (mean ± SD) ($N=103$)
2-Methyl-3-buten-2-ol[b]	4.25	115-18-4	NIST, STD	100.0	22.19 ± 1.19
3-Methyl-2-buten-1-ol[b]	8.70	556-82-1	NIST, STD	100.0	22.27 ± 1.07
3-Methyl-2-butenal	9.70	107-86-8	NIST, STD	100.0	16.22 ± 1.26
4-Heptanone	11.20	123-19-3	NIST, STD	100.0	21.03 ± 2.03
1-Methylthio-3-methylbut-2-eneb	11.40	NA		100.0	19.07 ± 2.38
2,5-Dimethyl pyrazine[b,c]	11.80	123-32-0	NIST, STD	100.0	25.89 ± 0.46
Trimethyl pyrazine[b,c]	13.82	14667-55-1	NIST, STD	100.0	20.94 ± 1.05
3-Ethylcyclopentanone	14.17	10264-55-8	NIST	100.0	15.71 ± 2.5
6-Methyl-5-hepten-2-one	14.35	110-93-0	NIST, STD	100.0	16.03 ± 1.24
2-Ethenyl-6-methyl pyrazine[b]	14.49	13925-09-2	NIST, PB	100.0	20.73 ± 2.62
2,5-Dimethyl-3-ethyl pyrazine[b]	15.57	13360-65-1	NIST, STD	100.0	22.93 ± 1.36
2-Methyl-6-(1-propenyl) pyrazine[b,c]	16.40	18217-81-7	NIST	100.0	22.5 ± 2.06
Acetophenone	17.04	98-86-2	NIST, STD	100.0	17.9 ± 1.31
Unknown (17.40)	17.40			100.0	14.92 ± 1.97
3,5-Dimethyl-2-propyl pyrazine	17.48	32350-16-6	NIST	100.0	18.99 ± 1.85
2,6-Dimethylcyclohexanol	17.99	5337-72-4	NIST	100.0	15.03 ± 1.55
2,5-Dimethyl-3-(2-methylpropyl) pyrazine	18.42	32736-94-0	NIST	100.0	18.3 ± 2.29
2-Acetyl-3,5-dimethyl pyrazine	18.78	54300-08-2	NIST, STD	100.0	17.34 ± 1.16
3-Butyl-2,5-dimethyl pyrazine	19.60	40790-29-2	NIST	98.0	14.88 ± 5.38
β-Cyclocitral	20.20	432-25-7	NIST, STD	98.0	13.75 ± 4.53
2,5-Dimethyl-3-(3-methylbutyl) pyrazine	21.00	18433-98-2	NIST	100.0	19.39 ± 2.33
Unknown (21.79)	21.79			100.0	18.73 ± 1.47
Unknown (21.96)	21.96			100.0	16.64 ± 2.47
Geranyl nitrile	25.07	5146-66-7	NIST	100.0	17.63 ± 3.5
3-Hydroxy-β-damascone	26.78	102488-09-5	NIST	100.0	14.94 ± 2.28

[a]Identification methods: NIST National Institute of Standards and Technology 2011 library of mass spectra; STD retention time match to authentic standard; PB retention time and mass spectral match to compound in peanut butter (see text)
[b]Also found in maned wolf urine by Goodwin et al. (2012)
[c]Also found in maned wolf urine by Childs-Sanford (2005)

that have not previously been identified in maned wolf urine or in other mammalian excretions. The compounds found to be common to all maned wolves in this study could prove important as a control or background signal in future bioassay research. It is possible that they confer nonreproductive signaling among conspecifics. Furthermore, the compounds that were shown to differ by sex are good candidates for semiochemical testing through bioassay research.

Acknowledgments We deeply appreciate the commitment of the staff at participating institutions (Jason Ahistus, Copper Aitken-Palmer, Ginger Eye, Tom Eyring, Don Goff, Priscilla Joyner, Jessica Kordell, Ken Lang, Paul Marinari, Jordan Maupin, Karen Meeks, MaryJo Stearns, and Linda Tomas) for ensuring consistent sample collection and for keeping our animals healthy. The GC–MS and its CombiPAL system were funded, maintained and made available by George Mason University's Shared Research Instrumentation Facility. Laboratory supplies were provided by the Maned Wolf Species Survival Plan and the George Mason College of Science Smithsonian Graduate Student Research Fellowship. Thank you to the George Mason University OSCAR program and Zuzanna Abdala for assisting in running the samples from Male 2660. MEK was supported by Teaching and Research Assistantships from George Mason University and by the Smithsonian Graduate Student Research Fellowship.

References

Adams GP, Ratto MH (2013) Ovulation-inducing factor in seminal plasma: a review. Anim Reprod Sci 136:148–156

Albone ER (1984) Mammalian semiochemistry: the investigation of chemical signals between mammals. Wiley, Chichester

Álvarez-Sánchez B, Priego-Capote F, Luque de Castro MD (2012) Study of sample preparation for metabolic profiling of human saliva by liquid chromatography-time of flight/mass spectrometry. J Chromatogr A 1248:178–181

Apps PJ (2013) Are mammal olfactory signals hiding right under our noses? Naturwissenschaften 100:487–506

Apps PJ, Mmualefe L, McNutt JW (2012) Identification of volatiles from the secretions and excretions of African wild dogs (*Lycaon pictus*). J Chem Ecol 38:1450–1461

Asa CS, Valdespino C (1998) Canid reproductive biology: an integration of proximate mechanisms and ultimate causes. Am Zool 38:251–259

Asa CS, Mech LD, Seal US (1984) The use of urine, feces and anal-gland secretions in scent marking by a captive wolf (*Canis lupus*) pack. Anim Behav 33:1034–1036

Asa CS, Mech LD, Seal US, Plotka ED (1990) The influence of social and endocrine factors on urine-marking by captive wolves (*Canis lupus*). Horm Behav 24:497–509

Asa CS, Bauman K, Callahan P, Bauman J, Volkmann DH, Jöchle W (2006) GnRH-agonist induction of fertile estrus with either natural mating or artificial insemination, followed by birth of pups in gray wolves (*Canis lupus*). Theriogenology 66:1778–1782

Asa CS, Bauman JE, Coonan TJ, Gray MM (2007) Evidence for induced estrus or ovulation in a canid, the island fox (*Urocyon littoralis*). J Mammal 88:436–440

Bedos M, Flores JA, Fitz-Rodríguez G, Keller M, Malpaux B, Poindron P, Delgadillo JA (2010) Four hours of daily contact with sexually active males is sufficient to induce fertile ovulation in anestrous goats. Horm Behav 58:473–477

Bondia-Pons I, Savolainen O, Törrönen R, Martinez JA, Poutanen K, Hanhineva K (2014) Metabolic profiling of goji berry extracts for discrimination of geographical origin by non-targeted liquid chromatography coupled to quadrupole time-of-flight mass spectrometry. Food Res Int 63:132–138

Bronson FH, Marsden HM (1964) Male-induced synchrony of estrus in deer mice. Gen Comp Endocrinol 4:634–637
Bronson FH, Whitten WK (1968) Oestrus-accelerating pheromone of mice: assay, androgen-dependency and presence in bladder urine. J Reprod Fertil 15:131–134
Burger BV (2005) Mammalian semiochemicals. Top Curr Chem 240:231–278
Burger BV, Smit D, Spies HSC, Schmidt C, Schmidt U, Telitsina AY (2001) Mammalian exocrine secretions XVI. Constituents of secretion of supplementary sacculi of dwarf hamster, *Phodopus sungorus*. J Chem Ecol 27:1277–1288
Burger BV, Visser R, Moses A, Le Roux M (2006) Elemental sulfur identified in the urine of cheetah, *Acinonyx jubatus*. J Chem Ecol 32:1347–1352
Burger BV, Viviers MZ, Bekker JPI, Le Roux M, Fish N, Fourie WB, Weibchen G (2008) Chemical characterization of territorial marking fluid of male Bengal tiger, *Panthera tigris*. J Chem Ecol 34:659–671
Buttery RG, Takeoka GR, Teranishi R, Ling LC (1990) Tomato aroma components: identification of glycoside hydrolysis volatiles. J Agric Food Chem 38:2050–2053
Buttery RG, Ling LC, Stern DJ (1997) Studies on popcorn aroma and flavor volatiles. J Agric Food Chem 45:837–843
Buttery RG, Orts WJ, Takeoka GR, Nam Y (1999) Volatile flavor components of rice cakes. J Agric Food Chem 47:4353–4356
Carlson DA, Gese EM (2008) Reproductive biology of the coyote (*Canis latrans*): integration of mating behavior, reproductive hormones, and vaginal cytology. J Mammal 89:654–664
Carter CS, Getz LL, Gavish L, McDermott J, Arnold P (1980) Male-related pheromones and the activation of female reproduction in the prairie vole (*Microtus ochrogaster*). Biol Reprod 23:1038–1045
Childs-Sanford S (2005) The captive maned wolf (*Chrysocyon brachyurus*): nutritional considerations with emphasis on management of cystinuria. Dissertation, University of Maryland
Cohen-Tannoudji J, Lavenet C, Locatelli A, Tillet Y, Signoret JP (1989) Non-involvement of the accessory olfactory system in the LH response of anestrous ewes to male odour. J Reprod Fertil 86:135–144
Cohen-Tannoudji J, Einhorn J, Signoret JP (1994) Ram sexual pheromone: first approach of chemical identification. Physiol Behav 56:955–961
Concannon PW (2011) Reproductive cycles of the domestic bitch. Anim Reprod Sci 124:200–210
De Amorim MTSP, Comel C, Vermande P (1982) Pyrolysis of polypropylene. J Anal Appl Pyrolysis 4:73–81
Delbarco-Trillo J, Burkert BA, Goodwin TE, Drea CM (2011) Night and day: the comparative study of strepsirrhine primates reveals socioecological and phylogenetic patterns in olfactory signals. J Evol Biol 24:82–98
DeMatteo KE, Porton IJ, Kleiman DG, Asa CS (2006) The effects of the male bush dog (*Speothos venaticus*) on the female reproductive cycle. J Mammal 87:723–732
Dietz JM (1985) *Chrysocyon brachyurus*. Mamm Species 234:1–4
Dodge JC, Kristal MB, Badura LL (2002) Male-induced estrus synchronization in the female Siberian hamster (*Phodopus sungorus sungorus*). Physiol Behav 77:227–231
Fan W, Xu Y, Zhang Y (2007) Characterization of pyrazines in some Chinese liquors and their approximate concentrations. J Agric Food Chem 55:9956–9962
Fluck AAJ, Mitchell W, Perry HM (1961) Composition of buchu leaf oil. J Sci Food Agric 12:290–292
Gelez H, Archer E, Chesneau D, Magallon T, Fabre-Nys C (2004) Inactivation of the olfactory amygdala prevents the endocrine response to male odour in anoestrous ewes. Eur J Neurosci 19:1581–1590
Gese EM, Ruff RL (1997) Scent-marking by coyotes, *Canis latrans*: the influence of social and ecological factors. Anim Behav 54:1155–1166
Goodwin TE, Rasmussen LEL, Schulte BA, Brown PA, Davis BL, Dill WM, Dowdy NC, Hicks AR, Morshedi RG, Mwanza D, Loizi H (2005) Chemical analysis of preovulatory female

African elephant urine: a search for putative pheromones. In: Mason B, Müller-Schwarze D (eds) Chemical signals in vertebrates 10. Springer, New York, pp 128–139

Goodwin TE, Songsasen N, Broederdorf LJ, Burkert BA, Chen CJ, Jackson SR, Keplinger B, Rountree ME, Waldrip ZJ, Weddell ME, Desrochers LP, Baker WK, Helsper JPFG (2012) Hemiterpenoids and pyrazines in the odoriferous urine of the maned wolf (*Chrysocyon brachyurus*). In: East ML, Dehnhard M (eds) Chemical signals in vertebrates 12. Springer, New York, pp 171–184

Gu Q, David F, Lynen F, Rumpel K, Dugardeyn J, van der Straeten D, Xu G, Sandra P (2011) Evaluation of automated sample preparation, retention time locked gas chromatography-mass spectrometry and data analysis methods for the metabolic study of *Arabidopsis* species. J Chromatogr A 1218:3247–3254

Guadayol JM, Caixach J, Ribé J, Cabañas J, Rivera J (1997) Extraction, separation and identification of volatile organic compounds from paprika oleoresin (Spanish type). J Agric Food Chem 45:1868–1872

Hagey L, MacDonald E (2003) Chemical cues identify gender and individuality in giant pandas (*Ailuropoda melanoleuca*). J Chem Ecol 29:1479–1488

Harris RL, Davies NW, Nicol SC (2012) Chemical composition of odorous secretions in the Tasmanian short-beaked echidna (*Tachyglossus aculeatus setosus*). Chem Senses 37:819–836

Hindle R, Noestheden M, Peru K, Headley J (2013) Quantitative analysis of naphthenic acids in water by liquid chromatography-accurate mass time-of-flight mass spectrometry. J Chromatogr A 1286:166–174

Hradecký P (1985) Possible pheromonal regulation of reproduction in wild carnivores. J Chem Ecol 11:241–250

IUCN RedList of Threatened Species (2015) International Union for Conservation of Nature, Gland, Switzerland. http://www.redlist.org. Accessed 25 Aug 2015

Iwata E, Wakabayashi Y, Kakuma Y, Kikusui T, Takeuchi Y, Mori Y (2000) Testosterone-dependent primer pheromone production in the sebaceous gland of male goat. Biol Reprod 62:806–810

Jemiolo B, Alberts J, Sochinski-Wiggins S, Harvey S, Novotny M (1985) Behavioral and endocrine responses of female mice to synthetic analogues of volatile compounds in male urine. Anim Behav 33:1114–1118

Jemiolo B, Harvey S, Novotny M (1986) Promotion of the Whitten effect in female mice by the synthetic analogs of male urinary constituents. Proc Natl Acad Sci U S A 83:4576–4579

Johns MA, Feder HH, Komisaruk BR, Mayer AD (1978) Urine-induced reflex ovulation in anovulatory rats may be a vomeronasal effect. Nature 272:446–448

Johnson AEM, Freeman EW, Colgin M, McDonough C, Songsasen N (2014) Induction of ovarian activity and ovulation in an induced ovulator, the maned wolf (*Chrysocyon brachyurus*), using GnRH agonist and recombinant LH. Theriogenology 82:71–79

Jordan NR, Golabek KA, Apps PJ, Gilfillan GD, McNutt JW (2013) Scent-mark identification and scent-marking behaviour in African wild dogs (*Lycaon pictus*). Ethology 119:644–652

Jorgenson JW, Novotny MV, Carmack M, Copland GB, Wilson SR, Katona S, Whitten WK (1978) Chemical scent constituents in the urine of the red fox (*Vulpes vulpes L.*) during the winter season. Science 199:796–798

Kean EF, Muller CT, Chadwick EA (2011) Otter scent signals age, sex, and reproductive status. Chem Senses 36:555–564

Kim K, Mall C, Taylor SL, Hitchcock S, Zhang C, Wettersten HI, Jones AD, Chapman A, Weiss RH (2014) Mealtime, temporal, and daily variability of the human urinary and plasma metabolomes in a tightly controlled environment. PLoS One 9:e86223

Kleiman DG (1972) Social behavior of the maned wolf (*Chrysocyon brachyurus*) and bush dog (*Speothos venaticus*): a study in contrast. J Mammal 53:791–806

Ku K-L, Lee R-S, Young CT, Chiou RY-Y (1998) Roasted peanut flavor and related compositional characteristics of peanut kernels of spring and fall crops grown in Taiwan. J Agric Food Chem 46:3220–3224

Liu X, Jin Q, Liu Y, Huang J, Wang X, Mao W, Wang S (2011) Changes in volatile compounds of peanut oil during the roasting process for production of aromatic roasted peanut oil. J Food Sci 76:C404–C412

Ma W, Miao Z, Novotny MV (1999) Induction of estrus in grouped female mice (*Mus domesticus*) by synthetic analogues of preputial gland constituents. Chem Senses 24:289–293

Maia OB, Gouveia AMG (2002) Birth and mortality of maned wolves *Chrysocyon brachyurus* (Illiger, 1811) in captivity. Braz J Biol 62:25–32

Marchlewska-Koj A, Pochron E, Sliwowska A (1990) Salivary glands and preputial glands of males as source of estrus-stimulating pheromone in female mice. J Chem Ecol 16:2817–2822

Marsden HM, Bronson FH (1964) Estrous synchrony in mice: alteration by exposure to male urine. Science 144:1469

Mártin J, Barja I, López P (2010) Chemical scent constituents in feces of wild Iberian wolves (*Canis lupus signatus*). Biochem Sys Ecol 38:1096–1102

Miller KV, Jemiolo B, Gassett JW, Jelinket I, Wiesler D, Novotny M (1998) Putative chemical signals from white-tailed deer (*Odocoileus virginianus*): social and seasonal effects on urinary volatile excretion in males. J Chem Ecol 24:673–683

Mills GA, Walker V (2000) Headspace solid-phase microextraction procedures for gas chromatographic analysis of biological fluids and materials. J Chromatogr A 902:267–287

Möller OM (1973) Progesterone concentrations in the peripheral plasma of the blue fox (*Alopex lagopus*) during pregnancy and the oestrous cycle. J Endocrinol 59:429–438

Mondain-Monval M, Dutourne B, Bonnin-Laffargue M, Canivenc R, Scholler R (1977) Ovarian activity during the anoestrus and the reproductive season of the red fox (*Vulpes vulpes L.*). J Steroid Biochem 8:761–769

Monfort SL, Wasser SK, Mashburn KL, Burke M, Brewer BA, Creel SR (1997) Steroid metabolism and validation of noninvasive endocrine monitoring in the African wild dog (*Lycaon pictus*). Zoo Biol 16:533–548

Morè L (2006) Mouse major urinary proteins trigger ovulation via the vomeronasal organ. Chem Senses 31:393–401

Murata K, Wakabayashi Y, Kitago M, Ohara H, Watanabe H, Tamogami S, Warita Y, Yamagishi K, Ichikawa M, Takeuchi Y, Okamura H, Mori Y (2009) Modulation of gonadotropin-releasing hormone pulse generator activity by the pheromone in small ruminants. J Neuroendocrinol 21:346–350

Murata K, Tamogami S, Itou M, Ohkubo Y, Wakabayashi Y, Watanabe H, Okamura H, Takeuchi Y, Mori Y (2014) Identification of an olfactory signal molecule that activates the central regulator of reproduction in goats. Curr Biol 24:681–686

Pal SK (2003) Urine marking by free-ranging dogs (*Canis familiaris*) in relation to sex, season, place and posture. Appl Anim Behav Sci 80:45–59

Parker MN (2010) Territoriality and scent marking behavior of African wild dogs in northern Botswana. Dissertation, The University of Montana

Poddar-Sarkar M, Brahmachary RL (2004) Putative chemical signals of leopard. Anim Biol 54:255–259

Radulović N, Đorđević A, Palić R, Ziatković B (2010) Essential oil composition of *Hypericum annulatum* Moris (Hypericaceae) from Serbia. J Essent Oil Res 22:619–624

Raymer J, Wiesler D, Novotny M, Asa C, Seal US, Mech LD (1986) Chemical scent constituents in urine of the wolf (*Canis lupus*) and their dependence on reproductive hormones. J Chem Ecol 12:297–314

Reiter LE (2012) Understanding poor reproduction in the maned wolf (*Chrysocyon brachyurus*) held *ex situ*. Thesis, George Mason University

Risticevic S, Pawliszyn J (2013) Solid-phase microextraction in targeted and nontargeted analysis: displacement and desorption effects. Anal Chem 85:8987–8995

Robledo N, Arzuffi R (2012) Influence of host fruit and conspecifics on the release of the sex pheromone by *Toxotrypana curvicauda* males (Diptera: Tephritidae). Environ Entomol 4:387D391

Rodden MD, Sorenson LG, Sherr A, Kleiman DG (1996) Use of behavioral measures to assess reproductive status in maned wolves (*Chrysocyon brachyurus*). Zoo Biol 15:565–585

Rodden MD, Baker B, Barbiers R, Bush M, Collette J, Goff D, Hammond B, Lindsey SL, McAloose D, Meeks K, Songsasen N, Sterns MJ (2007) Maned wolf husbandry manual. Maned Wolf Species Survival Plan

Saucier C, dos Polidoro AS, dos Santos AL, Schneider JK, Caramão EB, Jacques RA (2014) Comprehensive two-dimensional gas chromatography with mass spectrometry applied to the analysis of volatiles in artichoke (*Cynara scolymus* L.) leaves. Ind Crop Prod 62:507–514

Schaal B, Coureaud G, Langlois D, Giniès C, Sémon E, Perrier G (2003) Chemical and behavioural characterization of the rabbit mammary pheromone. Nature 424:68–72

Seal US, Plotka ED, Packard JM, Mech LD (1979) Endocrine correlates of reproduction in the wolf: serum progesterone, estradiol and LH during the estrous cycle. Biol Reprod 21:1057–1066

Service KM, Brereton RG, Harris S (2001) Analysis of badger urine volatiles using gas chromatography-mass spectrometry and pattern recognition techniques. Analyst 126:615–623

Sillero-Zubiri C, Hoffmann M, Macdonald DW (eds) (2004) Canids: foxes, wolves, jackals, and dogs: status survey and conservation action plan. IUCN/SSC Canid Specialist Group, Gland

Smith TE, Tomlinson AJ, Mlotiewicz JA, Abbot DH (2001) Female marmoset monkeys (*Callithrix jacchus*) can be identified from the chemical composition of their scent marks. Chem Senses 26:449–458

Smyth GK (2004) Linear models and empirical Bayes methods for assessing differential expression in microarray experiments. Stat Appl Genet Mol Biol 3, 3

Sohrabi M, Zhang L, Zhang K, Ahmetagic A, Wei MQ (2014) Volatile organic compounds as novel markers for the detection of bacterial infections. Clin Microbiol 3:151

Soini HA, Wiesler D, Koyama S, Féron C, Baudoin C, Novotny MV (2009) Comparison of urinary science of two related mouse species, *Mus spicilegus* and *Mus domesticus*. J Chem Ecol 35:580–589

Songsasen N, Rodden MD (2010) The role of the species survival plan in maned wolf *Chrysocyon brachyurus* conservation. Int Zoo Yearb 44:136–148

Songsasen N, Rodden M, Brown JL, Wildt DE (2006) Patterns of fecal gonadal hormone metabolites in the maned wolf (*Chrysocyon brachyurus*). Theriogenology 66:1743–1750

Songsasen N, Johnson AE, Reiter LE, Rodden MD (2014) Reproduction and factors affecting reproductive success in the maned wolf. In: Consorte-McCrea AG, Santos EF (eds) Ecology and conservation of the maned wolf: multidisciplinary perspectives. CRC Press, Boca Raton

Stojanović-Radić Z, Čomić L, Radulović N, Dekić M, Ranđelović V, Stefanović O (2010) Chemical composition and antimicrobial activity of *Erodium* species: *E. ciconium* L., *E. cicutarium* L., and *E. absinthoides* wild (Geraniaceae). Chem Pap 64:368–377

Tirillini B, Verdelli G, Paolocci F, Cicciloi P, Frattoni M (2000) The volatile organic compounds from the mycelium of *Tuber borchii* Vitt. Phytochemistry 55:983–985

Toftegaards CL, Moore C, Bradley AJ (1999) Chemical characterization of urinary pheromones in brown antechinus, *Antechinus stuartii*. J Chem Ecol 25:527–535

Valdespino C (2007) Physiological constraints and latitudinal breeding season in the Canidae. Physiol Biochem Zool 80:580–591

van den Berg RA, Hoefsloot HCJ, Westerhuis JA, Smilde AK, van der Werf MJ (2006) Centering, scaling and transformations: improving the biological information content of metabolomics data. BMC Genomics 7:142

Walker SL, Waddell WT, Goodrowe KL (2002) Reproductive endocrine patterns in captive female and male red wolves (*Canis rufus*) assessed by fecal and serum hormone analysis. Zoo Biol 21:321–335

West PR, Weir AM, Smith AM, Donley ELR, Cezar GG (2010) Predicting human developmental toxicity of pharmaceuticals using human embryonic stem cells and metabolomics. Toxicol Appl Pharmacol 247:18–27

Whitten WK, Bronson FH, Greenstein JA (1968) Estrus-inducing pheromone of male mice: transport by movement of air. Science 161:584–585

Whittle CL, Bowyer RT, Clausen TP, Duffy LK (2000) Putative pheromones of rutting male moose (*Alces alces*): evolution of honest advertisement? J Chem Ecol 26:2747–2762

Wirant SC, Halvorsen KT, McGuire B (2007) Preliminary observations on the urinary behaviour of female Jack Russell Terriers in relation to stage of the oestrous cycle, location, and age. Appl Anim Behav Sci 106:161–166

Wold S, Esbensen K, Geladi P (1987) Principal component analysis. Chemom Intell Lab 2:37–52

Wood WF, Terwillger MN, Copeland JP (2005) Volatile compounds from anal glands of the wolverine, *Gulo gulo*. J Chem Ecol 31:2111–2117

Wood WF, Copeland JP, Yates RE, Horsey IK, McGreevy LR (2009) Potential semiochemicals in urine from free-ranging wolverines (*Gulo gulo* Pallas, 1780). Biochem Syst Ecol 37:574–578

Wyatt T (2014) Pheromones and animal behavior: communication by smell and taste, 2nd edn. Cambridge University Press, Cambridge

Zhang J-X, Soini HA, Bruce KE, Wiesler D, Woodley SK, Baum MJ, Novotny MV (2005) Putative chemosignals of the ferret (*Mustela furo*) associated with individual and gender recognition. Chem Senses 30:727–737

Zhang J-X, Sun L, Zhang J-H, Feng ZY (2008) Sex- and gonad-affecting scent compounds and 3 male pheromones in the rat. Chem Senses 33:611–621

Chapter 32
Learning and Applications of Chemical Signals in Vertebrates for Human–Wildlife Conflict Mitigation

Bruce A. Schulte

32.1 Human–Wildlife Conflict

Human–wildlife conflict (HWC) encompasses all negative interactions between humans and wildlife (Messmer 2000). Conflict between humans and wildlife is economically costly and ubiquitous globally (Woodroffe et al. 2005). HWC is a density-dependent phenomenon that takes on two major forms for humans, namely property damage and personal safety (Madden 2004) with costs that can extend well beyond the specific conflict (Barua et al. 2013). When considered broadly, wildlife can be any form of animal life not domesticated by humans. Animals may consume the food crops cultivated by humans and their actions can be destructive to human machinations. HWC is estimated to be the leading source of both crop damage and, in the form of competition for land use, biodiversity loss globally (Yudelman et al. 1998; Hanski 2005). Wildlife also may attack humans directly (Treves and Karanth 2003) and be vectors of attack by disease-causing organisms (Joseph et al. 2013). For most of human evolution, people competed with wildlife for food via hunting and gathering, and served as prey for some carnivorous predators. Once humans became more sedentary as a result of plant cultivation and animal domestication, wildlife could impact property as well as personal safety.

For millennia, human populations were low relative to the available landscape (Klein Goldewijk et al. 2010), so animal intrusions into human developments were likely relatively rare. However, post the industrial revolution, human population growth has become exponential, and by sheer probability alone, the likelihood of HWC has increased in a density-dependent manner. Among many, three factors of relevance herein have acerbated the conflict. One, humans have occupied prime land for habitation and cultivation, driving species out of critical habitat.

B.A. Schulte (✉)
Department of Biology, Western Kentucky University,
Bowling Green, KY 42101-1080, USA
e-mail: bruce.schulte@wku.edu

B.A. Schulte et al. (eds.), *Chemical Signals in Vertebrates 13*,
DOI 10.1007/978-3-319-22026-0_32

Two, humans have grown plants in unnaturally high, monoculture densities, selecting for said plants to be highly nutritious while also selecting against their natural defenses. Three, humans have raised livestock and pets in the same habitat where the prey, medium- to large-sized herbivores, of sympatric predators once foraged. In essence, humans have established unprotected candy shops in the midst of a children's playground. This has led to some of the most economically detrimental forms of HWC, crop, and livestock depredation with potential associated threats for which social and biological factors have to be considered when developing mitigation schemes (Dickman 2010).

32.2 Population Level Management of HWC

Historically, the management approach to ameliorating HWC was largely at the level of the population, although killing particular problem animals has always been one attempt at a solution. One common management strategy targets the population size reduction of the problem species directly or indirectly. Direct population reduction occurs through such practices as hunting, lethal traps, or poisoning (Treves and Naughton-Treves 2005). These practices have a variety of potentially negative side effects including unforeseen trophic cascades, non-target species mortality, and bioaccumulation of toxins in the environment that can impact human health. For large-scale agricultural systems, insect pests comprise the most widespread and deleterious source of crop losses (Zalucki et al. 2012). Yet, global food security may be more dependent on smallholder farmers, and such systems are intimately associated with biodiversity conservation (Tscharntke et al. 2012). The loss of biodiversity in such systems can result in an increase in pest prevalence and concomitant economic and social strife. Traditional, large-scale approaches to pest control, such as widespread use of pesticides, are not viable alternatives for smallholder farmers. Therefore, an approach to pest management that is more broadly applicable, environmentally safe, effective, and efficient has been sought.

A second method to reduce population size is to target future growth through reproductive control (Fagerstone et al. 2010). Rather than reducing current population size in the short term, this process shifts the intrinsic rate of increase such that deaths exceed births and the population size decreases over time. Numerous problems, such as long testing periods, lack of investment in this often expensive process, and sometimes social resistance to the option, have reduced the widespread implementation of this practice (Rutberg 2013). A third method of population control is to translocate animals out of the region if suitable habitat is available in which to place them. The success of this method is often measured more by the reduction in population size and conflict at the source and not the survivorship and viability of the translocated animals (but see Whisson et al. 2012). If the moved animals do not survive well, then this becomes a delayed form of lethal population control. Increasingly, this method meets resistance for the reason that with time, it simply moves the problem elsewhere or delays the conflict until the moved animals return or the population increases again (Linnell et al. 1997; Conover 2002; Shivik et al. 2011).

Beyond reducing effective population numbers, another strategy to control animal populations involves the exclusion of the problem species through permanent obstructions such as fences (Ferguson and Hanks 2010). While such obstructions might work for animals such as medium to large vertebrates, they are less effective for small vertebrates (e.g., mice and rats) or invertebrates. Fences can be used in one of two ways—fencing around the human area (crops or livestock) or fencing around the wildlife area (park or refuge). In either case, fences can be expensive and require regular maintenance. They also are not specific to the problem species; therefore, all species that cannot get over, through or under the fence must deal with its presence. Thus, individuals involved in trying to reduce HWC, and specifically crop depredation, continue to search for means to keep problem animals out with minimal economic cost and continuous human labor, while also maintaining a viable ecosystem for animals that do not crop raid (Woodroffe et al. 2005).

The discovery of chemical signals that mediate intra- and inter-specific interactions (semiochemicals) led to the application of such chemicals for controlling the population size, movement, and distribution of pest species (Hokkanen 1991). Plants whose semiochemicals attract particular insect pests can be grown around crop fields (Reddy and Guerrero 2004). Intraspecific chemical signals, or pheromones, have been placed within traps to disrupt mating or reduce population sizes around crop fields (El-Sayed et al. 2006; Witzgall et al. 2010). An initial limitation to this practice was the lack of knowledge on the specific chemicals that served as chemical signals. Over time, the work of chemical ecologists and their collaborators have led to the identification of pheromones that could be used in traps or as lures. Major advantages of using pheromones over other chemicals are their lack of toxicity in the environment, low dosages, and their specificity, avoiding deleterious consequences for non-target species (Witzgall et al. 2010). However, the use of pheromones is not without its disadvantages. The same species may show geographic variation in its signals, which could affect the success of artificial chemicals used to modify pest behavior (Unbehend et al. 2014). In addition, because of the often rapid generation time of invertebrate pests, an evolutionary problem could occur by traps and lures selecting against the pheromone, leading to the modification of the natural signal through directional selection (Dumenil et al. 2014). In nature, such pheromones are adaptive by attracting one sex to the other or by aggregating individuals to a food source. No such benefits accrue to the individuals when responding to a pheromone used in a trap. This leads to the classic "arm-race" or Red Queen problem so well studied with predator–prey (parasite–host) interactions (Van Valen 1973; Liow et al. 2011). That is, individuals that can distinguish between the compound(s) being released by a conspecific versus a trap will do better and thus reduce the effectiveness of the trap. In addition, the extreme reduction of one invertebrate pest could create an opportunity for other species in that feeding guild to attack the crop. Because pheromones are species-specific, the trap or lure would only work for a single problem species.

Chemical signals also can be used to attract vertebrates. For mammalian carnivores, meat is often used but food is often not a viable signal for herbivores. In either case, food has the problem of spoiling; whereas, prey or conspecific chemical signals may retain their effectiveness for a longer duration (Spurr et al. 2004; Martin et al. 2014). While the use of artificially created chemical attractants is progressing for invertebrate pests (e.g., Bray et al. 2014; Higbee et al. 2014), much less progress has been achieved for vertebrate pests. The use of artificial lures for vertebrates may be limited by the paucity of known chemical signals with definitive attractant functions (Wyatt 2014).

Natural chemical signals also could function as an odor or scent fence (Hediger 1949). In nature, territorial scent marks are likely not the sole means of repelling potential intruders, but they can reduce the probability of intrusion (Gosling 1982, 1990). Pheromones such as territorial markers or perhaps more effectively, alarm signals, might be enlisted to reduce HWC at crops or livestock holdings. For insects, one problem with this approach can be the volatility of the alarm compounds. The challenge is determining means to release the compound only in the presence of an invading pest, have the pest go in the appropriate direction (i.e., away from the crop), and to replenish the supply of the chemical in a timely and cost-effective manner or have the chemicals last throughout the problem period (Bray et al. 2014). In this regard, territorial markers for vertebrate species might be useful. For example, many territorial mammals create scent marks around the perimeter of their territory, updating the marks periodically (Peters and Mech 1975). The scent marks can remain effective for many days, making their application for HWC a possibility. One such attempt was tried with North American beavers (*Castor canadensis*). Welsh and Müller-Schwarze (1989) placed beaver scent around viable but currently unoccupied sites to see if beavers would avoid settling in these regions. The scent marking fence worked but only when the beaver population in the region was low; that is, the scent fence was not a sufficient defense when the need for open habitat by beavers was at a premium. A more recent study with grey wolves (*Canis lupus*) used human placed scents with good success as a biofence, although the human effort involved was considerable (Ausband et al. 2013). Jackson et al. (2012) showed that conspecific scents could be used to alter the movements of African wild dogs (*Lycaon pictus*), but Anhalt et al. (2014) indicated that a simulated wolf pack using scents and sounds did not affect real wolf pack movements. It seems most likely that animal scent marks can serve to delineate territories but they do not by themselves repel all intruders (Gosling and Roberts 2001; Roberts 2012).

Thus, there are limitations with trying to draw animals of a problem species away from a target area such as a crop field and with fencing them out through either physical or chemical means. Yet, in tandem, these mechanisms may be more effective. For this approach, the local population size of the pest species is reduced but not eliminated and their avoidance of an area (e.g., crop field) is reinforced through reward. This is modification of a similar approach used for invertebrate pests in which chemical deterrents are coupled with chemical attractants or lures in a process known as a push–pull approach (Cook et al. 2007).

A sound understanding about the biology of the pest species forms the foundation of a push–pull strategy for integrated pest management (Cook et al. 2007). With push–pull, aversive stimuli, such as natural chemical signals, repel insects from crops and attractive odors lure them into another, viable area. Natural chemicals are appealing signals for push–pull because of their potential for slow, long-lasting release and high specificity without adverse side effects. Additionally, natural associations with biologically relevant chemical signals are likely to be better suited for push–pull management because behavioral responses are strengthened by generations of selective pressure. Several recent studies have shown that this approach is effective in repelling invertebrate pests from valuable crops toward trap plants (Khan et al. 2007, 2011; Hassanali et al. 2008). The push–pull strategy can be applied to vertebrate pests, although our understanding of chemical communication in vertebrates remains a limitation in using this approach. The need to test this strategy with vertebrate pests should provide even more impetus for the identification and study of vertebrate semiochemicals. However, in the meantime, the need to reduce HWC and specifically crop raiding by vertebrate species is immediate and imperative. Thus, other avenues to mitigate HWC are needed.

32.3 Individual Level Management of HWC

One such possibility exists by considering the identity of the problem animals, which is not a reasonable option when dealing with the far more numerous invertebrate pests. In a number of instances, the conflict with vertebrate wildlife species is not a population level issue but an individual issue. When a problem individual or group is identified, killing has been a common response, and translocation to a less human occupied landscape is another possibility (Treves and Karanth 2003; Goodrich and Miquelle 2005). Lethal control is often an undesired option (Treves et al. 2006). Several scenarios may preclude translocation from being a viable solution. No unoccupied habitat is available, such that the translocated animal is destined to die in the new habitat or leave it, and quite possibly return to the original region. By taking the animal away, the habitat is now open and another individual may occupy it, resulting in the same problem with a new occupant. Thus, it may be ideal to retain the individual in the region but reduce its impact. If the impact results from progeny, then contraception might be an option. However, if the occupant is itself the issue, then an understanding at the individual level might provide insight on solutions (e.g., Blumstein and Fernandez-Juricic 2010).

The field of conservation behavior strives to assist in the maintenance of biodiversity through an application of behavioral principles. When an HWC incident can be attributed to repeated offenses by one or a group of individuals, such as a predator attacking livestock or an herbivore raiding a crop field, then determining the reasons why this individual has become a problem can be useful. For instance, an animal that is injured or diseased may resort to potentially easier prey than hunting wildlife or foraging for native vegetation. In other cases, aspects of the behavioral repertoire or status may make the individual prone to HWC. A dominant individual or one that has disperse a long distance may be a high risk taker and thus human provisioned food sources may be included in its search image. Identifying the characteristics of animals that cause HWC could yield meaningful profiles, providing predictive power in when HWC will occur and by whom, and thereby facilitate preventive measures.

Réale et al. (2007) described five categories of personality traits (Table 32.1) that could be used to create such a profile. These personality axes are akin to the big five personality factors associated with humans (Table 32.1) (Digman 1990). An interesting parallel to these personality factors exists with the dimensions of feelings evoked by odors in humans (Table 32.1) (Chrea et al. 2009). These dimensions suggest that in humans odors might tap into basic personality characteristics. This is not unreasonable given that odors assist in the formation of deep and lasting memories with strong emotional components. In terms of HWC abatement, the idea would be to use odors to recall or form and then recall these strong memories. If the recall of memories leads to a change in behavior from what would have occurred otherwise, then the animal has displayed learning. Interestingly, learning can thwart the reduction of HWC. Through habituation, association, social, or other forms of learning, an animal ignores or overcomes the human-created disruption or obstacle used to keep the animal away from the livestock or crops (Shivik et al. 2003). Thus, birds learn to disregard the brightly colored scarecrow, predators stop avoiding motion-activated lights in the absence of other negative reinforcement, herbivores learn that the sounds of dogs barking without the presence of real dogs can go unheeded, and a variety of animals may figure out how to overcome or avoid a fence or other such impediment. However, other repellents base their functionality on learning by associating an unwanted behavior to an aversive stimulus or associating it to a cue of the impending aversive stimulus (Shivik et al. 2003; Schakner and Blumstein 2013).

The question posed herein is whether problem animals can be trained to avoid human-dominated areas and have this learning maintained through the judicious use of memory-evoking odors as has been studied in humans (Schab 1990). Research with humans has shown that odors can affect the learning process. Aversive learning can result in lower threshold detection of a compound (Ahs et al. 2013). Humans also can learn to discriminate between enantiomers when one is paired with an aversive stimulus; whereas, the enantiomers are unable to be discriminated in the absence of said stimulus (Li et al. 2008). This ability also can occur through pairing with positive reinforcement such as chocolate (Pool et al. 2014). In children (Chu 2008) and mice (Jones et al. 2005), an ambient odor leads to better performance. An odor can even affect an individual's attention. Seigneuric et al. (2010) demonstrated that humans shown a collage of objects focused more readily on the object for which they had been exposed previously to its odor (e.g., exposed to scent of an orange, then shown a collage of pictures of which one was an orange). Thus, odors have a cross-modal influence in sensory processing. In everyday life, humans rarely use the memory of odors to conduct tasks (Köster et al. 2014). Rather, familiar odors provide a sense of comfort or security and unusual odors can evoke a response but more to seek the source or make an association than to identify the odor itself.

Table 32.1 Comparison of the animal personality axes and in humans, the big five personality factors and the odor-evoked feelings

Animal personality[a]	Human big five[b]	Odor evoked feelings[c]
Activity	Openness to experience	Arousal
Exploration	Conscientiousness	Conscious recollection of emotional memories
Boldness	Extraversion	Well-being
Sociability	Agreeableness	Social interaction
Aggressiveness	Neuroticism	Danger prevention

Each factor or axis has its counterpart at the other end of the scale (e.g., bold-shy; exploration-avoidance)
[a]Réale et al. (2007)
[b]Digman (1990)
[c]Chrea et al. (2009)

32.4 Using Learning and Odors to Ameliorate HWC

To date, the reduction of HWC at the individual level has emphasized using aversive or fear-inducing stimuli, and often ones that have an evolutionary-based meaning (Schakner and Blumstein 2013). The benefit of this approach is that the signals are reinforced in nature. That is, predators, distasteful prey, and humans that might harm the animals are real threats for the animals that are causing conflict. However, several problems also exist. In some cases, the natural odor may only be relevant to a subset of animals. For example, a territorial marker may be intended for one sex, such that opposite sexed animals would not feel threatened, or it may only be functional at certain times of year if territoriality in the species is not year round. Another dilemma that arises is similar to a Batesian mimicry situation. In this form of mimicry a palatable model looks, smells, acts, or otherwise mimics an unpalatable model. Predators learn to avoid the model and thereby also avoid the mimic. For this system to be maintained, the model must be repulsive without killing the consumer. Furthermore, the relative frequency of models must exceed mimics; otherwise, the consumer is successful (consumes palatable prey) more often that it is not (attempts to eat unpalatable prey). In human-dominated landscapes in which defenseless crops or livestock are present, protecting these resources via auditory, visual, tactile, gustatory, or olfactory defenses that mimic a real threat can quickly lead to mimics greatly outnumbering models. Thus, the consumer that ignores the potential threat and raids the crops or attacks the livestock does better than the consumer that leaves the area hungry. One solution without leaving crops or livestock unprotected is to give the raiding animal aversive reinforcement to stay away without killing it (because then no learning can occur). Such aversion can be given repeatedly but this is often expensive and logistically difficult (Baruch-Mordo et al. 2011). Alternatively, the aversion can be severe enough that it creates a deep impression so that the consumer does not return regularly, doing damage each time. Unfortunately, in many situations the learning is not deeply rooted and consumers return repeatedly and often.

For example, taste aversion studies in terrestrial predators often indicate initial success but decreased effectiveness with time (Shivik 2006). Furthermore, highly aversive stimuli may lead with repeated exposure to animal welfare issues, be expensive, or have the potential to be lethal, which again prevents further learning. The challenge is to establish the learning and then prevent its decay in a cost-effective, humane, and non-lethal manner.

Schakner and Blumstein (2013) suggest associative learning that induces not just immediate fear (short-term flight away from stimulus) but anxiety (long-term staying away) will be an effective deterrent protocol. Cues that predict the fear (flight) behavior create anxiety, and it is this anxiety that promotes avoidance. For many marine mammals, sound may be the pertinent cue (Schakner and Blumstein 2013), but for terrestrial vertebrates that come into conflict with humans, odors may be more salient. Naturally occurring odors could be useful if they are always associated with a potentially negative interaction. In this case, using such odors as repellents would be akin to a Müllerian mimicry scenario in which only models exist. One example might be the current practice of using bee sounds and bee nest strung along flexible fences to repel African elephants (*Loxodonta africana*) from crop fields (King et al. 2007, 2011). Elephants learn about the negative consequences of disturbing bee hives while foraging on trees (Vollrath and Douglas-Hamilton 2002). The odor of bees could be coupled with their sounds and presence to reinforce further the deterrence signal. Further reinforcement is provided by associating the naturally aversive odor with the presence of humans and their actions to repel elephants (Webber et al. 2007; Hoare 2012).

Where a natural repellent is not present or feasible, an odor would be selected to couple with the aversion to the targeted species. A key feature of this approach is the use of odors that are affordable, do not have an already defined meaning to the animal, and are not present or at least not common in its environment. Thus, the selected odor for a specific species and problem can be associated with a fear-inducing stimulus through conditioning. Once the association is established, the odor can be used as a scent fence when needed. It would be important not to saturate the environment with this odor as the animal must be able to function normally in most of its habitat and only be stimulated into anxiety when it approaches a problem area (e.g., near livestock or crops). An odor that evokes the powerful memory of the fear-inducing experience will suffice to keep the animal away from the area without the animal experiencing fear. In social animals, it could be possible that such associations would be transferred to social group members (Olsson and Phelps 2007).

32.5 Conclusion

The reduction of HWC has invoked various forms of obstacles to deter animals from human areas and different types of operant conditioning in which a negative stimulus is supplied to animals that enter human areas. The pairing of odors with an aversive stimulus in a form of classical conditioning could lead to the formation of deep-rooted, emotive memories that create anxiety, resulting in animals avoiding the odor field. Several challenges exist for this idea. First, it must be determined that such an association can be created and that it is indeed maintained. Second, the logistics must be confronted of what odor and how to deploy it not only during the training period but thereafter for the animals that are causing problems. If these animals are long-lived, recall their learning when exposed to the odor, and can transfer this learning to conspecifics, then an efficient process has been created that reduces HWC and also helps to maintain local biodiversity.

References

Ahs F, Miller SS, Gordon AR, Lundström JN (2013) Aversive learning increases sensory detection sensitivity. Biol Psychol 92:135–141

Anhalt CM, Van Deelen TR, Schultz RN, Wydeven AP (2014) Effectiveness of a simulated pack to manipulate wolf movements. Hum-Wildl Interact 8:210–217

Ausband DE, Mitchell MS, Bassing SB, White C (2013) No trespassing: using a biofence to manipulate wolf movements. Wildl Res 40:207–216

Barua M, Bhagwat SA, Jadhav S (2013) The hidden dimensions of human–wildlife conflict: health impacts, opportunity and transaction costs. Biol Conserv 157:309–316

Baruch-Mordo S, Breck SW, Wilson KR, Broderick J (2011) The carrot or the stick? Evaluation of education and enforcement as management tools for human–wildlife conflicts. PLoS One 6, e15681

Blumstein DT, Fernandez-Juricic E (2010) A primer of conservation behavior. Sinauer Associates, Sunderland

Bray DP, Carter V, Alves GB, Brazil RP, Bandi KK, Hamilton JGC (2014) Synthetic sex pheromone in a long-lasting lure attracts the visceral leishmaniasis vector, *Luzaomyia longipalpis*, for up to 12 weeks in Brazil. PLoS One 8(3), e2723. doi:10.1371/journal.pntd.0002723

Chrea C, Grandjean D, Delplanque S, Cayeux I, Le Calvé B, Aymard L, Velazco MI, Sander D, Schere KR (2009) Mapping the semantic space for the subjective experience of emotional responses to odors. Chem Senses 34:49–62

Chu S (2008) Olfactory conditioning of positive performance in humans. Chem Senses 33:65–71

Conover ME (2002) Resolving human–wildlife conflicts: the science of wildlife damage management. CRC Press, Boca Raton

Cook SM, Kahn ZR, Pickett JA (2007) The use of push-pull strategies in integrated pest management. Annu Rev Entomol 52:375–400

Dickman AJ (2010) Complexities of conflict: the importance of considering social factors for effectively resolving human–wildlife conflict. Anim Conserv 13:458–466

Digman JM (1990) Personality structure: emergence of the five-factor model. Annu Rev Psychol 41:417–440

Dumenil C, Judd GJR, Bosch D, Baldessari M, Gemeno C, Grott AT (2014) Intraspecific variation in female sex pheromone of the codling moth *Cydia pomonella*. Insects 5:705–721

El-Sayed AM, Suckling DM, Wearing CH, Byers JA (2006) Potential of mass trapping for long-term pest management and eradication of invasive species. J Econ Entomol 99:1550–1564

Fagerstone KA, Miller LA, Killian G, Yoder CA (2010) Review of issues concerning the use of reproductive inhibitors, with particular emphasis on resolving human-wildlife conflict in North America. Integr Zool 1:15–30

Ferguson K, Hanks J (eds) (2010) Fencing impacts: a review of the environmental, social and economic impacts of game and veterinary fencing in Africa with particular reference to the

Great Limpopo and Kavango-Zambezi transfrontier conservation areas. Mammal Research Institute, Pretoria, Available at: www.wcs-ahead.org/gltfca_grants/grants.html
Goodrich JM, Miquelle DG (2005) Translocation of problem Amur tigers *Panthera tigris altaica* to alleviate tiger-human conflicts. Oryx 39:1–4. doi:10.1017/S0030605305001146
Gosling LM (1982) A reassessment of the function of scent marking in territories. Z Tierpsychol 60:89–118
Gosling LM (1990) Scent-marking by resource holders: alternative mechanisms for advertising the costs of competition. In: Macdonald DW, Muller-Schwarze D, Natynczuk SE (eds) Chemical signals in vertebrates, V. Oxford University Press, Oxford, pp 315–328
Gosling LM, Roberts SC (2001) Scent-marking by male mammals: cheat-proof signals to competitors and mates. Adv Study Behav 30:169–217
Hanski I (2005) Landscape fragmentation, biodiversity loss and the societal response. EMBO Rep 6:388–392
Hassanali A, Herren H, Khan ZR, Pickett JA, Woodcock CM (2008) Integrated pest management: the push-pull approach for controlling insect pests and weeds of cereals, and its potential for other agricultural systems including animal husbandry. Philos Trans R Soc B 363:611–621
Hediger H (1949) Saugetier – Territorien und ihre Markierung. Bijdr Tot de Dierk 28:172–184
Higbee BS, Burks CS, Larsen TE (2014) Demonstration and characterization of a persistent pheromone lure for the navel orangeworm, *Amyelois transitella* (Lepidoptera: Pyralidae). Insects 5:596–608
Hoare R (2012) Lessons from 15 years of human–elephant conflict mitigation: management considerations involving biological, physical and governance issues in Africa. Pachyderm 51:60–74
Hokkanen HMT (1991) Trap cropping in pest management. Annu Rev Entomol 36:119–138
Jackson CR, McNutt JW, Apps JJ (2012) Managing the ranging behavior of African wild dogs (*Lycaon pictus*) using translocated scent parks. Wildl Res 39:31–34
Jones SV, Heldt SA, Davis M, Ressler KJ (2005) Olfactory-mediate fear conditioning in mice: simultaneous measurements of fear-potentiated startle and freezing. Behav Neurosci 119:329–335
Joseph MB, Mihaljevic JR, Arellan AL, Kueneman JG, Preston DL, Cross PC, Johnson PTJ (2013) Taming wildlife disease: bridging the gap between science and management. J Appl Ecol 50:702–712
Khan ZR, Midega CAO, Wadhams LJ, Pickett JA, Mumuni A (2007) Evaluation of Napier grass (*Pennisetum purpureum*) varieties for use as trap plants for the management of African stemborer (*Busseola fusca*) in a push-pull strategy. Entomol Exp Appl 124:201–211
Khan Z, Midega C, Pittchar J, Pickett J, Bruce T (2011) Push-pull technology: a conservation agriculture approach for integrated management of insect pests, weeds and soil health in Africa. Int J Agric Sustain 9:162–170
King LE, Douglas-Hamilton I, Vollrath F (2007) African elephants run from the sound of disturbed bees. Curr Biol 17:R832–R833
King LE, Douglas-Hamilton I, Vollrath F (2011) Beehive fences as effective deterrents for crop-raiding elephants: field trials in northern Kenya. Afr J Ecol 49:431–439
Klein Goldewijk K, Beusen A, Janssen P (2010) Long term dynamic modeling of global population and built-up area in a spatially explicit way: HYDE 3.1. The Holocene 20:565–573
Köster EP, Møller P, Moject J (2014) A "misfit" theory of spontaneous conscious odor perception (MITSCOP): reflections on the role and function of odor memory in everyday life. Front Psychol 5:64. doi:10.3389/fpsyg.2014.00064
Li W, Howard JD, Parrish TB, Gottfried JA (2008) Aversive learning enhances perceptual and cortical discrimination of indiscriminable odor cues. Science 319:1842–1845
Linnell JDC, Aanes R, Swenson JE, Odden J, Smith ME (1997) Translocation of carnivores as a method for managing problem animals: a review. Biodiv Conserv 6:1245–1257
Liow LH, Van Valen L, Stenseth NC (2011) Red queen: from populations to taxa and communities. TREE 26:349–358

Madden F (2004) Creating coexistence between humans and wildlife: global perspectives on local efforts to address human–wildlife conflict. Hum Dimen Wildl 9:247–257

Martin JA, Marshall C, Belant JL, Cagle S, West BC (2014) New live-trapping method improves capture rates for nine-banded armadillos. Wildl Biol Pract 10:69–74

Messmer TA (2000) The emergence of human-wildlife conflict management: turning challenges into opportunities. Int Biodeterior Biodegrad 45:97–102

Olsson A, Phelps EA (2007) Social learning of fear. Nature Neurosci 10(9). doi:10.1038/nn1968

Peters RP, Mech LD (1975) Scent-marking in wolves. Am Sci 63:628–637

Pool E, Delplanque S, Porcherot C, Jenkins T, Cayeux I, Sander D (2014) Sweet reward increases implicit discrimination of similar odors. Front Behav Neurosci 8. doi:10.3389/fnbeh.2014.00158

Réale D, Reader SM, Sol D, McDougall PT, Dingemanse NJ (2007) Integrating animal temperament within ecology and evolution. Biol Rev 82:291–318

Reddy GVP, Guerrero A (2004) Interactions of insect pheromones and plant semiochemicals. Trends Plant Sci 9:254–261

Roberts SC (2012) On the relationship between scent-marking and territoriality in callitrichid primates. Int J Primatol 33:749–761

Rutberg AT (2013) Managing wildlife with contraception: why is it taking so long? J Zoo Wildl Med 44:S38–S46

Schab F (1990) Odors and the remembrance of things past. J Exp Psychol Learn Mem Cogn 16:648–655

Schakner ZA, Blumstein DT (2013) Behavioral biology of marine mammal deterrents: a review and prospectus. Biol Conserv 167:380–389

Seigneuric A, Durand K, Jiang T, Baudouin J-Y, Schaal B (2010) The nose tells it to the eyes: crossmodal associations between olfaction and vision. Perception 39:1541–1554

Shivik JA (2006) Tools for the edge: what's new for conserving carnivores? Bioscience 56: 253–259

Shivik JA, Treves A, Callahan P (2003) Nonlethal techniques for managing predation: primary and secondary repellents. Conserv Biol 17:1531–1537

Shivik JA, Ruid D, Willging RC, Mock KE (2011) Are the same bears repeatedly translocated from corn crops in Wisconsin? Ursus 22:114–119

Spurr EB, Ragg JR, O'Connor CE, Hamilton WJ, Moller H, Woolhouse AD, Morse CW, Morriss GA, Arnold GC, Clapperton BK (2004) Effect of concentration of anal gland scent lures on the capture rate of ferrets (*Mustela furo*) in winter and spring. N Z J Zool 31:227–232

Treves A, Karanth UK (2003) Human-carnivore conflict and perspectives on carnivore management worldwide. Conserv Biol 17:1491–1499

Treves A, Naughton-Treves L (2005) Evaluating lethal control in the management of human-wildlife conflict. In: Woodroffe R, Thirgood S, Rabinowitz A (eds) People and wildlife: conflict or coexistence? Cambridge University Press, Cambridge, pp 86–106

Treves A, Wallace RB, Naughton-Treves L, Morales A (2006) Co-managing human-wildlife conflicts: a review. Hum Dimen Wildl 11:383–396

Tscharntke T, Clough Y, Wanger TC, Jackson L, Motzke I, Perfecto I, Vandermeer J, Whitbread A (2012) Global food security, biodiversity conservation and the future of agricultural intensification. Biol Conserv 151:53–59

Unbehend M, Hänniger S, Vásquez GM, Juárez ML, Reisig D, McNeil JN, Meagher RL, Jenkins DA, Heckel DG, Groot AT (2014) Geographic variation in sexual attraction of *Spodoptera frugiperda* corn- and rice-strain males to pheromone lures. PLoS One 9(2), e89255. doi:10.1371/journal.pone.0089255

Van Valen L (1973) A new evolutionary law. Evol Theory 1:1–30

Vollrath F, Douglas-Hamilton I (2002) African bees to control African elephants. Naturwissenschaften 89:508–511

Webber AD, Hill CM, Reynolds V (2007) Assessing the failure of a community-based human-wildlife conflict mitigation project in Budongo Forest Reserve, Uganda. Oryx 41:177–184

Welsh RG, Müller-Schwarze D (1989) Experimental habitat scenting inhibits colonization by beaver, *Castor canadensis*. J Chem Ecol 15:887–893

Whisson DA, Holland GJ, Carlyon K (2012) Translocation of overabundant species: Implications for translocated individuals. J Wildl Manage 76:1661–1669

Witzgall P, Kirsch P, Cork A (2010) Sex pheromones and their impact on pest management. J Chem Ecol 36:80–100

Woodroffe R, Thirgood S, Rabinowitz A (eds) (2005) People and wildlife: conflict or coexistence? Cambridge University Press, Cambridge

Wyatt TD (2014) Pheromones and animal behavior: chemical signals and signatures, 2nd edn. Cambridge University Press, Cambridge

Yudelman M, Ratta A, Nygaard D (1998) Pest management and food production: looking to the future. Food, agriculture, and the environment discussion paper 25. International Food Policy Research Institute, Washington, DC

Zalucki MP, Shabbir A, Silva R, Adamson D, Shu-Sheng L, Furlong MJ (2012) Estimating the economic cost of one of the world's major insect pests, *Plutella xylostella* (Lepidoptera: Plutellidae): just how long is a piece of string? BioOne 105:1115–1129

Chapter 33
An Experimental Test of the Effect of Diet on Preen Wax Composition in New Zealand Silvereyes (*Zosterops lateralis*)

Laura Azzani, Justin L. Rasmussen, Steven P. Gieseg, and James V. Briskie

33.1 Introduction

Research in avian communication has increasingly focused on "odor signatures," defined as olfactory or chemical cues that enable individual recognition (see Bonadonna and Mardon 2013; Caro et al. 2015 for a review). Odor signatures, and olfactory signals in general, could function in territorial marking and defense (Rich and Hurst 1998; Hayward and Hayward 2010), individual recognition (Mardon et al. 2010), and mate choice (Balthazart and Schoffeniels 1979; Hirao et al. 2009). However, for odor signatures to function in communication, they need to be consistent (or at least predictable) over time within the same individual (Hauser 1997). Kwak et al. (2008) point out that individual characteristics (not just individual chemical signatures) are assumed to be relatively invariant over time if they function as a cue for individual recognition, and this stability would therefore be expected of olfactory signatures. Nevertheless, short-term fluctuations in body odor due to stress, diet, and individual condition are known (Valenta and Rigby 1968; Ferkin et al. 1997; Yamazaki et al. 2002).

A signature role for odors would be less likely if odors were easily, and directly, modified by environmental factors, such as changes in diet. If the odor profile of an animal closely reflects what it just ate rather than some intrinsic, individual characteristic, then communication of individual qualities based upon it could be impaired. For example, Kwak et al. (2008) demonstrated that mice were able to

Electronic supplementary material: The online version of this chapter (doi:10.1007/978-3-319-22026-0_33) contains supplementary material, which is available to authorized users. The R script that contains the code for the simulated analysis is stored as text file. To run it in R, download the file and change the extension from '.txt' to '.R'

L. Azzani (✉) • J.L. Rasmussen • S. Gieseg • J.V. Briskie
School of Biological Sciences, University of Canterbury, Private Bag 4800,
Christchurch, New Zealand
e-mail: laura.azzani@gmail.com

B.A. Schulte et al. (eds.), *Chemical Signals in Vertebrates 13*,
DOI 10.1007/978-3-319-22026-0_33

detect a change in diet of congeners by odor alone more readily than a difference in MHC-dependent odors. Nevertheless, variation in diet did not mask the expression of MHC-dependent odor types, as suggested by previous studies (Brown et al. 1996; Schellinck et al. 1997). Despite the increasing realization that odors may play an important role in social communication in birds (Caro and Balthazart 2010), the influence of diet on odors in birds, but also vertebrates in general, has been tested in only a few studies (non-avian vertebrates: Ferkin et al. 1997; Havlicek and Lenochova 2006; Kwak et al. 2008; Osada et al. 2011).

Most studies on the influence of diet on odor production in birds have been carried out by poultry researchers, aimed at investigating differences in taste and smell of meat and eggs produced by broilers fed different diets (Mielnik et al. 2002; Overland and Skrede 2012). Dietary lipids are readily assimilated and stored in fat deposits in birds, as well as influencing fatty acid composition of other tissues and organs (McCue et al. 2009), and so it is not surprising that researchers have confirmed a direct link between food intake and meat fat quality (Bou et al. 2005; Shin et al. 2011). However, the main source of odors in most birds are the preen waxes produced by the uropygial gland (Jacob et al. 1979; Mardon et al. 2010; Caro et al. 2015), which are likely to have a different physiological mechanism of production than that of body fat (Apandi and Edwards 1964; Sandilands et al. 2004). The uropygial gland has a high capacity for synthesizing lipids de novo (Noble et al. 1963; Tang and Hansen 1972; Buckner and Kolattukudy 1975; Jacob and Ziswiler 1982; Stevens 1996). However, the composition of preen wax is known to change with season in some species (Reneerkens et al. 2002), and it is possible that such changes could be at least partly triggered by concurrent seasonal changes in diet. Given that hormonal changes also occur seasonally, experiments that manipulate food intake independently of season are needed to determine whether diet alone affects changes in preen wax composition.

To date, the results of experimental tests of diet on preen wax composition have been variable. Thomas et al. (2010) found that diet affected preen wax composition in a feeding study on white-throated sparrows, *Zonotrichia albicollis*. The treatment consisted of an artificial diet enriched with sesame oil or fish oil; these oils contain fatty acids that differ in their average chain length and in their n3:n6 ratio. The authors found that birds fed the sesame oil diet had longer chain monoesters than those fed the fish oil diet. Apandi and Edwards (1964) found a similar effect of corn-oil supplementation on the composition of uropygial secretions of chickens. In contrast, Pan et al. (1979) found no effect of diet on preen wax composition in domestic poultry. An experimental study on the "make-up hypothesis" in tawny owls, *Strix aluco*, also indirectly suggests no effect of diet on preen wax composition, at least with respect to uropygial compounds that contribute to bill reflectance (Piault et al. 2008). In some species of sandpipers, one food-restriction experiment showed an effect of diet (Reneerkens et al. 2007), but unpublished results reported in the same paper suggest an absence of dietary effects on the composition of uropygial gland secretions (reported in Reneerkens et al. 2006). In short, there have been few experiments designed to test the influence of diet on preen wax composition. Even if the biochemical pathways between dietary intake and preen wax production are not direct, changes in energy budgets as a result of dietary changes could affect preen wax production.

Determining if the odors of wild birds are influenced by changes in diet would help explain differences in odors of birds occupying different habitats (Levy and Strain 1982) and the seasonal changes in preen wax composition reported for some birds (Reneerkens et al. 2002), which may or may not be partly influenced by seasonal changes in diet (Shaw et al. 2011). Alternatively, confirming that diet has little role in preen wax composition, would support a greater role for a genetic basis to odors (as shown in mice by Kwak et al. 2008), which may be used to reliably signal information such as gender or individual identity. To this end, we carried out a diet supplementation experiment on a wild population of silvereyes, *Zosterops lateralis*, in New Zealand. As progression in the breeding season and sex can affect preen wax composition in a number of species (Caro and Balthazart 2010), we also examined the effect of sampling date and sex on preen wax composition.

33.2 Materials and Methods

33.2.1 Study Site and Species

The experiment took place in a regenerating native forest on the east coast of the South Island, New Zealand (Kowhai Bush Reserve, 173°36′E, 42°23′S). The silvereye is a small (11–14 g) passerine bird that ranges throughout Australia and the South Pacific, although it only recently self-colonized New Zealand, with the first birds arriving in the early nineteenth century (Higgins et al. 2006). They have a generalist diet, feeding on a variety of insects, fruit and nectar (review in Higgins et al. 2006), although there have been no detailed studies of diet of silvereyes at the study site. Sexes are similar in size (mass (g): females = 12.9 ± 1.1, males = 12.4 ± 1.2) and coloration, and both parents perform incubation and feeding duties. From 22 November to 16 December 2011, birds were captured using mist nets (under banding permit 2008078 and research permit NM-34075-FAU). This coincides with the period of peak nesting for this species at the study site. Each bird was handled using a new pair of latex gloves and placed in a clean paper bag before processing. This precaution was taken to minimize contamination of the preen wax samples with oil from the hands of the handlers. Preen wax samples were collected by gently squeezing the uropygial gland with forceps with wax-coated tips; the extruded droplet was then collected on a clean stainless-steel inoculation loop and placed in a clean glass insert lodged inside a glass vial, sealed with Teflon-covered lids to minimize evaporation or contamination of samples. Vials were kept cool using freezer packs in the field for a maximum of 8 h, until later frozen. Samples were kept at −20 °C until analysis. All birds ($n = 107$) were banded with numbered aluminum bands to allow resampling of individuals as the season and experiment progressed.

33.2.2 Diet Supplementation Experiment

Birds were given *ad libitum* access to tallow (a commercially available mixture of drippings from beef, mutton, and pork fat; "Butchery Pure Dripping" brand) for 22 days, from 26 November to 17 December 2011. The fat was provided in a series of feeders set up in the study area. Silvereyes readily fed on the fat, and earlier work had confirmed that supplemental feeding significantly increased reserves of body fat in this species (Barnett and Briskie 2007). Consumption of fat was monitored to ensure the feeders were always stocked. No birds were caught or sampled during the first 5 days of the feeding experiment to allow time for the changes in the energy balance of silvereyes to influence their preen wax production. The time needed for diet to start affecting preen wax composition is unknown but a period of 5 days was deemed to be sufficient as the time-frame was comparable to that employed in one other study (Piault et al. 2008) and the feeding protocol used is known to change both the fat reserves and singing behavior of silvereyes within a 24 h period (Barnett and Briskie 2007). Sampling of birds occurred on the sixth and seventh day after the feeders were deployed (1 and 2 December 2011; sampling period 1). Birds were sampled again 11 days (6 December 2011; sampling period 2) and 21 and 22 days (16 and 17 December 2011; sampling period 3) after the feeders were deployed. In addition to these 3 time points, "pre-feeding" samples were also collected from birds caught within the 7-day period preceding the commencement of diet supplementation (no birds could be collected from 20 to 22 November because of adverse weather conditions, so the actual time-frame for the pre-feeding samples was 4 days). The temporal structure of the experiment was chosen to allow for the detection of any changes in preen wax composition that may be related to a change in diet and energy intake of silvereyes.

33.2.3 Sample Preparation and Analysis

Samples were dissolved in 100 μl of ethyl acetate, poured directly into the insert containing the inoculation loop and the preen wax. The vial was then re-capped, and vortexed for 60 s at 700 rpm to ensure dissolution of the preen wax. The original cap was then substituted with a chromatographic cap fitted with a single-use PTFE silicone septum. Samples were analyzed on a Shimadzu GC-2010 gas chromatograph, equipped with a Shimadzu AOC-20i+s auto-injector and a Varian CP-SIL 5 CB capillary column (25 m length × 320 μm internal diameter × 0.12 μm film thickness). Injection volume was 1 μl, with a 6:1 split ratio. Injection port temperature was set at 250 °C, the carrier gas was nitrogen with a total flow of 19.0 ml/min and a linear velocity of 36.7 cm/s. The FID detector operated at 320 °C, with a sampling interval of 40 msec. Oven temperature was programmed as follows: initial temperature 70 °C with a hold time of 4 min, then increased to 130 °C at a rate of 20 °C/min, and finally increased to 320 °C at 4 °C/min rate (hold time 15 min). Results were recorded on Shimadzu's GCSolution, version 2.3 (©Shimadzu 2002-2009) software.

33.2.4 Quality Control

Sequences of linear alkanes ranging from C6 to C40 (C7-C40 Saturated Alkane Mixture in hexane, 49452-U Supelco, Sigma-Aldrich), and palmytoil palmytate (C16-C16 ester) were injected at regular intervals during the analysis period to ensure consistency and monitor column performance. In addition, empty vials that were manipulated in the field and processed in the lab in the same manner as sample vials were analyzed to control for any background or environmental odors.

33.2.5 Post-processing

Traces with an absolute maximum intensity in the peak region inferior to 8000 uV were excluded from further analysis. Low trace intensity was most likely due to an insufficient quantity of preen wax having been drawn from the bird. As it was hard to standardize the quantity of preen wax extracted from each bird, we used the relative proportions of compounds rather than their absolute quantities. Observations were post-processed in R (R Core Team 2014), by aligning the retention times and areas for each peak across all observations. Peaks that were present on a chromatogram but fell below the detection limit of the instrument were given an arbitrary value of 0.000001. Only peaks that were above the detection limit for the instrument in at least 5 % of the samples were included in the analysis. The area of each peak was converted to its proportional contribution to total peak area in that sample. The proportions were then square-root transformed, to reduce the influence of large peaks (Borcard et al. 2011). These two steps correspond to the Hellinger transformation, and the Euclidean distance calculated on transformed data is identical to the Hellinger distance (Legendre and Birks 2012). This distance is metric and has proven efficient in separation of ecological datasets (Legendre and Gallagher 2001; Anderson and Willis 2003; Kindt and Coe 2005).

33.2.6 Statistical Analysis

Differences in the chemical profiles of birds among the different diet time points (encoded in a matrix of Hellinger distances) were analyzed using two multivariate techniques: (1) permutational MANOVA ("non parametric MANOVA," after

Anderson 2001, implemented in R, package vegan (Oksanen et al. 2013), function Adonis) and (2) Canonical Analysis of Principal Coordinates (CAP, Anderson and Willis 2003, implemented via FORTRAN program by M.J. Anderson). These two techniques share some similarities, but each is expected to perform slightly better in different situations, depending on the data structure and the correlation matrix of the dependent variables; PERMANOVA is expected to perform better when the dependent variables are not highly correlated, while for a dataset containing several abundant variables that are highly correlated, CAP can be more efficient at detecting significant changes in minor variables that are not correlated with the former (Anderson 2004). Adonis allows testing of multi-way hypotheses, whereas CAP, as implemented in FORTRAN, only allows to test one grouping factor at a time, so the joint effect of "diet treatment" and "sex" was analyzed in Adonis only. Both PERMANOVA and CAP possess some key advantages over other multivariate techniques: they allow any dissimilarity measure to be used, rather than being limited to metric distance measures, and, they do not require the response variables to meet stringent assumptions, such as multivariate normality, which are seldom satisfied in chemoecological data sets (Anderson and Willis 2003). Finally, CAP enables us to generate a visual representation of the constrained ordination (by plotting samples against axes that maximize the differences between the specified *a priori* groups, i.e. the canonical axes), which can be compared with a robust unconstrained ordination (i.e. NMDS) to explore the multivariate patterns in the dataset (Anderson and Willis 2003).

Effect of sampling date on preen wax composition was evaluated using both multivariate and univariate techniques. Because of the temporal structure of the diet supplementation experiment, and because data belonging to three of the four time points (i.e. pre-diet supplementation, sampling period 1, sampling period 3) were collected over a period of days, we also performed a PERMANOVA analysis, on the same Hellinger distances, to verify whether collection date, rather than feeding time point, was correlated with variation in chemical distances.

Recaptured birds were excluded from the above analyses ($n=92$) to avoid pseudoreplication: the samples used in the above analysis have all been collected from different individuals (Pre feeding: $n=9$; 6 days feeding: $n=22$; 11 days feeding: $n=26$; 21 days feeding: $n=35$). However, in the context of chemical profiles being used as individual signature, it is interesting to investigate the "consistency" of such signals (i.e. if individuals maintain a coherent signature in the face of dietary changes). To this end, we performed a separate analysis—'Response of individual birds to food supplementation'—testing for effect of diet on individual birds that were captured during more than one sampling period. Because of a smaller sample size ($n=15$ different individuals), we split the 30 replicate samples into two groups: "sampling period 1," comprised of birds first captured 5 days after the beginning of diet supplementation, and "later sampling periods," comprised of the same individuals recaptured 5 or 14–16 days later. To analyze multivariate responses from repeated samples, PERMANOVA was carried out using vegan function Adonis to obtain a value for the model pseudo-F ratio. Then, specifying a custom permutation scheme to take into account dependency of data points, a "trustworthy" P-value was obtained via a randomization test, with 1999 permutations restricted within individuals.

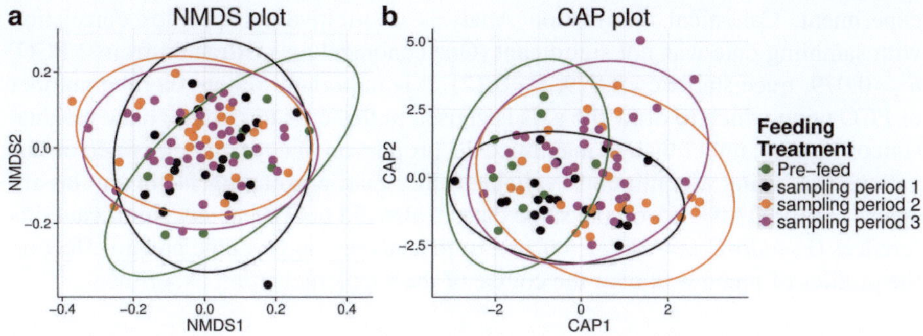

Fig. 33.1 (**a**) NMDS plot from unconstrained ordination and (**b**) CAP plot from constrained analysis of preen wax profiles from independent samples collected during the feeding experiment. Each point corresponds to one individual, i.e. one chemical profile

33.3 Results

33.3.1 Effect of Diet Treatment and Sex

Clustering based on the diet treatment did not appear to be supported either in the NMDS or in the CAP ordination (Fig. 33.1a, b). CAP analysis did not highlight any significant differences among pre-diet supplementation and the three diet time points. The squared correlations of the three canonical axes extracted from 7 PCO ($\delta^2_1=0.211$, $\delta^2_2=0.082$, and $\delta^2_3=0.025$) were not significantly higher than correlations obtained by random permutation of the observations (number of permutations = 9999, trace statistic = 0.319, $P=0.10$). "Leave-one-out allocation" of observations to pre-specified groups resulted in 36 % of cases being correctly classified, which is similar to the proportion of observations correctly classified by chance alone, i.e. 28 % (for details of this calculation, see Tabachnick and Fidell 2006, p. 404). A two-way PERMANOVA also yielded non-significant results for both feeding treatment (Pseudo-$F_{3,84}=1.47$, $P=0.082$) and sex (Pseudo-$F_{1,84}=1.37$, $P=0.19$). Sex × Treatment interaction was tested in a previous model, found to be non-significant ($P=0.97$), and therefore removed from the final model. Thus, there was no evidence that food supplementation or sex affected the preen wax profiles of silvereyes.

33.3.2 Effect of Collection Date

As our experiment was carried out over a period of several weeks, seasonal changes in preen wax composition may have affected the results. To determine if seasonal changes were apparent in our study, we examined the effect of collection date on preen wax composition. No change with date was detected during the course of the experiment. Canonical Correlation Analysis of multivariate profiles correlation with sampling date was not significant (One canonical axis extracted from 4 PCO, $\delta^2_1=0.079$, trace statistic$=0.079$, $P=0.12$). It is important to note that the number of PCO upon which to draw the axis is chosen in this case to minimize the residual sum of squares, rather than to maximize the proportion of correctly classified observations (i.e. using a continuous variable rather than a grouping factor to classify against). PERMANOVA on the same dataset also did not reveal any significant differences (Pseudo-$F_{1,90}=1.47$, $P=0.16$). Both analyses suggest date had no effect on the profiles of preen wax over the course of the supplementation experiment.

33.3.3 Response of Individual Birds to Food Supplementation

Birds sampled during feeding period 1 and recaptured during feeding period 2 or 3 did not exhibit any significant change in preen wax profile (repeated measures PERMANOVA: Pseudo-$F_{1,28}=0.68$, $P=0.36$). Preen wax profiles did not appear to mirror the duration of diet supplementation: the preen wax profiles were either unchanged (see example in Fig. 33.2) or the change was not unidirectional. This lack of any consistent change between sampling periods suggests the feeding regime was an unlikely explanation for such diffused changes (Fig. 33.3). In contrast, two

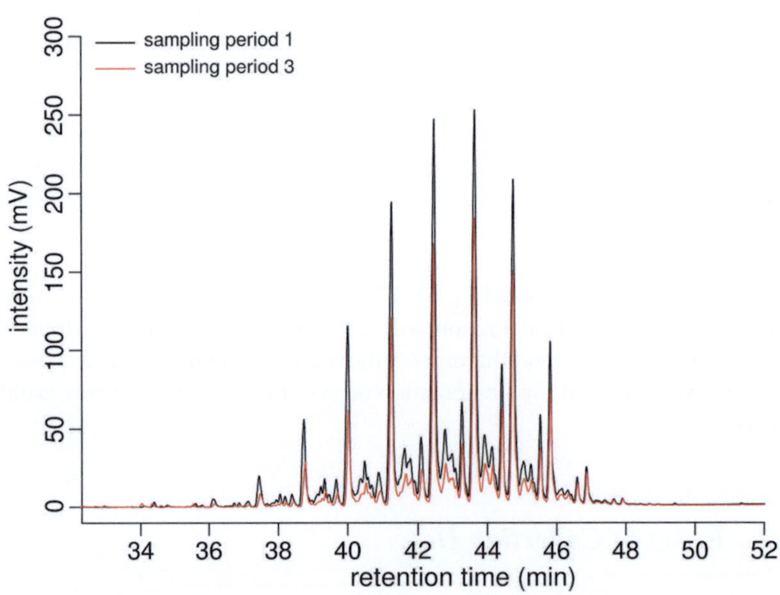

Fig. 33.2 Example of a silvereye's chemical profile. For graphical clarity, only a subsection of the chromatogram is displayed, centered around the ester section of the preen wax (Rt: 32–52 min). Pictured here are chromatograms obtained from a recaptured individual (SE50): the profile remains substantially unchanged between sampling period 1 compared to sampling period 3

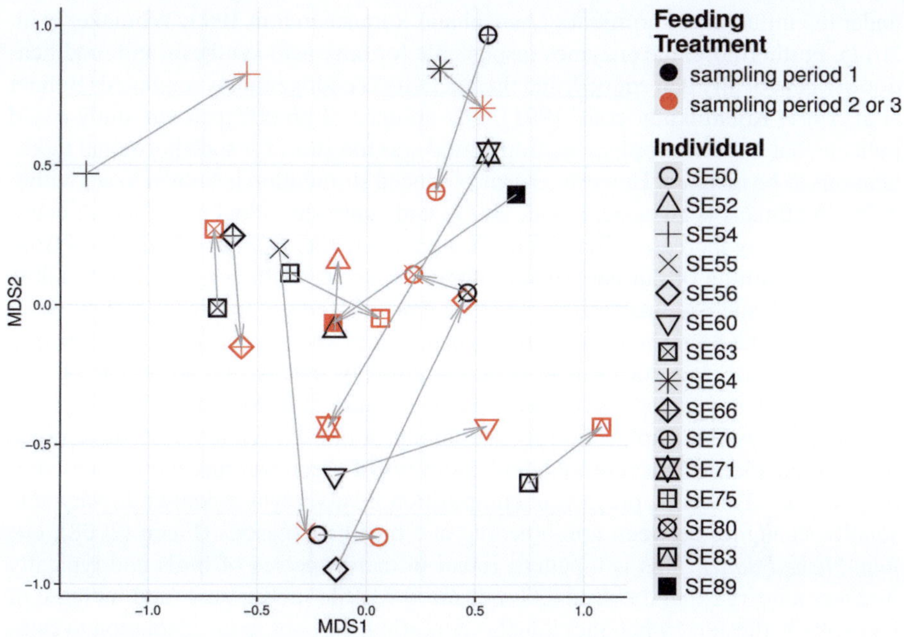

Fig. 33.3 NMDS plot, multivariate intra-individual (repeated measures) comparison: lines connect preen wax profiles from 15 individuals sampled in period 1 (black) and periods 2 or 3 (red)

artificial datasets produced by simulating a big or small effect of feeding on the preen wax profile resulted in both a significant repeated measures PERMANOVA and a NMDS plot that clearly highlighted the clustering of samples based on feeding status (see Online Supplementary material).

33.4 Discussion

Dietary fat supplementation did not affect preen wax composition in silvereyes, both in the population as a whole and within individuals. The remarkable consistency of chemical profiles of silvereyes in the face of sizeable changes in dietary intake would suggest preen wax composition is largely controlled endogenously, and thus our study supports the hypothesis that preen wax in silvereyes would qualify as a state signal that carries information about the physiological and genetic attributes of an individual (Bonadonna and Mardon 2013).

As direct incorporation of fat from food into the preen wax seems unlikely (Noble et al. 1963; Buckner and Kolattukudy 1975), it is still possible that changes in energetic intake could stimulate hormonal synthesis or release which would in turn influence the activity of the uropygial gland. The uropygial gland is particularly rich in receptors for steroid hormones and the activity of the gland is known to be under the influence of hormones (Asnani and Ramachandran 1993; Whittaker et al. 2011). Furthermore, the enzymes responsible for fatty acid synthesis and modification are genetically determined, and their activity is endogenously regulated (Bohnet et al. 1991; Kolattukudy et al. 1991). The absence of an effect in our study could indicate that the diet supplementation period was too short for such hormonal modifications to be instated. However, steroid-induced stimulation is known to act within hours in humans (Falkenstein et al. 2000), and acute steroidogenesis, which could be enhanced by increased dietary fats, occurs in a matter of hours (Miller and Bose 2011). The timing of our experiment, therefore, seems to be long enough to allow detection of such indirect effects. Given that our feeding protocol is known to change both the fat reserves and singing behavior of silvereyes within a 24 h period (Barnett and Briskie 2007), it also seems unlikely that we failed to induce any change in preen wax because of inadequate exposure to the treatment.

Another possible explanation for the failure of our experiment to change preen wax composition is that physiological pathways of preen wax production are completely fixed. However, preen wax composition in silvereyes is known to vary seasonally, changing between non-breeding and breeding seasons (Fluen (2008), *our unpublished data*). This is a pattern found in many species of birds and typically involves a decrease in the production of more volatile monoesters (and increase of less volatile diesters) when individuals are nesting, perhaps as an adaptation to camouflage nests from predators that use olfaction to locate their prey (Reneerkens et al. 2005). Seasonal changes in the preen wax of silvereyes indicates that preen wax production is not completely fixed, but instead that any changes are presumably the result of endogenous mechanisms (e.g. hormonal states) accompanying breeding, and not local changes in the type and quantity of food available.

Although we did not find any significant change in preen wax composition in silvereyes, it is possible that providing supplemental food may have increased the amount or volume of wax produced per unit time. An increase in the rate of preen wax production could then be used as a signal, with those individuals producing more wax (and presumably more persistent odor signals), conveying their better condition, in a manner analogous to studies showing high quality individuals having more colorful plumages, or more complex song repertoires (e.g. Hill 1991). Unfortunately, it was not feasible to reliably measure the volume of preen wax production in silvereyes in our experiment.

The lack of a change in preen wax with diet is consistent with the hypothesis that such compounds could be used as reliable chemosignals. Previous workers have found evidence that chemosignals are used by birds for species recognition (Bonadonna and Mardon 2010; Zhang et al. 2013; Krause et al. 2014), for kin recognition (Coffin et al. 2011; Krause et al. 2012), and for homing to the nest (Bonadonna and Bretagnolle 2002; Krause and Caspers 2012). All functions require a relatively stable odor profile, that can (a) reliably convey information about the genetic make-up of an individual or (b) be recognizable over time in order to be used as an effective cue for locating and returning to the nest. Equally, if avian chemosignals are to function as cues for reproduction, (i.e. convey information about the sexual maturity and/or sexual receptiveness of an individual) (Bohnet et al. 1991;

Hirao et al. 2009), the chemical profile has to be robust to transient changes such as those introduced by diet. Nevertheless, at present it is not known if the odors produced by preen wax are used by silvereyes in communication. There are no sexual differences in preen wax composition between males and females, as is found in other species. It bears noticing that the similarity in preen wax composition between the sexes in silvereyes could be related to the fact that both males and females share in incubation and brooding. However, allopreening behavior is common in silvereyes and occurs regularly between members of a pair, between siblings in a nest, and between parents and offspring (Kikkawa and Wilson 1983). Allopreening puts individuals in direct contact with the preen wax (and thus odors) of other individuals, and thus has the potential to be used for individual recognition and perhaps measures of quality.

The extent to which birds use chemical signals is species-specific, possibly related to the ecology of a species and to the greater or lesser extent to which it can use other senses (Martin 2012). Comparative studies have only been undertaken within Procellariiformes (Bonadonna and Bretagnolle 2002; Cunningham 2003), but other species also use olfactory cues for predator detection (Amo et al. 2011) although this ability is not universal (Johnson et al. 2011). It is possible that species which use endogenous olfactory cues as state signals show little modification of their profile in response to diet (for the reasons stated above), whereas species that do not rely on a signaling function could have a less "stable" profile, but there are too few studies at present to examine this hypothesis.

Our experiment only manipulated one aspect of the silvereye diet (i.e. fat intake) and we cannot rule out that other changes to diet might induce compositional changes in preen wax. Fat was chosen because of the effect it has on body mass and fat reserves, as well as on dawn chorus performance (Barnett and Briskie 2007) and breeding behavior in silvereyes (Barnett and Briskie 2010). It would be worthwhile manipulating the diet of silvereyes in other ways to determine whether the lack of change we observed is a general pattern across all types of diets. Lab studies have the advantage in that diet can be manipulated under controlled conditions as our use of wild birds meant it was not possible to control an individual's entire diet. However, field experiments have the undoubted value of mirroring what happens under natural diet variations. As our study is the first to determine whether diet can alter preen wax in a wild bird, further trials are needed to confirm that any diet-induced changes in the lab are likely to occur in the wild.

Finally, despite the fact that we did not detect any significant change in the profiles of preen wax in relation to either sex, date or diet supplementation, does not mean preen wax composition is identical between individuals. Both the analyses on independent samples and on recaptured birds showed that there is remarkable variation among individuals. Our results indicate this variation is not correlated to diet, sex or seasonality. However, our results do not reveal the factors responsible for the differences in preen wax composition among individuals. The next step is to determine whether the variation observed in chemometric analyses of preen wax can be perceived by the birds, and if so, what role this variation plays in either intra- or inter-specific communication.

Acknowledgements The authors wish to thank David Lloyd-Jones and Matt Weiland for help in the field, Jack Van Berkel, Ngaire Button and the Kaikoura Field Station for providing accommodation and research space during our field season, and the Royal Society of New Zealand Marsden Fund for funding the project. L.A. would also like to acknowledge the contribution of Kay Cichini, who wrote the code to implement the repeated measures PERMANOVA and made it publicly available on his blog. The original code can be found at http://thebiobucket.blogspot.co.nz/2011/04/repeat-measure-adonis-lately-i-had-to.html

References

Amo L, Visser ME, van Oers K (2011) Smelling out predators is innate in birds. Ardea 99:177–184

Anderson MJ (2001) A new method for non-parametric multivariate analysis of variance. Austral Ecol 26:32–46

Anderson MJ (2004) Canonical analysis of principal coordinates. A computer program by Marti J. Anderson. Department of Statistics, University of Auckland

Anderson MJ, Willis TJ (2003) Canonical analysis of principal coordinates: a useful method of constrained ordination for ecology. Ecology 84:511–525

Apandi M, Edwards HM (1964) Studies on the composition of the secretions of the uropygial gland of some avian species. Poult Sci 43:1445–1462

Asnani MV, Ramachandran AV (1993) Roles of adrenal and gonadal steroids and season in uropygial gland function in male pigeons, *Columba livia*. Gen Comp Endocrinol 92:213–224

Balthazart J, Schoffeniels E (1979) Pheromones are involved in the control of sexual behaviour in birds. Naturwissenschaften 66:55–56

Barnett CA, Briskie JV (2007) Energetic state and the performance of dawn chorus in silvereyes (*Zosterops lateralis*). Behav Ecol Sociobiol 61:579–587

Barnett CA, Briskie JV (2010) Silvereyes *Zosterops lateralis* increase incubation attentiveness in response to increased food availability. Ibis 152:169–172

Bohnet S, Rogers L, Sasaki G, Kolattukudy PE (1991) Estradiol induces proliferation of peroxisome-like microbodies and the production of 3-hydroxy fatty acid diesters, the female pheromones, in the uropygial glands of male and female mallards. J Biol Chem 266: 9795–9804

Bonadonna F, Bretagnolle V (2002) Smelling home: a good solution for burrow-finding in nocturnal petrels? J Exp Biol 205:2519–2523

Bonadonna F, Mardon J (2010) One house two families: petrel squatters get a sniff of low-cost breeding opportunities. Ethology 116:176–182

Bonadonna F, Mardon J (2013) Besides colours and songs, odour is the new black of avian communication. In: East ML, Dehnhard M (eds) Chemical signals in vertebrates, vol 12. Springer, New York, NY, pp 325–339

Borcard D, Gillet F, Legendre P (2011) Numerical ecology with R. Springer, New York, NY

Bou R, Guardiola F, Barroeta AC, Codony R (2005) Effect of dietary fat sources and zinc and selenium supplements on the composition and consumer acceptability of chicken meat. Poult Sci 84:1129–1140

Brown RE, Schellinck HM, West AM (1996) The influence of dietary and genetic cues on the ability of rats to discriminate between the urinary odors of MHC-congenic mice. Physiol Behav 60:365–372

Buckner JS, Kolattukudy PE (1975) Lipid biosynthesis in the sebaceous glands: synthesis of multibranched fatty acids from methylmalonyl-coenzyme A in cell-free preparations from the uropygial gland of goose. Biochemistry 14:1774–1782

Caro SP, Balthazart J (2010) Pheromones in birds: myth or reality? J Comp Physiol A Neuroethol Sensory, Neural, Behav Physiol 196:751–766

Caro SP, Balthazart J, Bonadonna F (2015) The perfume of reproduction in birds: chemosignaling in avian social life. Horm Behav 68:25–42

Coffin HR, Watters JV, Mateo JM (2011) Odor-based recognition of familiar and related conspecifics: a first test conducted on captive Humboldt penguins (*Spheniscus humboldti*). PLoS One 6:10–13

Cunningham GB (2003) A comparison of the olfactory abilities of three species of procellariiform chicks. J Exp Biol 206:1615–1620

Falkenstein E, Tillmann HC, Christ M, Feuring M, Wheling M (2000) Multiple actions of steroid hormones - a focus on rapid, nongenomic effects. Pharmacol Rev 52:513–556

Ferkin MH, Sorokin ES, Johnston RE, Lee CJ (1997) Attractiveness of scents varies with protein content of the diet in meadow voles. Anim Behav 53:133–141. doi:10.1006/anbe.1996.0284

Hauser MD (1997) The evolution of communication. MIT Press, Cambridge

Havlicek J, Lenochova P (2006) The effect of meat consumption on body odor attractiveness. Chem Senses 31:747–752

Hayward G, Hayward M (2010) Potential amplification of territorial advertisement markings by black-backed jackals (*Canis mesomelas*). Behaviour 147:979–992

Hill GE (1991) Plumage coloration is a sexually selected indicator of male quality. Nature 350:337–339

Higgins PJ, Peter JM, Cowling SJ (2006) Handbook of Australian, New Zealand & Antarctic birds. Volume 7, Part B, Dunnock to starlings. Oxford University Press, Melbourne. Pages 1745-1792

Hirao A, Aoyama M, Sugita S (2009) The role of uropygial gland on sexual behavior in domestic chicken *Gallus gallus domesticus*. Behav Processes 80:115–120

Jacob J, Ziswiler V (1982) Chapter 4 - The uropygial gland. In: Farner D, King J, Parkes K (eds) Avian Biology. Academic, Amsterdam, pp 199–324

Jacob J, Balthazart J, Schoffeniels E (1979) Sex differences in the chemical composition of uropygial gland waxes in domestic ducks. Biochem Syst Ecol 7:149–153

Johnson LS, Murphy SM, Parrish GW (2011) Lack of predator-odor detection and avoidance by a songbird, the house wren. J F Ornithol 82:150–157

Kikkawa J, Wilson JM (1983) Breeding and dominance among the Heron Island silvereyes *Zosterops lateralis chlorocephala*. Emu 83:181–198

Kindt R, Coe R (2005) Tree diversity analysis: a manual and software for common statistical methods for ecological and biodiversity studies. World Agroforestry Centre, Nairobi

Kolattukudy PE, Bohnet S, Sasaki G, Rogers L (1991) Developmental changes in the expression of S-acyl fatty acid synthase thioesterase gene and lipid composition in the uropygial gland of mallard ducks (*Anas platyrhynchos*). Arch Biochem Biophys 284:201–206

Krause ET, Caspers BA (2012) Are olfactory cues involved in nest recognition in two social species of estrildid finches? PLoS One 7:e36615. doi:10.1371/journal.pone.0036615

Krause ET, Krüger O, Kohlmeier P, Caspers BA (2012) Olfactory kin recognition in a songbird. Biol Lett 8:327–329

Krause ET, Brummel C, Kohlwey S, Baier MC, Müller C, Bonadonna F (2014) Differences in olfactory species recognition in the females of two Australian songbird species. Behav Ecol Sociobiol 68:1819–1827

Kwak J, Willse A, Matsumura K, Opiekun MC, Yi W, Preti G, Yamazaki K, Beauchamp GK (2008) Genetically-based olfactory signatures persist despite dietary variation. PLoS One 3:e3591. doi:10.1371/journal.pone.0003591

Legendre P, Birks HJB (2012) Chapter 8: From classical to canonical ordination. In: Birks HJB, Lotter AF, Juggins S, Smol JP (eds) Track. environmental change using lake sediments. Springer, Netherlands, pp 201–248

Legendre P, Gallagher ED (2001) Ecologically meaningful transformations for ordination of species data. Oecologia 129:271–280

Levy EM, Strain PM (1982) The composition of the preen gland waxes of some marine birds: a word of caution for chemotaxonomists. Comp Biochem Physiol Part B Comp Biochem 72:255–260

Mardon J, Saunders SM, Anderson MJ, Couchoux C, Bonadonna F (2010) Species, gender, and identity: cracking petrels' sociochemical code. Chem Senses 35:309–321

Martin GR (2012) Through birds' eyes: insights into avian sensory ecology. J Ornithol 153: S23–S48

McCue MD, Amitai O, Khozin-Goldberg I, McWilliams SR, Pinshow B (2009) Effect of dietary fatty acid composition on fatty acid profiles of polar and neutral lipid tissue fractions in zebra finches, *Taeniopygia guttata*. Comp Biochem Physiol A Mol Integr Physiol 154:165–172

Mielnik MB, Herstad O, Lea P, Nordal J, Nilsson A (2002) Sensory quality of marinated frozen stored chicken thighs as affected by dietary fish fat and vitamin E. Int J Food Sci Technol 37:29–39

Miller WL, Bose HS (2011) Early steps in steroidogenesis: intracellular cholesterol trafficking. J Lipid Res 52:2111–2135

Noble RE, Stjernholm RL, Mercier D, Lederer E (1963) Incorporation of propionic acid into a branched-chain fatty acid of the preen gland of the goose. Nature 199:600–601

Oksanen J, Blanchet G, Kindt R, Legendre P, Minchin P, Simpson G, Solymos P, Stevens MHW (2013) Vegan: community ecology package. R Package Version 2. 0–10

Osada K, Hanawa M, Tsunoda K, Izumi H (2011) Alteration of mouse urinary odor by ingestion of the xenobiotic monoterpene citronellal. Chem Senses 36:137–147

Overland M, Skrede A (2012) Fatty acid composition, oxidative stability and sensory quality of meat from broiler chicken fed autolysate from bacteria grown on natural gas. J Anim Physiol Anim Nutr (Berl) 96:739–746

Pan PR, Dilworth BC, Day EJ, Chen TC (1979) Effect of season of the year, sex, and dietary fats on broiler performance, abdominal fat, and preen gland secretion. Poult Sci 58:1564–1574

Piault R, Gasparini J, Bize P, Paulet M, McGraw KJ, Roulin A (2008) Experimental support for the makeup hypothesis in nestling tawny owls (*Strix aluco*). Behav Ecol 19:703–709

R Core Team (2014) R: a language and environment for statistical computing. R Core Team, Vienna

Reneerkens J, Piersma T, Damsté JSS (2002) Sandpipers (Scolopacidae) switch from monoester to diester preen waxes during courtship and incubation, but why? Proc R Soc B Biol Sci 269:2135–2139

Reneerkens J, Piersma T, Damsté JSS (2005) Switch to diester preen waxes may reduce avian nest predation by mammalian predators using olfactory cues. J Exp Biol 208:4199–4202

Reneerkens J, Piersma T, Damsté JSS (2006) Discerning adaptive value of seasonal variation in preen waxes: comparative and experimental approaches. Acta Zool Sin 52:272–275

Reneerkens J, Piersma T, Damsté JSS (2007) Expression of annual cycles in preen wax composition in red knots: constraints on the changing phenotype. J Exp Zool 307A:127–139

Rich T, Hurst J (1998) Scent marks as reliable signals of the competitive ability of mates. Anim Behav 56:727–735

Sandilands V, Powell K, Keeling L, Savory CJ (2004) Preen gland function in layer fowls: factors affecting preen oil fatty acid composition. Br Poult Sci 45:109–115

Schellinck HM, Slotnick BM, Brown RE (1997) Odors of individuality originating from the major histocompatibility complex are masked by diet cues in the urine of rats. Anim Learn Behav 25:193–199

Shaw CL, Rutter JE, Austin AL, Garvin MC, Whelan RJ (2011) Volatile and semivolatile compounds in gray catbird uropygial secretions vary with age and between breeding and wintering grounds. J Chem Ecol 37:329–339

Shin D, Narciso-Gaytán C, Park JH, Smith SB, Sánchez-Plata MX, Ruiz-Feria CA (2011) Dietary combination effects of conjugated linoleic acid and flaxseed or fish oil on the concentration of linoleic and arachidonic acid in poultry meat. Poult Sci 90:1340–1347

Stevens L (1996) Avian biochemistry and molecular biology. Cambridge University Press, Cambridge

Tabachnick BG, Fidell LS (2006) Using multivariate statistics, 5th edn. Allyn & Bacon, Inc., Needham Heights, MA

Tang BY, Hansen IA (1972) Lipogenesis in chicken uropygial glands. Eur J Biochem 31:372–377

Thomas RH, Price ER, Seewagen CL, MacKenzie SA, Bernards MA, Guglielmo CG (2010) Use of TLC-FID and GC-MS/FID to examine the effects of migratory state, diet and captivity on preen wax composition in white-throated sparrows *Zonotrichia albicollis*. Ibis 152:782–792

Valenta JG, Rigby MK (1968) Discrimination of the odor of stressed rats. Science 161:599–601

Whittaker DJ, Soini HA, Gerlach NM, Gerlac NM, Posto AL, Novotny MV, Ketterson ED (2011) Role of testosterone in stimulating seasonal changes in a potential avian chemosignal. J Chem Ecol 37:1349–1357

Yamazaki K, Boyse EA, Bard J, Curran M, Kim D, Ross SR, Beauchamp GK (2002) Presence of mouse mammary tumor virus specifically alters the body odor of mice. Proc Natl Acad Sci U S A 99:5612–5615

Zhang Y-H, Du Y-F, Zhang J-X (2013) Uropygial gland volatiles facilitate species recognition between two sympatric sibling bird species. Behav Ecol 24:1271–1278

Index

A
Acanthisitta chloris, 286
Aethia cristatella, 287
African elephants (*Loxodonta africana*), 64, 506
African painted dogs (APDs) (*Lycaon pictus*), 391
 behavior, 394
 description, 395–398
 fecal hormone metabolites, 394
 fecal sample collection and enzyme immunoassay analysis, 393
 at Lincoln Park Zoo, 392
 odor cues and behavioral measurements, 392–393
 wild dogs, 502
  abundances, 472
  behaviors of, 462
  chemical analysis, 464–465
  components code for pair, 465–469
  DUOs, 462, 463, 465, 468, 470
  potential correlation/multicollinearity problems, 472
  potential signaling components, 472–473
  principal components trail, 466
  sample collection, 463–464
  signal of pair-specificity, 467, 471
  statistical analysis, 465
  study site and population, 463
Agapornis fischeri, 254
Airborne volatiles by thermal desorption, 442–445
Alkaloid profiles, 327
Alkaloid resistance
 diversity, 325–326
 evolution, 324–327
Allobates, 307
Allomones
 anurans, 341
 in *Ranitomeya variabilis*, 346
Ameerega, 308, 317, 319, 326
Ameliorate HWC, 505–506
Analyses of similarity (ANOSIM), 92
AnalyzerPro deconvolution software, 418, 419
Anas sp., 287
Anogenital gland secretions (AGS), 364, 376
Anomaloglossus, 307
Anser anser, 212
Antechinus stuartii, 274
Antilope cervicapra (*see* Blackbuck)
Anurans
 allomones and synomones, 341
 kairomones in, 341
 primer pheromones in, 340–341
 releaser pheromones in, 340
 semiochemicals in, 339–340
 signature mixtures in, 341
Aphredoderus sayanus, 254
Appeasing pheromones, 385
Apteryx sp., 282
Arachis pintoi, 264
Archaic New Zealand frogs, chemical communication
 animal collection and odor sampling, 352–353
 bioacoustic signaling, 351–352
 chemical cue sources, 354–356

B.A. Schulte et al. (eds.), *Chemical Signals in Vertebrates 13*,
DOI 10.1007/978-3-319-22026-0

Archaic New Zealand frogs, chemical communication (*cont.*)
- perch site, 357–358
- self attraction *vs.* conspecific avoidance, 353, 355
- self/nonself recognition, 353–355
- social discrimination, 356–357
- testing conditions and analysis, 354
- tree crevice, 357–358

Aromobates nocturnus, 317
Aromobates saltuensis, 317
Artemisia, 173
Asian elephants (*Elephas maximus*), 64, 376, 416
- bioassay procedures, 234
- bioresponse levels, 235
- brevicomin enantiomers in, 236
- chemical analyses, 232
- data manipulation and analysis, 234
- frontalin
  - biochemical precursor, 238
  - blood, musth, 236–237
  - enantiomeric proportions for, 235–237
  - follicular phase females, 235
  - musth episodes of three male asian elephants, 238–239
  - precursor and testosterone levels, 237–238
- protocol I, 233
- protocol II, 233
- protocol III, 233–234
- temporal gland secretion samples, 231–232
- TGS sample, 237

Atelopus, 323
Australian prey species, 262

B

Baryphthengus martii, 308
Baseline drift, 421
Bioacoustic signaling, 351, 352
Biostimulation, 35
Birds
- chemical defenses in, 285–286
- chemical signals in, 177
- chromatographic profile, 294
- defensive secretion, 297
- foul-smelling secretion, 290
- interspecific brood parasitism, 288
- malodorous avian chemical defenses, 293–298
- malodorous secretion, 288–292
- non-evicting brood parasite, 289
- odorous defenses, 286–287
- principal component analysis, 295
- roles and importance of, 284–285
- sense of smell in, 282–284

Blackbuck
- aggressive behavior, 221–222
- dominance hierarchy, 219
- ecology of, 218
- fecal and interdigital gland scent marking, 221
- lekking, 219–220
- 2-methyl-4-heptanone, 223
- 2-methyl propanoic acid, 223
- pheromone binding proteins in, 224
- population, 217
- preorbital gland scent marking, 220–221
- regular estrous cycle, 217
- reproductive biology and odor communication, 217
- social organization, 218
- sources and identification, 222–223
- territoriality and territory size, 219
- testosterone influences dominance, 222
- urinary scent marking, 221

Blue petrels (*Halobaena caerulea*)
- airborne volatiles by thermal desorption, 442–445
- chemical analysis, 437
- cotton swab by solid-phase microextraction, 440, 450
  - chromatographic analysis, 441
  - field sampling, 441
  - sample preparation and extraction, 441
- cotton swab by thermal desorption, 442
- extraction methods, 454–456
- feather anlysis
  - chromatographic analysis, 440
  - by direct solid-phase thermal desorption, 439
  - extraction, 439
  - field sampling, 438, 439
  - sample preparation, 439, 440
- novel techniques for sampling odor, 452–454
- sample collection, 437, 451–452
- solvent extraction
  - feathers, 449
  - uropygial secretions, 445
- thermal desorption
  - airborne volatiles, 450
  - of feathers, 449
- uropygial secretions analysis, 437
  - chromatographic analysis, 438
  - sample preparation and extraction, 437–438

Bovine pheromones
estrous urine, 38
reproductive cycle modification
estrous synchrony, 34
female rats, 34
females on female cattle influence, 35
females on male cattle influence, 36
males on female cattle influence, 35
vaginal secretion, 39
Brachycephalus, 323
Burkholderia, 111
BVSTEP procedure, 51

C
Caecilians, 351
Callithrix jacchus, 122, 221
Caltech electronic nose, 13
Canids, 478, 480
Canine sex pheromones, 382
Canis lupus dingo, 262, 266
Canis lupus familiaris, 263, 266
Canonical analysis of principal coordinates (CAP), 516, 517
Cathartes aura L., 282
Ceryle rudis L., 285
Chemical communication
in birds, 106
songbird, 106
Chemical deterrents, 502
Chemical signals, 503
intraspecific, 501
natural, 502
through scent-marking, 478–479
study of, 12
Chemosensory detection, 401
Chemosensory modality, 351
Choresine, 286
Clamator glandarius L., 289, 296, 297
Colostethus panamensis, 317, 323
CombiPAL, 481
Conservation behavior, 503
Coracias garrulus L., 287
Coral reef fishes
dynamic ecosystems, 17
habitat recognition and settlement site selection, 19–20
olfactory cues and early life history, 18
olfactory preferences isolate, 18
predator identification, 20–21
Corticosterone, 403–404
Corvus corone corone L., 289
Cotton swab by solid-phase microextraction, 440
chromatographic analysis, 441
field sampling, 441
sample preparation and extraction, 441
Cotton swab by thermal desorption, 442
Cuckoos
chemical defenses in, 285–286
fledgling in defensive posture, 297
interspecific brood parasitism, 288
malodorous avian chemical defenses, 293–298
malodorous secretion, 288–292
non-evicting, 289
non-parasitized nests, 289
odorous defenses, 286–287
roles and importance of, 284–285
secretion, 290
sense of smell in, 282–284
Cuculus canorus L., 296
Culex longiareolata, 255
Culex mosquitoes, 249
Culex pipiens, 251, 253, 255
Culex quinquefasciatus, 254
Culex restuans, 251–253
Culex tarsalis, 255
Culiseta longiareolata, 248
Cyanistes caeruleus L., 286

D
Deconvolution, 418
algorithm, 483
AnalyzerPro, 418, 419
mammal odors (*see* Mammal odors, deconvolution of)
settings, 422, 423
Defensive behaviors, 401
Demersal reef fishes, 18
Dendrobates auratus, 306
Dendrobates histrionicus, 326
Dendrobates pumilio, 322
Dendrobates ventrimaculatus, 344
Dendrobatidae. *See* Poison frogs
Dendroctonus jeffreyi, 230
Dendrolagus bennettianus, 262
Dendrolagus lumholtzi, 262 (*see* Lumholtz's tree-kangaroos)
DFA. *See* Discriminant function analyses (DFA)
Dietary lipids, 512
Dimethyl sulfide (DMS), 185
Diplorhoptrum, 318

Discriminant function analyses (DFA), 465–467
DNA extraction, 49
Domestic cat (*Felis catus*)
- description, 402–403
- urine collection, 403

Dominant urine overmarks (DUOs), 462, 463, 465, 468, 470
Dumetella carolinensis L., 285

E

Electro-olfactogram (EOG), 192
Elephant musth
- abundance of medium, 72
- African elephant urinary proteins, 72
- chemical signals
  - GC-MS, 65
  - hypotheses and predictions, 66
  - male TGS, 64
  - mammalian males, 63
  - mammalian semiochemistry, 65
  - sexually dimorphic odors, 65
  - SPDE, 65
  - urine, 64
- compound abundances, 71
- culturing bacteria from, 69
- IACUC approval, 70
- identifying bacteria from, 70
- long-chain fatty acids, 72
- microbial metabolism, 68
- sequencing African elephant urinary proteins, 70
- SPDE/GC-MS, 68
- urinary fatty acids, 70
- urine collection, 67

Enterobacter cloacae, 296
Enterococcus faecalis, 296
Enterococcus phoeniculicola, 296
Environmental challenges
- chemical communication dealing with disruption, 13
- endangered species conservation and breeding, 15
- habitat imprinting, in fish, 14
- mitigating wildlife problems, 14

Enzyme immunoassay, 393
Epipedobates, 306, 308, 317, 319, 320, 325, 326
Erethizion dorsatum, 137
Estrous synchrony, 34
Euclidean distance, 515
European badgers (*Meles meles*)
- bacteriological analyses, 49
- chemical analyses, 48
- chemical composition, 51
- microbiota, 53
- statistical analyses, 50
- study animals and sample collection, 47
- subcaudal pouch, 46

Exocrine odor glands, 377

F

Falco sp., 291
Fat, 514
Fatty acid abundances, 73
Feather analysis
- chromatographic analysis, 440
- by direct solid-phase thermal desorption, 439
- field sampling, 438, 439
- sample preparation and extraction, 439, 440

Fecal glucocorticoid metabolites (FGM), 391, 392
Fecal hormone metabolites, 394
Fecal odors effect, 391
Fecal scent marking, 221
Felinine, 402
Fulmarus glacialis L., 282

G

Gambusia affinis, 248, 249, 252
Gas chromatography-mass spectrometry (GC-MS), 65, 138, 368
Giant panda (*Ailuropoda melanoleuca*)
- breeding management, 364
- chemical signals in, 363–365
  - animal and sample collection, 365–366
  - data analyses, 368–369
  - description, 375–377
  - determination of female and male reproductive status, 366
  - estrus related substances, 371–373
  - fatty acids, 371
  - GC-MS analysis, 368
  - gestagenes, 373–374
  - seasonal profile of reproductive hormones, 369–370
  - solid-phase microextraction, 366, 370
  - static headspace analyses, 366, 374–375
  - stir bar sorptive extraction, 367–368, 371
- excretions of estrogen, 365
- scent marking, 364

Gonadotropin releasing hormone (GnRH), 478
Grey wolves (*Canis lupus*), 502

H
Haemorhous mexicanus, 286
Hamilton's frog (*Leiopelma hamiltoni*), 352, 356–358
Head space solid-phase microextraction-gas chromatography–mass spectrometry, 481–482
Heart rate (HR), 386
Hellinger transformation, 515
Horseradish peroxidase (HRP) ligands, 393
Human–wildlife conflict (HWC)
  description, 499–500
  individual level management of, 503–504
  using learning and odors to ameliorate, 505–506
  population level management of, 500–503
  reduction of, 506
Hydrobates pelagicus, 167
Hydrodynamic processes, 23
Hyloxalus azureiventris, 307, 346
Hyloxalus erythromos, 307
Hystrix brachyura, 137
  behavior observations, 140–141
  chemical composition, 141–143
  cresol, benzaldehyde and methyl-cyclohexane present in urine samples, 141
  HPLC of ACN extract of urine, 142
  observations of behavior, 138, 139
  urine samples chemical analysis, 138–139
Hystrix indica, 137

I
Ifrita kowaldi, 286
Indole, 425
Interdigital gland scent marking, 221
Intraspecific chemical signals, 501

J
Junco hyemalis, 212, 285
Juvenile reef fishes
  nature of odors, 22
  predator identification, 20

K
Kairomones, 401
  in anurans, 341
  ovipositing mosquitoes, 248
  in *Ranitomeya variabilis*, 345

L
Lagostomus maximus, 221
Leiopelmatid frogs, 352
Lekking, 219–220
Lemur catta, 121
Lepomis cyanellus, 249, 252
L-felinine, 405–408
Ligands, 393
Lipids, 512
Lower limits of detection (LODs), 415
Loxodonta africana, 231
Lumholtz's tree-kangaroos
  animal facilities and study subject, 264–265
  antipredator response of, 275
  archetype-specific antipredator responses, 261, 264, 274
  behavioral categories, 267
  chemical cues in, 273
  data analyses, 266–267
  duration of comfort behaviors, 272
  evolution of, 263, 276
  experimental design, 265–266
  fecal material and diet, 266
  frequencies of vigilance, 272
  mean duration of comfort behaviors, 267–272
  mean frequencies and durations, 268–272
  movement behaviors, 276
  multi-predator environments, 261
  neophobia, 273
  novel odors, 273
  observations, 265
  relaxed predator selection, 262
  from rock-wallabies, 275
  sex, age and weight of study subjects, 265
  terrestrial predators of, 275
  urbanized and fragmented landscape, 263
  vigilance aids, 272
Luteinizing hormone (LH), 34
Lyctocoris elongates, 230

M
Macropus eugenii, 261
Mammal odors, deconvolution of, 416, 417, 428–429
  data deconvolution, 418–419
  description, 426–427, 430
  detecting trace components and separating overlapped peaks
    average separation, 419–420
    column temperature, 422–423
    component and false detections, 421, 423

Mammal odors, deconvolution of (*cont.*)
deconvolution settings, 419, 422
hidden peaks, 420, 427
trace component, 419, 421
inter-sample comparisons, 428
peak size differences, 424–426
resolution and peak capacity, 427–428
retention
issues, 428–429
relative shift, 426
retention window, 423–424
sample preparation and analysis, 418, 423
spectral distortion, 428
spectral match, 424
Maned wolf (*Chrysocyon brachyurus*)
biology, 477–478
ovulators, 478
scent-marking, 478–479
urinary volatile organic compounds, 479–480
urine sample analysis, 480
compounds differing by sex, 485–491
compounds found, 491, 492
data analysis, 482–484
head space solid-phase microextraction-gas chromatography–mass spectrometry, 481–482
qualitative, peak alignment and preliminary analysis, 484–485
Mannophryne, 307
Maternal pheromones, 385
Mating behavior, 148–149
Meadow vole
age classes, 123–124
aging, 129, 130
breeding season, 132
discrimination of sexes, 126–129
habituation phase, 124–125
materials and methods, 123
odor-related behaviors, 131
opposite-sex conspecifics, 126–128, 131
same-sex conspecifics, 126–130, 132
stimulus scent marks, 125
test phase, 125
Medetera bistriata, 230
Melopsittacus undulates, 285
Meriones unguiculatus, 121
Mesocricetus auratus, 121
Mesocricetus brandti, 122
Methyl-imidazolidinedione isomers, 465
Methyl paraben (MP), 382
Mice (*Mus musculus*) estrus, 479
Microbiota, 53–54
Microstigma rotundatum, 342
Microtus ochrogaster, 121
Microtus pennsylvanicus, phytoestrogens, 206 (*see also* Meadow vole)
animals study, 207–208
attractivity, 208–211
proceptivity, 209–211
treatment groups, 208
MO BIO PowerSoil®, 108
Molothrus oryzivorus, 288
Monotremes, 146
Morelia amethystina, 262, 266
Morpholine-imprinted Coho salmon, 14
Mozambique tilapia, urinary amino acids in
aqueous fraction, 193–194
data treatment and statistical analysis, 192–193
from dominant male tilapia, 193
electro-olfactogram, 192
experimental animals, 190–191
measurement, 191
olfactory sensitivity to, 194–196, 200–201
and social status, 196–199, 201
MPP, 483–485
Musk deer (*Moschus moschiferus*), 377
Mus musculus, 121
Musth *vs.* non-musth
male African elephant, 67
male Asian elephant, 69

N

Natural chemical signals, 502, 503
Natural repellent, 506
Neonatal exposure
to cat urine, 405, 407, 408
to L-felinine, 406, 408
of mouse pups to stimulus, 403
to odorants, 402
Nest odor in zebra finches
breeding conditions, 168
chemical analysis, 168–170
gas chromatography, 170, 171
nest variability, 174
self-produced substances, 173
statistical analysis, 170
substances, 172
volatile organic compounds, 169
Neuroendocrine system, 401
Neutral model analysis, 93
Nonmetric multidimensional scaling (NMDS), 516, 517, 519
North American beavers (*Castor canadensis*), 502

Notonecta irrorata, 248
Notonecta maculata, 248, 255
Notophthalmus viridescens louisianensis, 252
Novel stimulus, 404

O
Oceanodroma leucorhoa, 284
Odor
  nature, 22
    in coral reef, 22
    movement, 23
    perception in larval and juvenile fishes, 23
  production, diet on, 512
  role of, 20–21
    collecting larvae, 24
    in situ experiments and field validation, 26
    laboratory-based behavioral experiments, 25
    olfactory cue biochemistry analysis, 26
  sexually dimorphic, 65
  source of, 512
  with stimulus, 506
  of wild birds, 513
Odorants, 402
Odor signatures, 511
Olfactory communication, 46, 461
Olfactory cues
  life cycles of coral reef fishes, 18
  role of, 19–20
Olfactory effects, 479
Olfactory enrichment, 391
Olfactory habitat imprinting, 14
Olfactory signals, 511
Olfactory stimulation, 479
Oophaga, 319
Open field test, 404, 407
Operational taxonomic units (OTUs), 92
Opposite-sex conspecific, 209–212
Osteocephalus, 345
Ovipositing mosquitoes
  data analysis, 251
  fish experiments, 249–250, 252–253
  newt experiment, 250–251, 253
  predator treatment, 252
  study location, 249
Ovulation, 479

P
Passerculus sandwichensis, 284
PCA. *See* Principal component analysis (PCA)
Pelecanoides georgicus, 168
Pelecanoides urinatrix, 168
PERMANOVA technique, 516
Permutational tests for homogeneity of multivariate dispersions (PERMDISP), 92
Personality factors, 504, 505
Pest management, 502, 503
Petrel chicks, 167
Petrogale spp., 275
Petroica australis, 287
Pheromone binding protein (PBP), 224
Pheromones, 351, 501, 502
Phodopus campbelli, 122
Phoeniculus purpureus, 296
Phyllobates aurotaenia, 326
Phyllobates terribilis, 318, 326
Phytoestrogens
  affecting gonadal hormone titers, 207
  animals study, 207–208
  attractivity, 207
    results, 210–211
    scent donors and subjects, 208
    testing procedure, 208–209
  estrogen-like compounds, 205
  opposite-sex conspecific, 209–212
  proceptivity, 207
    results, 211
    scent donors and subjects, 209
    testing procedure, 210
  on reproductive physiology, 205, 206
  on sexual behavior, 206
  soy-derived isoflavones, 205
  treatment groups, 208
Pica pica L., 289
Pitohui dichrous, 286
PIXEL FLUX software, 386
Plasma corticosterone, 403, 408
Plectropterus gambensis L., 286
Poison frogs, 339. *See also Ranitomeya variabilis*
  alkaloid profiles, 327
  alkaloid resistance
    diversity, 325–326
    evolution, 324–327
  autoresistance as chemical defense component, 323
  compounds, 309–310
  defensive/anti-predator properties, 309–310
  diet and defense, 317–318
  environmental source, 309–310
  evolutionary ecology, 306–317
  insects, bacteria, and fungi, 315–316

Poison frogs (*cont.*)
  muscular and neuronal receptors and channels, 311–314
  pharmacokinetics perspective, 319–322
  tetrodotoxin, 323–324
  toxicity and unpalatability, 318–319
Polar compounds, 428
Polyclonal antiserum, 393
Polypedates, 323
Predator odors, 401
Predator–prey relationships, 401
Preen gland bacterial communities
  male and female juncos, 111
  methods, 107–109
Preen wax composition, 512
Preorbital gland scent marking, 220–221
Preputial gland, 419
Primer pheromones
  in anurans, 340–341
  in *Ranitomeya variabilis*, 344
Principal component analysis (PCA), 484, 485
Principal coordinates analyses (PCoA), 112
Pseudocheirus peregrinus, 276
Pseudomonas, 111
Pseudophryne, 328
Psychotria, 318
Push–pull approach, 502, 503
Pyrazines, 484

R

Rana catesbeiana, 321
Rana pipiens, 326
Ranitomeya variabilis, 341–342
  allomones/synomones in, 346
  kairomones in, 345
  primer pheromones in, 344
  releaser pheromones in, 342–344
  semiochemicals in, 342, 343
  signature mixtures in, 345
Relative retention time (RRT), 138
Releaser pheromones
  in anurans, 340
  in *Ranitomeya variabilis*, 342–344
Repellent, 506
Reproductive signals, 145–146
Rhadinaea decorate, 308
Rheobates, 307

S

Salamanders, 351
Sandpipers pipers, 512
Saporito, 325
Sarcophilus harrisii, 277
Scent-marking, 364
  chemical signaling through, 478–479
  territorial, 461, 462
  urine, 478
Scent pouch bacterial communities
  age-related diversity, 96
  age-related variation, 99
  bacterial community analyses, 92
  development, 88–89
  fermentation hypothesis, 87
  IACUC approval, 94
  neutral model analysis, 93
  sample collection, 90
  sample preparation and next-generation sequencing, 91
  sequence analysis, 91
  striped hyenas, 89
  taxonomic characterization, 94
  variation in, 97
Semiochemical communication, in dogs
  antibiotics, 384
  appeasing pheromones, 385
  bacterial flora, 384
  repellents, 385
  sex pheromones
    canine, 382
    investigation, 385–386
    suspected structure, 383–384
  vaginal flora, 385
Semiochemicals, 15, 501
  in anurans, 339–340
  ovipositing mosquitoes, 248
  in *Ranitomeya variabilis*, 342
Serinus canaria L., 283
Sex pheromones, 351
Short-beaked echidna
  chemical profiles and mating activity, 155–157
  data selection and statistical analyses, 150–152
  estimating timing of pregnancy, 150
  hibernation, reproductive state and odor cues, 152–154, 157–160
  inter-annual differences in female, 152
  intra-annual differences in chemical profiles, 154–155
  mating activity, 150
  mating behavior, 146–148
  odor cues and sexual conflict, 160–161
  PERMANOVA analyses, 154, 156
  reproductive female echidnas, 155, 157
  reproductive signals, 145–146

sample collection and chemical analyses, 149
study site and animals, 148–149
temperature records, 153
temporal changes in signaling activity, 145
Sialia sialis L., 286
Signature mixtures
in anurans, 341
in *Ranitomeya variabilis*, 345
Silvereyes (*Zosterops lateralis*), preen wax composition
collection date, 517–518
diet supplementation experiment, 514
effect of diet treatment and sex, 517
individual birds to food supplementation, 518–519
non-breeding and breeding seasons, 520
post processing, 515
quality control, 515
sample preparation and analysis, 514–515
statistical analyis, 515–516
steroid hormones, 519–520
study site and species, 513
Silverstoneia, 307
Social behavior, 351
Solenopsis, 318
Solid phase dynamic extraction (SPDE), 65
Solid-phase microextraction (SPME), 290, 366, 367, 370, 371, 373, 481, 482
Somateria mollissima L., 287
Static headspace analyses (SHS), 366, 374
Stir bar sorptive extraction (SBSE), 367–369, 371–373
Stoddart, M., 11
Streptopelia senegalensis L., 285
Sturnus unicolor, 285
Subcaudal gland secretion (SGS), 47, 56–57
Symbiotic bacteria, 105
Synergistic effects, 12
Synomones
anurans, 341
in *Ranitomeya variabilis*, 346

T
Tachyglossus aculeatus setosus (*see* Short-beaked echidna)
Taeniopygia guttata (*see* Zebra finches)
Tamias striatus, 121
Terminal restriction fragment length polymorphism (T-RFLP) analysis, 49
Territorial scent marking, 461, 462
Testosterone influences dominance, 222
Tetrodotoxin (TTX), 323–324
Thamnophis sirtalis, 324
Thin layer chromatography (TLC), 139
Thylacinus cynocephalus, 263, 276
Thylacoleo carnifex, 263, 276
Tiger (*Panthera tigris*), 376
Timed artificial insemination (TAI) protocols, 33
Tonguing, 383
Trachypithecus phayrei crepusculus, 213
Trichosurus vulpecula, 275

U
Universal carnivore signals, 401
Upupa epops L., 286
Urinary amino acids, in mozambique tilapia
aqueous fraction, 193–194
data treatment and statistical analysis, 192–193
from dominant male tilapia, 193
electro-olfactogram, 192
experimental animals, 190–191
measurement, 191
olfactory sensitivity to, 194–196, 200–201
and social status, 196–199, 201
Urinary scent marking, 221, 478
Urinary volatile organic compounds, 479–480
Uropygial gland, 436, 437, 512
Uropygial secretions analysis
chromatographic analysis, 438
field sampling, 437
sample preparation and extraction, 437–438

V
Vespa orientalis L., 285
Volatile organic compounds (VOCs)
antagonist interactions, 285
by birds, 286
colony deter mosquitoes, 287
juvenile parasite secretion, 290
paradoxical interactions, 293–294
maned wolves, 479–480
Voltage-gated sodium channels (VGSCs), 323, 324
Vomeronasal organ (VNO) neurons, 415
Vulpus vulpus, 276

Z
Zebra finches
dimethyl sulfide, 185
experimental test arena, 180

Zebra finches (*cont.*)
fledgling, 178
food-related odors, 185
foraging-related context, 179
latency for leaving the start box, 182
nest odor in
breeding conditions, 168
chemical analysis, 168–170
gas chromatography, 170, 171
nest variability, 174
self-produced substances, 173
statistical analysis, 170
substances, 172
volatile organic compounds, 169
nonsocial foraging context, 184
number of errors, 183, 184
olfactory cues for social communication, 184
olfactory fingerprints, 179
olfactory foraging experiment, 178–181
Spearman correlations, 182
statistical analysis, 181
time to find the food, 183

MIX
Papier aus verantwortungsvollen Quellen
Paper from responsible sources
FSC® C105338

If you have any concerns about our products,
you can contact us on
ProductSafety@springernature.com

In case Publisher is established outside the EU,
the EU authorized representative is:
Springer Nature Customer Service Center GmbH
Europaplatz 3, 69115 Heidelberg, Germany

Printed by Libri Plureos GmbH
in Hamburg, Germany